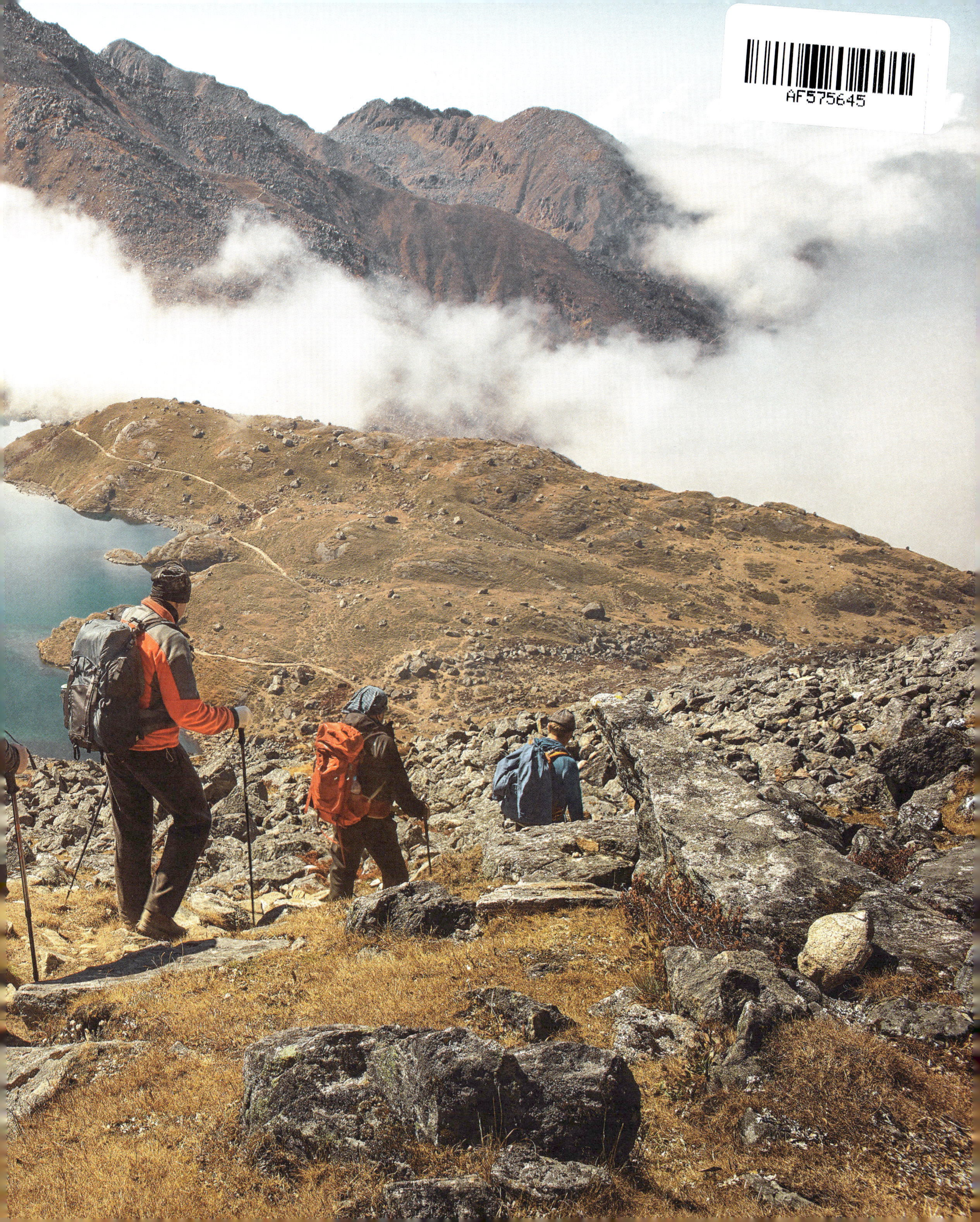

THE GREAT

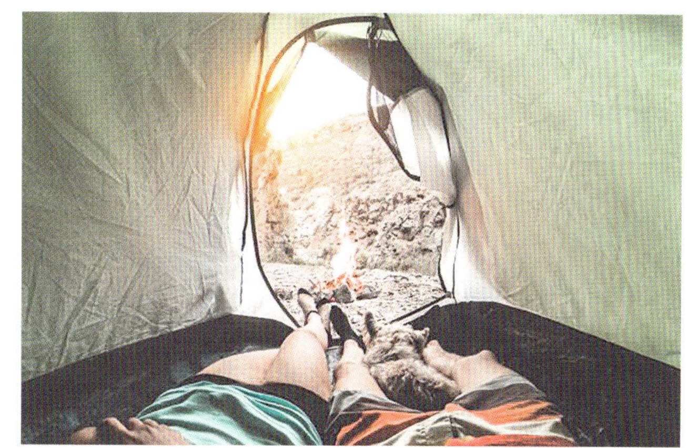

OUTDOORS

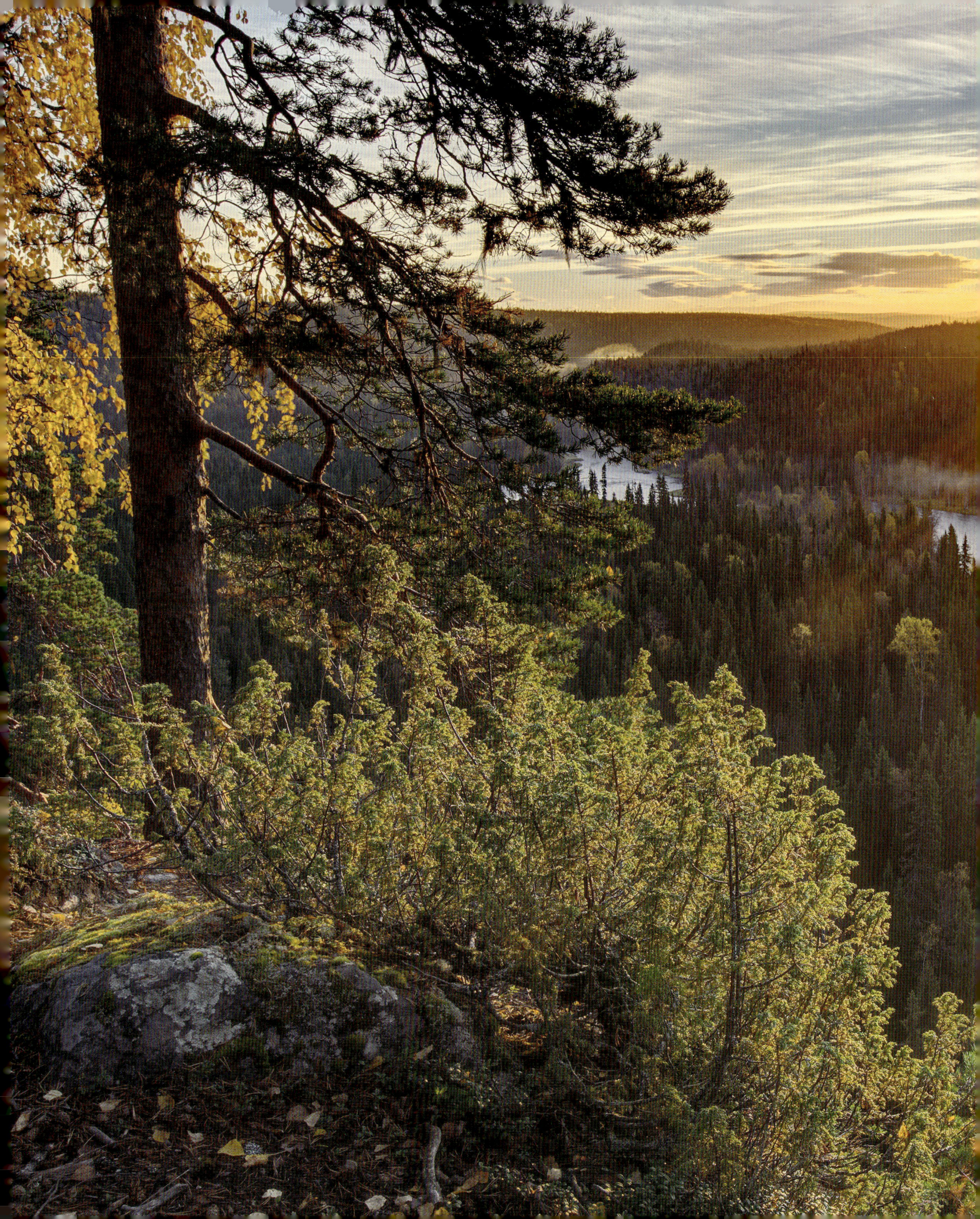

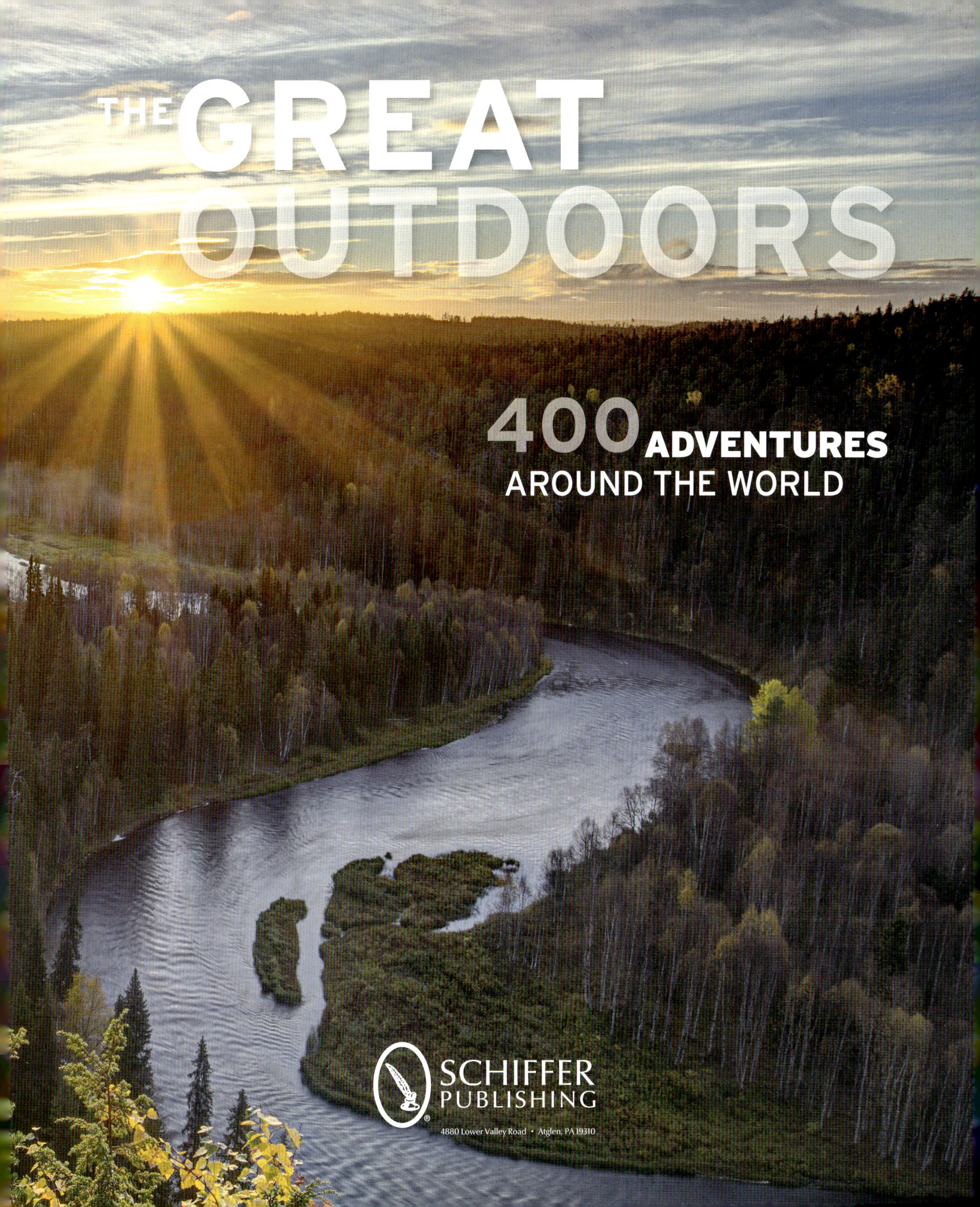
THE GREAT OUTDOORS
400 ADVENTURES
AROUND THE WORLD
SCHIFFER
PUBLISHING
4880 Lower Valley Road • Atglen, PA 19310

CONTENTS

GERMANY, AUSTRIA, SWITZERLAND

WESTERN EUROPE

NORTHERN EUROPE

EASTERN EUROPE

SOUTHERN EUROPE

AFRICA

ASIA MINOR / ARABIA

ASIA

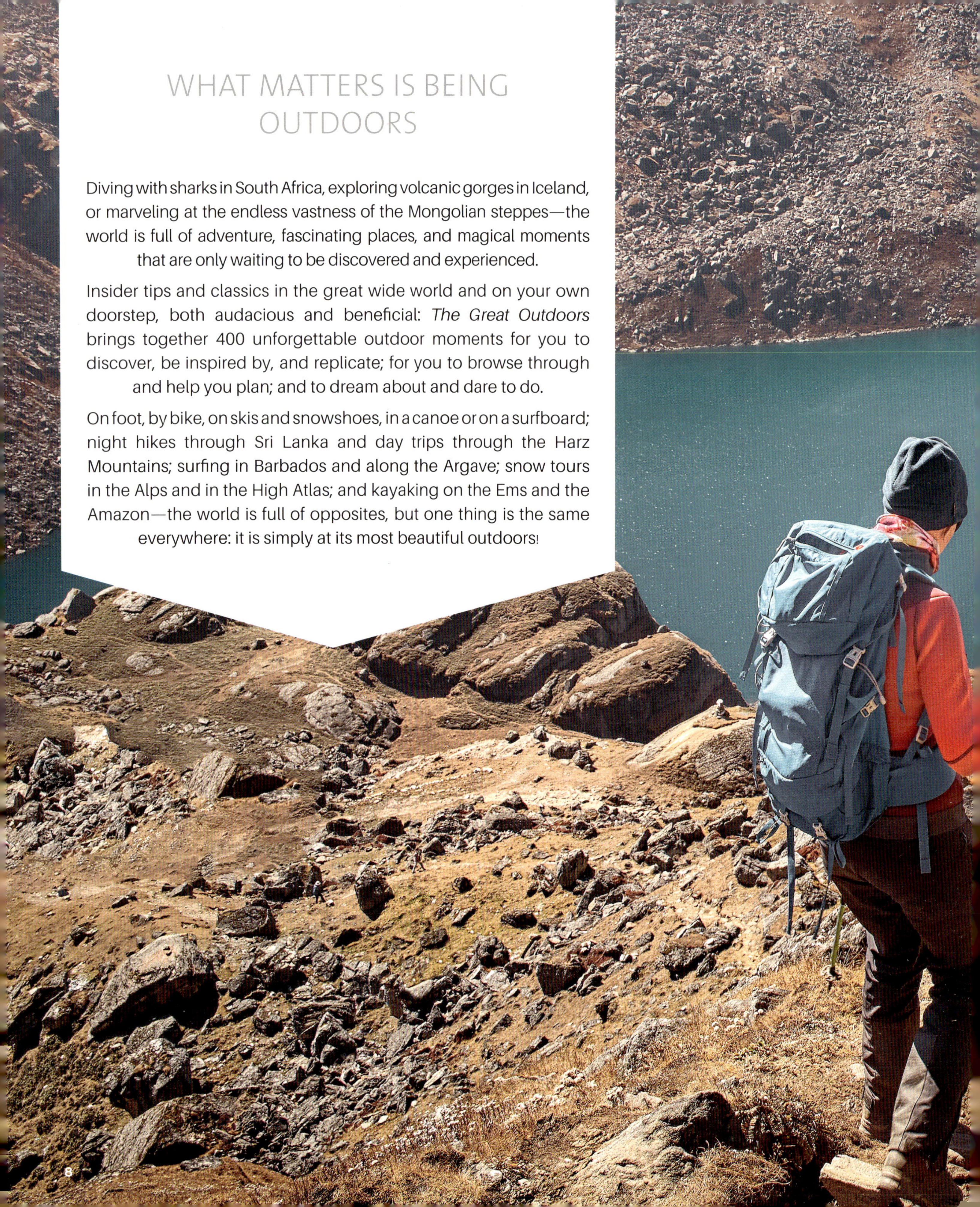

WHAT MATTERS IS BEING OUTDOORS

Diving with sharks in South Africa, exploring volcanic gorges in Iceland, or marveling at the endless vastness of the Mongolian steppes—the world is full of adventure, fascinating places, and magical moments that are only waiting to be discovered and experienced.

Insider tips and classics in the great wide world and on your own doorstep, both audacious and beneficial: *The Great Outdoors* brings together 400 unforgettable outdoor moments for you to discover, be inspired by, and replicate; for you to browse through and help you plan; and to dream about and dare to do.

On foot, by bike, on skis and snowshoes, in a canoe or on a surfboard; night hikes through Sri Lanka and day trips through the Harz Mountains; surfing in Barbados and along the Argave; snow tours in the Alps and in the High Atlas; and kayaking on the Ems and the Amazon—the world is full of opposites, but one thing is the same everywhere: it is simply at its most beautiful outdoors!

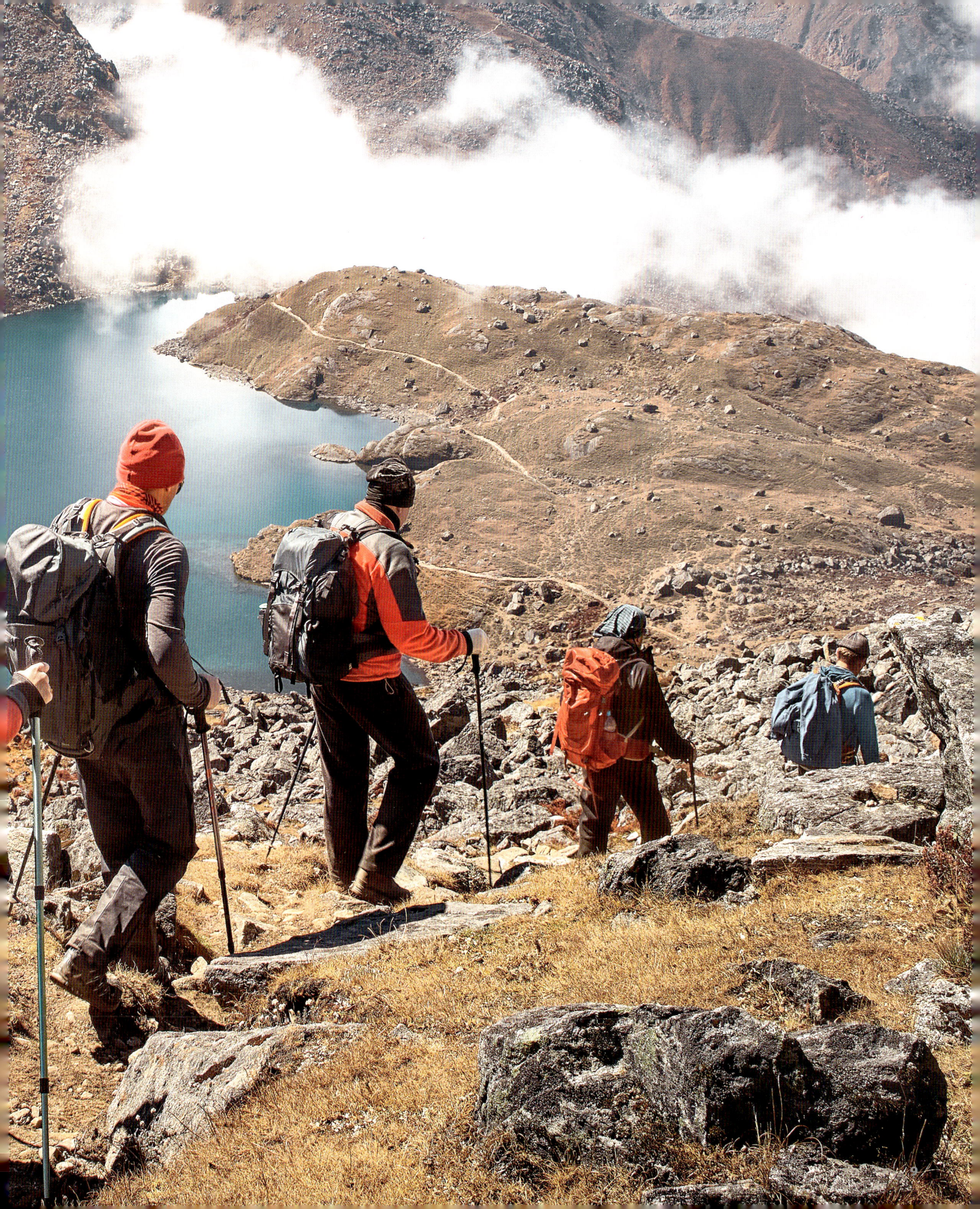

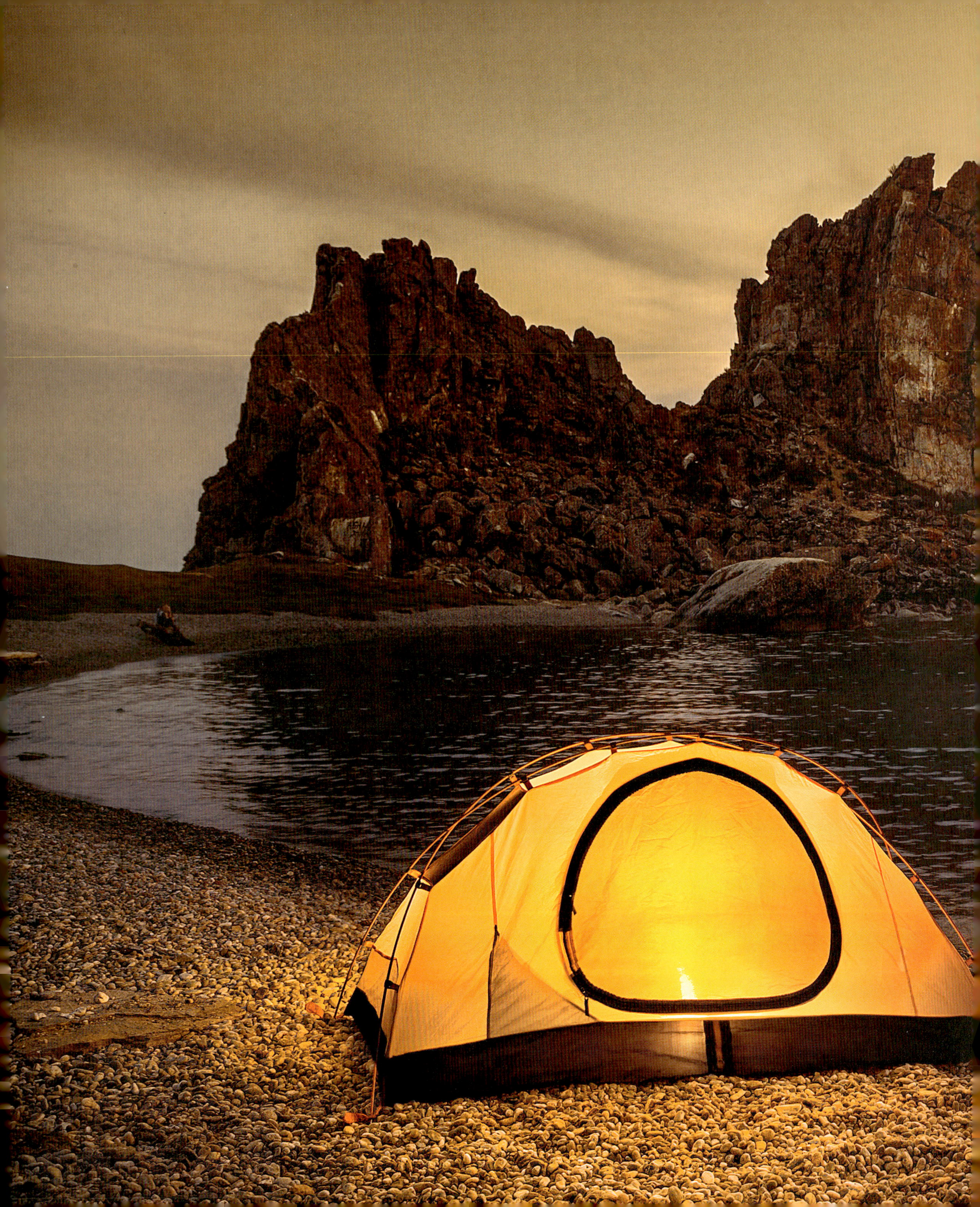

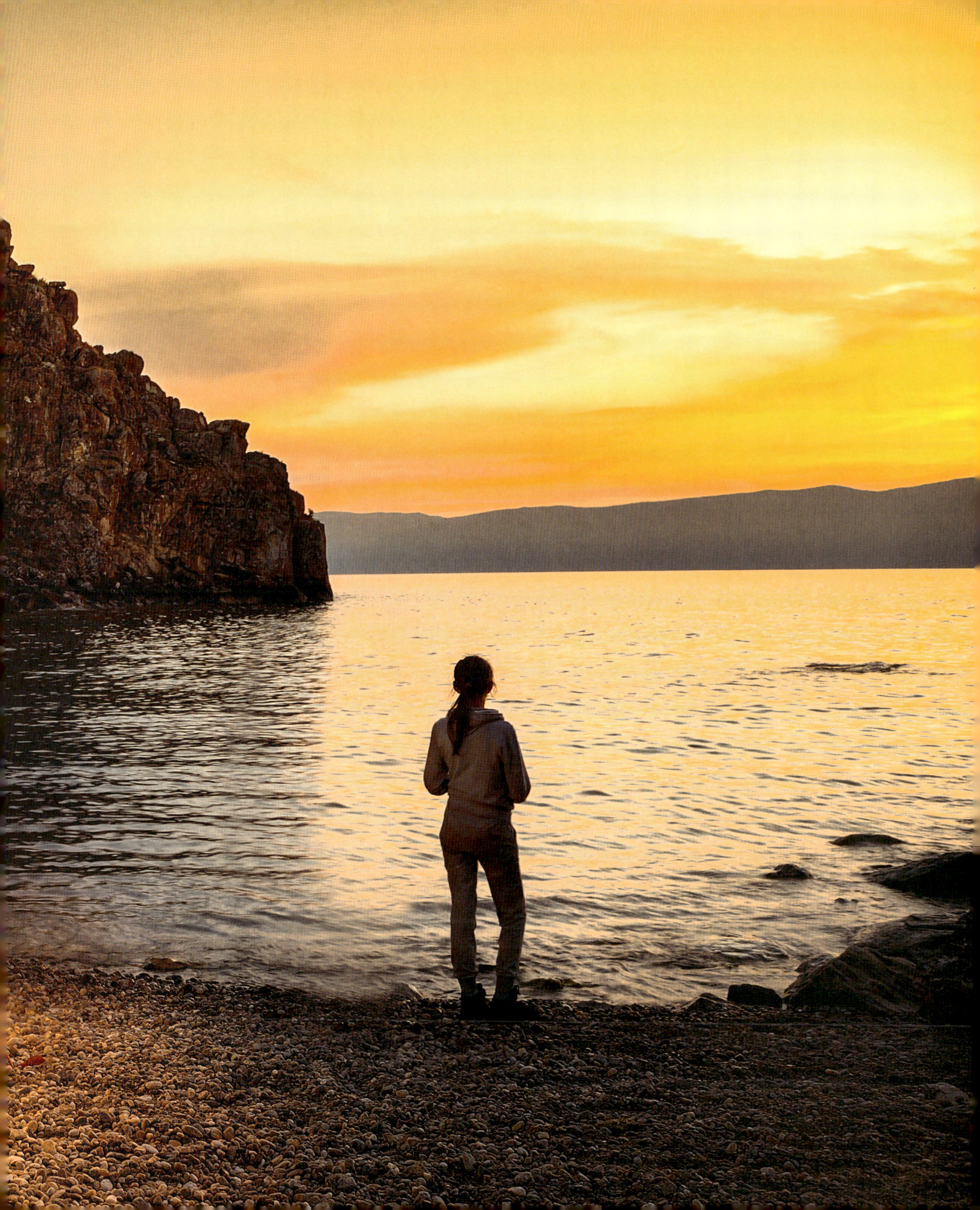

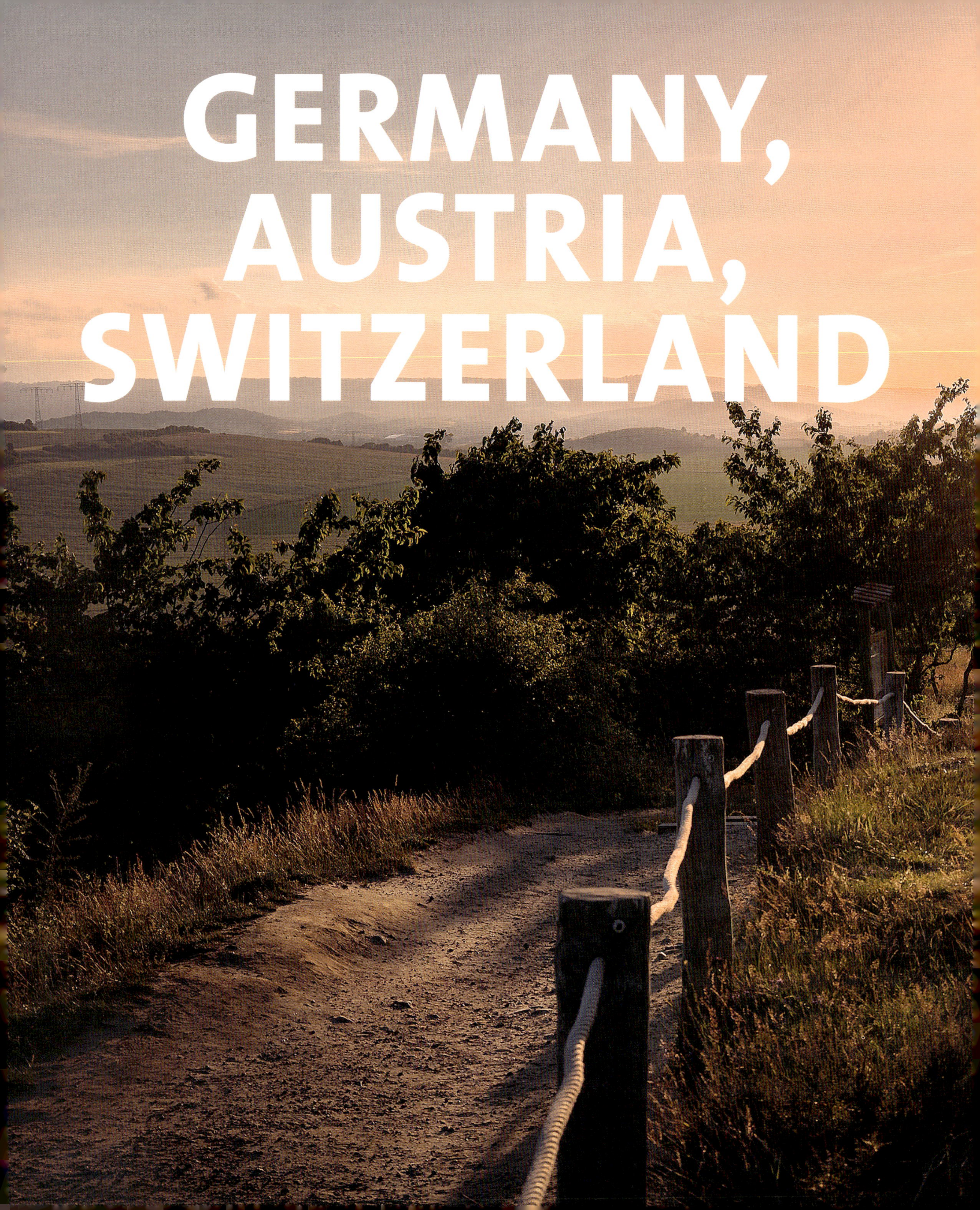

GERMANY, AUSTRIA, SWITZERLAND

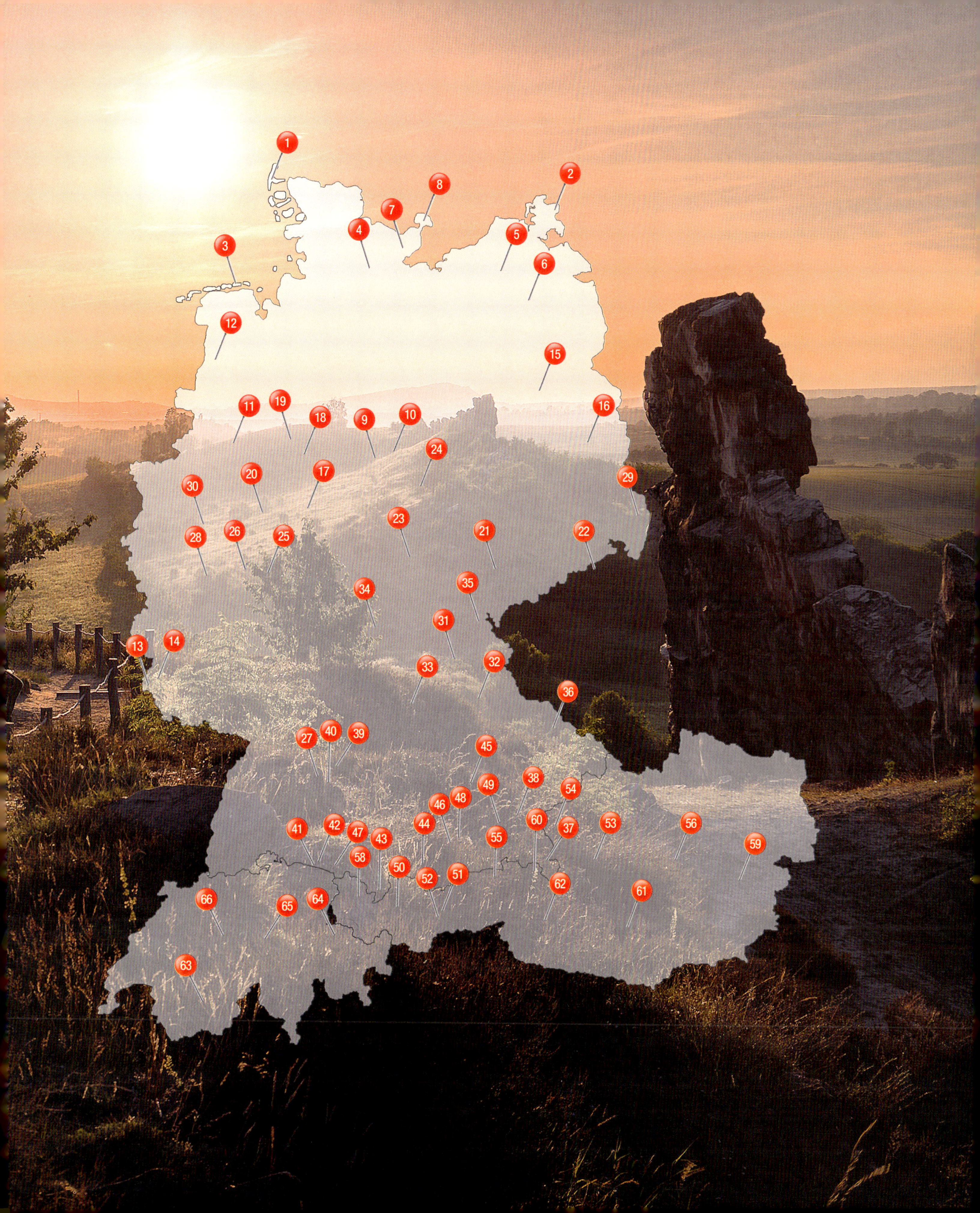
1
2
3
4
5
6
7
8
9
10
11
12
13
14
15
16
17
18
19
20
21
22
23
24
25
26
27
28
29
30
31
32
33
34
35
36
37
38
39
40
41
42
43
44
45
46
47
48
49
50
51
52
53
54
55
56
58
59
60
61
62
63
64
65
66

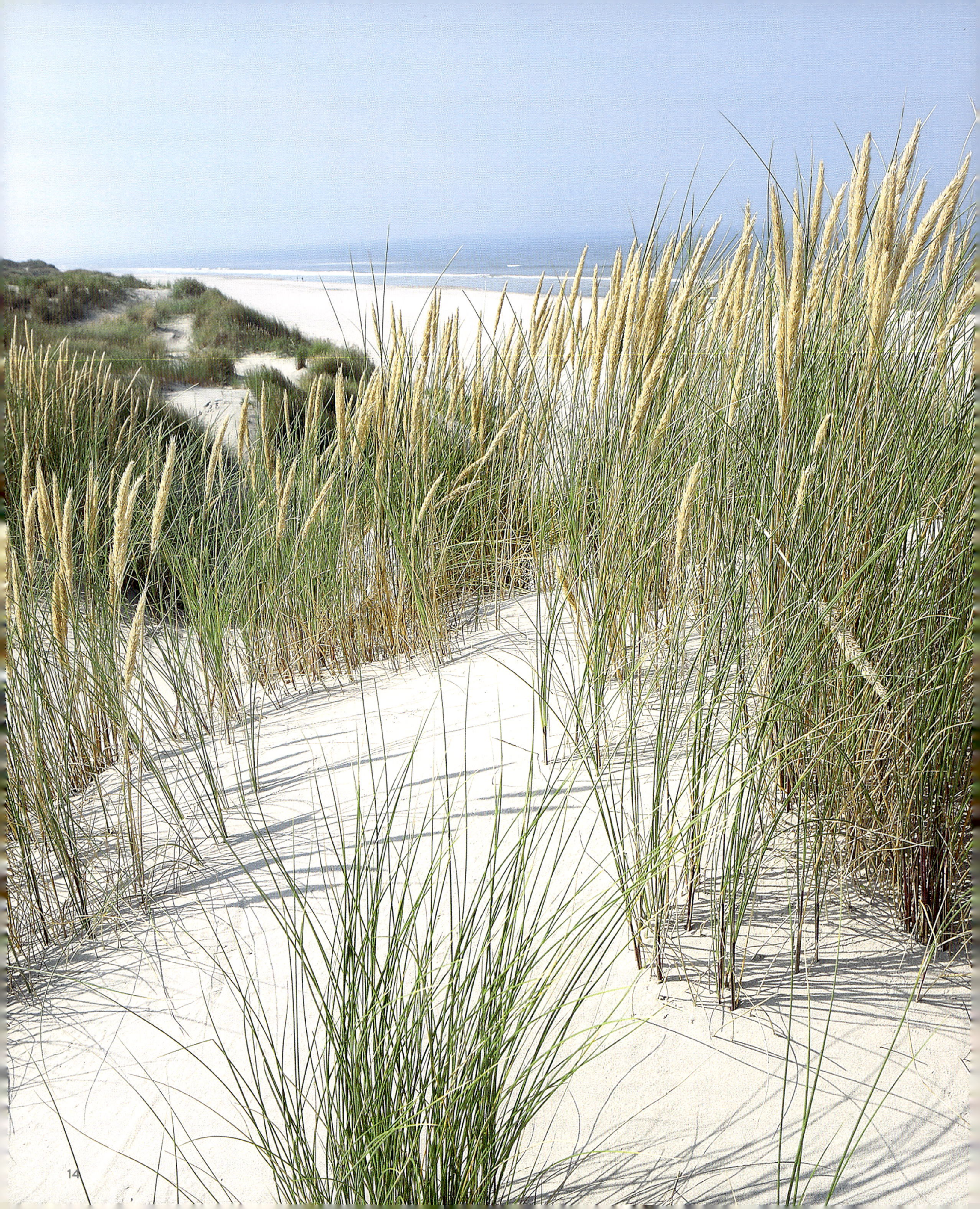

1. DUNES HIKE ON SYLT ISLAND

The northern tip of the island of Sylt, also known as "the elbow," is an extensive dune landscape where you can hike in peace. There are no cars driving here, and the usual tourist hustle and bustle is far away. Here, between the eastern and western lighthouses, you are at the northernmost point in Germany and can let the fresh wind blow the cobwebs away. You are walking, so to speak, between the tranquil Wadden Sea on one side and the often-stormy-and-unpredictable North Sea on the other, western side of the island. And since this headland is very narrow, you can easily switch from one side to the other and admire the differences in the fauna and flora on both sides. Salty air and the cries of seagulls accompany the hiker. And in the small port of List you can enjoy freshly harvested Sylt oysters.

Size: 2.75 mi. (4.5 km) long by 0.75 mi. (1.2 km) wide, an area of about 1.25 sq. mi. (3 km^2)
Information: any good travel guide about Sylt

2. HIKING ON THE BALTIC SEA ISLANDS

Let the wind blow through your hair, enjoy the maritime landscape, and let your gaze wander into the distance . . . The island of Rugen offers—besides its famous chalk cliffs—broad, fine white-sand beaches that are about 16–22 yards (15–20 m) wide. The pine forest often starts right behind them. You can find a good combination of culture and nature on the narrow heath between Binz and Mukran, with 6 mi. (10 km) of fine sandy beach and the museums in the Prora, a former Hitler-era building complex. You can range some 40 mi. (65 km) along the coast of the island of Fehmarn, past lighthouses such as Flugge and Staberhuk. And on the Danish island of Bornholm, you can admire the tallest Danish waterfall, the Døndalfald, at 24 ft. (22 m) high—in seven hours you can cross the entire island on well-marked trails.

Time it takes: a variety of hikes of different lengths

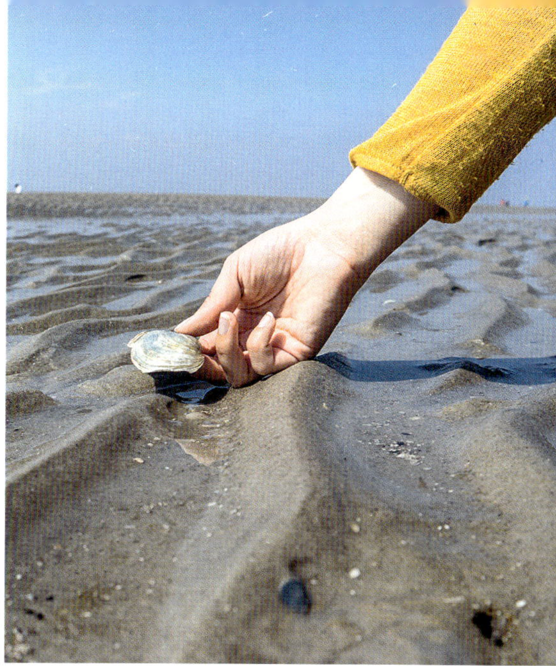

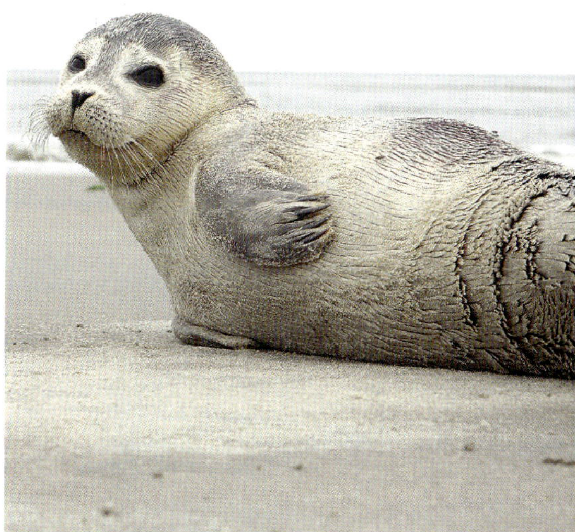

3. MUDFLAT HIKING THROUGH THE WADDEN SEA

The North Sea's Wadden Sea, at 3,475 sq. mi. (9,000 km^2), is the most extensive in the world. Twice a day the water retreats, leaving behind a vast mud landscape crisscrossed by creeks where waterfowl enthusiastically and successfully search for food. Nonfeathered bipeds should definitely get to know this unique landscape by taking a mudflat hike! On Spiekeroog Island, for example, you walk a 5.5 mi. long (9 km) outbound hike in about four hours and take a traditional fishing cutter for the return trip. The 4.33 mi. (7 km) island of Baltrum, which is ideal for "mudflat beginners," can be hiked in three hours. Langeoog offers various hikes ranging between 6 and 14 mi. (10 and 22 km) long; the shortest one takes about four hours. On Norderney, it takes three hours to hike the 5 to 7 mi. (8 to 11.5 km) route, and on Minsener Oog, four to five hours for 5 mi. (8 km).

Time it takes: a variety of hikes of different lengths

4. HIKING ON THE ALSTER HIKING TRAIL

The Alster Hiking Trail, part of the Via Baltica or North German Way of St. James, runs along the Alster River and Alster Lakes from its source in Schleswig-Holstein to its outlet into the Elbe River in Hamburg. The Alster's source region and the moors to the east are protected nature conservancy landscapes but can be explored on footpaths. Most of the hiking route runs right along the river and is also excellent for cycling. Along the way you come across cultural monuments here and there, such as the ruins of Henneberg Castle or the former lockkeeper's house near Poppenbuttel. On forest trails and stairway paths, through parkland and across bridges and locks, the hike leads away from Kayhude right into Hamburg's city center, and from there to the port landing stages.

Length: 23 mi. (37 km)

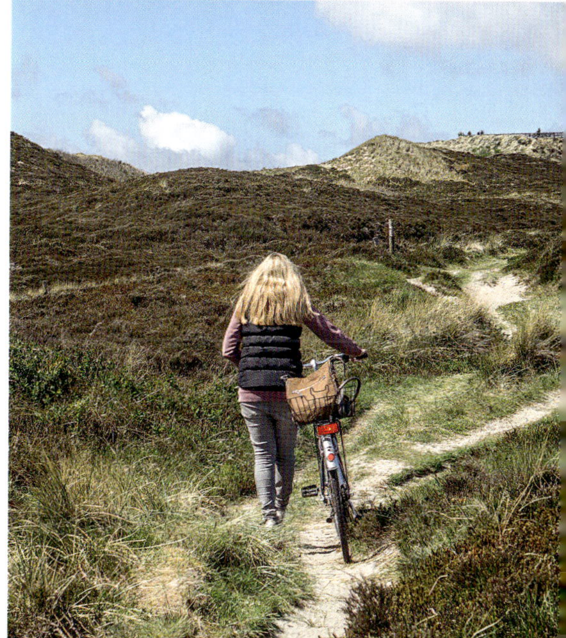

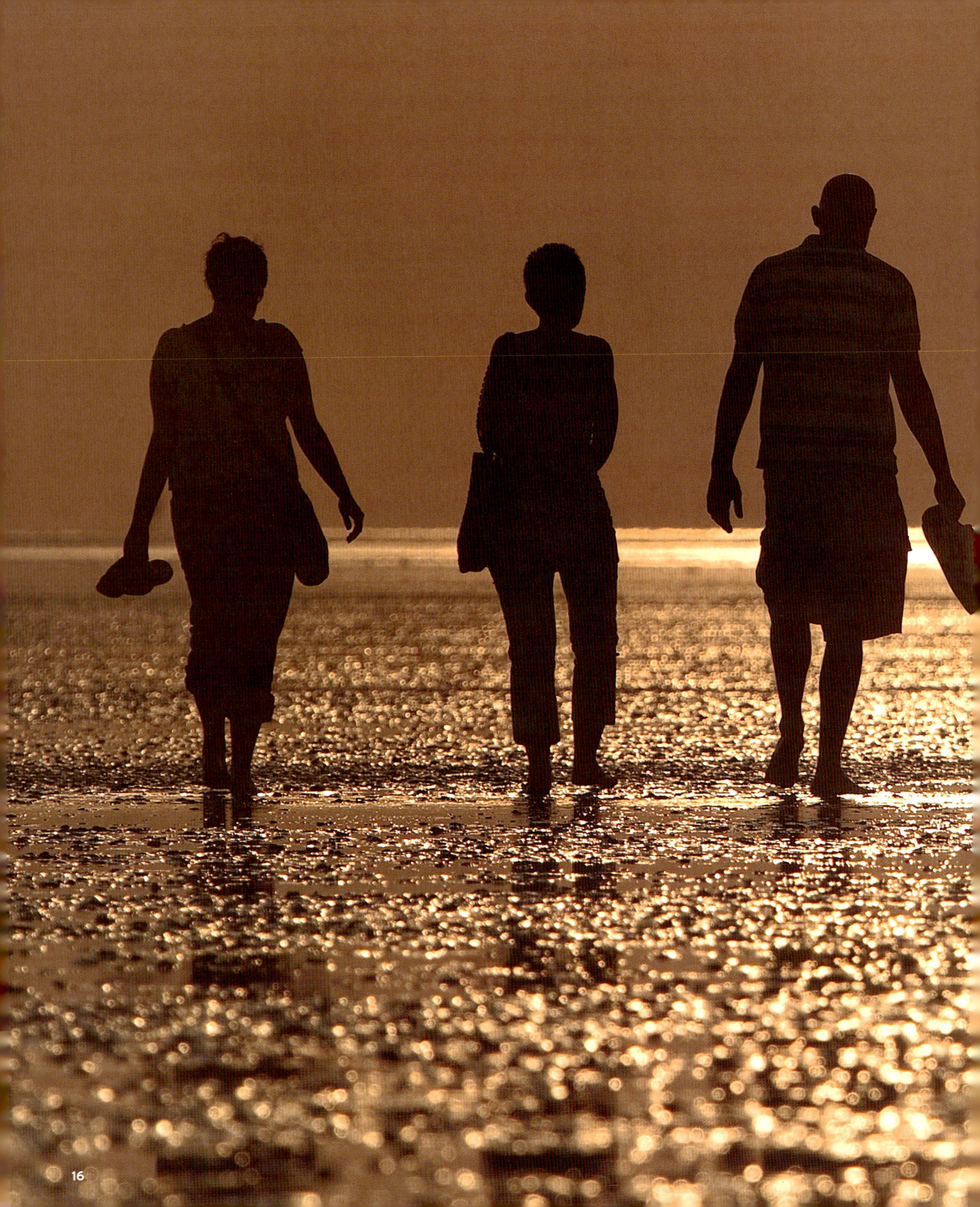

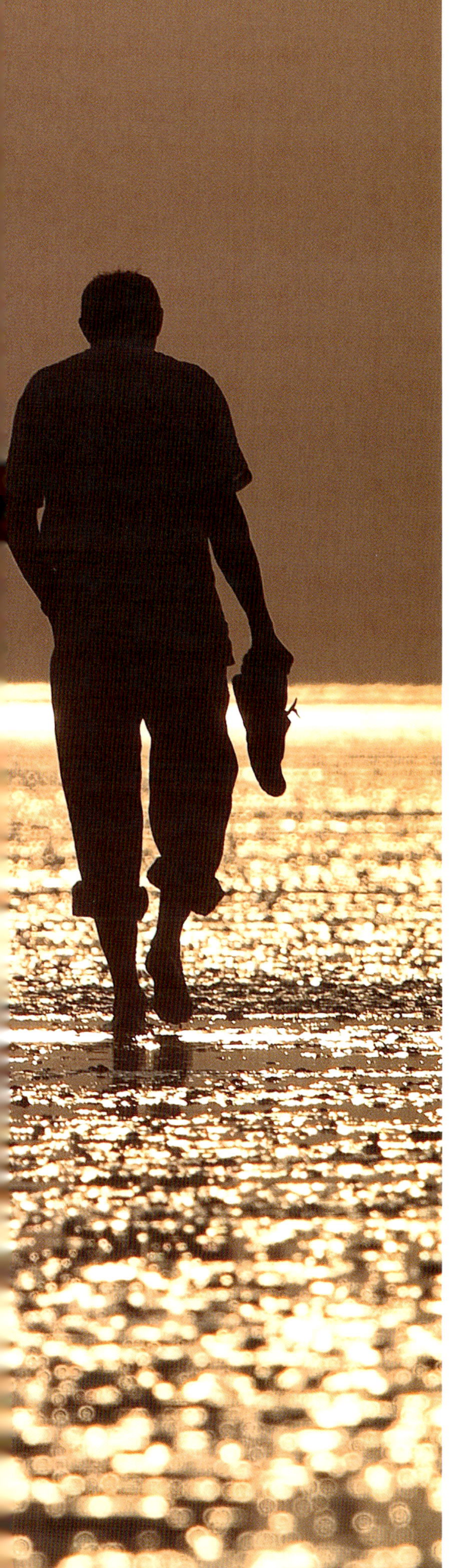

5. THE MECKLENBURG BIKE TOUR

Get on your bike and off you go on a Mecklenburg bike tour! Northern Germany's largest bicycle tour operator has a portfolio including more than 200 organized bicycle trips along national and international bike paths throughout Europe, but especially along the beautiful coastal landscapes of the state of Mecklenburg-Western Pomerania. For example, around Germany's largest island, Rugen; on the car-free island of Hiddensee; and across the Usedom Island landscape to the German-Polish border. There is also an excellent network of routes for bikers who love water biking, including beyond "Meck-Poms"; for example, through the "open country" between the East Frisian cities of Emden and Papenburg, as well as a Kiel Canal tour from Kiel to Brunsbuttel, and combined boat/bike tours between Hamburg and Copenhagen.

Time it takes: organized tours lasting several days
Starting point: depends on the route
Elevation difference: largely bike-friendly terrain
Information: www.mecklenburger-radtour.de

6. HOUSEBOATING ON THE MECKLENBURG SEENPLATTE LAKES

By houseboat through the lake paradise of Müritz & Co. In the town of Waren, the boat slowly slips its way into Müritz National Park, which presides over the Meck-Pom aquatic landscapes with its entrancing nature and a multibranched inland water system. With around 100 lakes and a myriad of smaller bodies of water, the protected Mecklenburgische Seenplatte Lake District alone covers an area of 124 sq. mi. (322 km^2), which corresponds to two-thirds of Germany's largest, the mighty Lake Constance. Sea eagles and ospreys, kingfishers, rails, swans, and herons are the flagship species of a rich bird life; beavers set up their quarters here and there; and there are even European bison grazing by the Kölpinsee, on the Nossentiner Heide (heath), as they did in ancient times. The water system extends to Potsdamer Platz in Berlin!

Boat rental: for 2 to 12 people
Conditions: no boat license required
Total distance: depending on route planning
Day charter: possible
Information: www.locaboat.com, www.nautal.de

7. EXPLORE LÜBECK ON FOOT

It is easy to walk around the Old Town of Lübeck—located on an island—along the inner banks of its waterways. From the Holstentor (Holsten Gate), symbol of the city, you go to the Museumshafen (Museum Harbor). The brick Marienkirche (St. Mary's Church) stands between the Holstentor and Rathausmarkt (Town Hall marketplace). The Breite Strasse takes you by the Buddenbrook House, where Thomas Mann's family once lived. The Gunter Grass and Willy Brandt houses, connected by a garden, can be reached via Pfaffenstrasse. The city is proud of its three Nobel Prize laureates. Beautifully restored townhouses present themselves along the often-still-cobblestoned streets of the Old Town. It's worth taking a look through the narrow passageways to the back courtyards: many of these houses have already been restored. These very narrow passageways were there to ensure that it was possible to carry coffins out of the courtyards in the event of a person's death.

Length / time it takes: once around the island is about 3–4 mi. (5–6 km); walk through the Old Town, 1–2 hours
Size: 100-hectare island surrounded by the Trave River and canals.

8. BY BIKE TO FEHMARN ISLAND

Ready for the islands? You don't always have to choose going by water or by air to get there. Fehmarn Island, in the Baltic Sea, Germany's third-largest island, can be easily reached by bike. The bike trip starts from the town Heiligenhafen, in eastern Holstein, which lies on the tip of the Wagrien Peninsula, then along the flat Baltic coast and by way of a small detour through the interior to Grossenbrode. From there on, the way to go is to always keep following your nose directly toward the sea—and across the Fehmarnsund Bridge, which is used by both road and rail traffic, to one of the most beautiful islands in the Baltic Sea. The capital of the island is Burg, which is the ideal destination for this original route, or as a starting point for many other bike and hiking tours.

Starting point: Heiligenhafen
Trail's end: Burg on Fehmarn Island
Information: www.outdooractive.com/de/radtour/ostseekueste-schleswig-holstein/fehmarn-von-burg-nach-heiligenhafen/

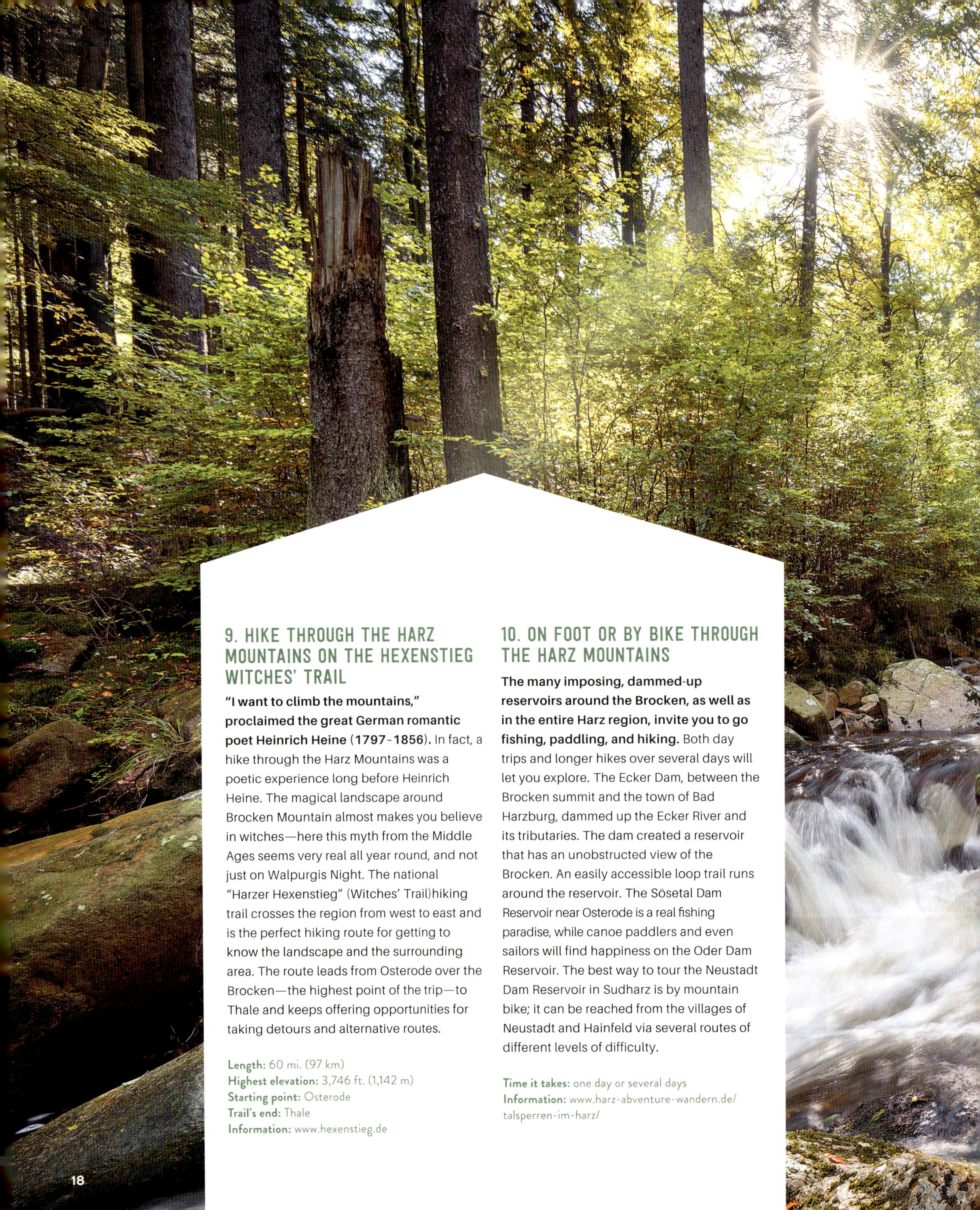

9. HIKE THROUGH THE HARZ MOUNTAINS ON THE HEXENSTIEG WITCHES' TRAIL

"I want to climb the mountains," proclaimed the great German romantic poet Heinrich Heine (1797–1856). In fact, a hike through the Harz Mountains was a poetic experience long before Heinrich Heine. The magical landscape around Brocken Mountain almost makes you believe in witches—here this myth from the Middle Ages seems very real all year round, and not just on Walpurgis Night. The national "Harzer Hexenstieg" (Witches' Trail)hiking trail crosses the region from west to east and is the perfect hiking route for getting to know the landscape and the surrounding area. The route leads from Osterode over the Brocken—the highest point of the trip—to Thale and keeps offering opportunities for taking detours and alternative routes.

Length: 60 mi. (97 km)
Highest elevation: 3,746 ft. (1,142 m)
Starting point: Osterode
Trail's end: Thale
Information: www.hexenstieg.de

10. ON FOOT OR BY BIKE THROUGH THE HARZ MOUNTAINS

The many imposing, dammed-up reservoirs around the Brocken, as well as in the entire Harz region, invite you to go fishing, paddling, and hiking. Both day trips and longer hikes over several days will let you explore. The Ecker Dam, between the Brocken summit and the town of Bad Harzburg, dammed up the Ecker River and its tributaries. The dam created a reservoir that has an unobstructed view of the Brocken. An easily accessible loop trail runs around the reservoir. The Sösetal Dam Reservoir near Osterode is a real fishing paradise, while canoe paddlers and even sailors will find happiness on the Oder Dam Reservoir. The best way to tour the Neustadt Dam Reservoir in Sudharz is by mountain bike; it can be reached from the villages of Neustadt and Hainfeld via several routes of different levels of difficulty.

Time it takes: one day or several days
Information: www.harz-abventure-wandern.de/talsperren-im-harz/

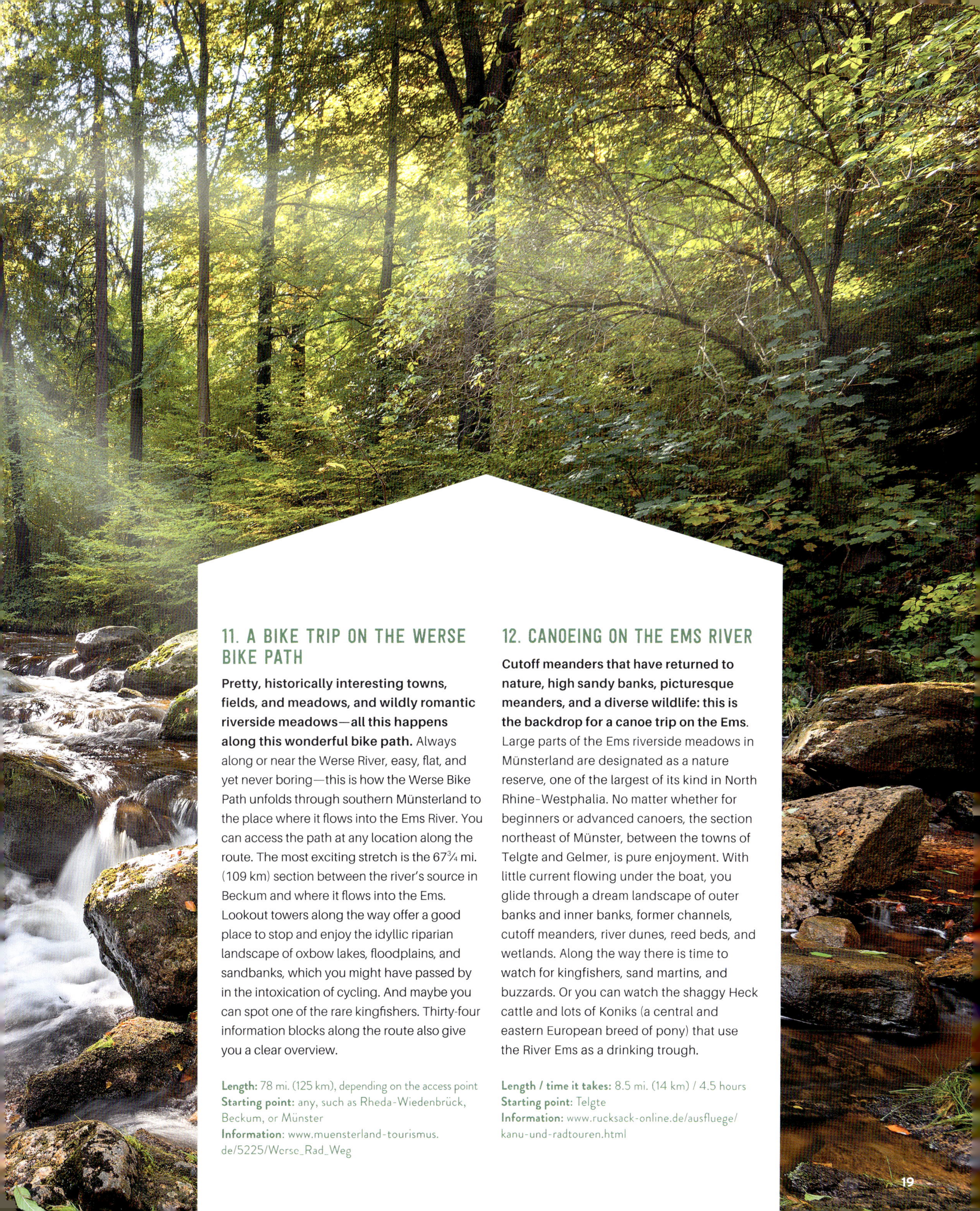

11. A BIKE TRIP ON THE WERSE BIKE PATH

Pretty, historically interesting towns, fields, and meadows, and wildly romantic riverside meadows—all this happens along this wonderful bike path. Always along or near the Werse River, easy, flat, and yet never boring—this is how the Werse Bike Path unfolds through southern Münsterland to the place where it flows into the Ems River. You can access the path at any location along the route. The most exciting stretch is the 67¾ mi. (109 km) section between the river's source in Beckum and where it flows into the Ems. Lookout towers along the way offer a good place to stop and enjoy the idyllic riparian landscape of oxbow lakes, floodplains, and sandbanks, which you might have passed by in the intoxication of cycling. And maybe you can spot one of the rare kingfishers. Thirty-four information blocks along the route also give you a clear overview.

Length: 78 mi. (125 km), depending on the access point
Starting point: any, such as Rheda-Wiedenbrück, Beckum, or Münster
Information: www.muensterland-tourismus.de/5225/Werse_Rad_Weg

12. CANOEING ON THE EMS RIVER

Cutoff meanders that have returned to nature, high sandy banks, picturesque meanders, and a diverse wildlife: this is the backdrop for a canoe trip on the Ems. Large parts of the Ems riverside meadows in Münsterland are designated as a nature reserve, one of the largest of its kind in North Rhine-Westphalia. No matter whether for beginners or advanced canoers, the section northeast of Münster, between the towns of Telgte and Gelmer, is pure enjoyment. With little current flowing under the boat, you glide through a dream landscape of outer banks and inner banks, former channels, cutoff meanders, river dunes, reed beds, and wetlands. Along the way there is time to watch for kingfishers, sand martins, and buzzards. Or you can watch the shaggy Heck cattle and lots of Koniks (a central and eastern European breed of pony) that use the River Ems as a drinking trough.

Length / time it takes: 8.5 mi. (14 km) / 4.5 hours
Starting point: Telgte
Information: www.rucksack-online.de/ausfluege/kanu-und-radtouren.html

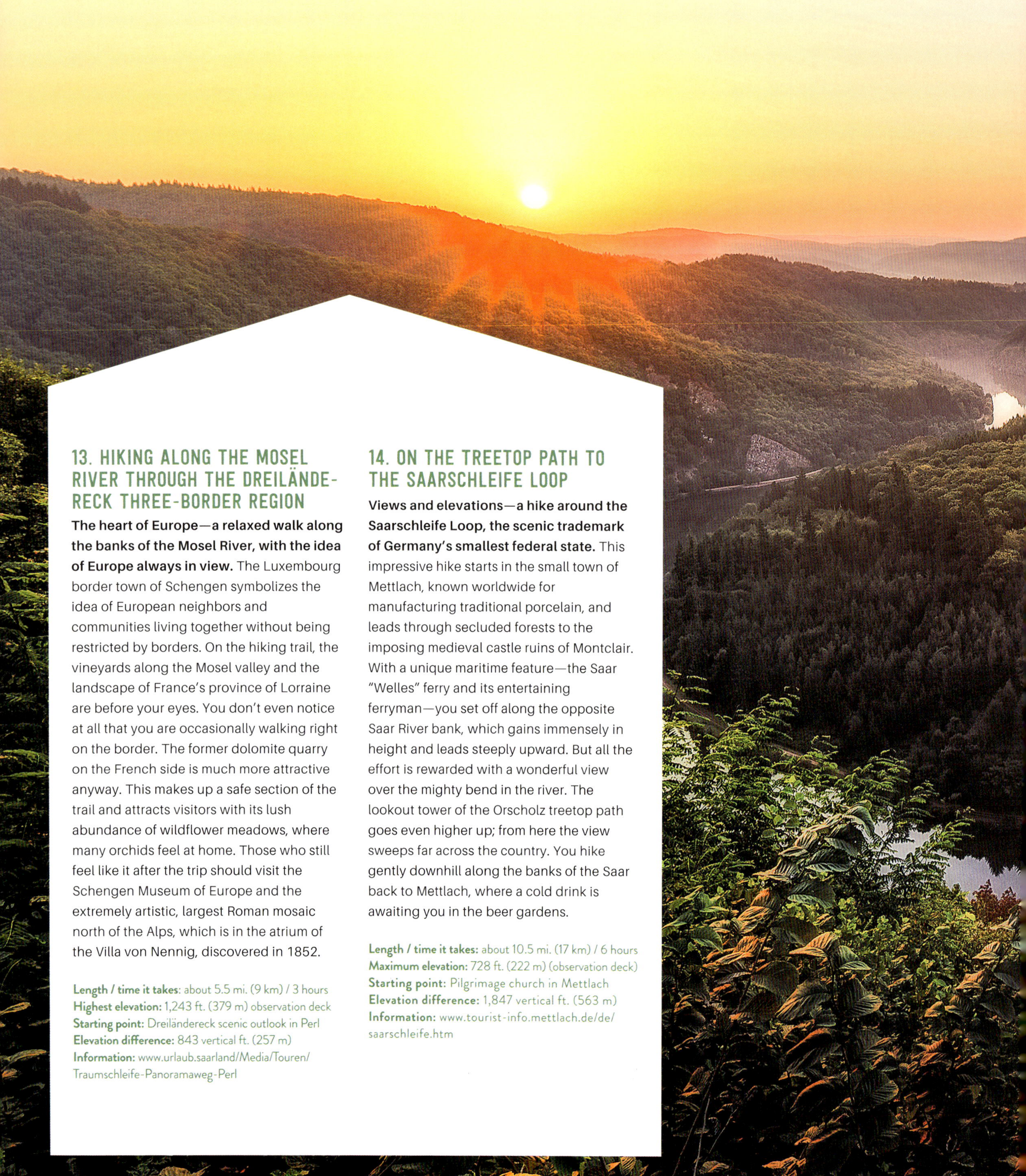

13. HIKING ALONG THE MOSEL RIVER THROUGH THE DREILÄNDE-RECK THREE-BORDER REGION

The heart of Europe—a relaxed walk along the banks of the Mosel River, with the idea of Europe always in view. The Luxembourg border town of Schengen symbolizes the idea of European neighbors and communities living together without being restricted by borders. On the hiking trail, the vineyards along the Mosel valley and the landscape of France's province of Lorraine are before your eyes. You don't even notice at all that you are occasionally walking right on the border. The former dolomite quarry on the French side is much more attractive anyway. This makes up a safe section of the trail and attracts visitors with its lush abundance of wildflower meadows, where many orchids feel at home. Those who still feel like it after the trip should visit the Schengen Museum of Europe and the extremely artistic, largest Roman mosaic north of the Alps, which is in the atrium of the Villa von Nennig, discovered in 1852.

Length / time it takes: about 5.5 mi. (9 km) / 3 hours
Highest elevation: 1,243 ft. (379 m) observation deck
Starting point: Dreiländereck scenic outlook in Perl
Elevation difference: 843 vertical ft. (257 m)
Information: www.urlaub.saarland/Media/Touren/Traumschleife-Panoramaweg-Perl

14. ON THE TREETOP PATH TO THE SAARSCHLEIFE LOOP

Views and elevations—a hike around the Saarschleife Loop, the scenic trademark of Germany's smallest federal state. This impressive hike starts in the small town of Mettlach, known worldwide for manufacturing traditional porcelain, and leads through secluded forests to the imposing medieval castle ruins of Montclair. With a unique maritime feature—the Saar "Welles" ferry and its entertaining ferryman—you set off along the opposite Saar River bank, which gains immensely in height and leads steeply upward. But all the effort is rewarded with a wonderful view over the mighty bend in the river. The lookout tower of the Orscholz treetop path goes even higher up; from here the view sweeps far across the country. You hike gently downhill along the banks of the Saar back to Mettlach, where a cold drink is awaiting you in the beer gardens.

Length / time it takes: about 10.5 mi. (17 km) / 6 hours
Maximum elevation: 728 ft. (222 m) (observation deck)
Starting point: Pilgrimage church in Mettlach
Elevation difference: 1,847 vertical ft. (563 m)
Information: www.tourist-info.mettlach.de/de/saarschleife.htm

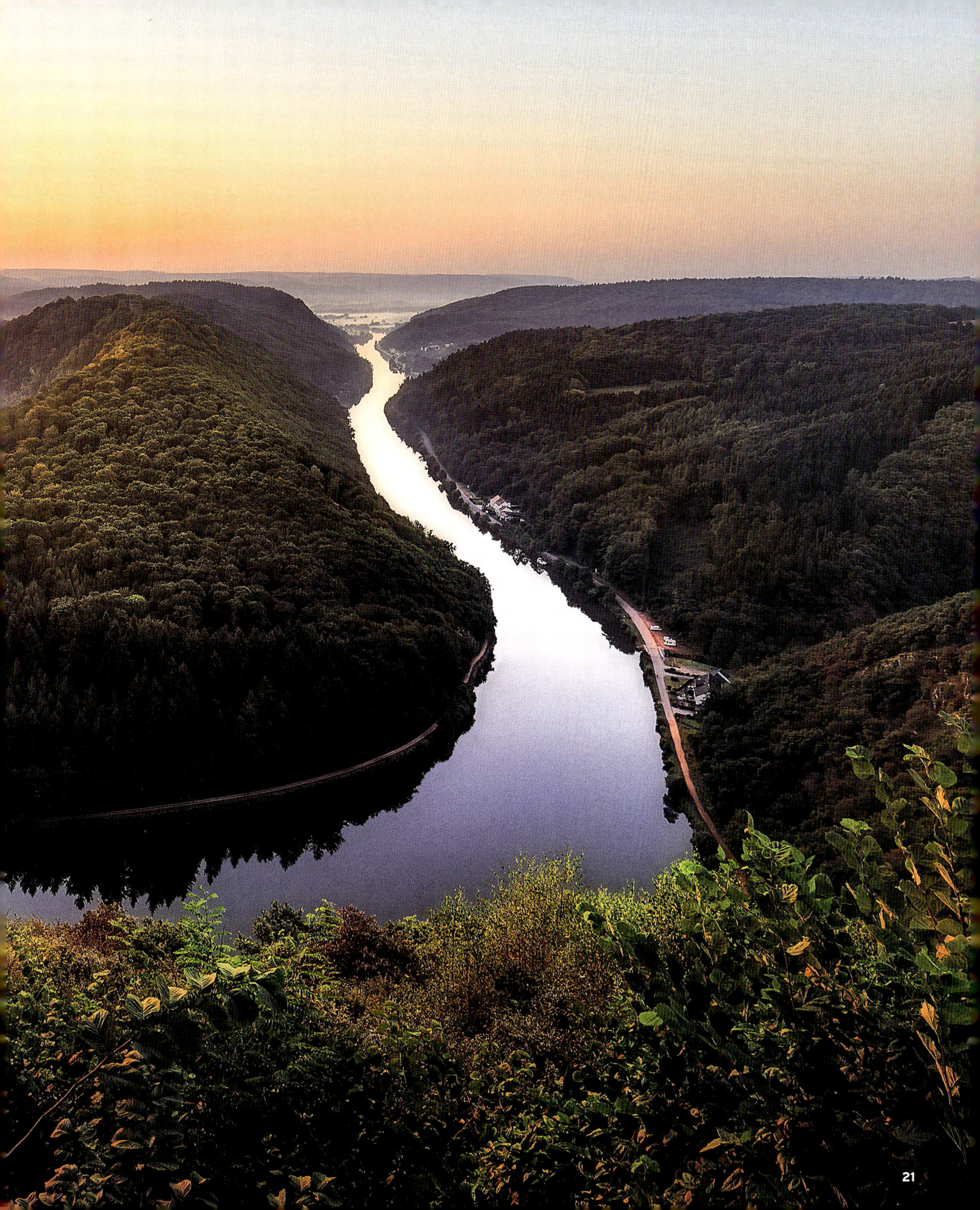

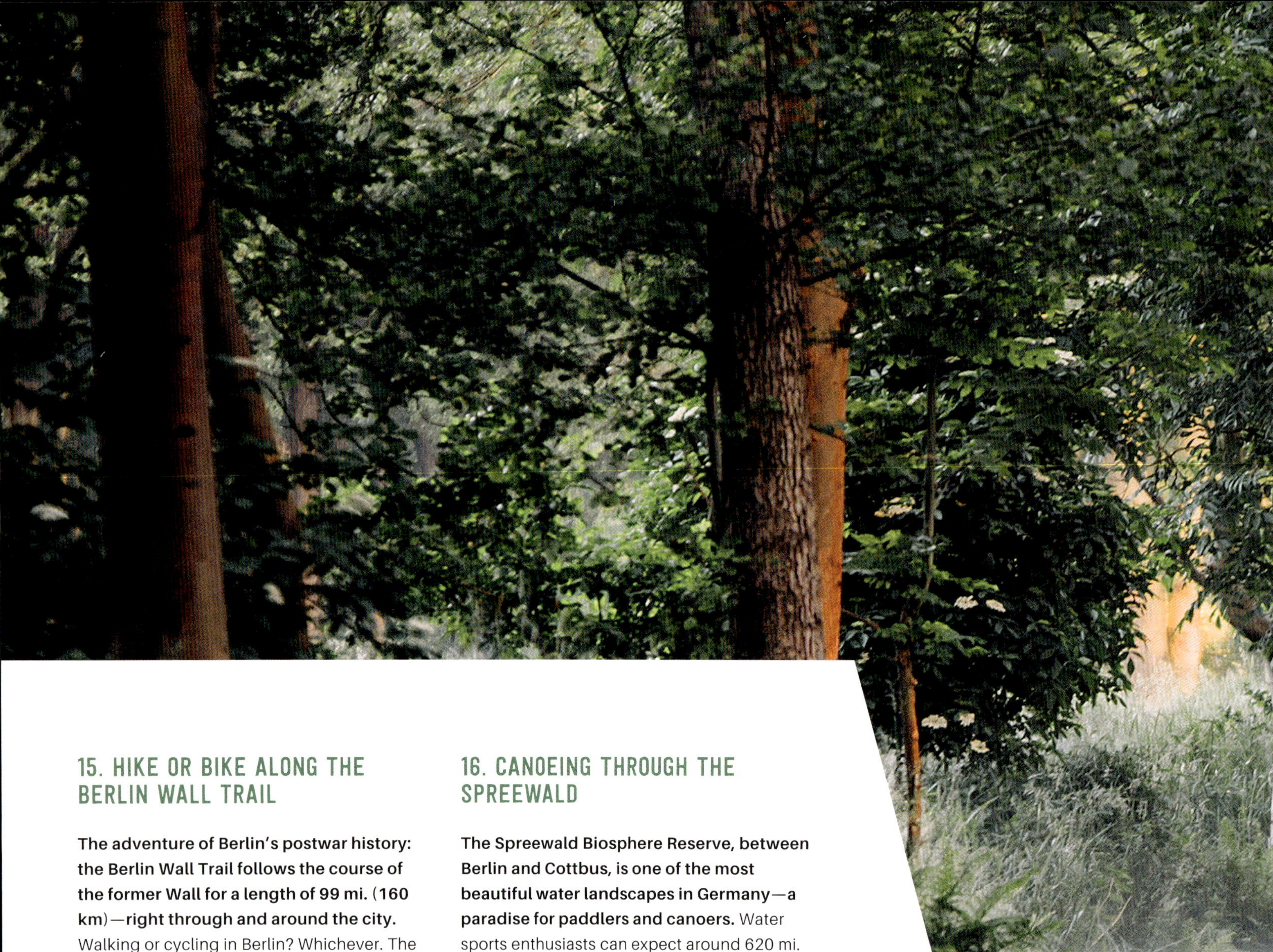

15. HIKE OR BIKE ALONG THE BERLIN WALL TRAIL

The adventure of Berlin's postwar history: the Berlin Wall Trail follows the course of the former Wall for a length of 99 mi. (160 km)—right through and around the city. Walking or cycling in Berlin? Whichever. The route alternates between the former customs road in West Berlin and the old military column patrol road laid out by the East German border troops along the former East German border installations—moments of horror included. The route combines sections steeped in history, where you can still find remnants of the wall, with lovely scenic routes to the north, south, and west of the city. Pure Berlin in all its facets. Photos and texts along the way are constant reminders of events that happened at these locations, while twenty-nine steles commemorate the dead and the events at these places.

Length / time it takes: 99 mi. (160 km) / 14 partial stages of 3–13 mi. (5–21 km); each stage 1–6 hours
Starting point: Veltheimstrasse, Hermsdorf
Information: www.berlin.de/mauer/mauerweg/

16. CANOEING THROUGH THE SPREEWALD

The Spreewald Biosphere Reserve, between Berlin and Cottbus, is one of the most beautiful water landscapes in Germany—a paradise for paddlers and canoers. Water sports enthusiasts can expect around 620 mi. (1,000 km) of navigable streams right before the gates of Germany's capital city, plus a fantastic, at times untouched, natural landscape—a paradise. Almost an insider tip: the largely undiscovered Unterspreewald area, which is often passed by on the way to the central Oberspree forest. That guarantees absolute peace and quiet—at least on weekdays. The trip from Schlepzig to Gross Wasserburg offers pleasant insight into how much this part of the Spreewald has retained its natural quality. Sometimes you feel as if you are in the primeval forest, and the strong current in the Puhlstrom River even creates a sense of adventure.

Length / time it takes: 13.5 mi. (22 km) / 1 day
Starting point: Schlepzig or Unterspreewald
Information: www.reiseland-brandenburg.de/poi/spreewald/kanutouren

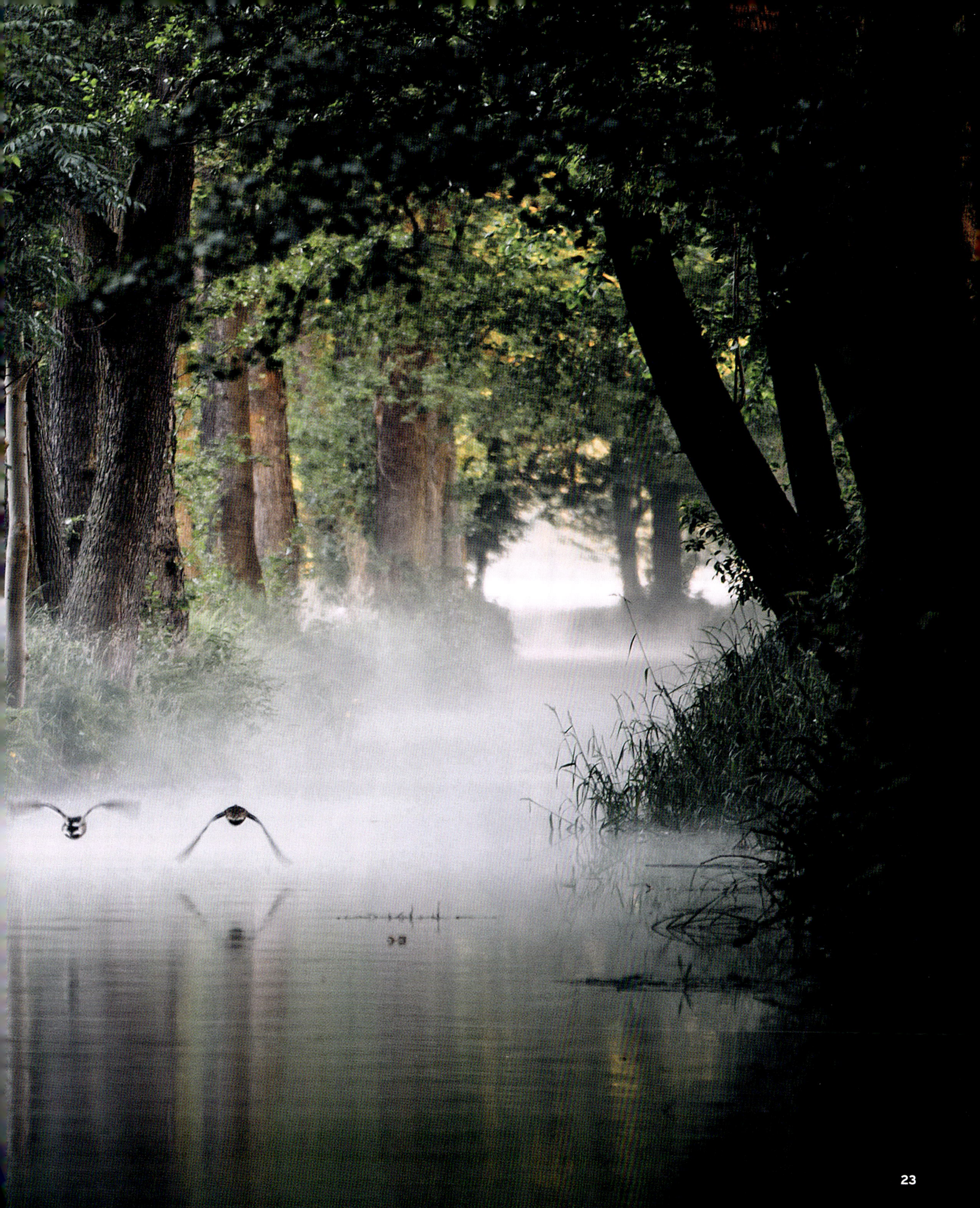

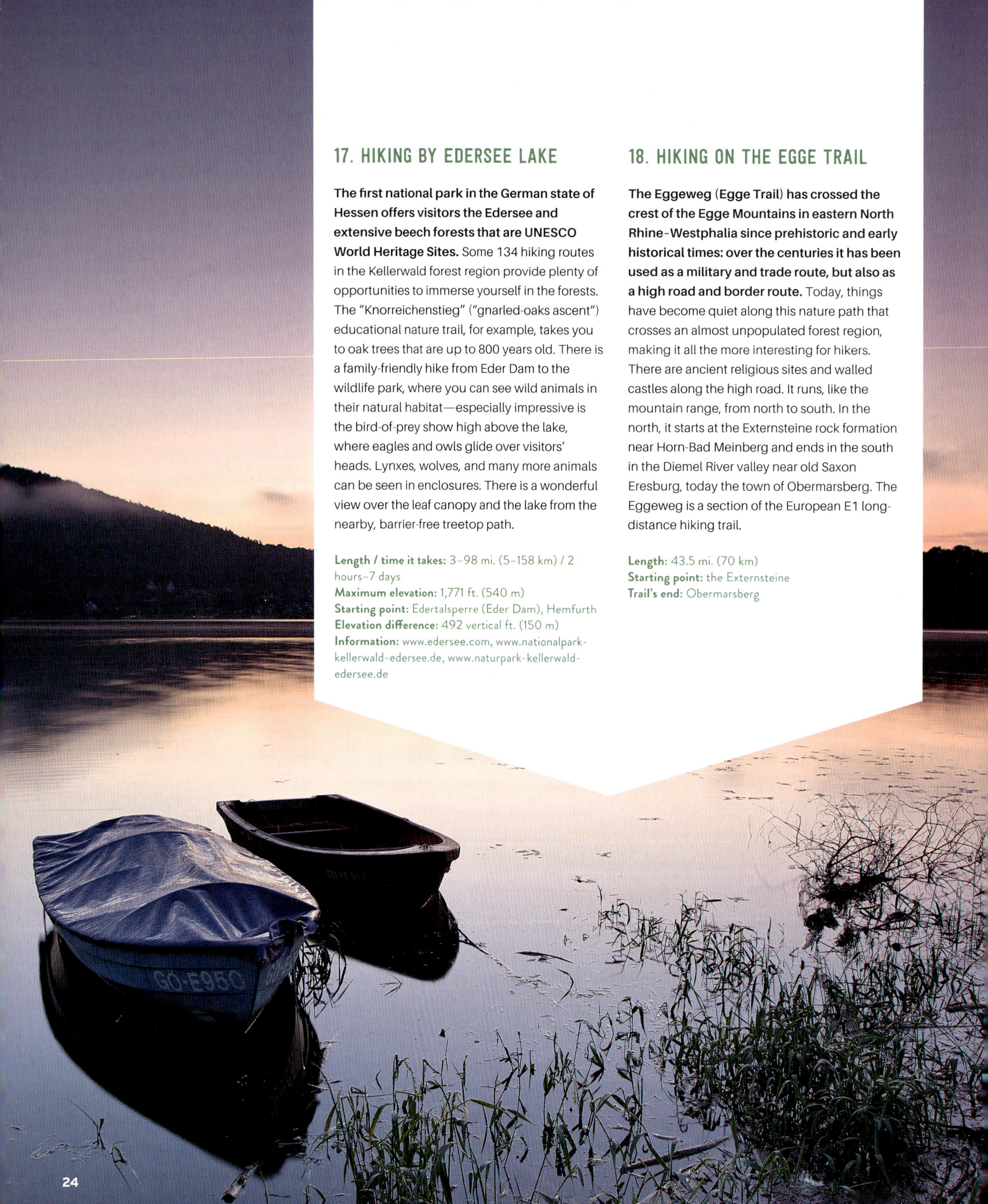

17. HIKING BY EDERSEE LAKE

The first national park in the German state of Hessen offers visitors the Edersee and extensive beech forests that are UNESCO World Heritage Sites. Some 134 hiking routes in the Kellerwald forest region provide plenty of opportunities to immerse yourself in the forests. The "Knorreichenstieg" ("gnarled-oaks ascent") educational nature trail, for example, takes you to oak trees that are up to 800 years old. There is a family-friendly hike from Eder Dam to the wildlife park, where you can see wild animals in their natural habitat—especially impressive is the bird-of-prey show high above the lake, where eagles and owls glide over visitors' heads. Lynxes, wolves, and many more animals can be seen in enclosures. There is a wonderful view over the leaf canopy and the lake from the nearby, barrier-free treetop path.

Length / time it takes: 3–98 mi. (5–158 km) / 2 hours–7 days
Maximum elevation: 1,771 ft. (540 m)
Starting point: Edertalsperre (Eder Dam), Hemfurth
Elevation difference: 492 vertical ft. (150 m)
Information: www.edersee.com, www.nationalpark-kellerwald-edersee.de, www.naturpark-kellerwald-edersee.de

18. HIKING ON THE EGGE TRAIL

The Eggeweg (Egge Trail) has crossed the crest of the Egge Mountains in eastern North Rhine-Westphalia since prehistoric and early historical times: over the centuries it has been used as a military and trade route, but also as a high road and border route. Today, things have become quiet along this nature path that crosses an almost unpopulated forest region, making it all the more interesting for hikers. There are ancient religious sites and walled castles along the high road. It runs, like the mountain range, from north to south. In the north, it starts at the Externsteine rock formation near Horn-Bad Meinberg and ends in the south in the Diemel River valley near old Saxon Eresburg, today the town of Obermarsberg. The Eggeweg is a section of the European E1 long-distance hiking trail.

Length: 43.5 mi. (70 km)
Starting point: the Externsteine
Trail's end: Obermarsberg

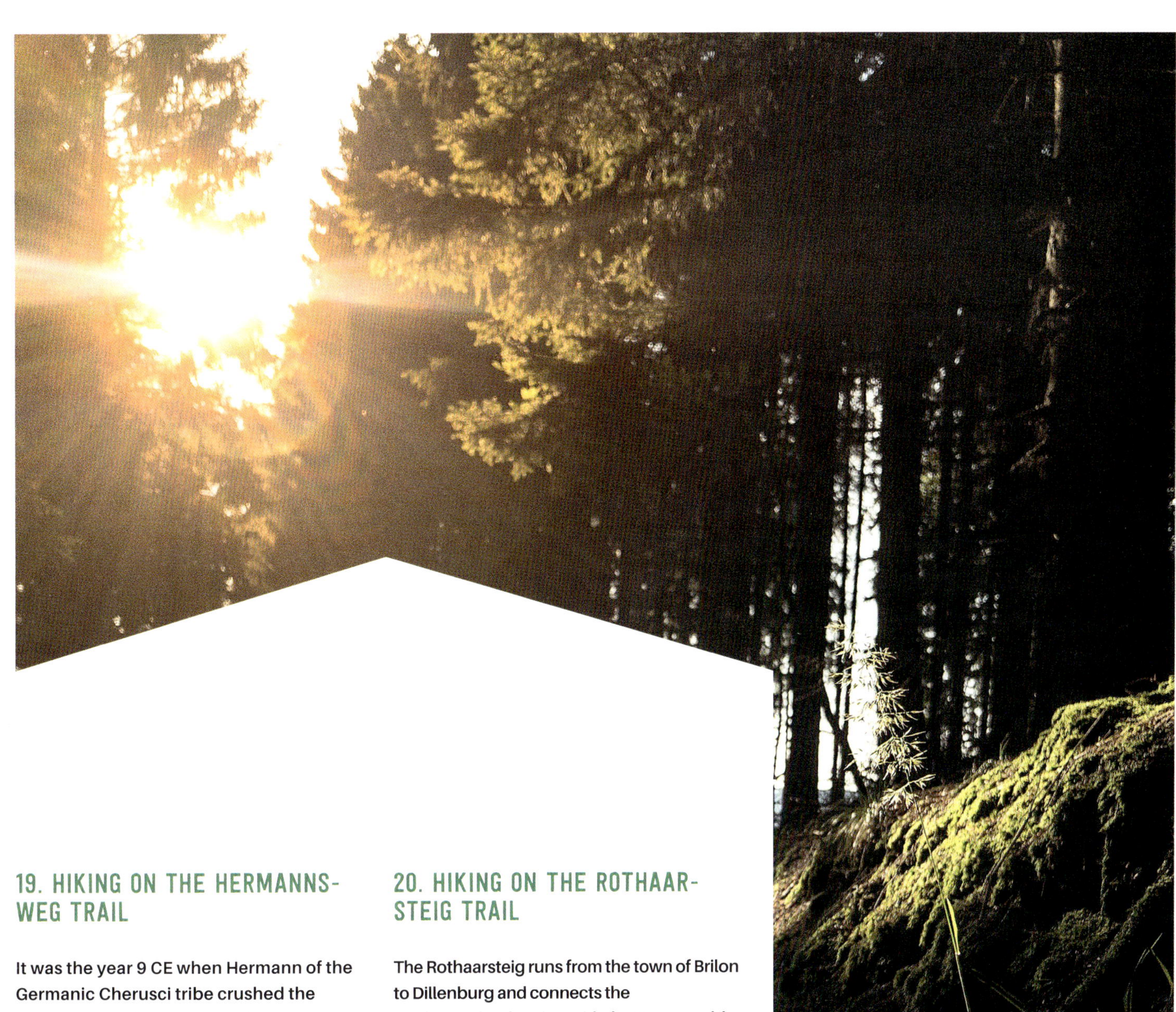

19. HIKING ON THE HERMANNS-WEG TRAIL

It was the year 9 CE when Hermann of the Germanic Cherusci tribe crushed the Romans in the battle of the Teutoburg Forest. The Hermann Monument, one of the many sights that line this high route from the town of Rheine to the Lippische Velmerstot hilltop, commemorates these events. The sights include the bizarre Dörenther Klippen rock formations, as well as the Externsteine and the Adlerwarte Berlebeck wildlife center in Detmold. The Hermannsweg is one of the most beautiful high-elevation trails in Germany and runs over the crest of the Teutoburg Forest, where it leads over heights of 328 to 1,300 ft. (100–400 m) to Horn-Bad Meinberg and ends on the 1,446 ft. high (441 m) Lippische Velmerstot rock formation.

Length: 97 mi. (156 km) / 8 stages

20. HIKING ON THE ROTHAAR-STEIG TRAIL

The Rothaarsteig runs from the town of Brilon to Dillenburg and connects the Hochsauerland region with the Westerwald region. It thus crosses the densely wooded center of Germany on trails close to nature. Through quiet forests and past idyllic meadows, it leads over the crest of the Rothaar Mountains. The trail passes by Bruchhauser Steine rock formation near Bruchhausen and the sources of various rivers, including the Ruhr, Eder, Sieg, Lahn, and Dill Rivers, among the beautiful stage points. Pretty towns such as Willingen and Bad Berleburg also lie along the route. In addition to the main trail, there are many shorter branch trails, such as the charming 5.5 mi. (14 km) valley route through the Grubensiepen River gorge or the mountain branch trail from Dillenburg to the southern Siegerland region, with its many scenic outlooks, which is about 20.5 mi. (53 km) long.

Length: 60.5 mi. (157 km) / 12 stages
Information: www.rothaarste

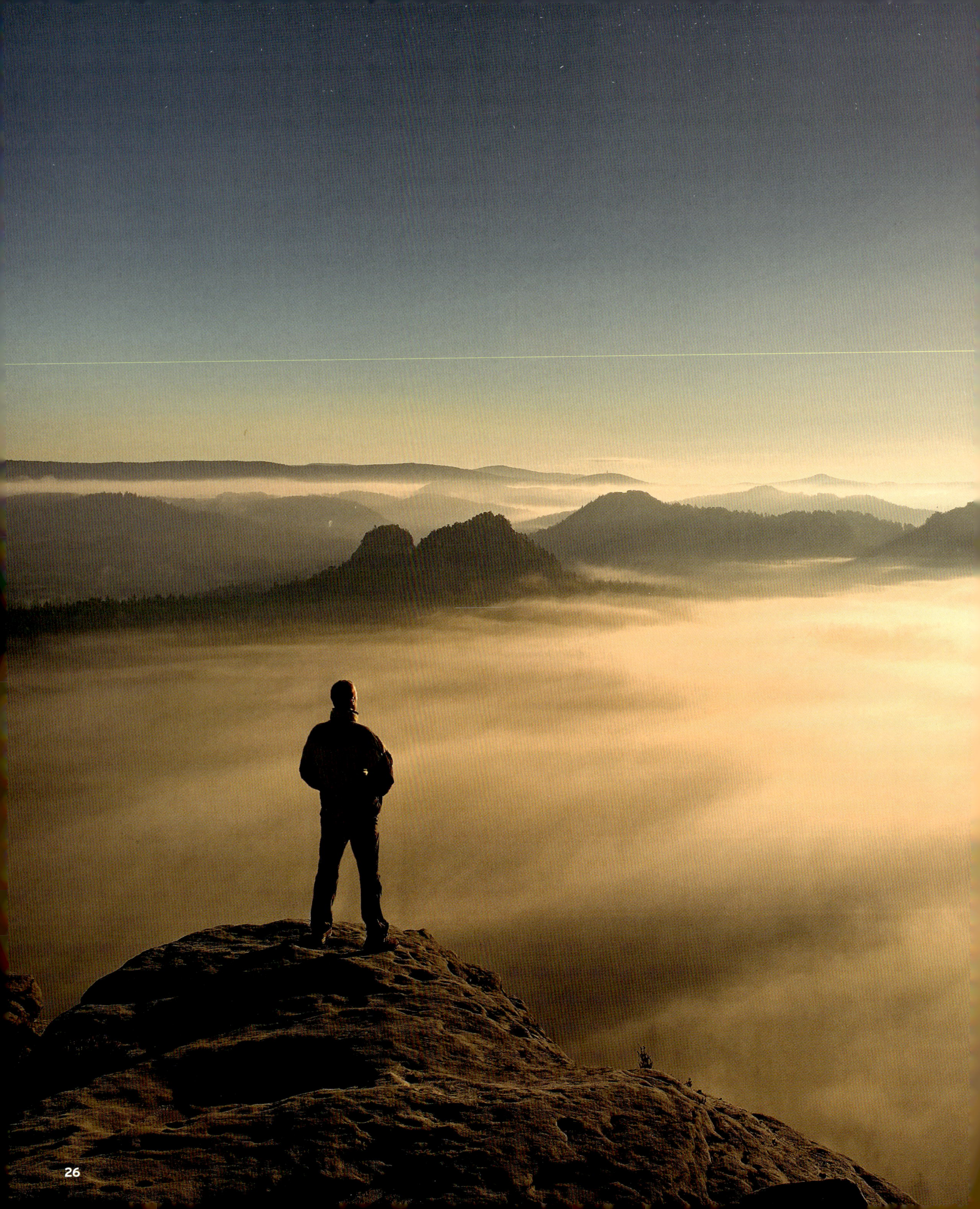

21. HIKING ON THE VOGTLAND PANORAMA TRAIL

The trail starts in the tranquil town of Greiz, in fact at something superlative: at 256 ft. (78 m), the Göltzsch Viaduct is the highest brick bridge in the world, and it also has a very impressive visual impact, with its rows of arches stacked one above the other. Along the way, stream and river valleys, summer meadows blooming with bright arnica and wild orchids, and quiet forest clearings with European jays and great spotted woodpeckers all invite you to discover and relax: the way is the goal. Special highlights: the lovely town of Plauen, the popular Pöhl Dam, the historical Albert spa in Bad Elster, the famous spas of Marienbad and Franzensbad across the Czech border, or the Vogtländ "Musikwinkel" ("musical corner") around the town of Markneukirchen. The trail was one of the first to be recognized as a "[high-]quality trail."

Length: 137 mi. (220 km) / 10 stages
Starting point/trail's end: Göltzschtal Viaduct

22. HIKE THROUGH THE PICTURESQUE ELBE SANDSTONE MOUNTAINS

As early as the eighteenth century, travelers in general and painters such as Caspar David Friedrich, and Ludwig Richter in particular, were fascinated by the grotesque rock formations, wildly romantic gorges, and sublime table mountains of the Saxon Switzerland. The Malerweg (Painter's Path) was laid out following the historical route across the Elbe Sandstone Mountains, a hiking route from Liebethal to Pirna. Many Romantic painters recorded their impressions on canvas; for example, Caspar David Friedrich's painting *The Wanderer above the Sea of Fog* was created along this route. The trail follows ways left in their natural condition through a varied and diverse landscape to impressive scenic outlooks. This hiking route brings together everything that Saxon Switzerland has to offer: gorges, steep paths, caves, and friendly people.

Length: 70 mi. (112 km) / 8 stages

23. HIKING THE RENNSTEIG RIDGE TRAIL THROUGH THE THURINGIAN FOREST

"There is peace over all the mountaintops; you barely feel a breath stirring in the treetops." *The Wanderer's Nightsong*, by Johann Wolfgang von Goethe, was written on Kickelhahn Mountain, north of the Rennsteig Ridge Trail. There is still peace in the broad Thuringian Forest. It is worth it to occasionally leave the ridge to see more than the forest and the beautiful views, because there is a lot of culture to the left and right of the trail. The starting point in Hörschel is on the Werra River—practically within sight of Wartburg Castle, where Martin Luther translated the Bible. From Hörschel there is an alternative route to the Wartburg and farther through the Dragon Gorge to the Hohe Sonne stage destination on the Rennsteig. Anyone can incorporate many destinations into the hike until they reach the final destination at Blankenstein an der Saale—or cover the route on a 17 mi. (27 km) bike path.

Length / time it takes: footpath: 104 mi. (168 km) / 5–8 days
Length / time it takes: bike path: 121 mi. (195 km) / 3–4 days
Highest elevation: 3,225 ft. (983 m)
Starting point: Hörschel
Trail's end: Blankenstein on the Saale
Elevation difference: 9,842 vertical ft. (3,000 m)
Information: www.Thüringer-Wald.com

24. HIKING ON BARBAROSSAWEG TRAIL

The Barbarossaweg Trail lies in the northern part of the federal states of Hessen and Thuringia and is one of the most beautiful and oldest hiking trails in Thuringia. The Barbarossa Trail was named after Emperor Friedrich I, also commonly known as Emperor Barbarossa or "Redbeard." The Barbarossaweg connects the legendary Kyffhäuser range of hills, with its many cities, monasteries, and castles, where the emperor lived and worked during the twelfth century. There are countless historical sights to be found along the route. In addition, the Ars Natura artists' initiative has teamed up to confront hikers with modern art along their march. It has installed around 150 works of art, especially along the stretch between Bad Wildungen and Waldkappel. The division of Germany after World War II severed the hiking trail; in the early 1990s it was redesigned as a long-distance hiking trail.

Length: 205 mi. (330 km)
Starting point: Korbach
Trail's end: Kelbra-Tilleda

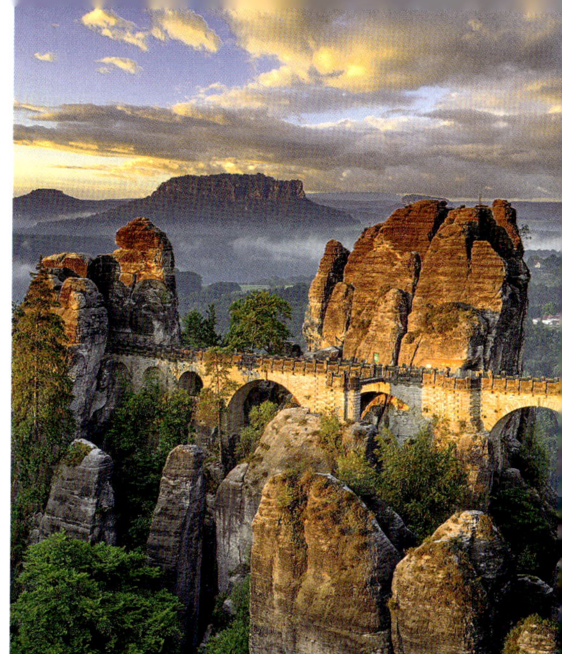

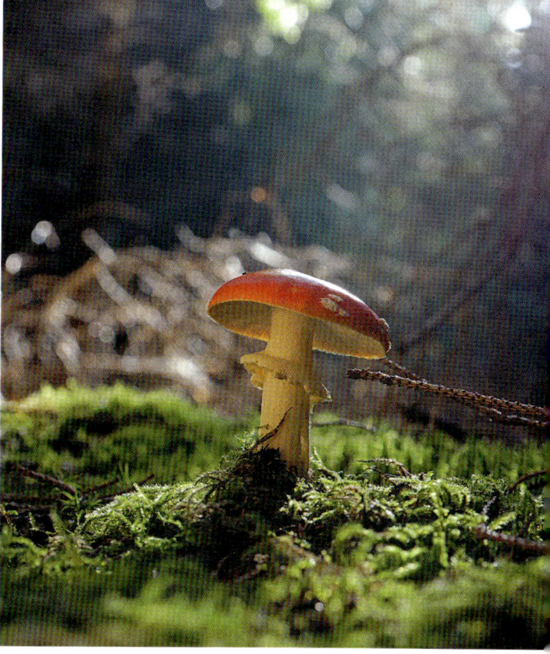

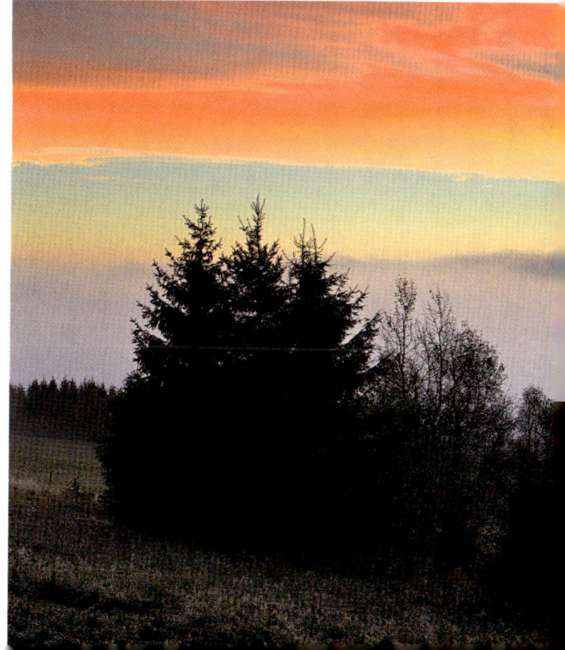

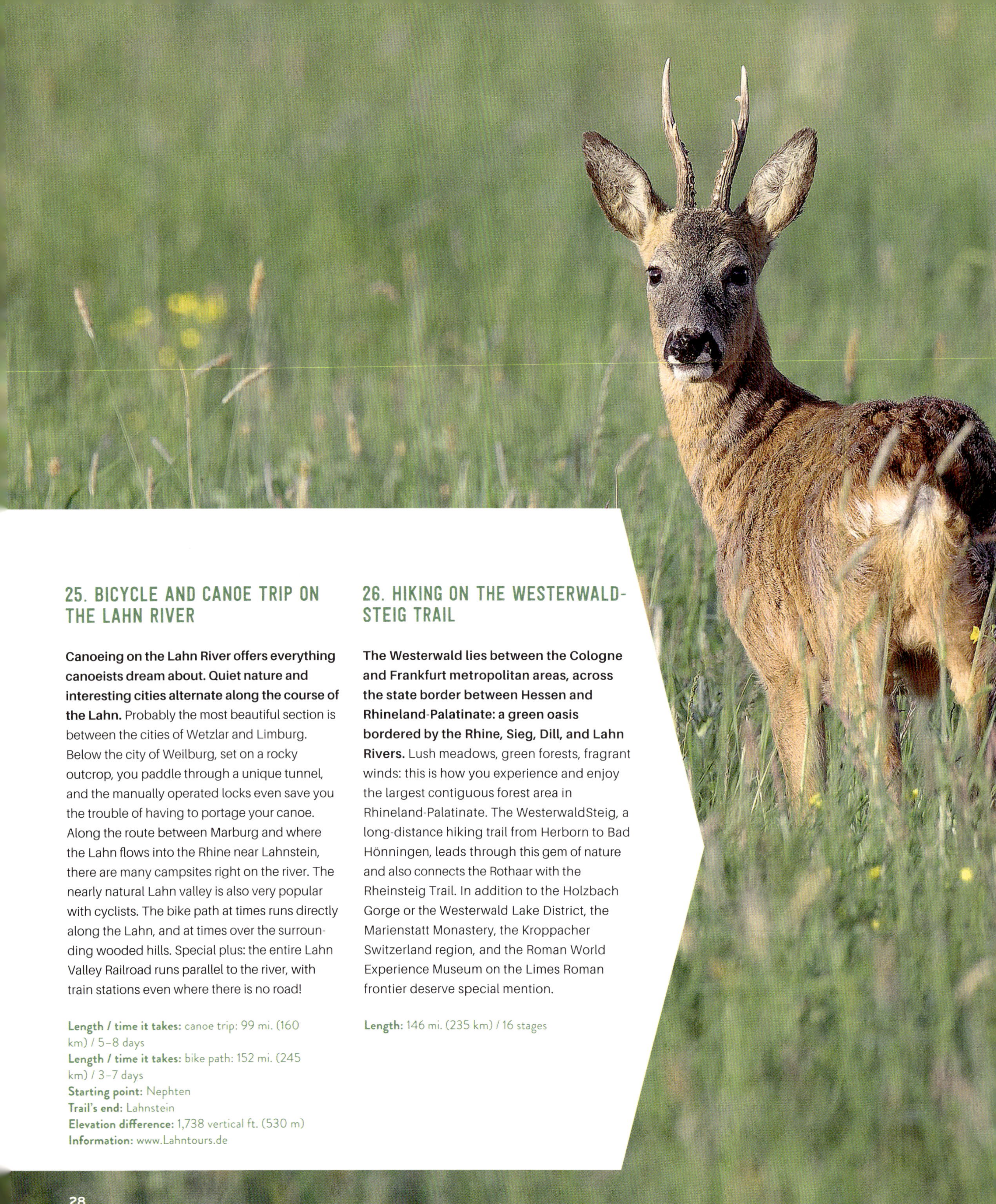

25. BICYCLE AND CANOE TRIP ON THE LAHN RIVER

Canoeing on the Lahn River offers everything canoeists dream about. Quiet nature and interesting cities alternate along the course of the Lahn. Probably the most beautiful section is between the cities of Wetzlar and Limburg. Below the city of Weilburg, set on a rocky outcrop, you paddle through a unique tunnel, and the manually operated locks even save you the trouble of having to portage your canoe. Along the route between Marburg and where the Lahn flows into the Rhine near Lahnstein, there are many campsites right on the river. The nearly natural Lahn valley is also very popular with cyclists. The bike path at times runs directly along the Lahn, and at times over the surrounding wooded hills. Special plus: the entire Lahn Valley Railroad runs parallel to the river, with train stations even where there is no road!

Length / time it takes: canoe trip: 99 mi. (160 km) / 5–8 days
Length / time it takes: bike path: 152 mi. (245 km) / 3–7 days
Starting point: Nephten
Trail's end: Lahnstein
Elevation difference: 1,738 vertical ft. (530 m)
Information: www.Lahntours.de

26. HIKING ON THE WESTERWALD-STEIG TRAIL

The Westerwald lies between the Cologne and Frankfurt metropolitan areas, across the state border between Hessen and Rhineland-Palatinate: a green oasis bordered by the Rhine, Sieg, Dill, and Lahn Rivers. Lush meadows, green forests, fragrant winds: this is how you experience and enjoy the largest contiguous forest area in Rhineland-Palatinate. The WesterwaldSteig, a long-distance hiking trail from Herborn to Bad Hönningen, leads through this gem of nature and also connects the Rothaar with the Rheinsteig Trail. In addition to the Holzbach Gorge or the Westerwald Lake District, the Marienstatt Monastery, the Kroppacher Switzerland region, and the Roman World Experience Museum on the Limes Roman frontier deserve special mention.

Length: 146 mi. (235 km) / 16 stages

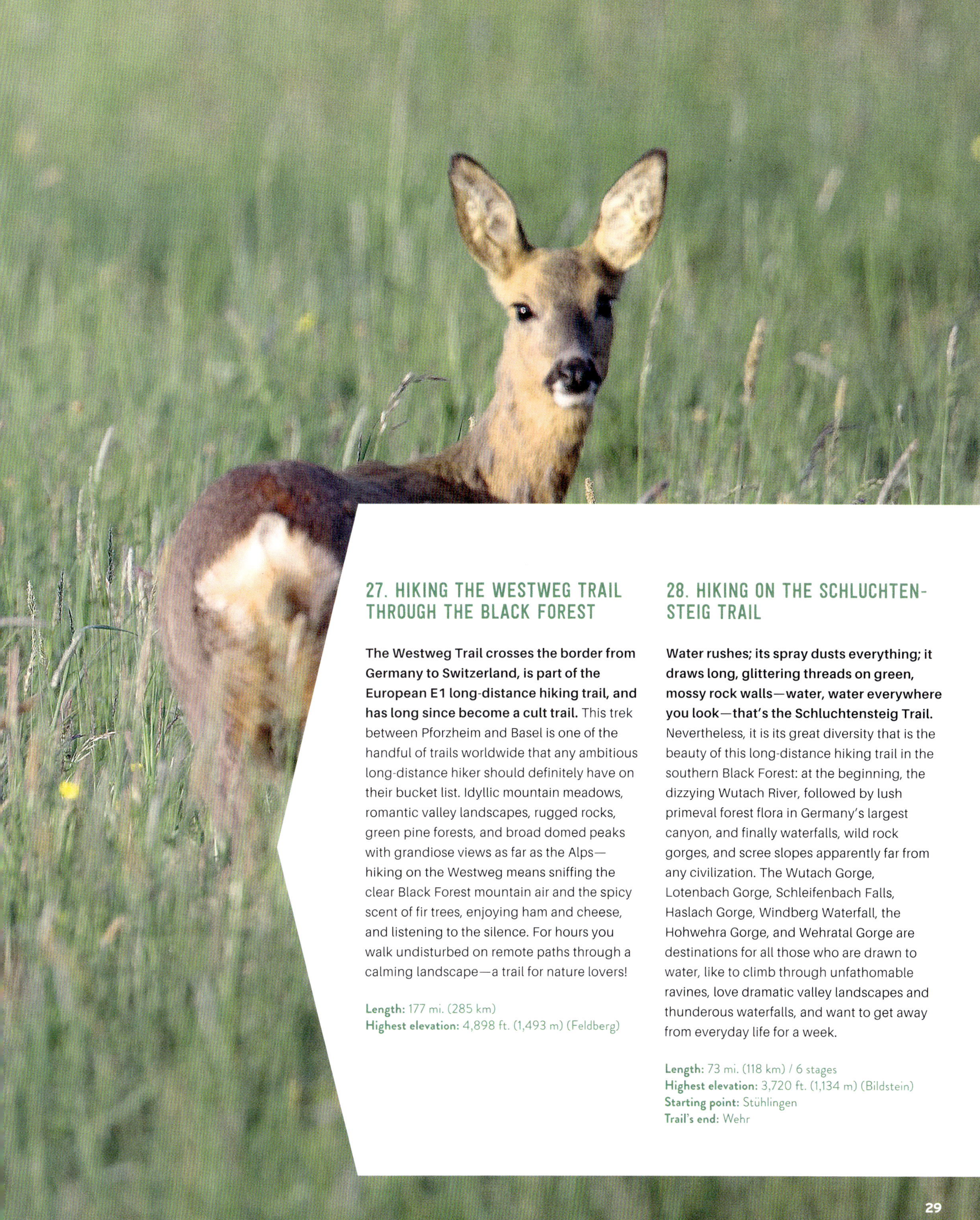

27. HIKING THE WESTWEG TRAIL THROUGH THE BLACK FOREST

The Westweg Trail crosses the border from Germany to Switzerland, is part of the European E1 long-distance hiking trail, and has long since become a cult trail. This trek between Pforzheim and Basel is one of the handful of trails worldwide that any ambitious long-distance hiker should definitely have on their bucket list. Idyllic mountain meadows, romantic valley landscapes, rugged rocks, green pine forests, and broad domed peaks with grandiose views as far as the Alps—hiking on the Westweg means sniffing the clear Black Forest mountain air and the spicy scent of fir trees, enjoying ham and cheese, and listening to the silence. For hours you walk undisturbed on remote paths through a calming landscape—a trail for nature lovers!

Length: 177 mi. (285 km)
Highest elevation: 4,898 ft. (1,493 m) (Feldberg)

28. HIKING ON THE SCHLUCHTEN-STEIG TRAIL

Water rushes; its spray dusts everything; it draws long, glittering threads on green, mossy rock walls—water, water everywhere you look—that's the Schluchtensteig Trail. Nevertheless, it is its great diversity that is the beauty of this long-distance hiking trail in the southern Black Forest: at the beginning, the dizzying Wutach River, followed by lush primeval forest flora in Germany's largest canyon, and finally waterfalls, wild rock gorges, and scree slopes apparently far from any civilization. The Wutach Gorge, Lotenbach Gorge, Schleifenbach Falls, Haslach Gorge, Windberg Waterfall, the Hohwehra Gorge, and Wehratal Gorge are destinations for all those who are drawn to water, like to climb through unfathomable ravines, love dramatic valley landscapes and thunderous waterfalls, and want to get away from everyday life for a week.

Length: 73 mi. (118 km) / 6 stages
Highest elevation: 3,720 ft. (1,134 m) (Bildstein)
Starting point: Stühlingen
Trail's end: Wehr

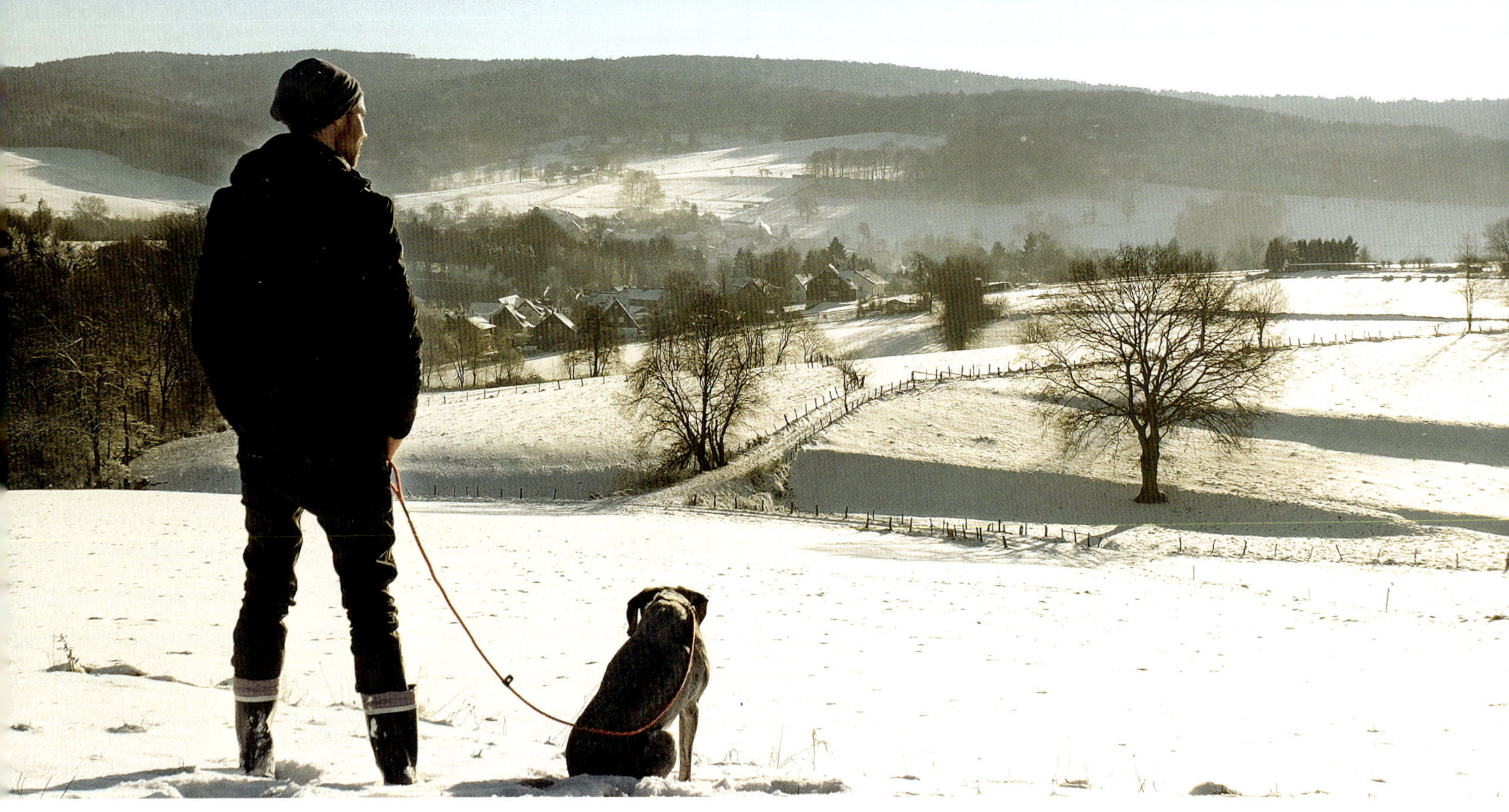

29. ON THE GERMAN UNITY HIKING TRAIL

This interesting interior German long-distance hiking trail leads from Görlitz, which is considered Germany's easternmost city, to the westernmost; namely, Aachen. It runs along other marked long-distance hiking trails, such as the northern Kammweg (Ridge Trail) in Oberlausitz (Upper Lusatia historical region), or the Rennsteig in the Thuringian Forest. On one section, between Bad Godesberg on the Rhine and Aachen, it is identical to the European E8 long-distance hiking trail (Ireland-Black Sea). The idea for this trail was born during a social get-together in 1990. The unusual thing about it is that, unlike the majority of the major long-distance hiking trails, it does not traverse the landscape in a north-south direction but connects the whole of Germany together from east to west—as a sign of German unity.

Length: about 670 mi. (1,080 km)
Information: einheitsweg.de

30. HIKING THROUGH THE BERGISCHES LAND MOUNTAIN REGION

Along the Wupper River—from the source of the Wipper (later the Wupper)—to the Rhine, more than 200 bridges cross the Wupper. The Wuppertal suspension railroad is the international showpiece of the Bergisches land region, but its most important lifeblood is the river, which is a good 72 mi. (116 km) long. Dammed in several places, it serves as a regional drinking-water reservoir for the entire area. Many hiking trails meander idyllically along its banks. The small village of Beyenburg, on the former border between Westphalia and the Rhineland, is romantically reflected in the surface of the water. Waterwheels drove the small grinding mills and were the foundation of the industrial world fame of its industries, and not only the Solingen bladesmiths. The Wupper River meanders contemplatively below the striking Burg Castle, the representative ancestral seat of the Duchy of Berg, only to pass the monumental steel skeleton of the impressive Mungsten Bridge shortly afterward.

Length / time it takes: 50 mi. (80 km) / 3–4 days
Highest elevation: 1,378 ft. (420 m)
Starting point: Wipperfürth
Elevation difference: 1,854 vertical ft. (565 m)
Information: www.bergisches-wanderland.de

31. HIKING IN THE HERSBRUCKER ALB

The Höhenglücksteig is considered the most demanding and most beautiful *via ferrata* (in Italian, literally "iron path") climbing route in Germany outside the Alps. Since there are opportunities to bypass passages everywhere, beginners as well as advanced climbers can climb this via ferrata together—and if you don't want to climb at all, you can enjoy the action from the hiking trail below. There is a "Via Ferrata Bambini" (children's iron way) at the entrance, where beginners and children can have fun. The approach through beech forest and over rocky humps in the forest is already a pleasure. Although the rocky ledges are not very high, they always offer beautiful views. Since the trail is divided into three sections, linked by forest trails, it is easy to get off the via ferrata before the third and most difficult part.

Time it takes: 3 hours
Starting point: Hiking-trail parking lot between Hegendorf and Neutras or Hirschbach, the "Goldener Hirsch" inn
Highest elevation: 1,929 ft. (588 m)
Elevation difference: 164 vertical ft. (50 m)
Information: www.fraenkische-schweiz.com, www.via-ferrata.de/klettersteige/topo/hoehenglueckssteig

32. CAVING TOUR THROUGH THE BISMARCK GROTTO

Narrow passages, large halls, and lots of darkness—a cave tour is a special kind of nature experience. Some of the most impressive cave systems in Europe can be found in the Franconian Alb range; one of them is the legendary Bismarck Grotto—one of the largest caves in Franconian Switzerland. Even entering the karst cave makes this action into something special: you have to rappel down to the dark, uncertain depths by using a climbing harness. From there begins a mystical journey through the underworld. Besides some tricky climbing spots, keeping oriented is one of the main issues in this deeply fissured, widely branched, and multistory column cave system. The rope-secured exit through the "corkscrew" also doesn't leave any room for claustrophobia.

Time it takes: 3 to 6 hours
Cave length: 3,937 ft. (1,200 m)
Cave depth: 170 ft. (52 m)
Starting point: Rinnenbrunn.
No entry: from October 1 to April 15
Information: www.leinen-los.de

33. HIKING IN FRANCONIAN SWITZERLAND

Franconian Switzerland offers an unbelievable number of well-marked trails through wildly romantic valleys where rocks tower over half-timbered houses. The landscape is crisscrossed by the Wiesent River, where canoeists and kayakers frolic. The Riesenburg Cave ruins lie high above the Wiesent. A set of stairs lead steeply up here, with impressive views of whatever still remains standing of the cave; a rock bridge spans the ascent. At the village of Engelhardsberg, the trail goes across fields and back down into the forest, where the steep paths, often over rocky ground, lead to the Hohe Kreuz and the Oswalds Cave. It descends from there, with a wonderful view, to the village of Muggendorf, where refreshments await you by the Wiesent for your return trip. The hike takes you on enchanting forest trails to the Quakenschloss and the Adlerstein, and back to the Riesenhöhle.

Length / time it takes: 6.2 mi. (10 km) / 3–5 hours
Starting point: Doos hiking-trail parking lot, loop trail
Highest elevation: 1,742 ft. (531 m)
Elevation difference: 1,607 vertical ft. (490 m)
Information: www.fraenkische-schweiz.com

34. THE BAVARIAN FOREST: CANOE TRIP ON THE REGEN RIVER

The Regen—this river west of the town of Zwiesel is the great classic among canoe routes. Now perfectly developed, this popular boating trail has everything a canoeist's heart desires: canoe rentals, rest stops and tent sites, mooring points, portage points with stairs, information boards . . . All of this makes the trip on the river a completely unproblematic canoe tour on the most beautiful river for hiking in the Upper Palatinate. The Regen winds through the Bavarian Forest region in a beautiful landscape: sometimes far away from any roads, sometimes flowing at a brisk pace, sometimes lazily, and through many natural river pools. The water of the Regen is clean, and it offers many places for swimming. In summer, a paradise for children—in winter, for the lonely trapper in his Indian canoe.

Length / time it takes maximum 67 mi. (107 km) / 5 days
Level of difficulty: WW 1
Starting point: Blaibacher See (lake) near Bad Kötzting
Information: www.kajakschule-oberland.de

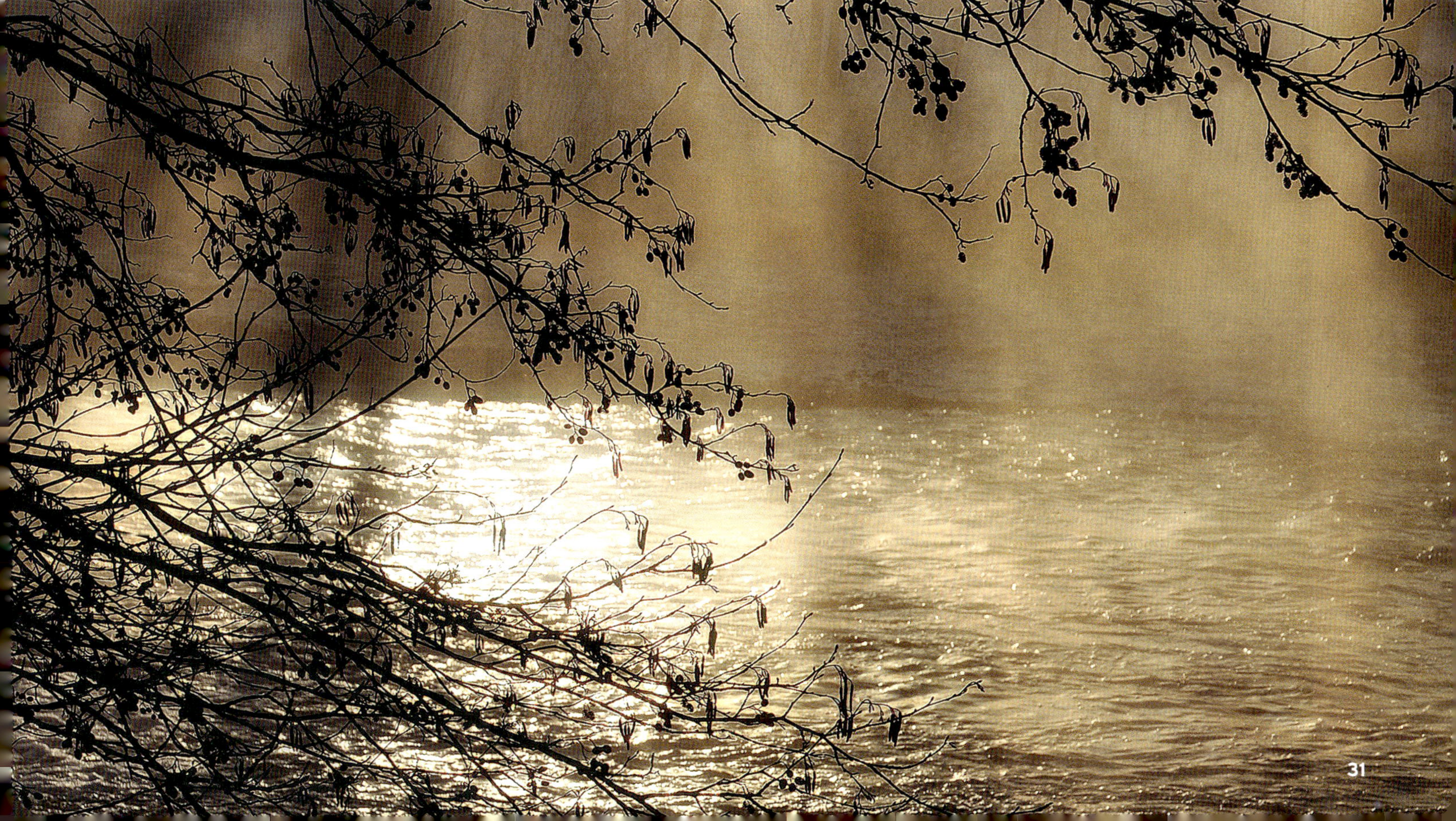

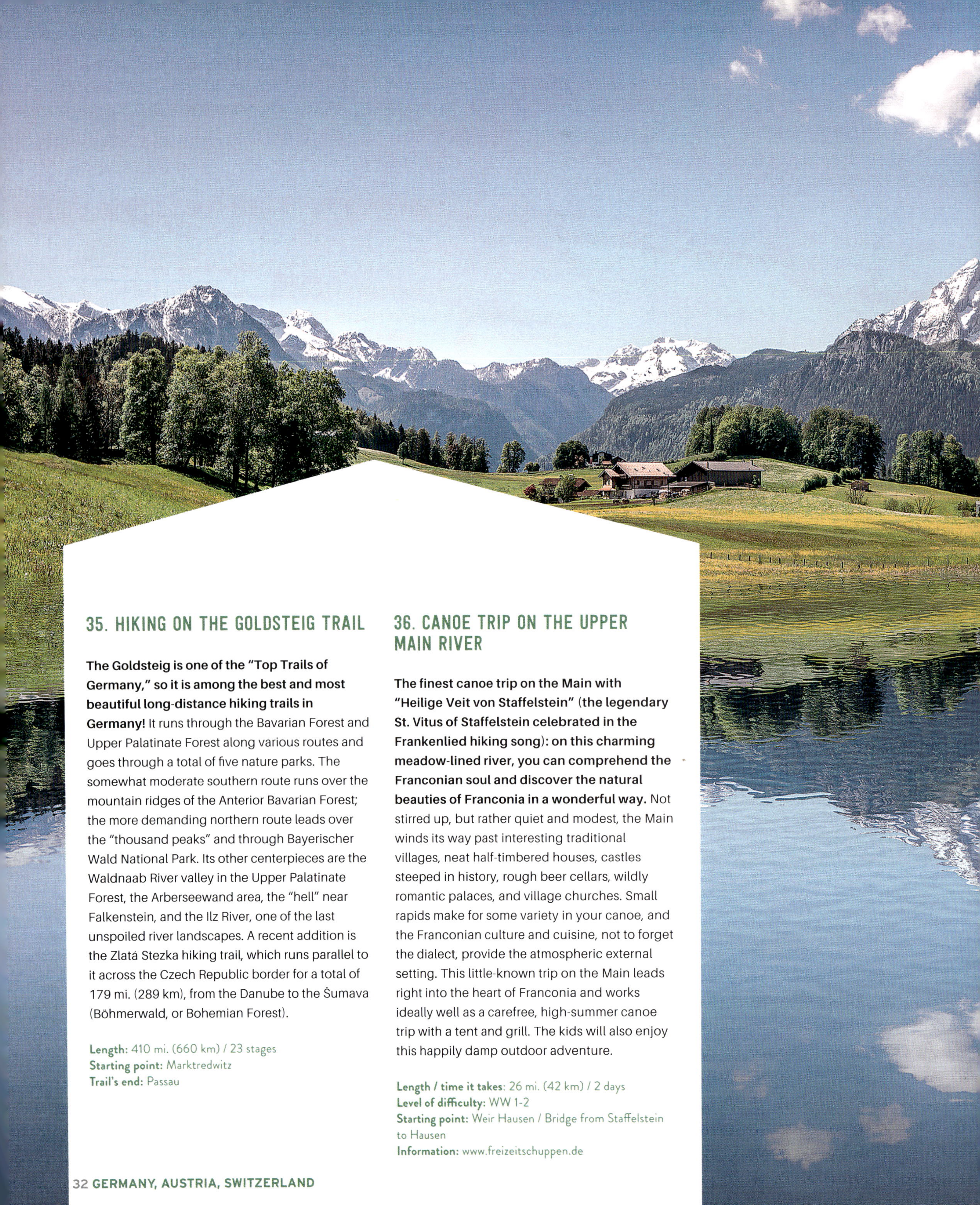

35. HIKING ON THE GOLDSTEIG TRAIL

The Goldsteig is one of the "Top Trails of Germany," so it is among the best and most beautiful long-distance hiking trails in Germany! It runs through the Bavarian Forest and Upper Palatinate Forest along various routes and goes through a total of five nature parks. The somewhat moderate southern route runs over the mountain ridges of the Anterior Bavarian Forest; the more demanding northern route leads over the "thousand peaks" and through Bayerischer Wald National Park. Its other centerpieces are the Waldnaab River valley in the Upper Palatinate Forest, the Arberseewand area, the "hell" near Falkenstein, and the Ilz River, one of the last unspoiled river landscapes. A recent addition is the Zlatá Stezka hiking trail, which runs parallel to it across the Czech Republic border for a total of 179 mi. (289 km), from the Danube to the Šumava (Böhmerwald, or Bohemian Forest).

Length: 410 mi. (660 km) / 23 stages
Starting point: Marktredwitz
Trail's end: Passau

36. CANOE TRIP ON THE UPPER MAIN RIVER

The finest canoe trip on the Main with "Heilige Veit von Staffelstein" (the legendary St. Vitus of Staffelstein celebrated in the Frankenlied hiking song): on this charming meadow-lined river, you can comprehend the Franconian soul and discover the natural beauties of Franconia in a wonderful way. Not stirred up, but rather quiet and modest, the Main winds its way past interesting traditional villages, neat half-timbered houses, castles steeped in history, rough beer cellars, wildly romantic palaces, and village churches. Small rapids make for some variety in your canoe, and the Franconian culture and cuisine, not to forget the dialect, provide the atmospheric external setting. This little-known trip on the Main leads right into the heart of Franconia and works ideally well as a carefree, high-summer canoe trip with a tent and grill. The kids will also enjoy this happily damp outdoor adventure.

Length / time it takes: 26 mi. (42 km) / 2 days
Level of difficulty: WW 1-2
Starting point: Weir Hausen / Bridge from Staffelstein to Hausen
Information: www.freizeitschuppen.de

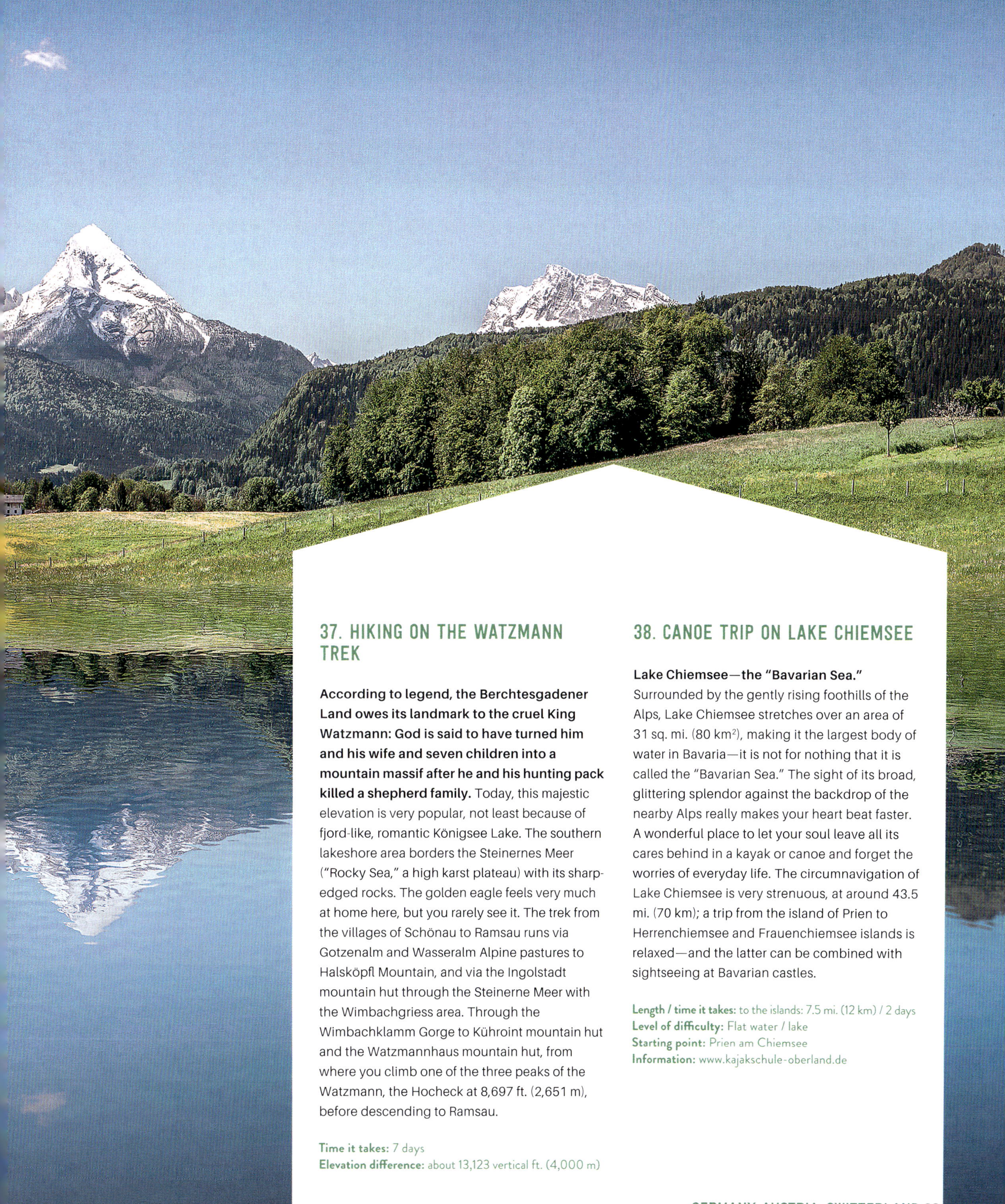

37. HIKING ON THE WATZMANN TREK

According to legend, the Berchtesgadener Land owes its landmark to the cruel King Watzmann: God is said to have turned him and his wife and seven children into a mountain massif after he and his hunting pack killed a shepherd family. Today, this majestic elevation is very popular, not least because of fjord-like, romantic Königsee Lake. The southern lakeshore area borders the Steinernes Meer ("Rocky Sea," a high karst plateau) with its sharp-edged rocks. The golden eagle feels very much at home here, but you rarely see it. The trek from the villages of Schönau to Ramsau runs via Gotzenalm and Wasseralm Alpine pastures to Halsköpfl Mountain, and via the Ingolstadt mountain hut through the Steinerne Meer with the Wimbachgriess area. Through the Wimbachklamm Gorge to Kühroint mountain hut and the Watzmannhaus mountain hut, from where you climb one of the three peaks of the Watzmann, the Hocheck at 8,697 ft. (2,651 m), before descending to Ramsau.

Time it takes: 7 days
Elevation difference: about 13,123 vertical ft. (4,000 m)

38. CANOE TRIP ON LAKE CHIEMSEE

Lake Chiemsee—the "Bavarian Sea." Surrounded by the gently rising foothills of the Alps, Lake Chiemsee stretches over an area of 31 sq. mi. (80 km^2), making it the largest body of water in Bavaria—it is not for nothing that it is called the "Bavarian Sea." The sight of its broad, glittering splendor against the backdrop of the nearby Alps really makes your heart beat faster. A wonderful place to let your soul leave all its cares behind in a kayak or canoe and forget the worries of everyday life. The circumnavigation of Lake Chiemsee is very strenuous, at around 43.5 mi. (70 km); a trip from the island of Prien to Herrenchiemsee and Frauenchiemsee islands is relaxed—and the latter can be combined with sightseeing at Bavarian castles.

Length / time it takes: to the islands: 7.5 mi. (12 km) / 2 days
Level of difficulty: Flat water / lake
Starting point: Prien am Chiemsee
Information: www.kajakschule-oberland.de

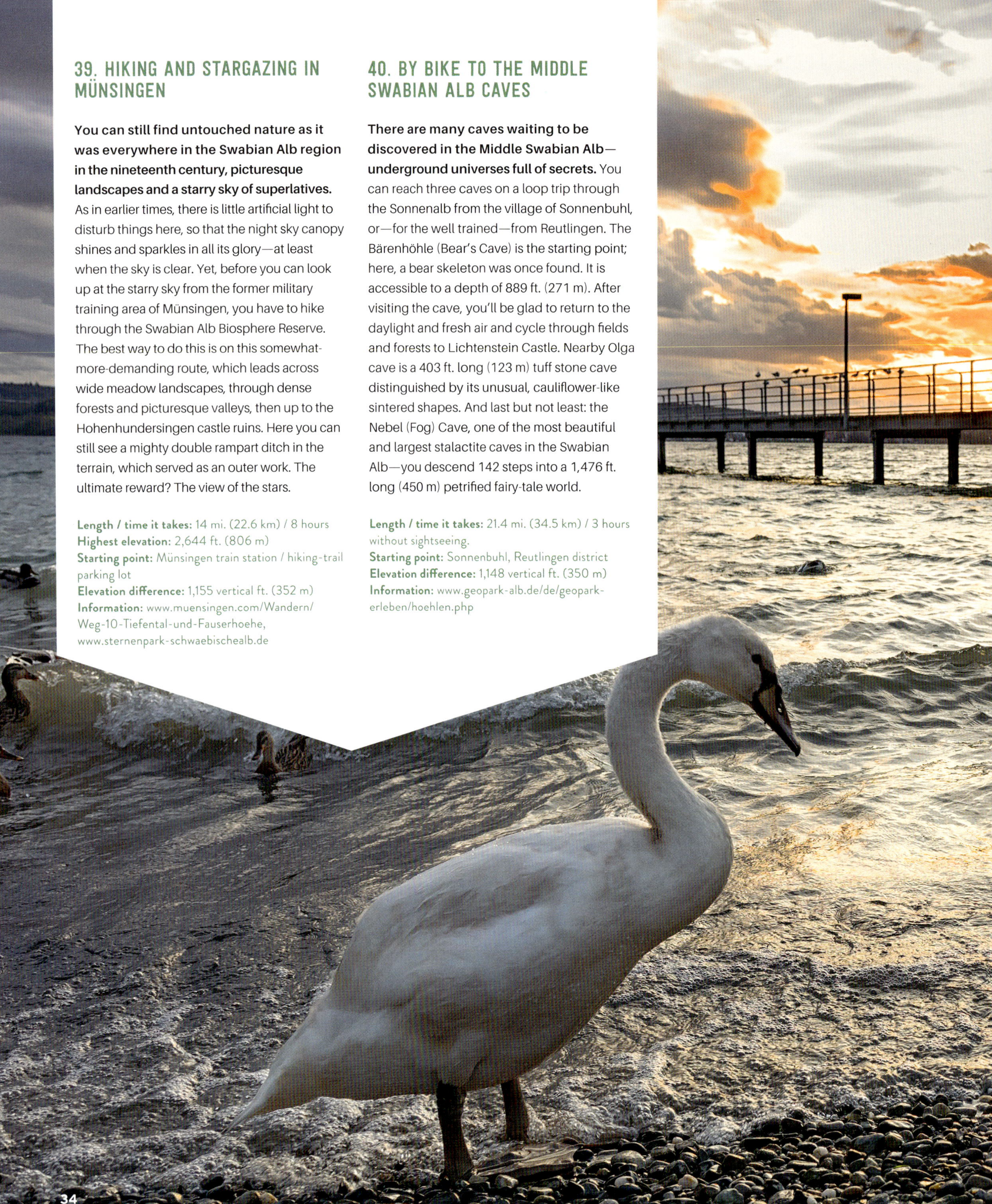

39. HIKING AND STARGAZING IN MÜNSINGEN

You can still find untouched nature as it was everywhere in the Swabian Alb region in the nineteenth century, picturesque landscapes and a starry sky of superlatives. As in earlier times, there is little artificial light to disturb things here, so that the night sky canopy shines and sparkles in all its glory—at least when the sky is clear. Yet, before you can look up at the starry sky from the former military training area of Münsingen, you have to hike through the Swabian Alb Biosphere Reserve. The best way to do this is on this somewhat-more-demanding route, which leads across wide meadow landscapes, through dense forests and picturesque valleys, then up to the Hohenhundersingen castle ruins. Here you can still see a mighty double rampart ditch in the terrain, which served as an outer work. The ultimate reward? The view of the stars.

Length / time it takes: 14 mi. (22.6 km) / 8 hours
Highest elevation: 2,644 ft. (806 m)
Starting point: Münsingen train station / hiking-trail parking lot
Elevation difference: 1,155 vertical ft. (352 m)
Information: www.muensingen.com/Wandern/Weg-10-Tiefental-und-Fauserhoehe, www.sternenpark-schwaebischealb.de

40. BY BIKE TO THE MIDDLE SWABIAN ALB CAVES

There are many caves waiting to be discovered in the Middle Swabian Alb—underground universes full of secrets. You can reach three caves on a loop trip through the Sonnenalb from the village of Sonnenbuhl, or—for the well trained—from Reutlingen. The Bärenhöhle (Bear's Cave) is the starting point; here, a bear skeleton was once found. It is accessible to a depth of 889 ft. (271 m). After visiting the cave, you'll be glad to return to the daylight and fresh air and cycle through fields and forests to Lichtenstein Castle. Nearby Olga cave is a 403 ft. long (123 m) tuff stone cave distinguished by its unusual, cauliflower-like sintered shapes. And last but not least: the Nebel (Fog) Cave, one of the most beautiful and largest stalactite caves in the Swabian Alb—you descend 142 steps into a 1,476 ft. long (450 m) petrified fairy-tale world.

Length / time it takes: 21.4 mi. (34.5 km) / 3 hours without sightseeing.
Starting point: Sonnenbuhl, Reutlingen district
Elevation difference: 1,148 vertical ft. (350 m)
Information: www.geopark-alb.de/de/geopark-erleben/hoehlen.php

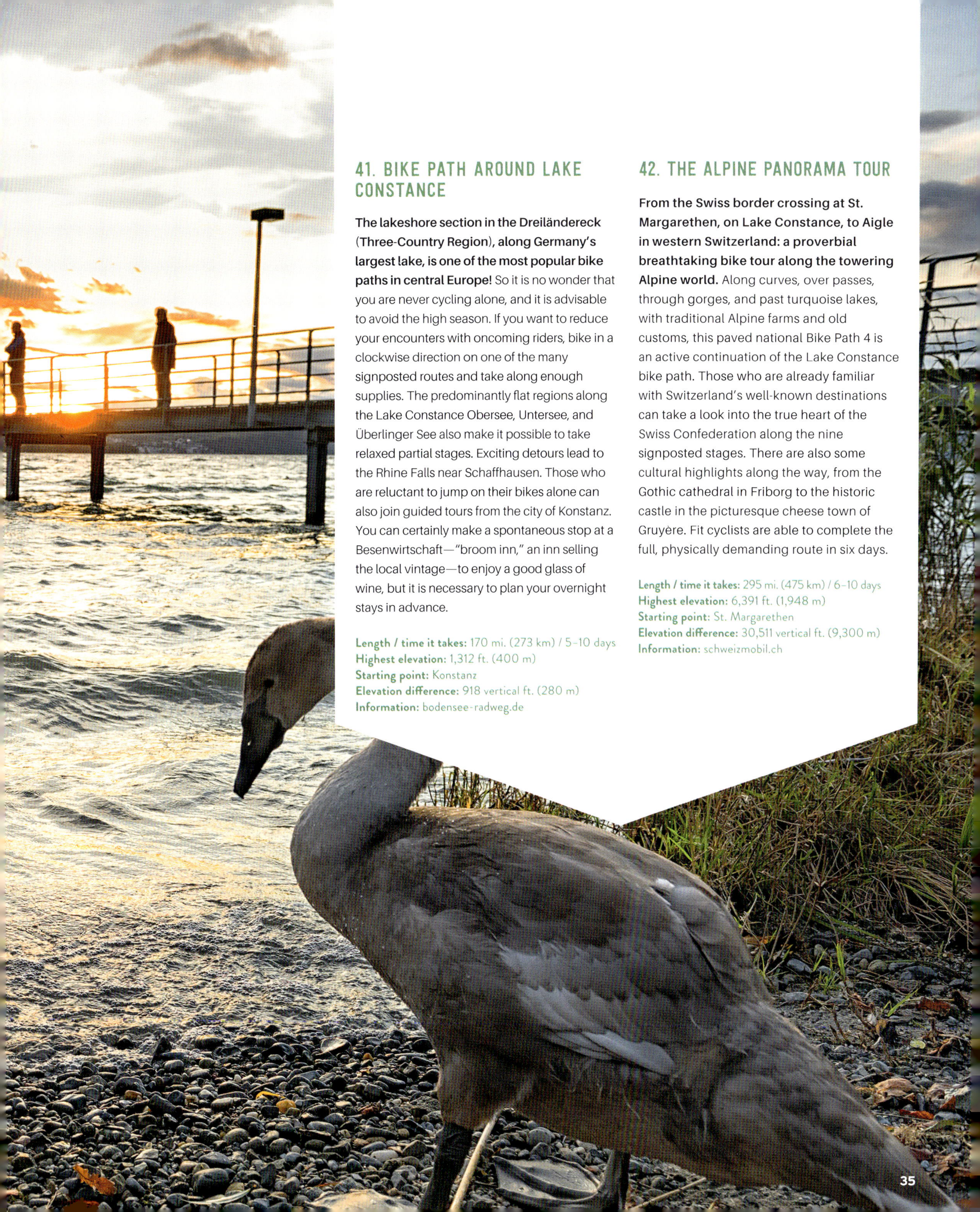

41. BIKE PATH AROUND LAKE CONSTANCE

The lakeshore section in the Dreiländereck (Three-Country Region), along Germany's largest lake, is one of the most popular bike paths in central Europe! So it is no wonder that you are never cycling alone, and it is advisable to avoid the high season. If you want to reduce your encounters with oncoming riders, bike in a clockwise direction on one of the many signposted routes and take along enough supplies. The predominantly flat regions along the Lake Constance Obersee, Untersee, and Überlinger See also make it possible to take relaxed partial stages. Exciting detours lead to the Rhine Falls near Schaffhausen. Those who are reluctant to jump on their bikes alone can also join guided tours from the city of Konstanz. You can certainly make a spontaneous stop at a Besenwirtschaft—"broom inn," an inn selling the local vintage—to enjoy a good glass of wine, but it is necessary to plan your overnight stays in advance.

Length / time it takes: 170 mi. (273 km) / 5–10 days
Highest elevation: 1,312 ft. (400 m)
Starting point: Konstanz
Elevation difference: 918 vertical ft. (280 m)
Information: bodensee-radweg.de

42. THE ALPINE PANORAMA TOUR

From the Swiss border crossing at St. Margarethen, on Lake Constance, to Aigle in western Switzerland: a proverbial breathtaking bike tour along the towering Alpine world. Along curves, over passes, through gorges, and past turquoise lakes, with traditional Alpine farms and old customs, this paved national Bike Path 4 is an active continuation of the Lake Constance bike path. Those who are already familiar with Switzerland's well-known destinations can take a look into the true heart of the Swiss Confederation along the nine signposted stages. There are also some cultural highlights along the way, from the Gothic cathedral in Friborg to the historic castle in the picturesque cheese town of Gruyère. Fit cyclists are able to complete the full, physically demanding route in six days.

Length / time it takes: 295 mi. (475 km) / 6–10 days
Highest elevation: 6,391 ft. (1,948 m)
Starting point: St. Margarethen
Elevation difference: 30,511 vertical ft. (9,300 m)
Information: schweizmobil.ch

43. HIKING ON HEILBRONNER WEG TRAIL

If you hanker after rocks, exertion, and wonderful views, there is no way of avoiding the Heilbronner Weg Trail in the Allgäu Mountain region—but if you get giddy and aren't sure-footed, it is better to stay down below. The ascent begins in Oberstdorf and gradually climbs through the Trettach River valley, then goes steeply up to the Kemptner mountain hut, which is ideal for an overnight stay. From here, the E5 highway leads south to Venice, and the actual Heilbronner Weg Trail begins at the foot of Mädelegabel ("girl's fork") Mountain. Via the Bockkarkopf and the Wilder Mann mountains, the trail goes to Steinschartenkopf peak, where a horizontal iron ladder spans a cleft in the narrow ridge and provides some thrills. Via ferrata equipment is not essential, but the trail is certainly not for beginners.

Time it takes: 11 hours / 2 days
Starting point: Oberstdorf
Trail's end: Einödsbach
Elevation difference: 4,921 vertical ft. (1,500 m)
Information: www.oberstdorf.de, www.via-ferrata.de/klettersteige/topo/heilbronnerweg-allgaeu

44. HIKES AROUND AND ON ZUGSPITZE MOUNTAIN

This adventurous multiday hike leads to one of the most beautiful places in the Wetterstein Mountains: King Ludwig II had a villa, the King's House on Schachen, built on the Schachen massif; an Alpine botanical garden is laid out next to it. An impressive first hike goes from Garmisch-Partenkirchen through the wild Partnach Gorge and via Kälbersteig and Schachenweg up to the Schachen villa. On the second day, hike along the Partnach mountain stream into the heart of the Wetterstein Mountains. The trail continues via Bock mountain hut to the picturesque Reintalanger mountain hut and the nearby source of the Partnach. Then, in a change with good prospects, hike along the south side of the mountains and through the "Gatterl" gap to Austria, to Ehrwalder Alm. If you want, you can climb the Zugspitze from Knorr mountain hut (three hours).

Length / time it takes: 19.25 mi. (31 km) / 3 days
Starting point: Garmisch-Partenkirchen
Trail's end: Ehrwald
Elevation difference: 9,186 vertical ft. (2,800 m)

45. CANOEING ON THE ISAR RIVER

With rucksack and pack through nature and "nude" swimmers: Munich's cult river, much praised, much beloved, and much celebrated in song, is and remains a very special canoeing experience. The winding Isar meanders in its wide gravel bed from Sylvensteinsee, a lake on the edge of the Alps, through pristine riverside meadows, forests, and towns such as Lenggries, Bad Tölz, Wolfratshausen, and Grunwald, to the heart of Munich—from pure nature to the center of the Bavarian state capital. The many gravel banks on both sides along the riverbank invite you to relax, swim, and have a snack. In the Pupplinger Au and Ascholdinger Au sections of the river, you will paddle past many nude swimmers in the summer. As a Munich native—a Münchner—would say, "Mei—halt nature pure"—"Whatever—keep nature pure," and casually paddle their canoe, heavily loaded with tent and provisions, farther down the river.

Length / time it takes: 45 mi. (72 km) / 3 days
Level of difficulty: WW 1-2 (3)
Starting point: Sylvensteinsee, near Lenggries
Information: www.kajakschule-oberland.de

46. MTB TOURS AROUND LAKE WALCHENSEE

Picture-book Bavaria: the Walchensee, shining in gorgeous blue green, is so beautiful it is almost kitschy, and knows how to inspire you with its fantastic location, embedded in an incredible mountain panorama. Besides, its water is also used for environmentally friendly energy generation: the Walchensee power plant supplies around 300 million kilowatt hours of electricity per year. On some weekends and during vacation time, the area around Germany's deepest Alpine lake is terribly crowded. But quiet days in spring or autumn are wonderful for cycling around it on a mountain bike: the special light, the turquoise of the lake, and the blue of the Bavarian sky. At the end of a successful outdoor day by the lake, there is Schweinshaxe (pork knuckle) with dumplings and sauerkraut, plus a Halbe (half liter of beer)—wonderful!

Length / time it takes: 16 mi. (26 km) / 2 hours
Highest elevation: 2,623 ft. (801 m)
Starting point: Urfeld
Information: www.walchensee-kochelsee.de

47. HIKING ON MAXIMILIANSWEG TRAIL

Following in the footsteps of Maximilian II of the Wittelsbach dynasty, who was king of Bavaria from 1848 to 1864, the Maximiliansweg Trail leads from the city of Lindau via Füssen, through the Allgäu Alps and the Alpine foothills, to Berchtesgaden, offering an extraordinary Alpine panorama and fascinating natural landscapes. If you follow the entire route, you will be on the trail for five weeks. The trail generally runs along the European E4 long-distance hiking trail and connects to the Austrian 04 Voralpenweg trail at both ends. Some sections are very demanding and require being sure-footed and experienced in the mountains. Along the trail you will pass top sights, including the royal castles of Neuschwanstein, Hohenschwangau, and Linderhof.

Length / time it takes: 224–236 mi. (360–380 km) / five weeks
Elevation difference: a total of 66–984 vertical ft. (20–300 m).

48. SNOWSHOEING IN THE SCHWARZENBACH VALLEY

Snowshoeing through the white splendor: roaming over the powdery snow on these oversized "flat feet" is a very special experience. And that in a snow-covered mountain valley, secluded and full of romance, with a wonderful view of the classic Bavarian Voralpen Mountains, the Benediktenwand, Achselköpfle, and Brauneck. Getting there is also relaxed: the trip on the BOB railroad, the Bavarian Oberlandbahn, takes just an hour from Munich. You put on your snowshoes quickly, and off you go through the wintry landscape into the Schwarzenbach valley: this winter hike in unspoiled nature guarantees rest and relaxation from the hectic everyday life of a big city. At the end of the loop trail, warm delicacies await you at the Pfaffensteff Inn in Wegscheid. Heart and soul—what more could you wish?

Length / time it takes: 5.15 mi. (8.3 km) / 2 hours
Highest elevation: 2,231 ft. (680 m)
Starting point: Wegscheid, near Lenggries
Elevation difference: 410 vertical ft. (125 m)
Information: www.lenggries.de

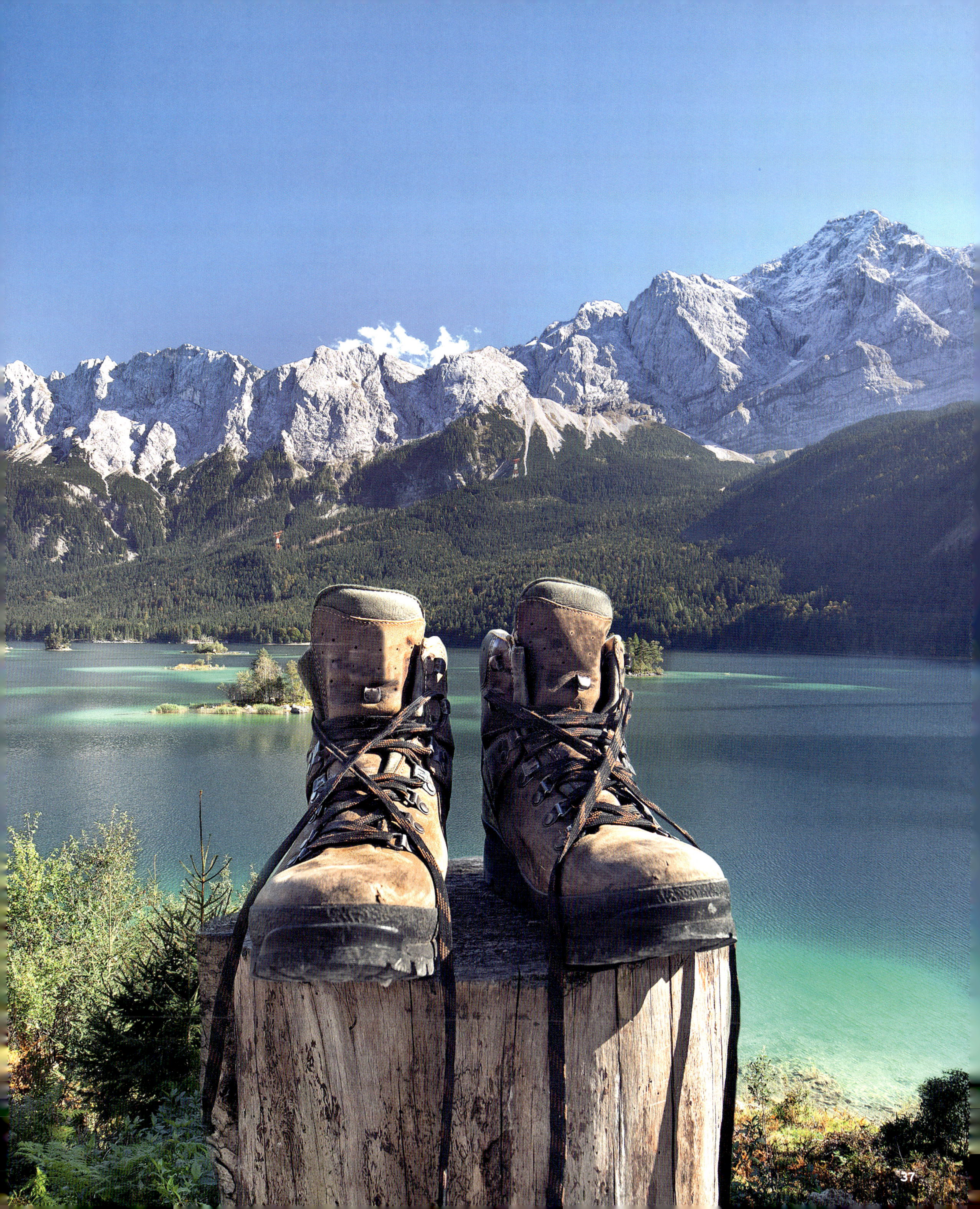

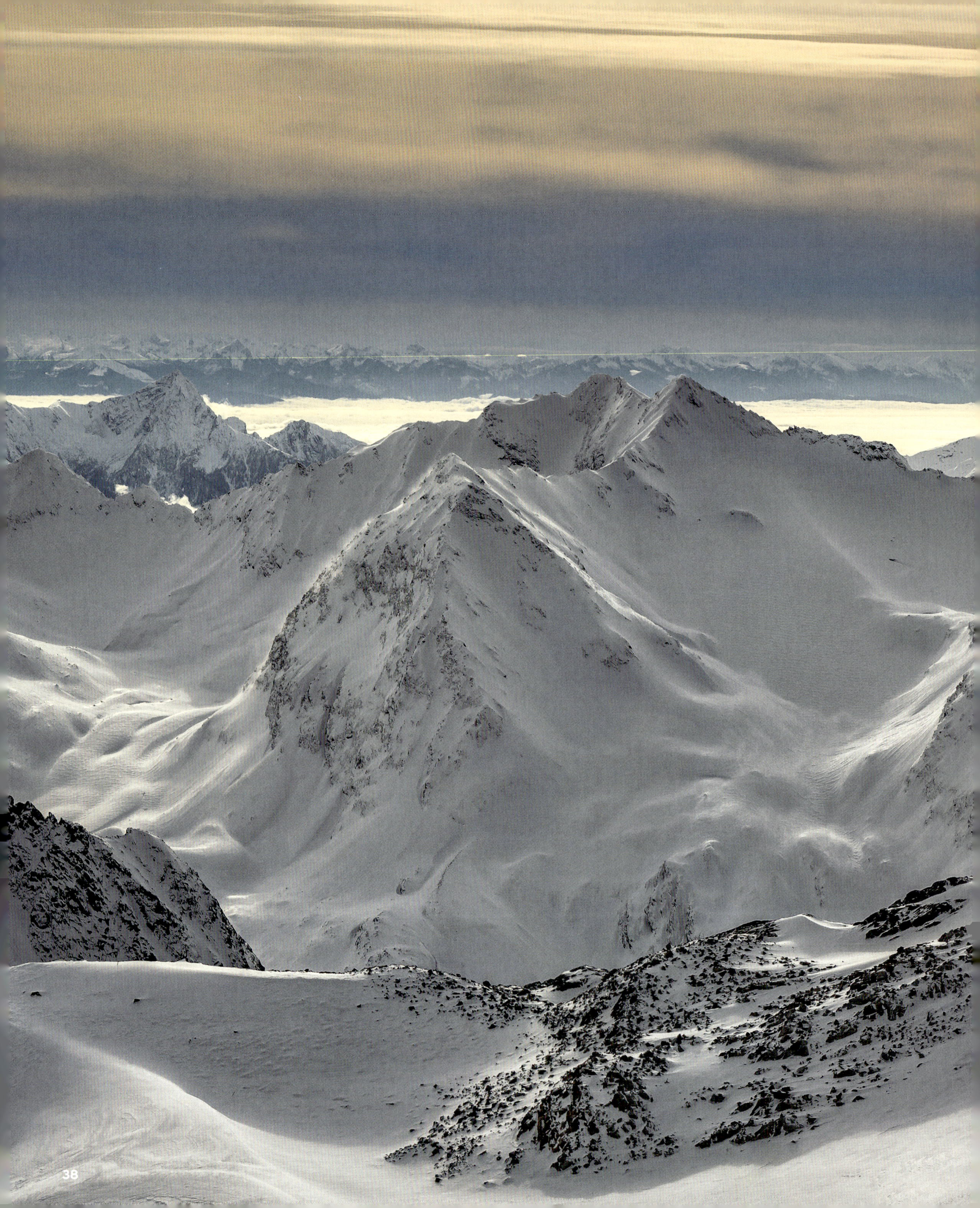

49. ON THE INN RIVER BIKE PATH THROUGH TYROL

Cycling through Tyrol—a relaxed bike tour from the Swiss border to Upper Bavaria. The almost 323 mi. long (520 km) Inn River already shows its characteristic green color in the narrow upper Inn valley basin, and it barely changes along the length of the bike path. The river, which, like the valley, is surrounded by high mountains, keeps getting wider along almost the entire length of the route. It has a steady, high-flow velocity that leaves cyclists amazed. What is special about this route, which runs consistently along the riverbank, is the fact that it feels as if it is constantly running downhill. Nevertheless, individual sections of the bike path can be quite hard on the legs. For example, making a detour to the carnival museum in the town of Imst or to Stams monastery. While you can reach the ski jump on the Bergisel above Innsbruck by elevator, trying out the stand on the jump ramp will raise a sweat on your forehead, despite the fantastic view.

Length / time it takes: 114 mi. (183 km) / 3–4 days
Highest elevation: 3,494 ft. (1,065 m)
Starting point: Maloja Pass, Switzerland
Trail's end: outlet into the Danube, Upper Bavaria
Elevation difference: 1,857 vertical ft. (566 m)
Information: innradweg.com

50. ALPINE CROSS-COUNTRY SKIING THROUGH THE TYROLEAN OBERLAND

Touring on skis—cross-country skiing with Alpine skis, garnished with a few biathlon skills. In winter, a layer of snow, often several feet deep, carefully protects the sleeping beauties of the Tschey, one of the most beautiful alpine meadows, which in early summer turns into a veritable sea of blossoms with countless butterflies. The skier enjoys the gentle, white, hilly landscape against a mighty mountain backdrop, because the ascents are not too steep and the descents not too hurried. You don't need groomed *loipe* cross-country ski trails, but simply ski cross-country on alpine skis covered with mohair climbing skins. You take the skins off when it is time to ski downhill again. Equipped with a bow and arrow, you can also practice for a biathlon as an alternative. There are some plastic animals from the summer archery course set up as targets between trees and next to hay barns. This is when a steady hand is the order of the day.

Length / time it takes: 5–6 mi. (8–10 km) / 3 hours
Highest elevation: 3,494 ft. (1,065 m)
Starting point: Berghof Pfunds in Greit
Elevation difference: 2,001 vertical ft. (610 m)
Information: www.tiroler-oberland.
com**Höhenunterschied:** 610 Hm
Info: www.tiroler-oberland.com

51. THE WILDE-WASSER-WEG LOOP HIKING TRAIL

Water as a phenomenon—powerful, full of life, and refreshing, it fascinates hikers in the Stubai Mountains. Here is the idea behind the Wilde-Wasser-Weg (Wild Water Trail): to show the evolution of a stream until it becomes a bubbling mountain river by following its course along easy as well as more-difficult trails. It begins with the murmuring brooks, which, starting from blue-green mountain lakes and glaciers, absorb brooks from other springs, gradually develop into rushing rivers, and finally tumble into the valley as mighty waterfalls, such as the Grawa Waterfall. The trail is divided into three stages of different levels of difficulty, so that less advanced hikers can also experience the fascination of water.

Length / time it takes: 6.5 mi. (10.5 km) / 5.5 hours
Highest elevation: 7,188 ft. (2,191 m)
Starting point: The WildeWasserArena on the Ruetz River bank, at the quarry at Ranalt
Elevation difference: 3,871 vertical ft. (1,180 m)
Information: www.stubai.at/aktivitaeten/wandern/wildewasserweg

52. LAKE HIKE FROM THE DRESDNER MOUNTAIN HUT

Panoramic views, lakes, lonely moorland—the Stubai High Trail to Mutterberger Lake offers all you could want of mountain scenery. Start off in the cable car and then up to the Dresdner mountain hut (if you walk briskly, you will make the ascent in one and a half hours). From there, the Stubai High Trail leads to some lakes in fantastic places. The route is spectacular, varied, and never boring. A mountain hiker will forget about the time and, halfway along the trail, unexpectedly comes upon the beautifully situated Mutterbergersee lake. It is one of the most beautiful mountain lakes in the eastern Alps. Always in view: the triumvirate of the Stubai Alps—Wilder Freiger, Wilder Pfaff, and Zuckerhutl. Time for a cozy picnic on the lakeshore. The trail continues through a picturesque raised-bog landscape and finally down to the Mutterberg Alm.

Length / time it takes: 5 mi. (8.3 km) / 4 hours
Highest elevation: 8,278 ft. (2,523 m)
Starting point: Stubai Glacier ski lift parking lot
Elevation difference: Ascent, 1,601 vertical ft. (488 m); descent, 3,337 vertical ft. (1,017 m)
Information: www.almenrausch.at/touren

53. THE "CRADLE OF VIA FERRATAS"

It is possible to access twenty via ferratas from Ramsau. The via ferrata to the Dachstein summit, built in 1843, is the oldest in the Alps! This classic stretches over ridges, peaks, and rock towers high above the valley and should be used only with via ferrata equipment. To avoid dealing with a very large difference in elevation when ascending and descending, it is advisable to make the hike along the loop route from the Guttenberg mountain hut (7,040 ft. [2,146 m]). Then you approach around the Koppenkar to Edelgriesshöhe peak. The many steep, iron-secured climbs over the pinnacles and ridges provide for plenty of adrenaline rush. Just as you are relaxing on a rock tower, you look into an almost vertical chimney on the rear side.

Time it takes: 11 hours / 2 days
Starting point: Guttenberg mountain hut
Highest elevation: 8,648 ft. (2,636 m)
Lowest point: 7,040 ft. (2,146 m)
Information: www.ramsau.com/de/via ferrata/via ferrata

54. MTB TOUR THROUGH SALZBURGER LAND

Mountain biking is becoming more and more popular in the Alps. For those who like a short and snappy downhill but shy away from the climb, Wagrain is the right place: the cable car takes you uphill, and from there, you cycle downhill through the Wagrain bike park. For beginners, the ride from the middle station over forest paths and tree roots is certainly enough, while experts can continue to the upper station to test their limits on the downhill run with wall rides, rollers and tables, and lots of jumps. If you like things a bit more relaxed, you can also take yourself and your bike comfortably in the direction of the Edelweiss Alm and get some fresh air and enjoy the view of the mountains.

Time it takes: 11 hours / 2 days
Starting point: Wagrain valley cable car station
Highest elevation: 6,070 ft. (1,850 m)
Lowest point: 2,756 ft. (840 m)
Information: www.wagrain-kleinarl.at/de/aktivitaeten/mountainbiken-e-bike mountainbike-routen

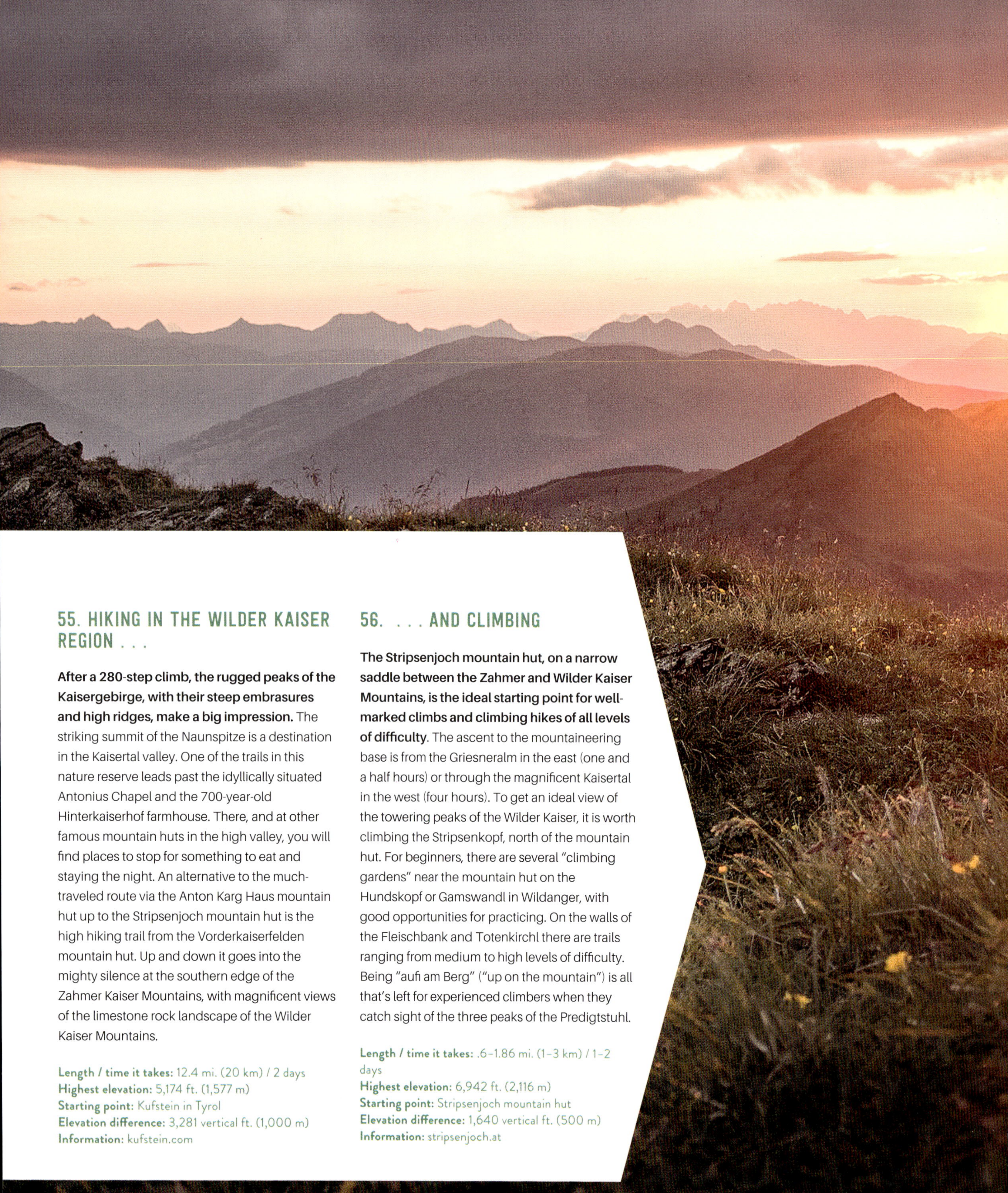

55. HIKING IN THE WILDER KAISER REGION . . .

After a 280-step climb, the rugged peaks of the Kaisergebirge, with their steep embrasures and high ridges, make a big impression. The striking summit of the Naunspitze is a destination in the Kaisertal valley. One of the trails in this nature reserve leads past the idyllically situated Antonius Chapel and the 700-year-old Hinterkaiserhof farmhouse. There, and at other famous mountain huts in the high valley, you will find places to stop for something to eat and staying the night. An alternative to the much-traveled route via the Anton Karg Haus mountain hut up to the Stripsenjoch mountain hut is the high hiking trail from the Vorderkaiserfelden mountain hut. Up and down it goes into the mighty silence at the southern edge of the Zahmer Kaiser Mountains, with magnificent views of the limestone rock landscape of the Wilder Kaiser Mountains.

Length / time it takes: 12.4 mi. (20 km) / 2 days
Highest elevation: 5,174 ft. (1,577 m)
Starting point: Kufstein in Tyrol
Elevation difference: 3,281 vertical ft. (1,000 m)
Information: kufstein.com

56. . . . AND CLIMBING

The Stripsenjoch mountain hut, on a narrow saddle between the Zahmer and Wilder Kaiser Mountains, is the ideal starting point for well-marked climbs and climbing hikes of all levels of difficulty. The ascent to the mountaineering base is from the Griesneralm in the east (one and a half hours) or through the magnificent Kaisertal in the west (four hours). To get an ideal view of the towering peaks of the Wilder Kaiser, it is worth climbing the Stripsenkopf, north of the mountain hut. For beginners, there are several "climbing gardens" near the mountain hut on the Hundskopf or Gamswandl in Wildanger, with good opportunities for practicing. On the walls of the Fleischbank and Totenkirchl there are trails ranging from medium to high levels of difficulty. Being "aufi am Berg" ("up on the mountain") is all that's left for experienced climbers when they catch sight of the three peaks of the Predigtstuhl.

Length / time it takes: .6–1.86 mi. (1–3 km) / 1–2 days
Highest elevation: 6,942 ft. (2,116 m)
Starting point: Stripsenjoch mountain hut
Elevation difference: 1,640 vertical ft. (500 m)
Information: stripsenjoch.at

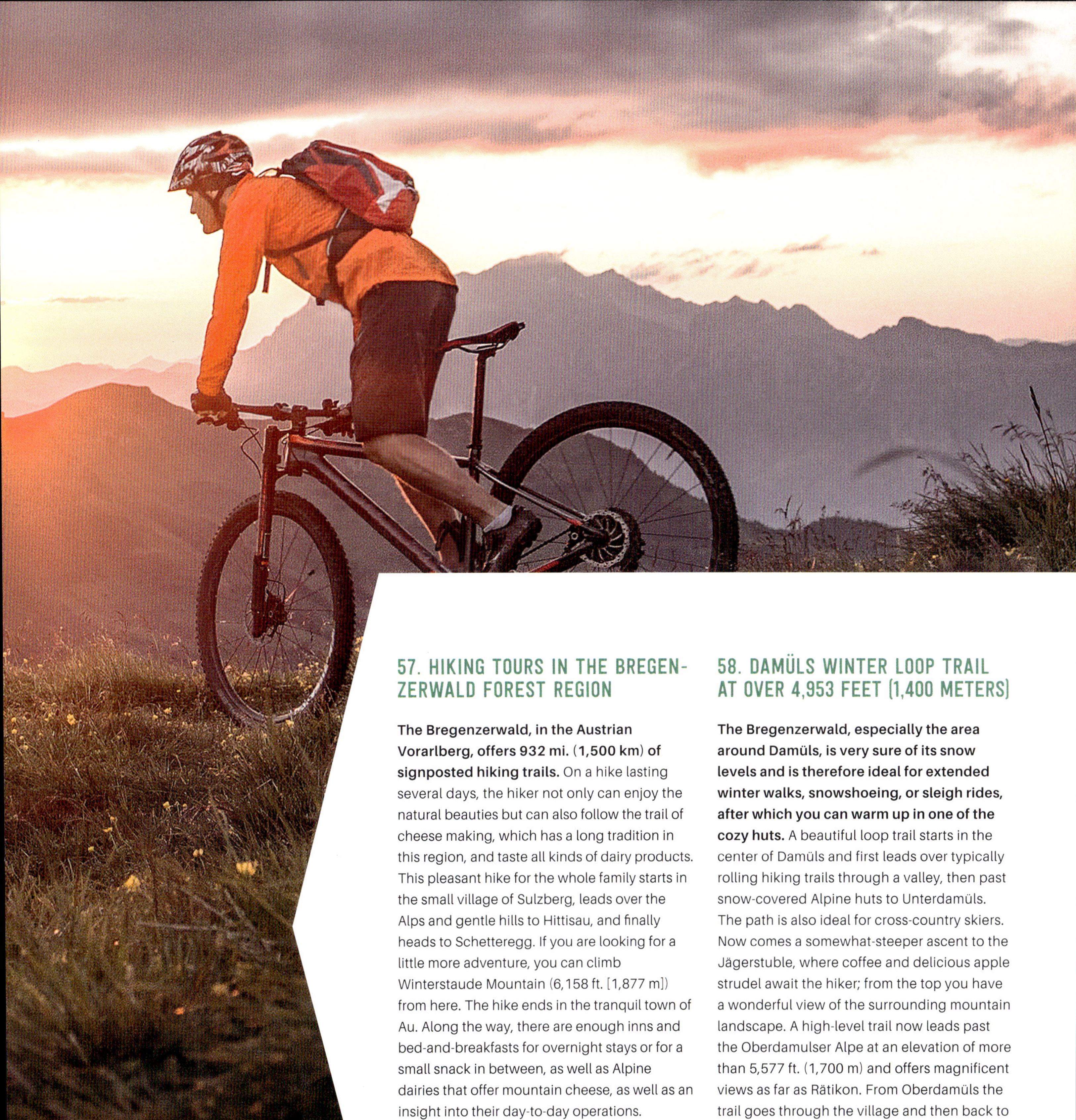

57. HIKING TOURS IN THE BREGENZERWALD FOREST REGION

The Bregenzerwald, in the Austrian Vorarlberg, offers 932 mi. (1,500 km) of signposted hiking trails. On a hike lasting several days, the hiker not only can enjoy the natural beauties but can also follow the trail of cheese making, which has a long tradition in this region, and taste all kinds of dairy products. This pleasant hike for the whole family starts in the small village of Sulzberg, leads over the Alps and gentle hills to Hittisau, and finally heads to Schetteregg. If you are looking for a little more adventure, you can climb Winterstaude Mountain (6,158 ft. [1,877 m]) from here. The hike ends in the tranquil town of Au. Along the way, there are enough inns and bed-and-breakfasts for overnight stays or for a small snack in between, as well as Alpine dairies that offer mountain cheese, as well as an insight into their day-to-day operations.

Length / time it takes: 47 mi. (75.8 km) / 3 days
Highest elevation: 2,677 ft. (816 m)
Starting point: Sulzberg
Elevation difference: 6,562 vertical ft. (2,000 m)

58. DAMÜLS WINTER LOOP TRAIL AT OVER 4,953 FEET (1,400 METERS)

The Bregenzerwald, especially the area around Damüls, is very sure of its snow levels and is therefore ideal for extended winter walks, snowshoeing, or sleigh rides, after which you can warm up in one of the cozy huts. A beautiful loop trail starts in the center of Damüls and first leads over typically rolling hiking trails through a valley, then past snow-covered Alpine huts to Unterdamüls. The path is also ideal for cross-country skiers. Now comes a somewhat-steeper ascent to the Jägerstuble, where coffee and delicious apple strudel await the hiker; from the top you have a wonderful view of the surrounding mountain landscape. A high-level trail now leads past the Oberdamulser Alpe at an elevation of more than 5,577 ft. (1,700 m) and offers magnificent views as far as Rätikon. From Oberdamüls the trail goes through the village and then back to the starting point.

Length / time it takes: 5 mi. (8 km) / 3 hours
Highest elevation: about 5,577 ft. (1,700 m)
Starting point: Damüls
Elevation difference: 951 vertical ft. (290 m)

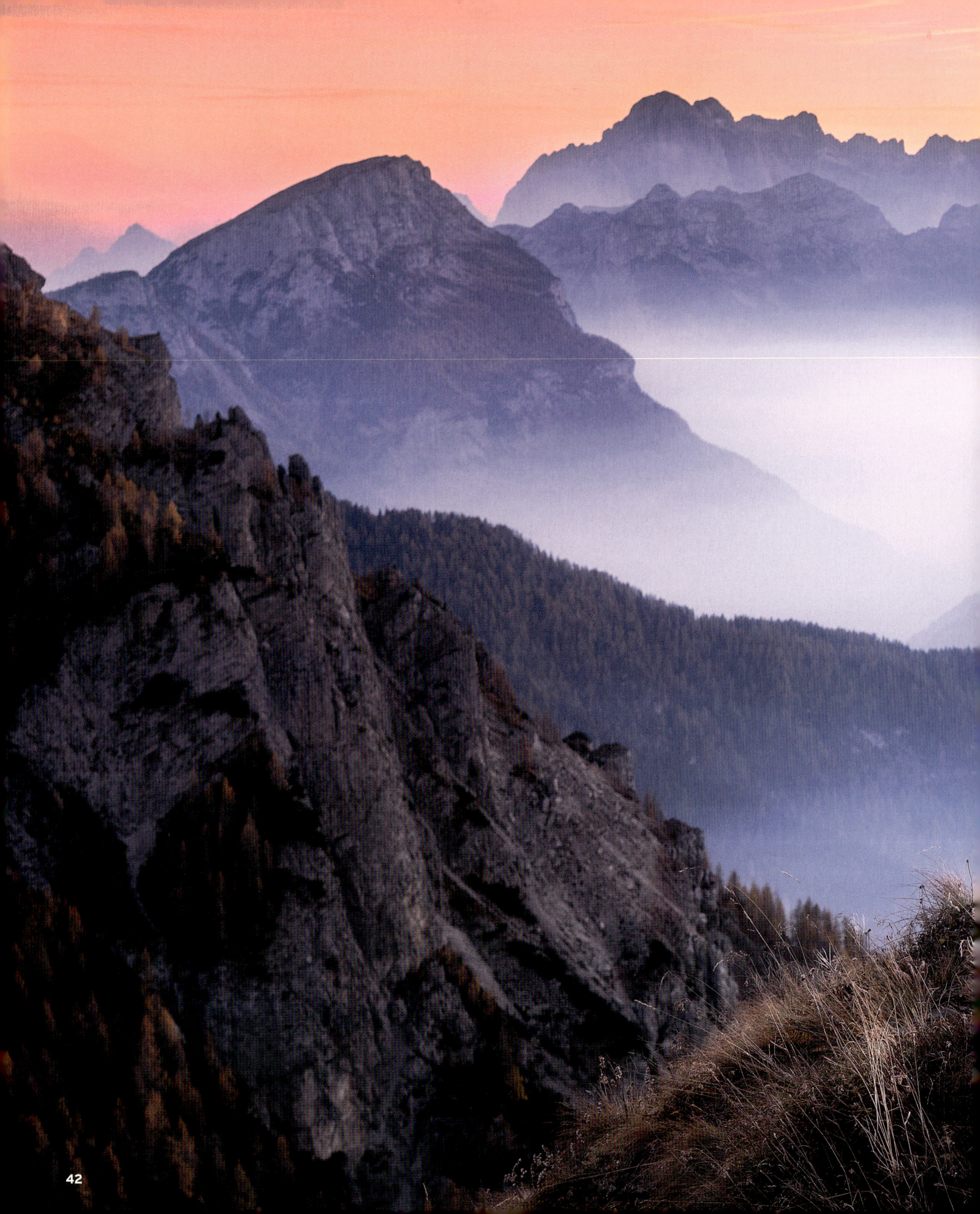

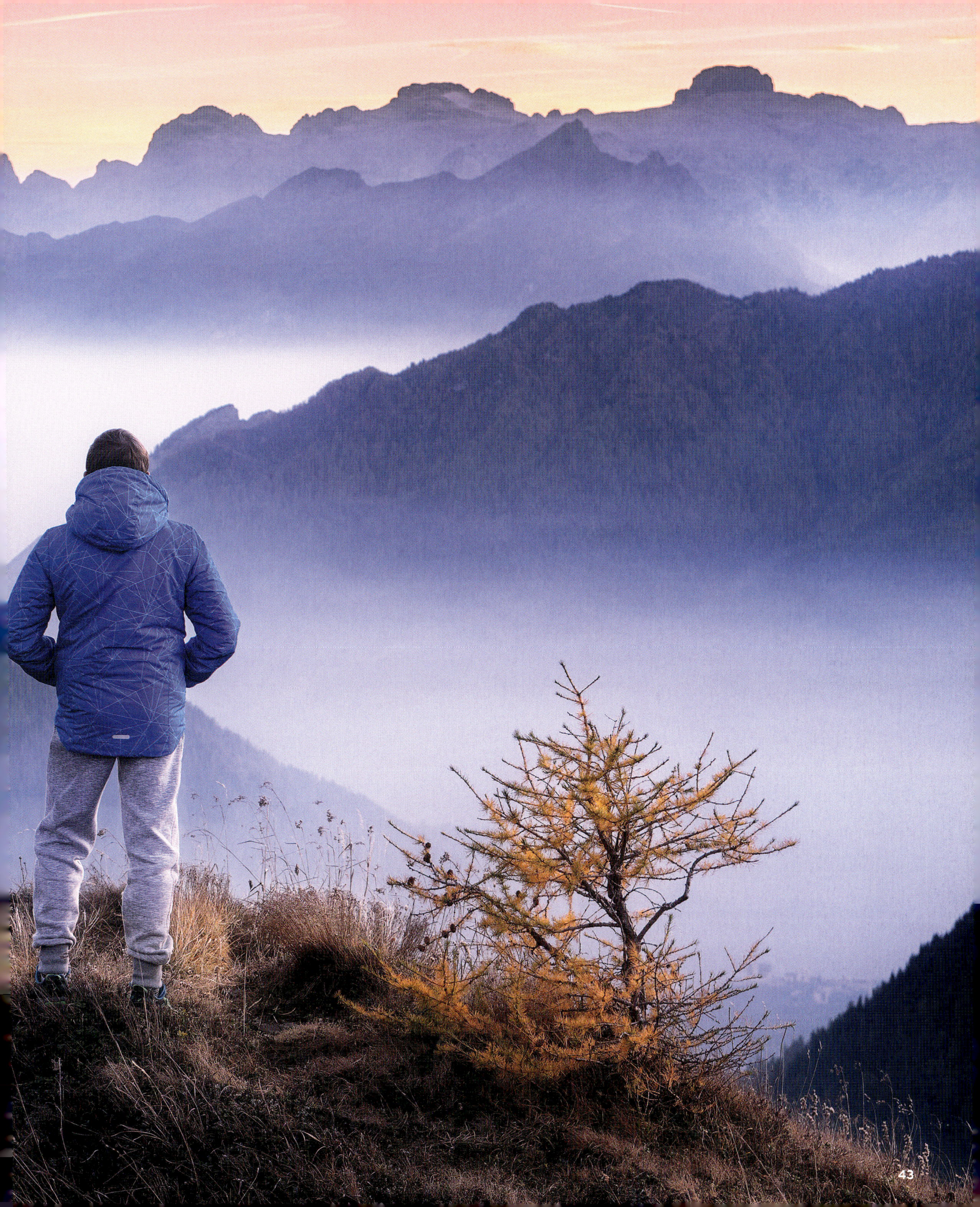

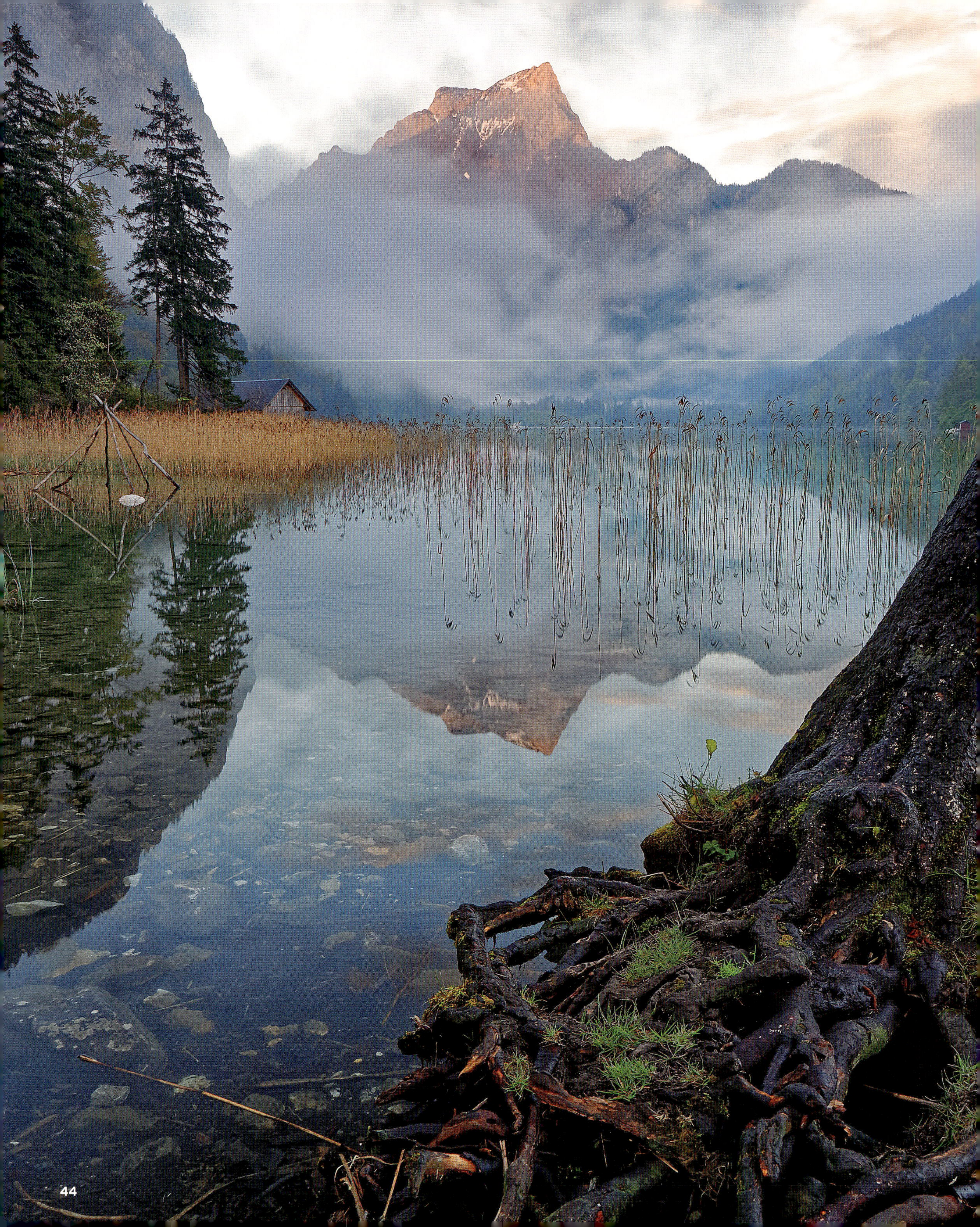

59. HIKE FROM THE WALDVIERTEL REGION TO STYRIA

The starting point of the Ostösterreichischen Grenzlandweg (East Austrian Borderland Trail) is 3,336 ft. high (1,017 m) Nebelstein Mountain in Lower Austria. It then runs along Austria's northern and eastern borders with the Czech Republic, Slovakia, Hungary, and Slovenia, through the hilly granite and gneiss highlands of the Bohemian Massif and Austria's Weinviertel wine district. After crossing Vienna, where you should definitely take a two-to-three-day break, the trail continues through the Burgenland lowlands and finally through the Styrian hill country, until it then finds its end at Bad Radkersburg. The Grenzlandweg Trail is also part of the European E8 long-distance hiking trail (Ireland–Ukraine) from Nebelstein to Wolfsthal, and from Vienna to Hochstrass it runs along the route of the E4 (Portugal–Cyprus).

Length: about 435 mi. (700 km)
Starting point: Nebelstein
Trail's end: Bad Radkersburg

60. HIKING ON PILGRIMAGE ROUTES THROUGH AUSTRIA

The Mariazell pilgrimage route system runs through the Austrian federal states of Vienna, Lower Austria, Burgenland, Upper Austria, Salzburg, Carinthia, and Styria. The system comprises traditional pilgrimage routes: the Vienna Mariazellerweg 06 (Via Sacra, about 78 mi. [125 km]), the Lower Austria Mariazellerweg 06 (about 158 mi. [255 km]), and the Burgenland Mariazellerweg 06 (about 90 mi. [145 km]). This also explains the great variety of scenery along the route. Klagenfurt, Linz-Pöstlingberg, Nebelstein and Perchtoldsdorf, Eisenstadt, and Salzburg, as well as Eibiswald, serve as starting points. The Mariazell pilgrimage route is part of the European E6 long-distance hiking trail (not yet fully established: Finland–Turkey); the west branch corresponds to the Voralpenweg (Trail 04).

Length: about 326 mi. (525 km)
Starting point: various
Trail's end: Mariazell
Information: www.mariazellerwege.at

61. HIKING THE EISENWURZEN TRAIL TO CARINTHIA

This long-distance hiking trail connects the northernmost point of Austria (Rottal on the Czech border) with the southernmost (the Seebergsattel on the border with Slovenia). The first section is easy to walk and runs through the Waldviertel to the valley of the Danube and Ybbs Rivers. Part two of the trail begins in Waidhofen with some moderately difficult and demanding sections. The trail is named after the Eisenwurzen Nature Park, the area around the lower Enns valley, where many small iron-processing companies settled during the late nineteenth and early twentieth centuries. The Via Alpina runs alongside this trail from Ingering II to the Oberst-Klinke mountain hut area. The last section is also more demanding and leads from Judenburg in the Mur valley, over the Drava River, and to the Seeberg saddle.

Length: about 342 mi. (550 km)
Highest elevation: Zirbitzkogel, 7,861 ft. (2,396 m)

62. HIKING ON THE WIENER HÖHENWEG TRAIL

The Wiener Höhenweg (Vienna High Trail) does not lie before the gates of Vienna, as the name would suggest, but rather in one of the most pristine and probably most impressive mountain landscapes of the Hohe Tauern range. The mountain trail runs as a multiday, high-alpine hike through Hohe Tauern National Park in the Schobergruppe range and connects several mountain hut bases. As part of the European E6 long-distance hiking trail (it will one day connect Finland with Turkey), the trail leads from the Winklerner via Wangenitzsee, Adolf-Nossberger, Elberfelder, Glorer, and Salm mountain huts to the Glocknerhaus inn. Experiencing this sublime mountain world far from everyday life slows things down for you in an incredible way. Hikers must have Alpine sure-footedness, not suffer from vertigo, have good orientation (fog!), and be in very good physical condition.

Length / time it takes: 22 mi. (35 km) / 5 stages, at 4–6 hours of walking per day
Elevation difference: 2,461 vertical ft. (750 m)
Requirements: mountain safety and experience; physically fit

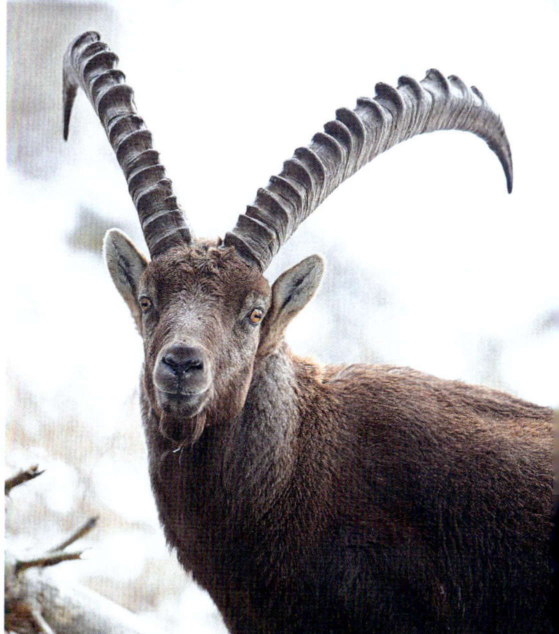

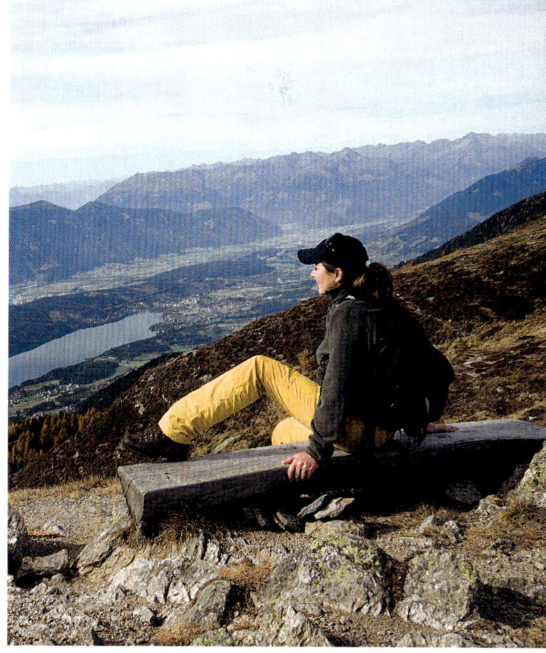

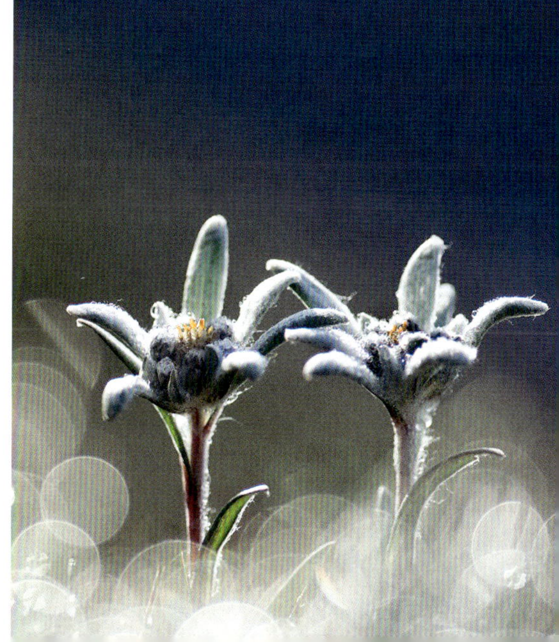

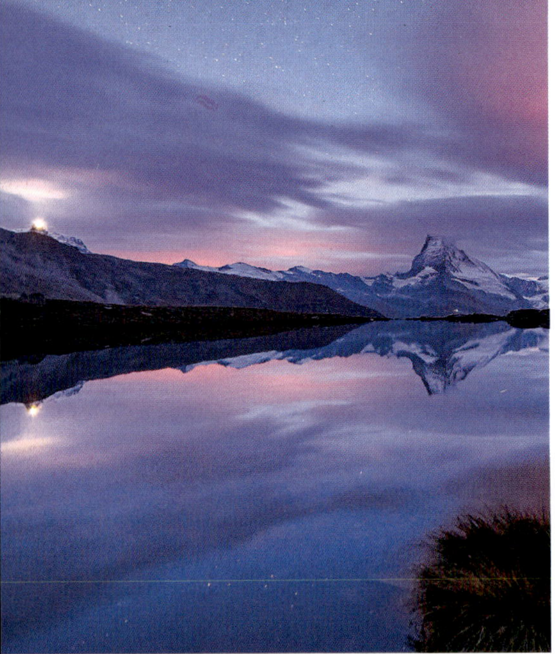

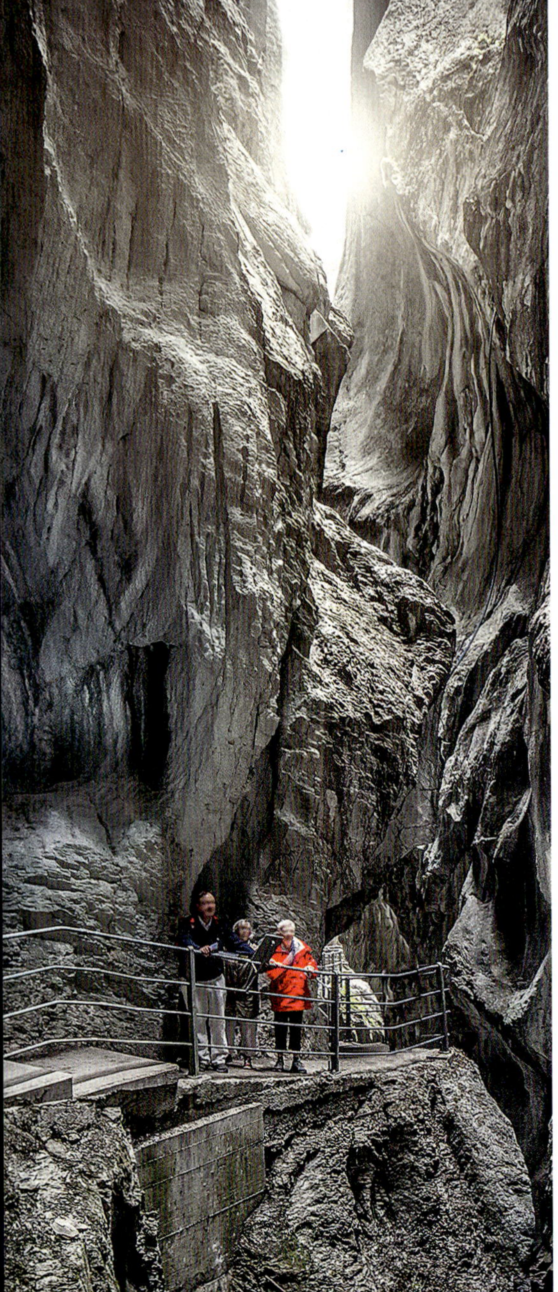

63. THE LONGEST SUSPENSION FOOTBRIDGE IN THE WORLD

Not for the faint of heart or for anyone who is afraid of heights: the view from the 1,621 ft. long (494 m) Charles Kuonen suspension bridge into the 279 ft. deep (85 m) valley goes as far as a rockfall area. The Europaweg Trail, which opened in 1997, with its spectacular view of the Matterhorn, is considered by many to be the most beautiful Alpine trail for demanding mountain hikers. The Europa mountain hut, at 7,431 ft. (2,265 m) above the municipality of Randa, is open only in spring and summer, like the rest of the trail. As soon as the first rays of sunshine reach the majestic mountain peaks, inquisitive nature lovers take the short route to the pedestrian rope bridge over the Mattertal valley. It was named for its main sponsor and was completed only in 2017. At dizzying heights, hikers experience unforgettable moments amid the peaks.

Length / time it takes: 21 mi. (33.7 km) / 2 days
Highest elevation: 8,907 ft. (2,715 m)
Starting point: Grächen
Elevation difference: 3,599 vertical ft. (1,097 m)
Information: europaweg.ch

64. SLEDDING FUN ON THE PREDA-TO-BERGÜN RUN

A different kind of rapid-curve descent: the longest illuminated "sled run" in Europe leads below the towering viaducts through an enchanting winter landscape in a UNESCO World Heritage area. The well-developed pass road between the villages of Preda and Bergun is closed in winter, but the two towns along the famous Rhaetian Railway route to St. Moritz are well connected by train anyway. This is how they came to this great idea for the whole family, right in the middle of Parc Ela. You simply borrow a two-runner vehicle in one of the two villages, sled down along all the curves, take the train back up again, and do this several times during a day. Who said sledding is only for kids?

Length / time it takes: 3.75 mi. (6 km) / one day
Highest elevation: 5,873 ft. (1,790 m)
Starting point: Preda
Elevation difference: 1,312 vertical ft. (400 m)
Information: schlittel-bahnorama.ch, schlitteln-berguen.ch

65. HIKING THROUGH THE ROSENLAUI GLACIER GORGE

The thundering waters of the Reichenbach River, bubbling and smacking, pour below the Rosenlaui glacier into a gorge that Johann Wolfgang von Goethe hiked through after it was discovered. Only gradually have isolated travelers begun to populate this little-known corner of Switzerland by descending through a mountain forest. The region is a UNESCO World Natural Heritage Site. The paved trail, with its breathtaking views, leads along refreshing waterfalls, bizarre rock formations, and romantic grottoes. There is an adventure pass named in memory of the famous hiker; it combines travel with entry to the glacier gorge and a "Goethe lunch" in the historic Hotel Rosenlaui at the entrance to the gorge.

Length / time it takes: 1,880 ft. (573 m) / half a day
Highest elevation: 6,562 ft. (2,000 m)
Starting point: Rosenlaui
Elevation difference: 508 vertical ft. (155 m)
Information: rosenlauischlucht.ch

66. THE JUNGFRAU MARATHON

The impressive panorama of the Eiger, Mönch, and Jungfrau Mountain troika is in view—it was no coincidence that this run was once awarded the title of the "most beautiful marathon in the world." Between Lake Thun and Lake Brienz in Interlaken, some 4,000 runners set off on a Saturday at the beginning of September on an initially moderate course, in awe of the already visible mountain world of the famous triumvirate. From Lauterbrunnen, with the Trummelbach Falls, the trail climbs steeply upward to the mountain village of Wengen and on to the finish line at Kleine Scheidegg at 6,890 ft. (2,100 m). Flag wavers, alphorn players, and a bagpipe provide a fascinating and atmospheric backdrop along the route. The sports event for handicapped athletes, the Jungfrau-Pararace, is held the day before.

Length / time it takes: 26.22 mi. (42.195 km) / 2 days
Highest elevation: 7,234 ft. (2,205 m)
Starting point: Interlaken
Elevation difference: 6,000 vertical ft. (1,829 m)
Information: jungfrau-marathon.ch

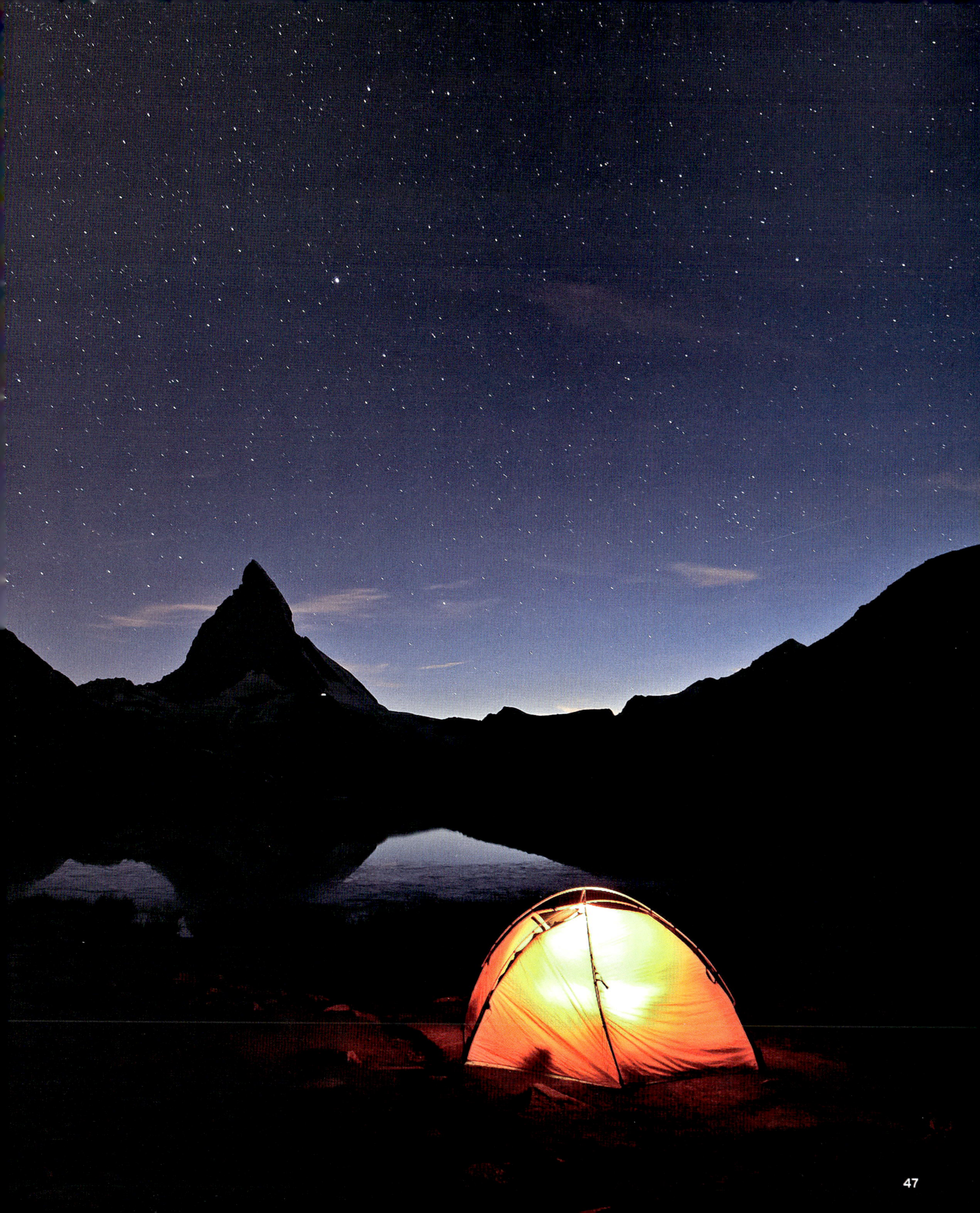

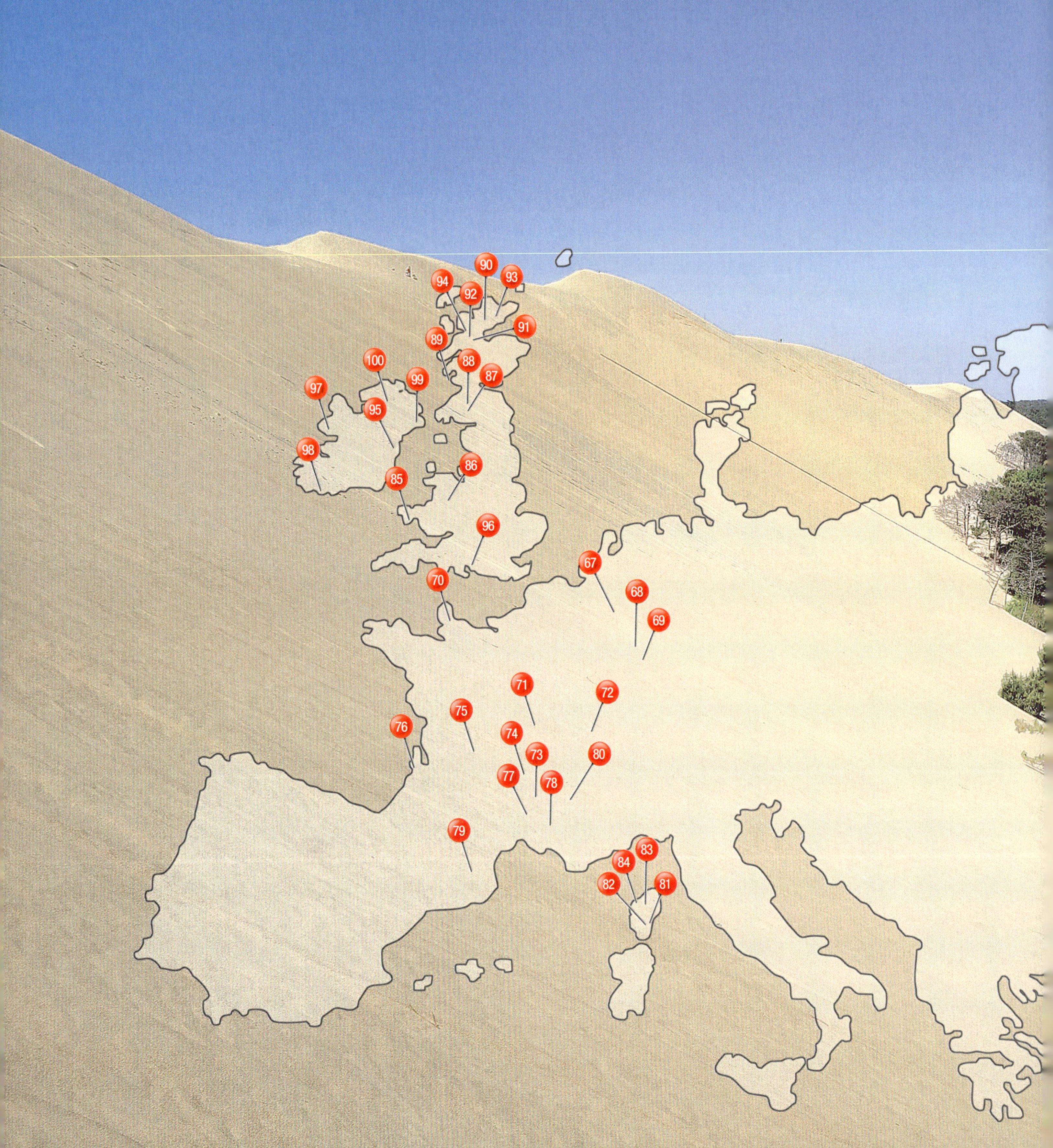

90
94
93
92
91
89
100
88
87
99
97
95
98
86
85
96
67
70
68
69
71
72
75
76
74
73
80
77
78
79
83
84
81
82

WESTERN EUROPE

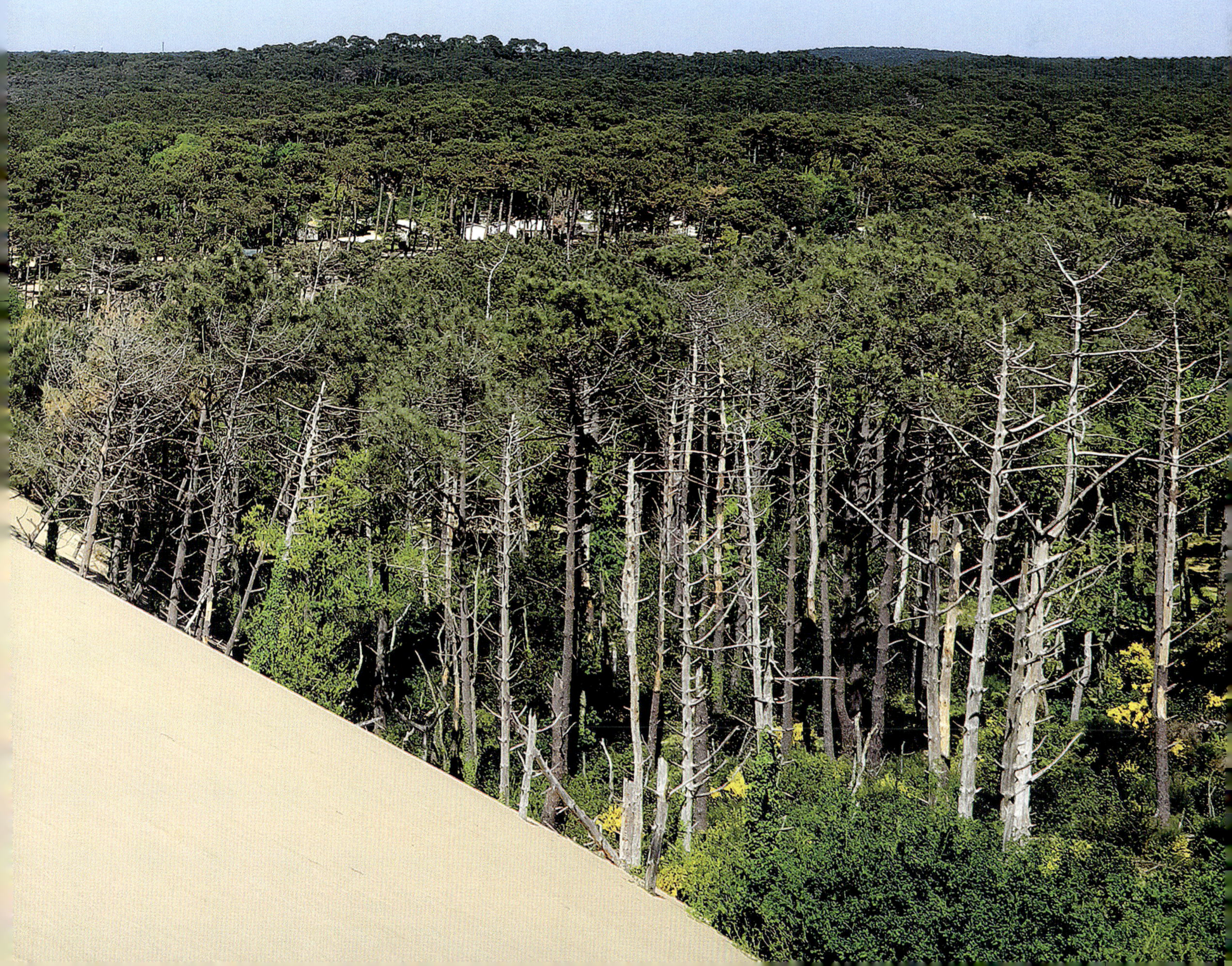

67. HIKING IN THE MULLERTHAL REGION

Valleys, forests, rocks, rivers, and waterfalls—the Grand Duchy of Luxembourg has one of the densest and most attractive hiking-trail networks in Europe. The trails take you through beautiful landscapes. The Mullerthal Trail, awarded a "Leading Quality Trail—Best of Europe" designation, is the beacon among the hiking trails of the Mullerthal region—Luxembourg's Little Switzerland. Via three loops, it goes through bizarre rock formations and past castles and palaces and offers magnificent panoramic views. You can hike it in six individual stages between 11 and 16 mi. (18 and 25 km) long. One of the most beautiful highlights is the "Schiessentümpel" waterfall near the village of Mullerthal. It lies behind the Heringer Millen brasserie, where you can also fortify yourself for your continued hike.

Length / time it takes: 70 mi. (112 km) / 5–7 days
Highest elevation: 1,312 ft. (400 m)
Starting point: Echternach
Elevation difference: 492–656 ft. (150–200 m)
Information: www.mullerthal-trail.lu

68. NATURE TRAIL IN THE NORTHERN VOSGES

On the Franco-German border, the Palatinate Forest and the Northern Vosges have been a cross-border UNESCO biosphere reserve since 1998. The region features densely forested areas, strange sandstone rock formations, around 100 castles and castle ruins, and many streams and small rivers that wind through the valleys. The best way to discover the natural treasures of the "Pays de la Petite-Pierre" is along the Loosthal educational nature trail. Spotting an eagle owl among the crevices is an especially impressive experience. The largest nocturnal bird of prey in Europe, it is particularly active at dusk and at night and is therefore difficult to spot during the day due to the perfect camouflage of its plumage. But searching for it is worth the time.

Length / time it takes: 2.5 mi. (4 km) / 1 day
Highest elevation: 1,280 ft. (390 m)
Starting point: Gorna forester's lodge
Elevation difference: 164 vertical ft. (50 m)
Information: www.parc-vosges-nord.fr

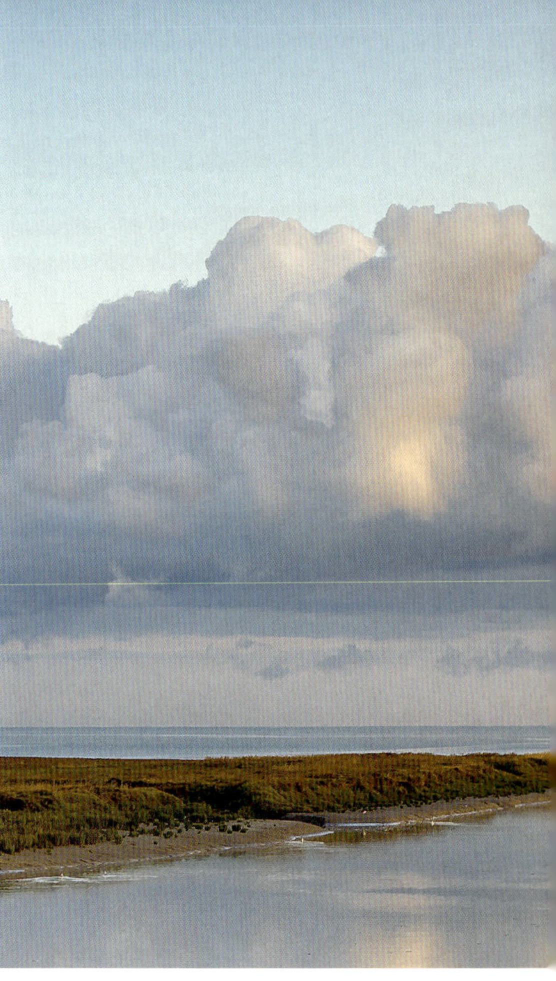

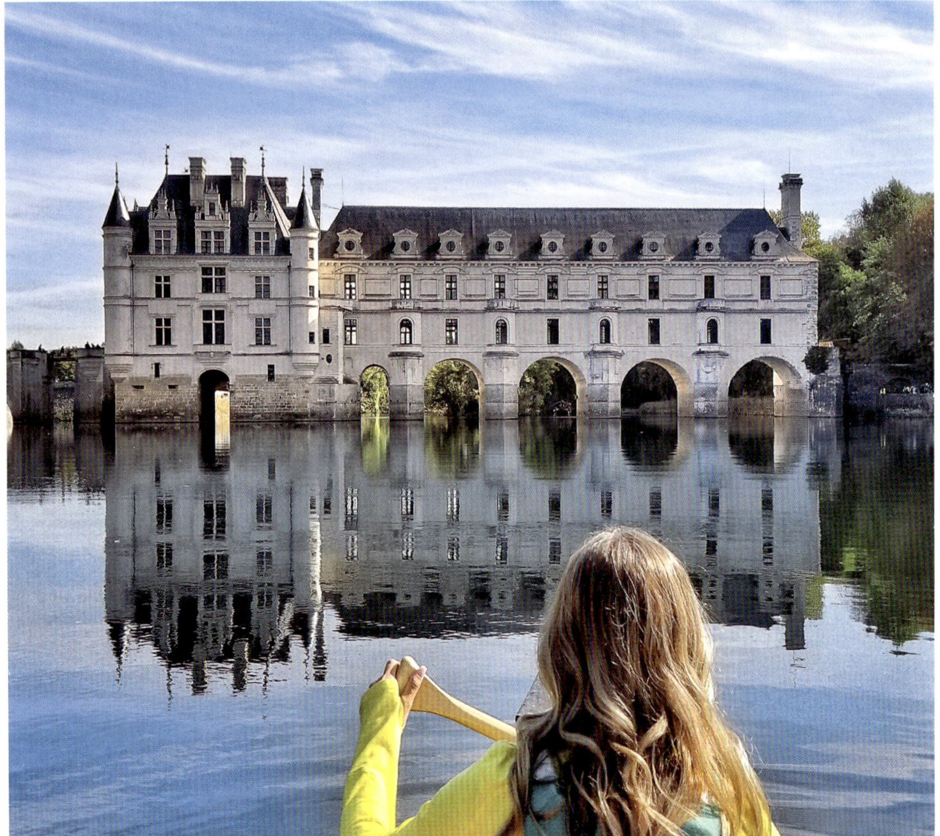

69. BY BIKE TO MONT-SAINT-MICHEL

Everyone knows this destination, the Mont-Saint-Michel monastery on its hill, but not this route: via the Véloscénie long-distance bike path. You could start from Paris or farther on from Versailles, but the last 106 mi. (171 km) from the small town of Alençon are the most beautiful and the most rural. These miles lead through largely deserted parts of the Normandy hinterland, far away from sensory overload, metropolises, and industry. Up and down it goes, over lush green ranges of hills, through densely forested areas, and past horse and cattle pastures and ivy-covered walls. The villages are slumbering, roses and hydrangeas grow next to houses, and the facades have colored wooden shutters. The route is accompanied by the crystal-clear air, which begins to taste of salt the closer it gets to the legendary Mont-Saint-Michel. It rises from the sea like an apparition.

Length / time it takes: 280 mi. (450 km) / about 1 week
Starting point: Paris
Trail's end: Mont-Saint-Michel
Information: www.veloscenie.com

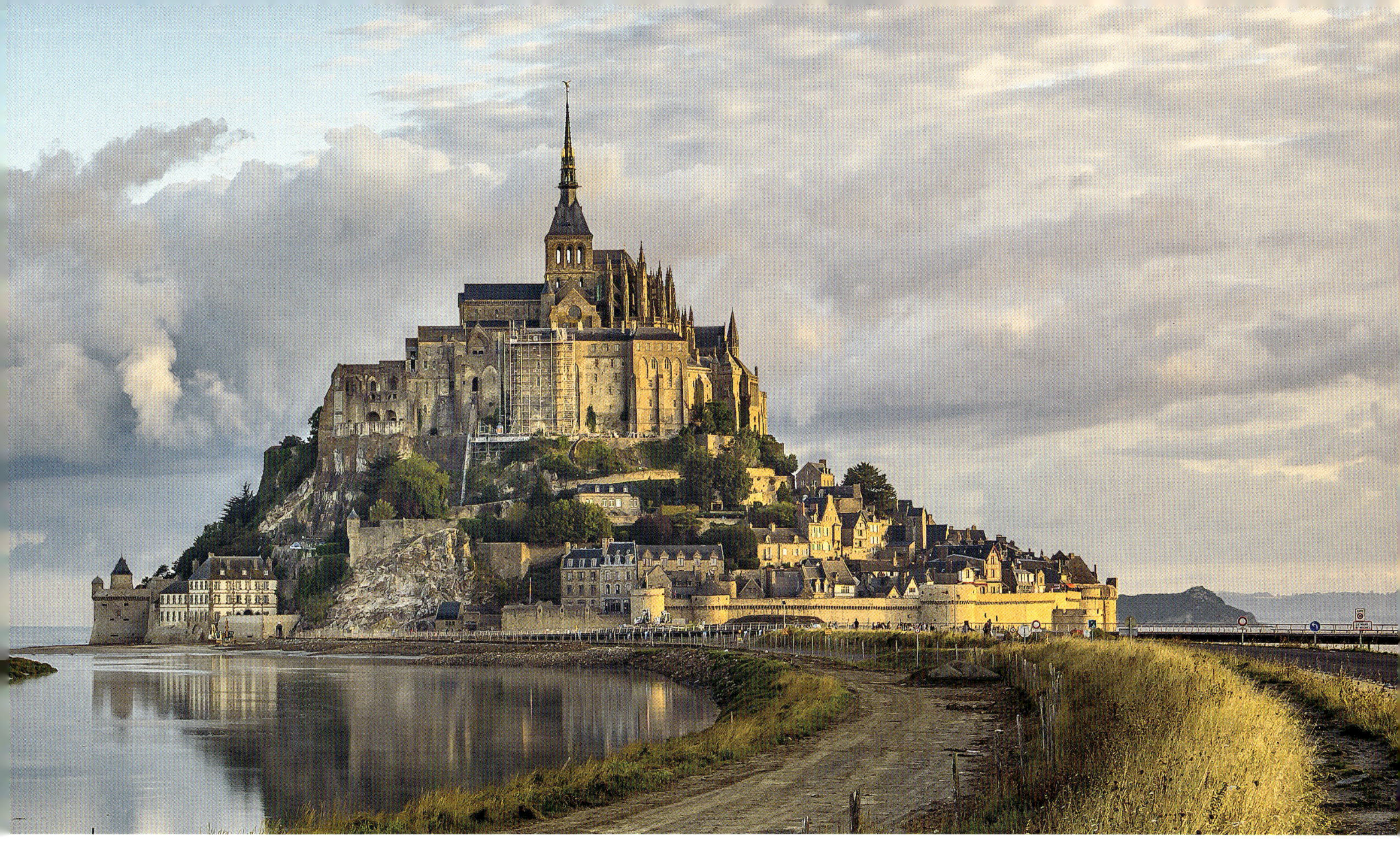

70. ALONG THE SENTIER DES DOUANIERS TRAIL

All roads lead, well, in this case not to Rome, but to the well-known abbey and hill island of Mont-Saint-Michel. But this time on foot: along the GR 223, the "Sentier des Douaniers," in English the "Customs Officers' Trail" (277 mi. [446 km] long), through Normandy. The fragmented coastline of the French département of La Manche sets the course and is particularly beautiful in spring, when nature glows in all imaginable colors against the background of the deep-blue sea. And in the second half of August, the heather on Cap de la Hague is in full bloom. All this naturally wants to be irrigated—and, in fact, from above. Especially in Normandy, you are never safe from precipitation. But the rocky coast, the beaches, the towns, the harbors, a piece of the world's cultural heritage (the Vauban towers in Saint-Vaast-la-Hougue), and everyone you encounter along the way will balance everything out.

Length / time it takes: 277 mi. (446 km) / at least 22 days
Starting point: Carentan
Trail's end: Mont-Saint-Michel
Information: www.manche-tourismus.com

71. CANOE TRIP ON THE LOIRE RIVER

A wilderness feeling in the heart of Europe! For canoeists, especially for newcomers, the Loire unfolds like a paradise among the long-distance rivers. Although you are consistently going downstream on the river from Decize and Cosne-Cours-sur-Loire, this week of canoeing is no surefire success. In fact, it is a real challenge: along wide loops through lonely strips of land and sidearms, as well as over small rapids. Steep embankments form the framework; you can see sandbanks with uprooted trees and the exuberant green of meadows, pastures, and alluvial forests. Small towns such as Nevers and La Charité-sur-Loire take you temporarily back to civilization, but in many places the silence is a balm for the soul. The deceleration rate is on average 2–3 mi. (4–5 km) per hour.

Length / time it takes: 61 mi. (98 km) / 7 days
Starting point: Decize
Information: organized paddling tours, such as at www.rucksack-reisen.de

72. BY FOOT THROUGH THE FRENCH JURA MOUNTAINS

The Échappée Jurassienne is the longest hiking trail through the French Jura mountain range. During the warmer season, you can really get into your stride here between the plains and the mountain range highlands: from Dole via Lons-le-Saunier and the winter sports resort of Les Rousses to Saint-Claude. The landscape features forested areas, steep slopes, and narrow valleys just as much as mountain lakes and waterfalls. Of historical note are the salt mines of Salins and the Royal Saltworks of Arcet-Senans, which are a UNESCO World Heritage Site, and the Benedictine abbey in Baume-les-Messieurs, one of the most beautiful villages in this area of France. Along the way you can also immerse yourself in regional culinary treasures by tasting the famous Comté cheese with a full-bodied red wine.

Length / time it takes: 186 mi. (300 km) / at least 16 days
Starting point: Dole.
Information: www.jura-tourism.com

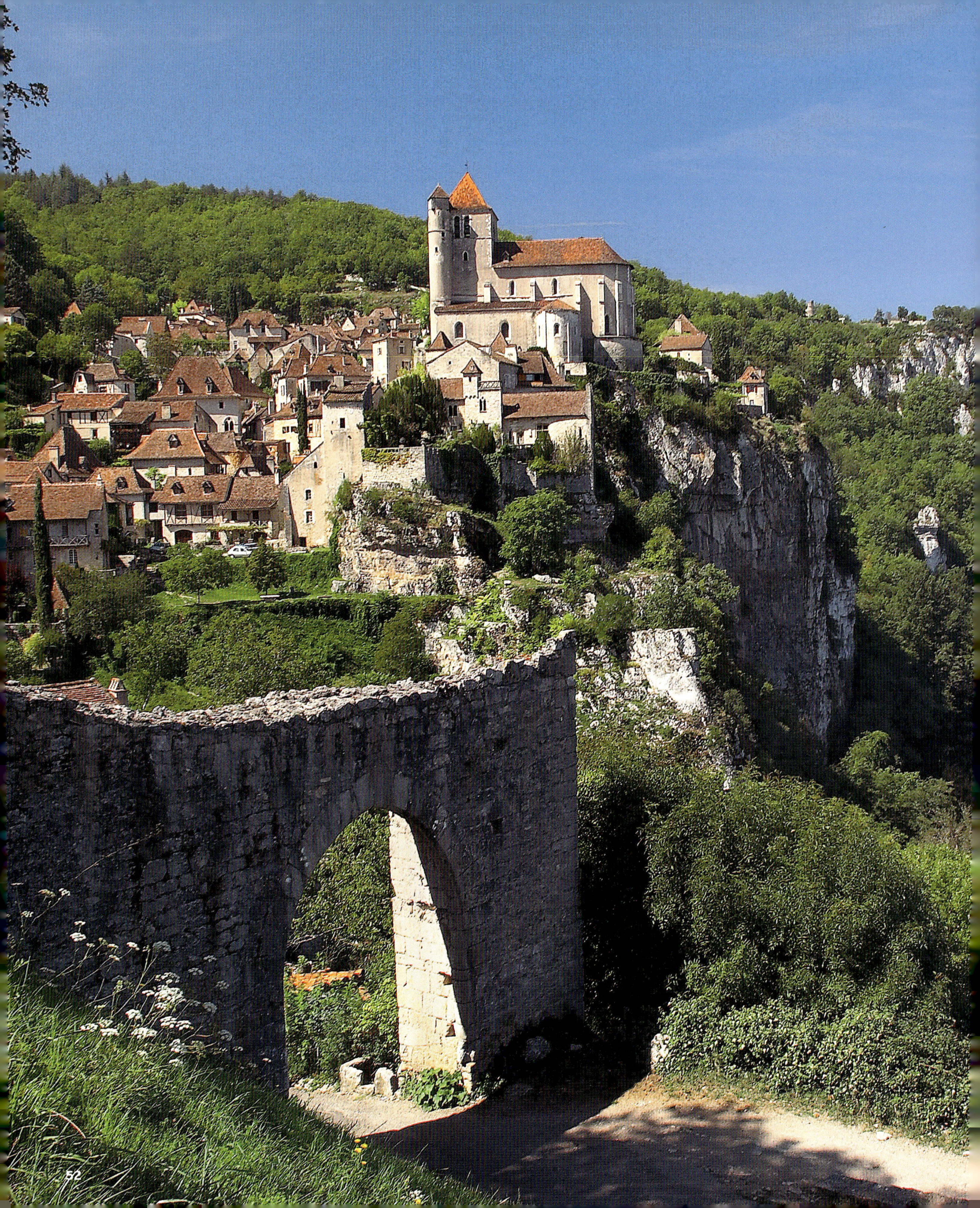

73. ON THE VIARHÔNA BIKE PATH

Suddenly the water comes to life, just before the day's goal of Le Pouzin. Tiny waves splash against the riverbank. Who is behind it? There are beavers here, but today it is muskrats, which ultimately disappear into the undergrowth. The next day there is more wildlife, such as herons and lizards. If you want to experience this and much more, pedal along the ViaRhôna long-distance bike path, named after the course of the Rhône River, which it follows for most of its route. You could start at Lake Geneva and then head toward Lyon, but if you want to cherry-pick the most-famous places, you have to tackle a 124 mi. (200 km) stretch south of Lyon from Vienne to Pont-Saint-Esprit. Simply fantastic: close to the water, past vineyards, pear and apple orchards, and picturesque villages. If you want to, you can continue cycling to the Mediterranean.

Length / time it takes: 506 mi. (815 km) / at least 12 days
Information: de.viarhona.com

74. HIKING ON VIA PODIENSIS PILGRIMAGE ROUTE

The final glances back: to the houses of Le Puyen-Velay, to Corneille Rock, to the remains of a volcanic cone, and to the cathedral. Then the word is "Saint James, we're coming!" Since the Middle Ages, the town of Le Puyen-Velay in the Auvergne region has been the starting point for one of the four most-important pilgrimage routes through France: the Via Podiensis, which stretches far away, more than 435 mi. (700 km) toward the Pyrenees. You pass through the highlands around Aubrac, with farms, grazing cattle, gorse bushes, ferns, and low stone walls. Stops along the way are Espalion, Conques (with its abbey), Figeac, Cahors (with its weirs crossing the Lot River), Moissac (with its fairy-tale cloister), Condom, the fortified village of Larressingle, and Aire-sur-l'Adour on the banks of the Adour River. In the small village of Ostabat-Asme, at the foot of the Pyrenees, the Via Podiensis converges with the Via Turonensis and Via Lemovicensis routes of the Camino de Santiago (Way of St. James).

Length / time it takes: more than 435 mi. (700 km) / about 5 weeks
Information: camino-europe.eu/de/eu/fr/jakobswege/france-via-podiensis-le-puy-moissac/

75. ON THE FLOW VÉLO LONG DISTANCE BIKE PATH

Class contrasts: the Flow Vélo long-distance bike path takes you from the interior of the département of Dordogne to the Atlantic. It starts in the Dordogne, then goes through the départements of Charente and Charente-Maritime. Impressions of nature are interspersed with injections of culture in the towns of Angoulême, Saintes, and Rochefort, where, at the instigation of Louis XIV, the famous naval arsenal and royal rope makers were established during the seventeenth century at a bend in the Charente River. Along the way, a visit to a distillery in the town of Cognac will immerse you in the high-proof world of brandy distilling. The end of the trail for the Flow Vélo is—after taking a ferry—the Atlantic island of Île d'Aix, a flat speck of forest and sand and massive fortress walls. France's dethroned emperor Napoleon spent his last days on French soil here in 1815.

Length / time it takes: 1,805 mi. (2,905 km) / 5–6 days
Starting point: Thiviers
Information: www.lafowvelo.com

76. CLIMBING THE DUNE DU PILAT

What a sandy giant! The Dune du Pilat rises more than 328 ft. (100 m) high, extends 1,640 ft. (500 m) wide, and is almost 2 mi. (3 km) long. You clamber up, facing the yellow-white wall as you ascend the dunes. You can't even quickly shake it off your ankles—that takes a certain amount of effort. At some point you automatically take off your shoes or sandals. In the morning, the sand is still cool and fine grained due to the wind; later on in the day it is riddled with the many traces of those who have gone before you. The panorama is magnificent at all times: far below, the blue of the Atlantic, which reaches into Arcachon Bay; behind it a green sea of pine. So just sit down and enjoy it. By the way, the exact heights of the dune and all its humps and bumps fluctuate, depending on the winds and natural drifting of the sand.

Elevation: 328 ft. (100 m)
Width: 1,640 ft. (500 m)
Information: www.ladunedpilat.com

77. PADDLING THROUGH THE ARDÈCHE GORGES

The upper reaches of the Ardèche River feature extreme whitewater and can be recommended only for very experienced paddlers—if the water levels are appropriate. Things are calmer on the lower reaches, which impress paddlers with a gorge that is around 19 mi. (30 km) long. The fascinating, scenic Gorges de l'Ardèche provide a fantastic backdrop for a relaxed canoeing experience and take you past gravel ridges, over rapids, and along spectacular loops in the river. A natural stone bridge spans the river near the village of Vallon-Pont-d'Arc. Passing under the imposing Pont d'Arc is a highlight of the tour. In summer the Ardèche is a very popular destination; in spring and autumn it is much quieter. There are plenty of boat rental companies that provide everything you need for a trip, including a pickup service.

Length / time it takes: 4–20 mi. (7–31 km) depending on the conditions / 2–6 hours
Starting point: Vallon-Pont-d'Arc
Information: www.ardeche-guide.com

78. HIKING THROUGH COLORADO PROVENÇAL, RUSTREL

The colorful, shimmering landscape of the Colorado Provençal de Rustrel stands out in contrast to the surrounding green forest. Earlier ocher mining and erosion have given the ocher quarries their distinctive relief surfaces. The first part of the trail leads under shady trees until you reach a first scenic outlook on a hill. From here you can let your gaze wander over this extraordinary natural spectacle. Impressive sections and formations have been given sonorous names, such as the pointed rock needles called "fairy chimneys," which were shaped by wind and weather, or the ocher-colored "Sahara," which welcomes the visitor in a completely different way than the light sand of the "White Desert." Hiking trails of various lengths lead through this fascinating and colorful landscape, whose tones range from white to deep red.

Length / time it takes: 7 mi. (11 km) / 3 hours
Highest elevation: 1,870 ft. (570 m)
Starting point: parking lot on the D22 east of Rustrel
Elevation difference: 2,231 vertical ft. (680 m)
Information: www.rustrel.fr

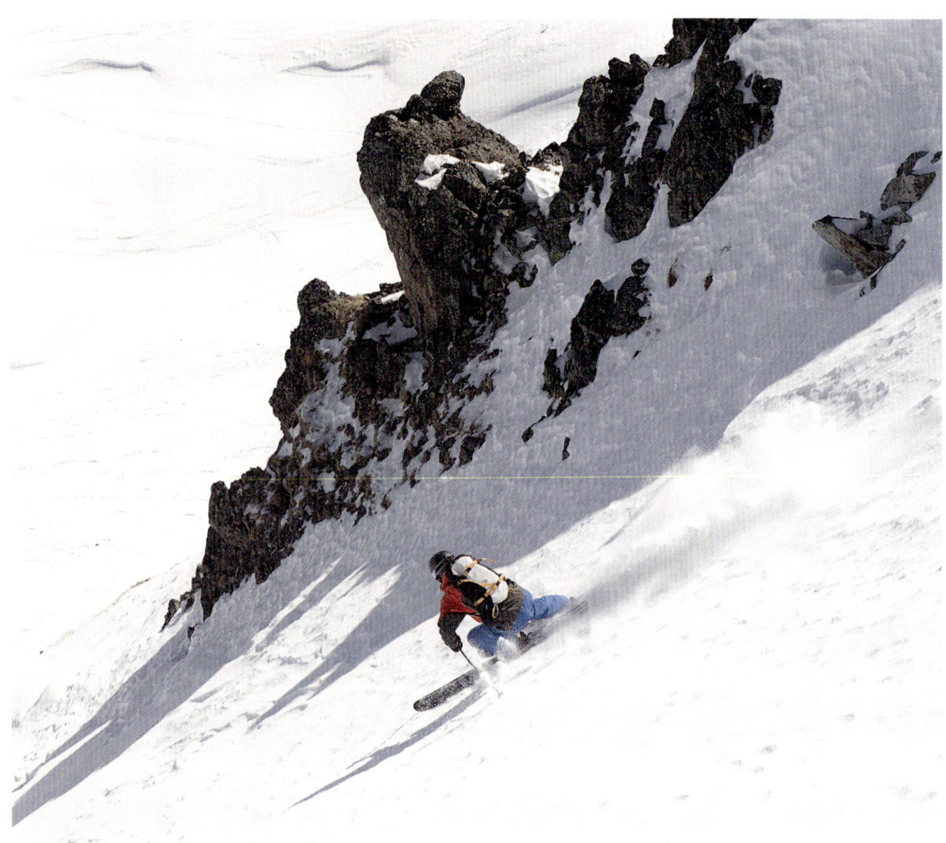

81. HIKING THROUGH FOREST PARC ALONG THE VIA FERRATA

Over the smooth rocks and up the via ferrata, the trail leads over the Solenzara River—although there are numerous climbing centers in Corsica, this one on the Solenzara, in the Bavella valley, is something special. The climbing routes—at various levels of difficulty—lead across the fantastic rocky landscape of the Bavella massif that begins here. The highlight is the via ferrata: iron brackets are driven into the rocks, and you shimmy up on them. The view of the wide river valley from above is gigantic! And to top it off, there is a ride on the zipline across the Solenzara. If you dare to look down, you will find wonderful swimming pools you can plunge into after all your exertion . . .

Time it takes: 3–4 hours
Starting point: D268 in the Bavella valley, next to the U Rosmarinu campsite
Requirements: No fear of heights; good physical condition
Information: www.corse-canyoning-parc.com

79. BY FOOT FROM THE ATLANTIC TO THE MEDITERRANEAN

The GR 10 hiking trail is a sought-after destination for experienced long-distance hikers. This demanding trail creates a connection between the Atlantic and the Mediterranean along the Pyrenees Ridge. It goes without saying that you have to be in top physical condition and have a good head for heights. June to October are considered the most-stable months for weather, but you should always be prepared for sudden weather changes. In the central Pyrenees and Ariège Pyrenees, the GR 10 leads over breathtaking high mountain passes. Then the situation relaxes again as you hike on summer pastures, shepherd's paths, and forest trails. Panoramas of the summits and valleys alternate; lakes and waterfalls, as well as distant villages, fill your photo memory cards. And the sea at the end—at this or that end—is the reward for all your efforts.

Length / time it takes: 5,285 mi. (8,505 km) / at least 60 days
Highest elevation: 8,970 ft. (2,734 m)
Starting point: Hendaye
Trail's end: Banyuls-sur-Mer

80. LA GRAVE—THE TRUE FREE-RIDE MECCA

Hardly anyone knows it, but there is no better place for free riding. Extreme skiers have chosen La Grave, a small village in the French Alps, as their homeland. The almost 13,123 ft. high (4,000 m) La Meije Mountain towers over La Grave, and there is only an old, completely inefficient gondola to take you up. There are no marked downhill runs; as a result, there are 6,890 vertical ft. (2,100 m) of pure skiing adventure on the gigantic north face of La Meije. From glaciers to wide-open spaces, into steep, narrow gullies, then into the forest—the countless downhill-skiing options offer all types of free riding. "Mountain as playground" or "serious fun"—if you come up here, you should be a good skier. You will not regret it and might even find a home here, like author Norbert Blank. La Grave is Mecca, the Holy Grail, and the Land of Milk and Honey all rolled into one.

Length / time it takes: one run 5.28 mi. (8.5 km) / 2 hours
Highest elevation: 11,811 ft. (3,600 m)
Elevation difference: 7,054 vertical ft. (2,150 m)
Starting point: Meije glacier cable car, La Grave
Information: www.skierslodge.com

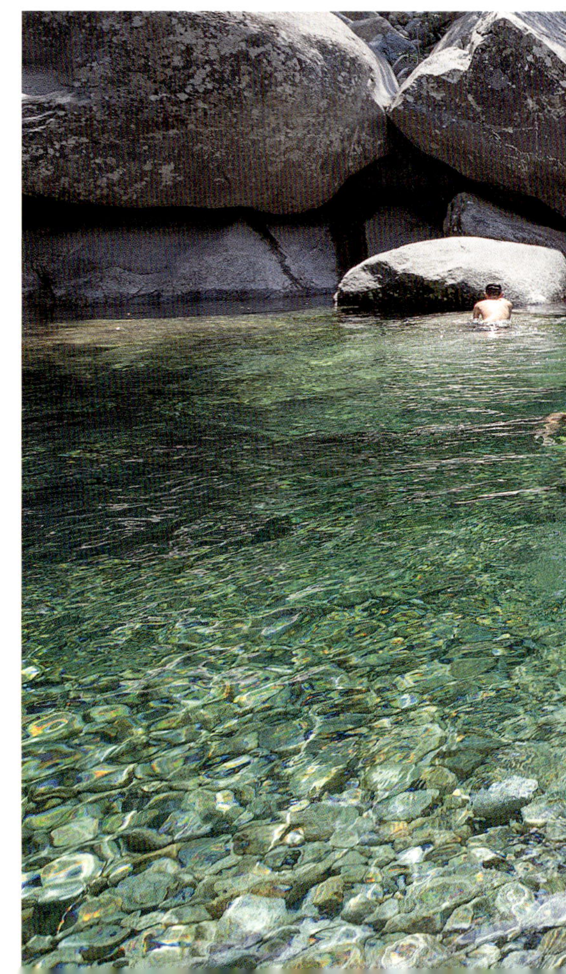

82. TRIP ALONG FIUMICELLI RIVER

Hiking, climbing, swimming, jumping—the Fiumicelli River valley is a unique natural playground. What starts out to be a still-easy wade through knee-deep water quickly turns out to be a demanding river hike with all kinds of shenanigans: the basins become deeper, the steep walls higher, the obstacles more massive, and the boulders in the riverbed more insurmountable. Cascades follow; you can climb up some of them but will have to go around some others. Small, sandy coves entice you to take a break and picnic. From the rocks, you can jump into the pools that open up here and there. After 1.86 mi. (3 km), a hiking trail crosses the river. It leads back to the road, and you have to walk back a bit on this to the starting point at the bridge.

Length / time it takes: 1.86 mi. (3 km); river, 1 mi. (1.7 km); hiking trail, 1.37 mi. (2.2 km); road / half a day
Starting point: Pont de Fiumicelli

83. HIKING AROUND LAKE NINO

Wild horses graze on the marshy shores; in the background the peaks of the Cinto Massif dominate the scenery. In front of the massif, in the middle of the green meadows, lies the deep-blue mountain lake at an elevation of 5,741 ft. (1,750 m). This heavenly place is the goal of a hike along a brook, through a pine forest, and steeply uphill over slabs of rock and boulders. There is plenty of variety along the way, including a secluded sheepfold, places for climbing, and a sensational panoramic view of the plateau from the Bocca à Stazzona pass. An alternative way back would be to follow the GR 20 long-distance hiking trail from the lake until you reach a piece of woodland. After 1,312 ft. (400 m), turn right and leave the GR 20, hike to Fontaine Caroline fountain, then go 6.5 mi. (2 km) along the road to the starting point.

Length / time it takes: 6.2 mi. (10 km) / 4–6 hours; return via GR 20 +2.5 mi. (+4 km)
Highest elevation: 5,764 ft. (1,757 m)
Starting point: Maison Forestière de Poppaghia (on the D84)
Elevation difference: 2,493 vertical ft. (760 m, +65 ft. [+20 m])

84. CANOEING ON THE FANGO RIVER

Pond turtles stretch their heads toward the sun, water lilies enchant the eye, and herons circle around. To top off this seemingly unreal scene created by the local fauna, you paddle Indian style in a canoe through the tributaries of the Fango River. It is something very special to experience the many branches in this nature reserve this way. At the Torra di Galeria, a former Genoese tower, the Fango flows into the sea—on one side is the beach, on the other the protected nature reserve. Below the Genoese building is the place where you can board the single or double canoes provided by the "Delta du Fango en Canoë" canoe supplier, and off you go into the wilderness . . .

Time it takes: starting from 1 hour
Starting point: Torra di Galeria on the D351
Information: www.delta-du-fango.com

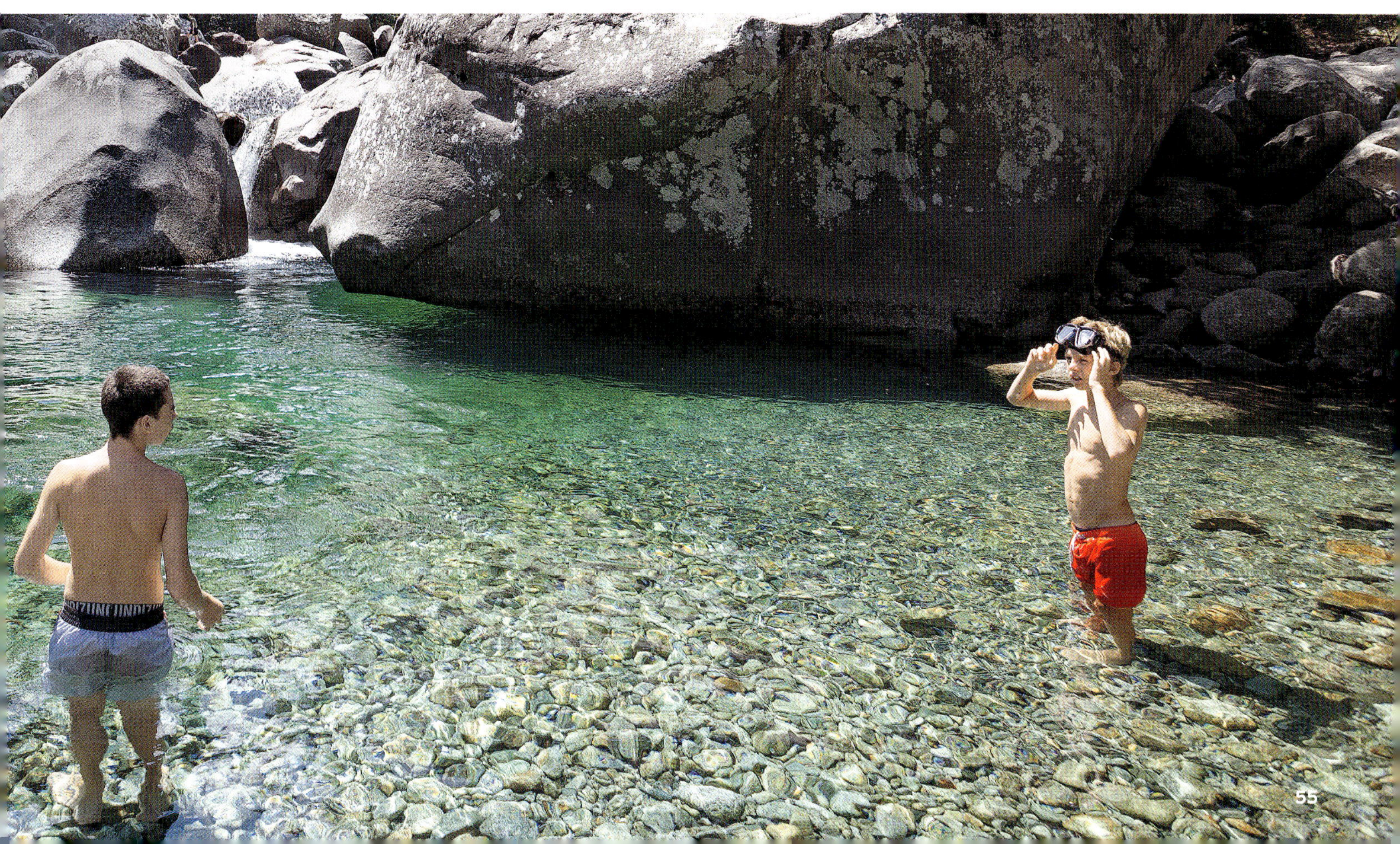

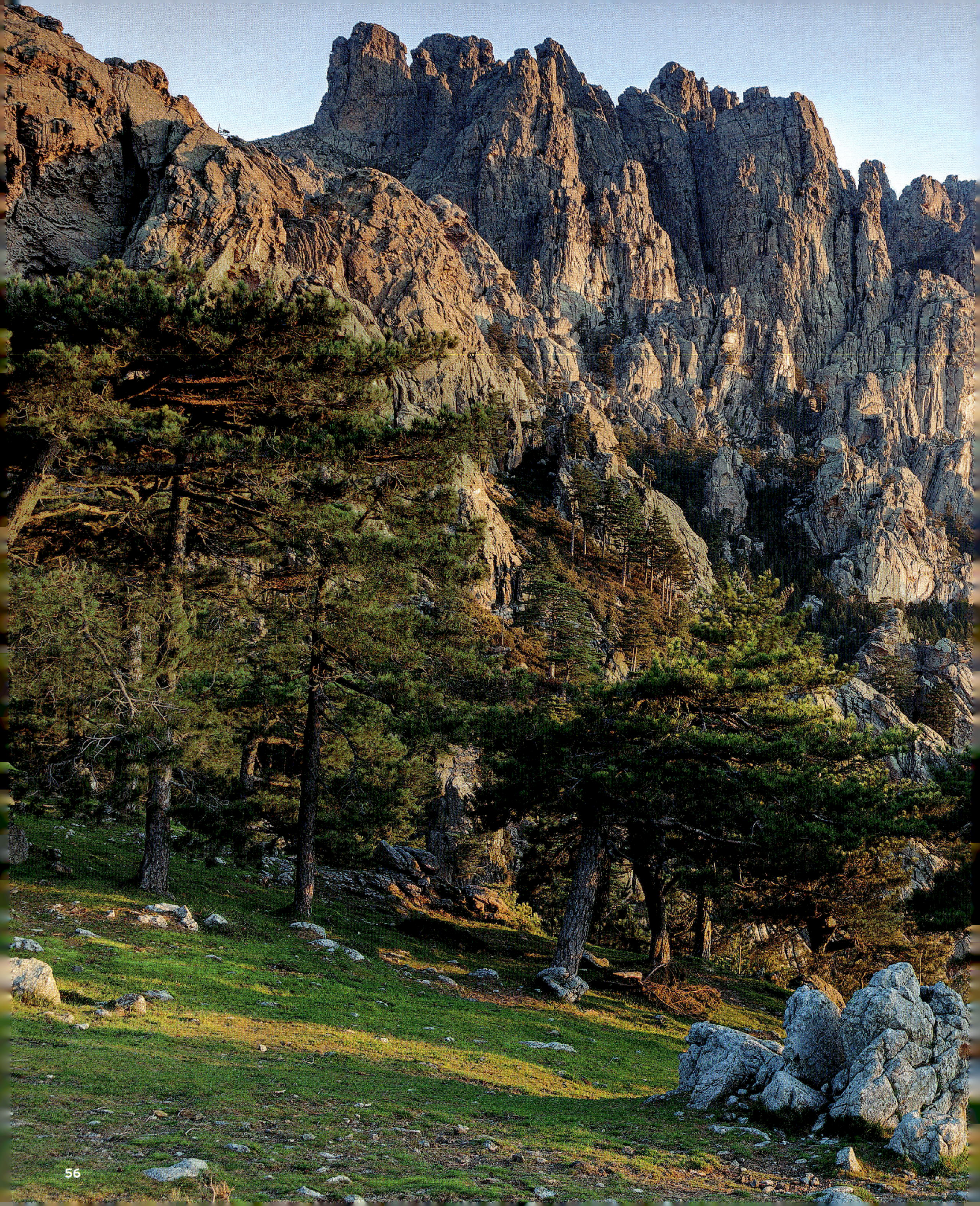

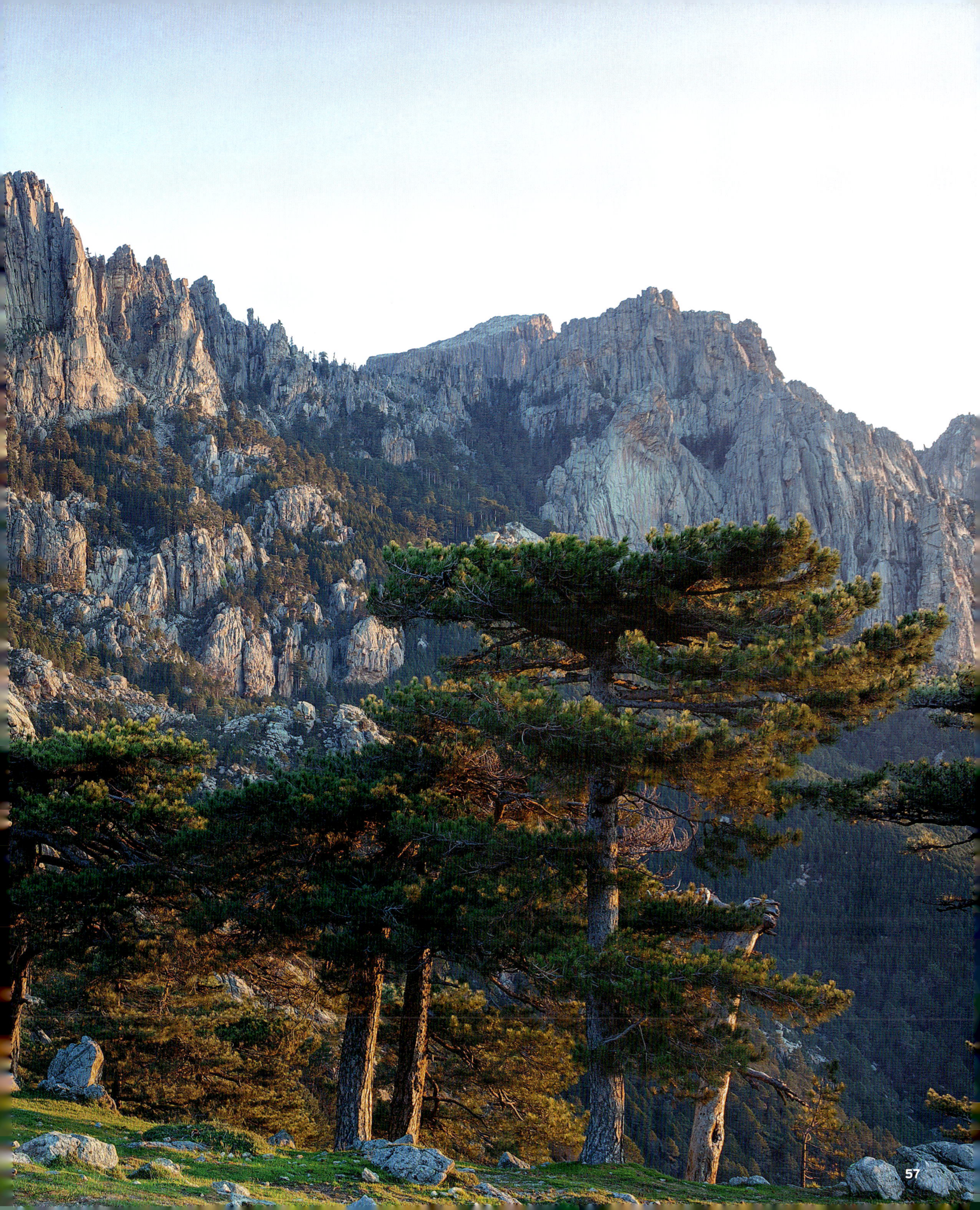

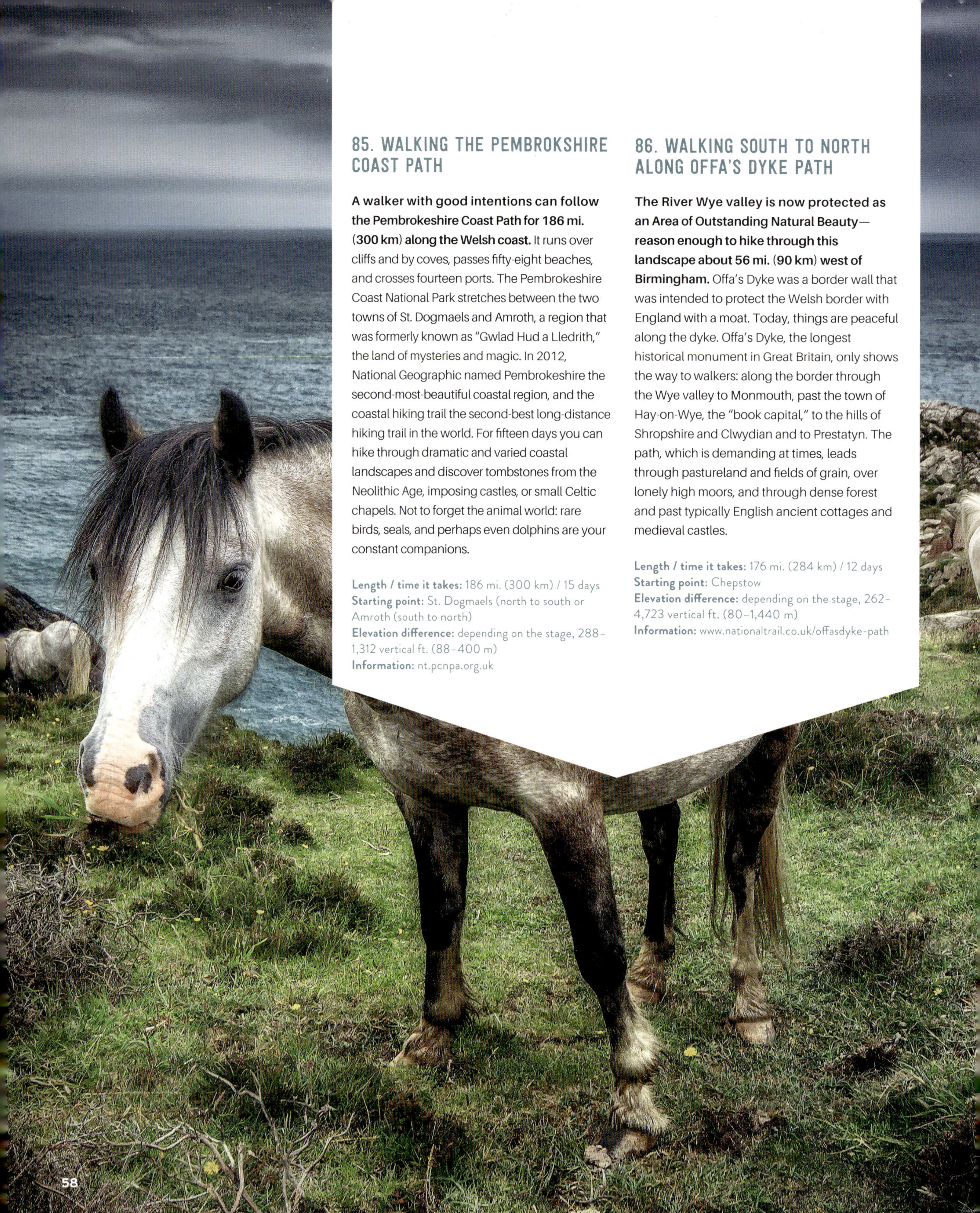

85. WALKING THE PEMBROKESHIRE COAST PATH

A walker with good intentions can follow the Pembrokeshire Coast Path for 186 mi. (300 km) along the Welsh coast. It runs over cliffs and by coves, passes fifty-eight beaches, and crosses fourteen ports. The Pembrokeshire Coast National Park stretches between the two towns of St. Dogmaels and Amroth, a region that was formerly known as "Gwlad Hud a Lledrith," the land of mysteries and magic. In 2012, National Geographic named Pembrokeshire the second-most-beautiful coastal region, and the coastal hiking trail the second-best long-distance hiking trail in the world. For fifteen days you can hike through dramatic and varied coastal landscapes and discover tombstones from the Neolithic Age, imposing castles, or small Celtic chapels. Not to forget the animal world: rare birds, seals, and perhaps even dolphins are your constant companions.

Length / time it takes: 186 mi. (300 km) / 15 days
Starting point: St. Dogmaels (north to south or Amroth (south to north)
Elevation difference: depending on the stage, 288–1,312 vertical ft. (88–400 m)
Information: nt.pcnpa.org.uk

86. WALKING SOUTH TO NORTH ALONG OFFA'S DYKE PATH

The River Wye valley is now protected as an Area of Outstanding Natural Beauty—reason enough to hike through this landscape about 56 mi. (90 km) west of Birmingham. Offa's Dyke was a border wall that was intended to protect the Welsh border with England with a moat. Today, things are peaceful along the dyke. Offa's Dyke, the longest historical monument in Great Britain, only shows the way to walkers: along the border through the Wye valley to Monmouth, past the town of Hay-on-Wye, the "book capital," to the hills of Shropshire and Clwydian and to Prestatyn. The path, which is demanding at times, leads through pastureland and fields of grain, over lonely high moors, and through dense forest and past typically English ancient cottages and medieval castles.

Length / time it takes: 176 mi. (284 km) / 12 days
Starting point: Chepstow
Elevation difference: depending on the stage, 262–4,723 vertical ft. (80–1,440 m)
Information: www.nationaltrail.co.uk/offasdyke-path

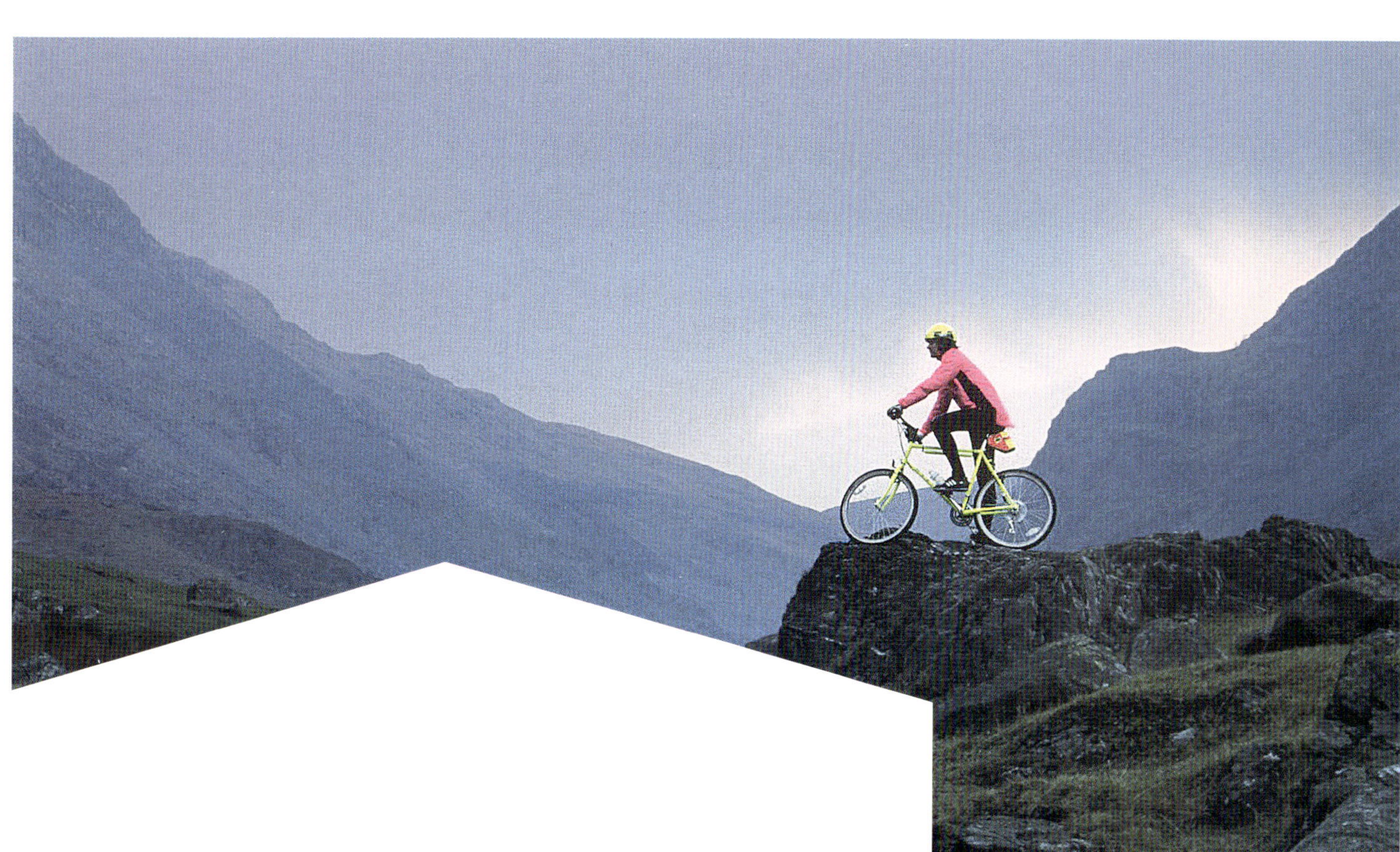

87. WALKING IN THE FOOTSTEPS OF PETER RABBIT

Everyone has probably already met them—the typical animal figures from the stories of Beatrix Potter. *Miss Potter*, the movie with Renée Zellweger about the life and work of the author and artist, is also atmospheric. Her home country, the Lake District in the county of Cumbria, delights visitors with its curving lakes in a gently rolling landscape of unspoiled nature and slate houses. A network of 2,175 mi. (3,500 km) of hiking trails, which extend up to the highest regional mountain, Scafell Pike (3,209 ft. [978 m]), offers plenty of variety. Windermere and Ullswater are the most-important lakes. The Cumbria Way, the best-known footpath, stretches from the town of Ulverston via Keswick to Carlisle. Definitely stop by the picturesque village of Grasmere with its Gingerbread Shop, and at Hill Top, once Beatrix Potter's farm.

Length / time it takes: 70 mi. (112 km) / 6–10 days
Highest elevation: 2,159 ft. (658 m)
Starting point: Windermere
Elevation difference: 10,010 vertical ft. (3,051 m)
Information: walklakes.co.uk

88. CLIMBING IN THE MINES IN CUMBRIA

Slate is abundant up here in the North of England; therefore its matte gray defines the architectural style of the entire region. For almost 300 years, the Honister Slate Mine near Keswick has been producing Westmorland Green Slates from the 450-million-year-old material. The adventures began twenty years ago, and the mine became an event. Since then, climbing and rappeling in one of the largest active mines in Cumbria has meant pure adrenaline. The via ferrata climbing route follows the original route taken by the miners; the more demanding Xtreme tour has won several awards. Or how about the Infinity Bridge, 1,969 ft. (600 m) above the often-rainy valley? The icing on the cake is the climbing tour—unique in England—on the walls of the mines to the highest point underground.

Time it takes: 1 day
Highest elevation: 2,126 ft. (648 m)
Starting point: Keswick
Information: honister.com

89. WALKING ON THE ISLE OF ARRAN

The Isle of Arran is also a very popular destination for Scots. It has all the landforms of the mainland and, thanks to the Gulf Stream, at the same time has a mild climate that often brings good weather. You have to be sure-footed to manage some of the ridge passages during this moderately difficult hike from south to north, and the ascent to Goat Fell, the highest mountain on the island, requires stamina. Damp weather and poor visibility make finding the path a challenge, and the rocky sections can become dangerously slippery. The moors and jagged rocks of the North especially create unforgettable impressions, and with a little luck you will see golden eagles and seals. There are numerous campsites and B&Bs along the route, and Brodick Castle and the Arran whiskey distillery near Lochranza round out the experience.

Length / time it takes: 42 mi. (68 km) / 3 days
Highest elevation: 2,867 ft. (874 m)
Starting point: Kilmory
Elevation difference: 12,467 vertical ft. (3,800 m)

90. WALKING OVER RANNOCH MOOR

This almost uninhabited, 81 sq. mi. (130 km²) moor lies on a 1,312 ft. high (400 m) plateau. Thanks to the bogs and the waterways that run through the area and the many small lakes and ponds that are features of the landscape, this easy hike can become a wet experience in some places. The starting point is the village of Dalwhinnie, lying almost at the geographical center of Scotland. It is also home to the country's highest-elevation malt whiskey distillery, at 1,069 ft. (326 m). The walk, which takes you through generally unspoiled landscapes, is flanked by numerous mountains towering more than 3,281 ft. (1,000 m) high, and you can find simple but very atmospheric bothies—huts for walkers—again and again along the route.

Length / time it takes: 43.5 mi. (70 km) / 4 days
Highest elevation: 2,132 ft. (650 m)
Starting point: Dalwhinnie
Elevation difference: 3,937 vertical ft. (1,200 m)

91. TREKKING ALONG THE WEST HIGHLAND WAY

It is not surprising that the infrastructure for walkers along the West Highland Way is extremely good, because it is one of the most famous and most frequently used footpaths in Scotland. Thanks to the well-developed and well-signposted trails, this long-distance hike is perfect for beginners. The hike begins at a striking stone pillar in the pedestrian zone of Milngavie, a suburb of Glasgow, and leads through beautiful Scottish Highland landscapes. The trail runs past lakes and impressive peaks and is lined with cozy hostels and rustic pubs up to the trail's end at Fort William. Fort William has good bus and train connections back to Glasgow.

Length / time it takes: 96 mi. (154 km) / 8 days
Highest elevation: 1,804 ft. (550 m)
Starting point: Glasgow
Elevation difference: 11,483 vertical ft. (3,500 m)

92. BAGGING ALL THE MUNROS

If you want to call yourself a "Munroist," you have to climb ("bag") all 282 Munros, any mountain in Scotland higher than 3,000 ft. (914.4 m). The highest peak is Ben Nevis, near Fort William in the northwestern Highlands, at 4,412 ft. (1,345 m). The air masses coming from the Atlantic tend to hang around this former volcano and result in fog and rain for 300 days a year; its Gaelic name means something like "mountain with its head in the clouds." Ben Lomond (3,196 ft. [974 m]), directly on Loch Lomond, is suitable for beginners. A well-developed and moderately rising hiking trail takes you up. If you want to enjoy wonderful views over Rannoch Moor up to the peaks of Glencoe, you should climb Schiehallion, the "fairy mountain of the Caledonians" (3,553 ft. [1,083 m]), which glitters because of the quartz.

Highest elevation: 4,412 ft. (1,345 m)
Information: www.walkhighlands.co.uk/munros/munros-height

93. WALKING THROUGH THE CAIRNGORM MOUNTAINS

There are multiday loop trails in Scotland's largest national park for experienced hikers. Five of the ten highest mountains in the country are in the Cairngorms, which form the largest contiguous mountain group in Great Britain. Compared to the West of Scotland, you can expect less precipitation in the Northeast, but it is significantly windier and cooler, and snowfall is by no means uncommon in the summer due to the plateau's elevation. The small village of Muir is a good starting point; from here you can hike over the four peaks of Cairn Toul, Carn na Criche, Braeriach, and Ben Macdui. The hikes go through lonely areas and at times through pathless terrain, so orientation can become a challenge. You should bring your own tent along, since there are only two huts, with no staff, along the way.

Length / time it takes: 39 mi. (63 km) / 4 days
Highest elevation: 4,294 ft. (1,309 m)
Starting point: Muir
Elevation difference: 11,483 vertical ft. (3,500 m)

94. ICE CLIMBING ON BEN NEVIS

Britain's highest mountain rises in Scotland. Ben Nevis, or simply "the Ben" as it is also affectionately called, has developed into a true El Dorado for ice climbers. In winter, when high levels of precipitation result in reliable ice formation, the mountain shows its true colors. Numerous routes of varying degrees of difficulty lead up the ridge on the mountain's 1,640 ft. high (500 m) and 2 mi. wide (3 km) northeast face. The best time to tackle ice climbing is between February and April. At the foot of the north face at 2,231 ft. (680 m), the CIC mountain hut is not only Britain's highest-elevation staffed mountain hut, it is also a good starting point for tours. If you don't want to go out on your own, you can book winter climbing courses here.

Highest elevation: 4,413 ft. (1,345 m)
Starting point: CIC mountain hut
Elevation difference: 3,281 vertical ft. (1,000 m)
Information: www.westcoast-mountainguides.co.uk/courses/cic-hut-winter-climbing-course/

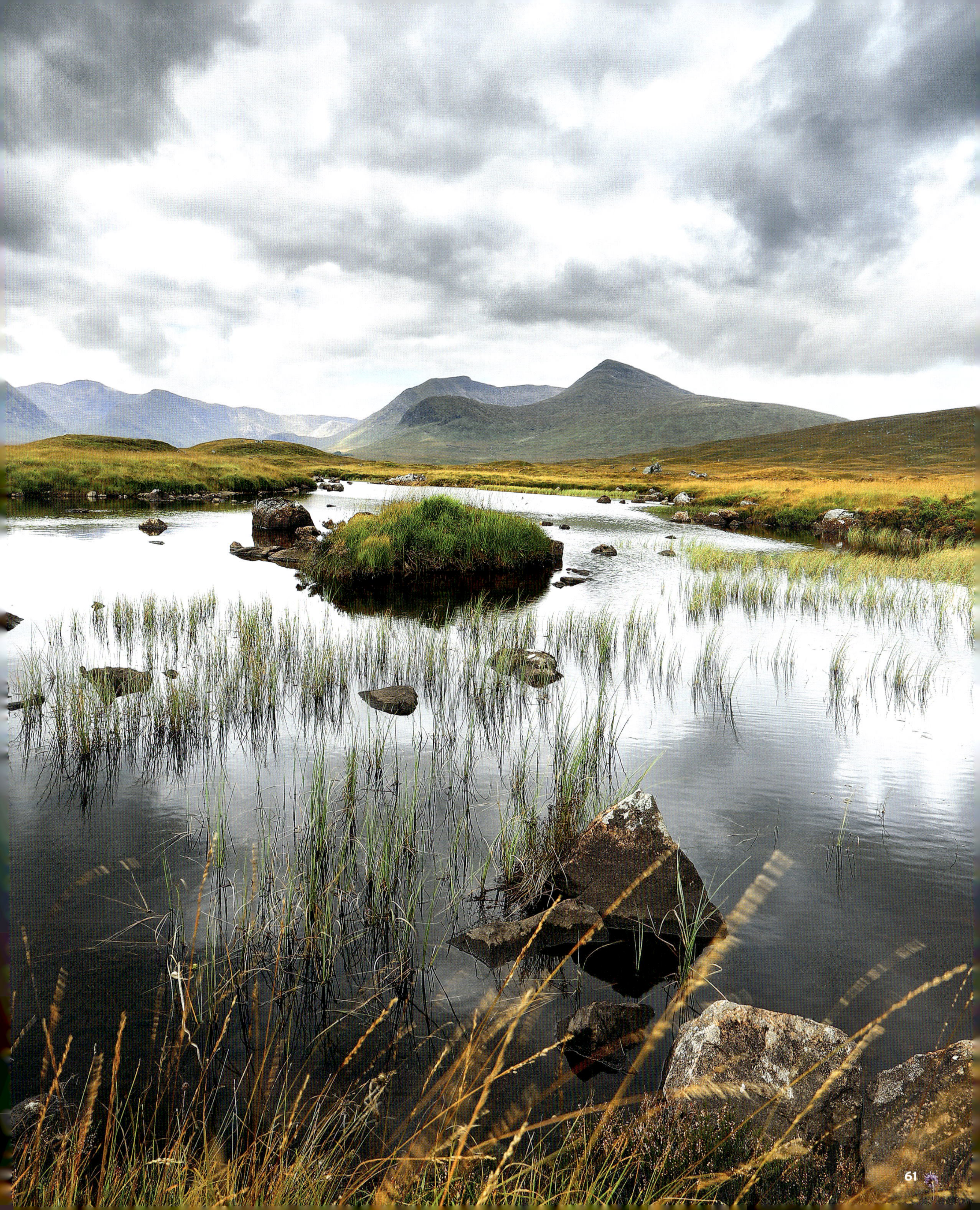

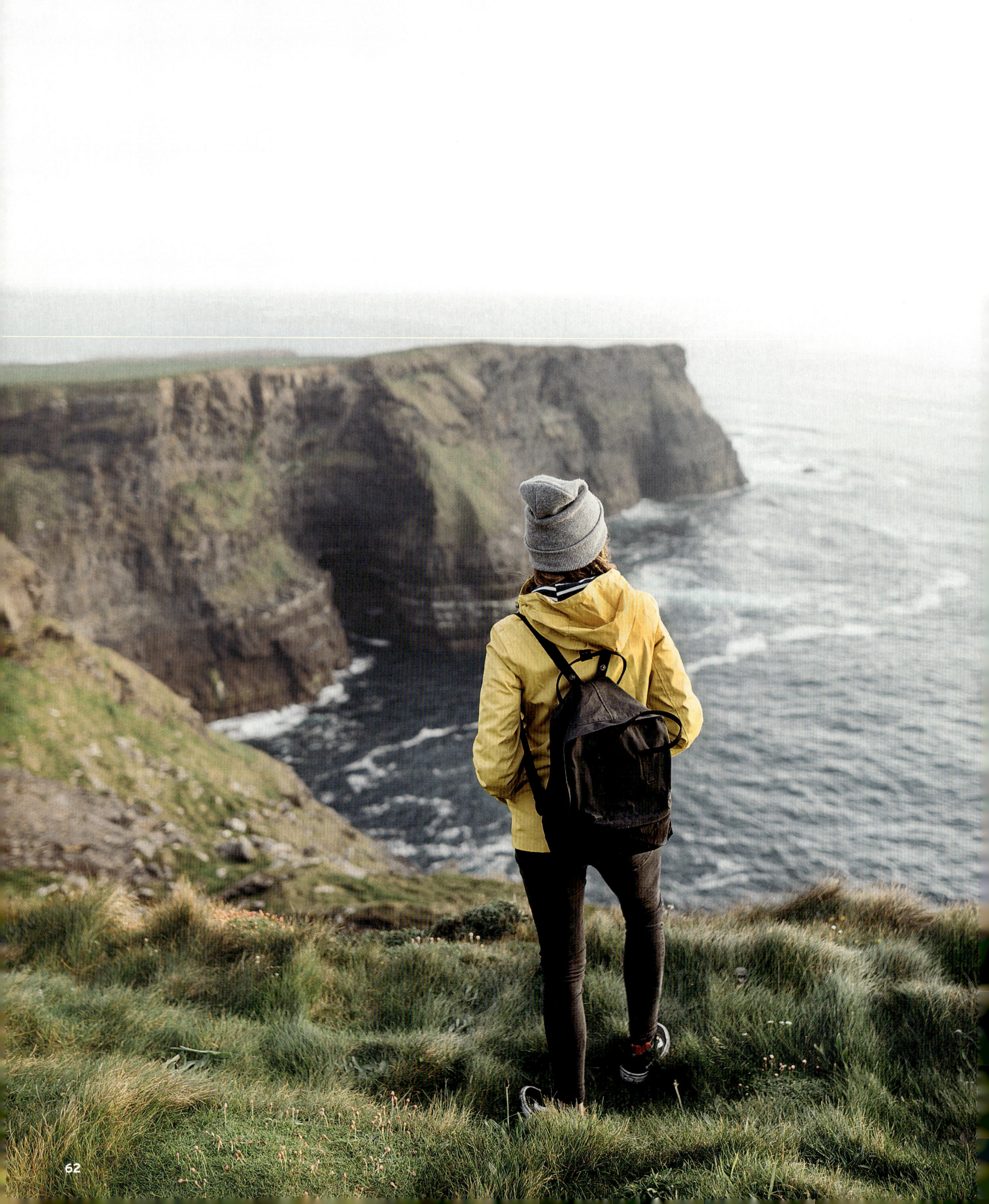

95. THE WICKLOW WAY

The Wicklow Way was the first signposted hiking trail in Ireland and starts south of Dublin. Over a distance of 81 mi. (130 km), it leads through the rolling hills of the Wicklow Mountains, covered by moors. Especially in summer, the slopes are covered with blooming purple heather and make for a very special experience. The monastery dedicated to Saint Kevin in Glendalough is definitely worth a visit. The almost 1,000-year-old and 108 ft. high (33 m) round stone tower and a Celtic high cross make the monastery one of the most famous in Ireland. If you don't want to walk all the way, you can choose from a variety of hikes of different degrees of difficulty and various stage lengths. Information and directions are available at the National Park Visitor Centre in the Glendalough Valley.

Length / time it takes: 81 mi. (130 km) / 7 days
Starting point: Marlay Park, Dublin
Trail's end: Clonegal

96. THE DINGLE WAY

In 2006, the Dingle Peninsula was described by National Geographic as "the most beautiful place on earth." The starting point and trail's end for the 111 mi. (179 km) loop is Tralee, the capital of County Kerry. The walking trail follows a large part of the coastline and leads past the slopes of the Slieve Mish sandstone mountain range and over Mount Brandon. The wide, sandy beaches in the north around the village of Castlegregory present a contrast to the rugged landscape on the southwestern tip of the Dingle Peninsula. History buffs can look forward to the many historical monuments along the way, such as the menhirs and Ogham standing stones. In Dingle there is a statue for a very special honorary citizen of this small town: Fungie the dolphin has been living in the secluded bay since 1983 and has become a real attraction since he first appeared.

Length / time it takes: 111 mi. (179 km) / 8 days
Starting point: Tralee
Marking: yellow arrow on a black background

97. WALKING TO THE CLIFFS OF MOHER

Over lush meadows and past grazing sheep, the walking path initially leads at ground level to the Cliffs of Moher Coastal Walk. From here the path becomes narrower and runs close to the world-famous cliffs. Countless birds find nesting places in the cracks and crevices of the steep cliff walls, which are up to 394 ft. (120 m) high and at times fall away vertically. From time to time you have to climb over a fence, but you should never leave the beaten path or step too close to the abyss. In wet weather and strong winds, you might consider taking a parallel-running alternative route. The Cliffs of Moher are one of Ireland's most popular destinations, and depending on the season, expect that there will be a lot of visitors. Since most of them remain close around the visitor center, it is worth taking a break before or after passing the center to enjoy the breathtaking landscape in peace and quiet.

Length / time it takes: 12 mi. (19 km) / 5 hours
Highest elevation: 686 ft. (209 m)
Starting point: Liscannor
Elevation difference: 1,739 vertical ft. (530 m)

98. NIGHT SWIMMING IN LOUGH HYNE

Embedded in a beautiful landscape during the day, saltwater Lough Hyne lake, which is linked to the Atlantic Ocean, has an almost magical fluorescence at night. This natural phenomenon is fascinating for nocturnal swimmers. Swimming in this ghostly sparkling lake after sunset is a unique, solemn experience. The biofluorescence works just as glow worms do, which is why you shouldn't swim when the moon is full—and quickly turn off the lights of the car when you arrive! The maritime nature reserve is also worth a trip in daylight. You can paddle a kayak to Castle Island, with the ruins of Cloghan Castle, or hike to Knockoumah Hill scenic lookout. There are also guided nighttime kayak tours!

Time it takes: half a day / half a night
Starting point: 4 mi. (6 km) from Baltimore, parking lot
Water temperature: around 57°F (14°C) in summer
Requirements: Neoprene swimwear is an advantage
Information: www.baltimore.ie/lough-hyne; kayak tour vendor: www.atlanticseakayaking.com

99. WALKING IN THE MOURNE MOUNTAINS

In the land of Narnia—the legendary Mourne Mountains in County Down are an ideal hiking area. Including the 2,788 ft. high (850 m) summit of Slieve Donnard, they represent the highest mountain in Northern Ireland. But what mountains they are: rising gently, then rising wildly again, enveloped by clouds of gorse, sometimes lovely, sometimes sublime, and almost always seeming remote. You don't have to be a mountaineer to climb them. Gentle hiking trails lead through enchanted forests and to lakes, such as in the Silent Valley. They have already been used as the setting for the film adaptations of *Game of Thrones* and *Narnia*. The Morne Way hiking trail leads from Newcastle on the coast to Rostrevor on Lough Carlingford, through the heart of this fascinating mountain world around Slieve Donnard. It is one of the most beautiful routes in Northern Ireland.

Length / time it takes: 26 mi. (42 km) / 2 days
Highest elevation: 2,461 ft. (750 m)
Starting point: Newcastle
Elevation difference: 2,132 vertical ft. (650 m)
Information: www.visitmournemountains.co.uk

100. CANOEING THROUGH THE FERMANAGH LAKELANDS

Fermanagh, the canoer's paradise—in the largest navigable lake district in Europe, the center of Lough Erne features sights that could have come from a picture book over the centuries—a journey through time in the age-old cultural history of Ireland. On Devinish Island you can visit an early Christian settlement from the sixth century, on White Island you are in the Celtic Bronze Age, and on Boa Island you are close to the 2,000-year-old stone deity called Janus, who was already there before there were any Celtic high crosses. On the Lough Erne Canoe Trail, the small coves, the narrow channels with a gentle current, and the countless islands make Lough Erne perfect for beginner to advanced canoers. There is also nature camping on the island of Lusty Beg.

Length / time it takes: 31 mi. (50 km) / 3–5 days
Starting point: Enniskillen
Information: www.canoeni.com/canoe-trails/lough-erne

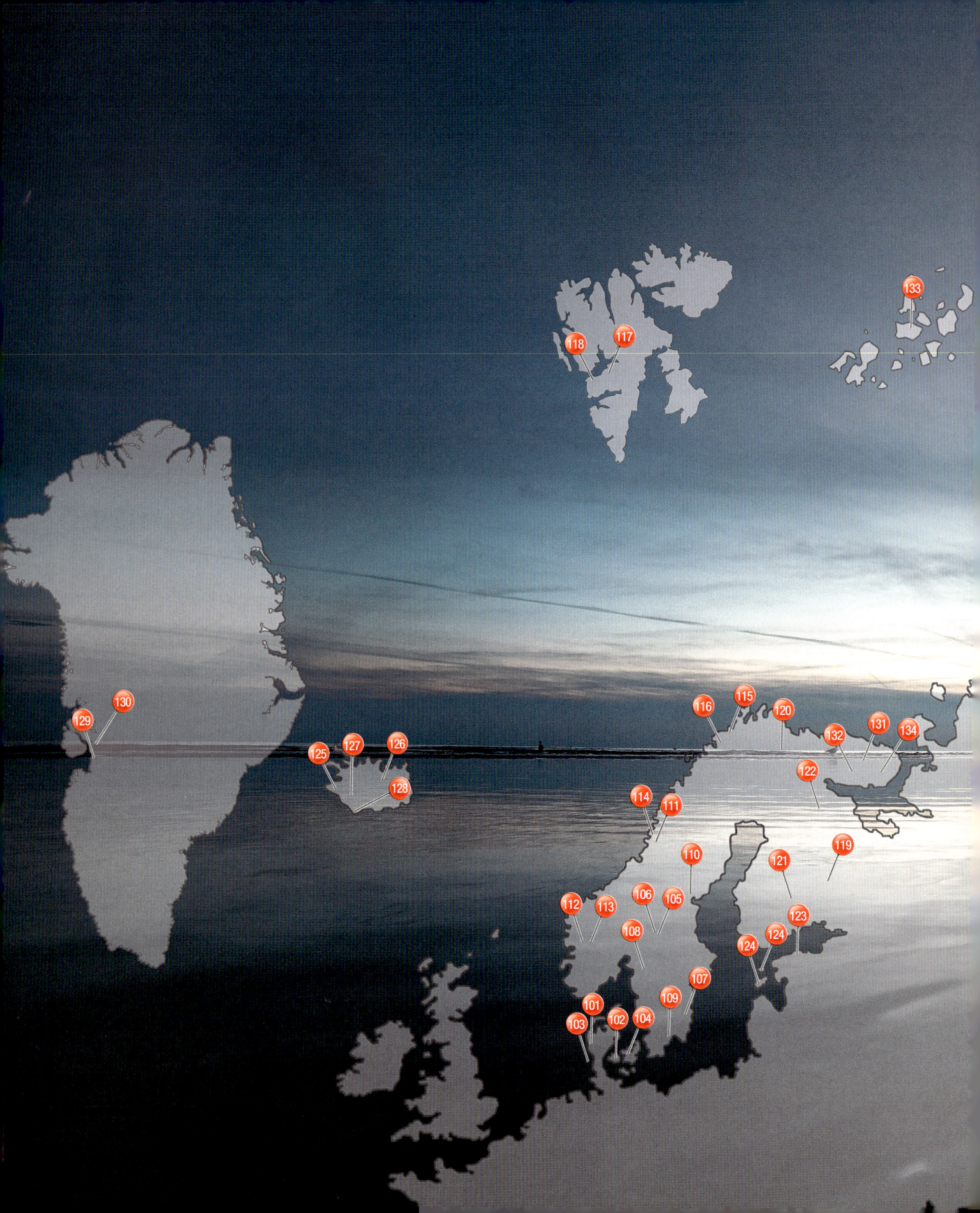
133
117
118
130
129
115
116
120
131
134
132
125
127
126
128
122
114
111
119
110
121
106
105
112
113
108
123
124
124
107
101
109
103
102
104

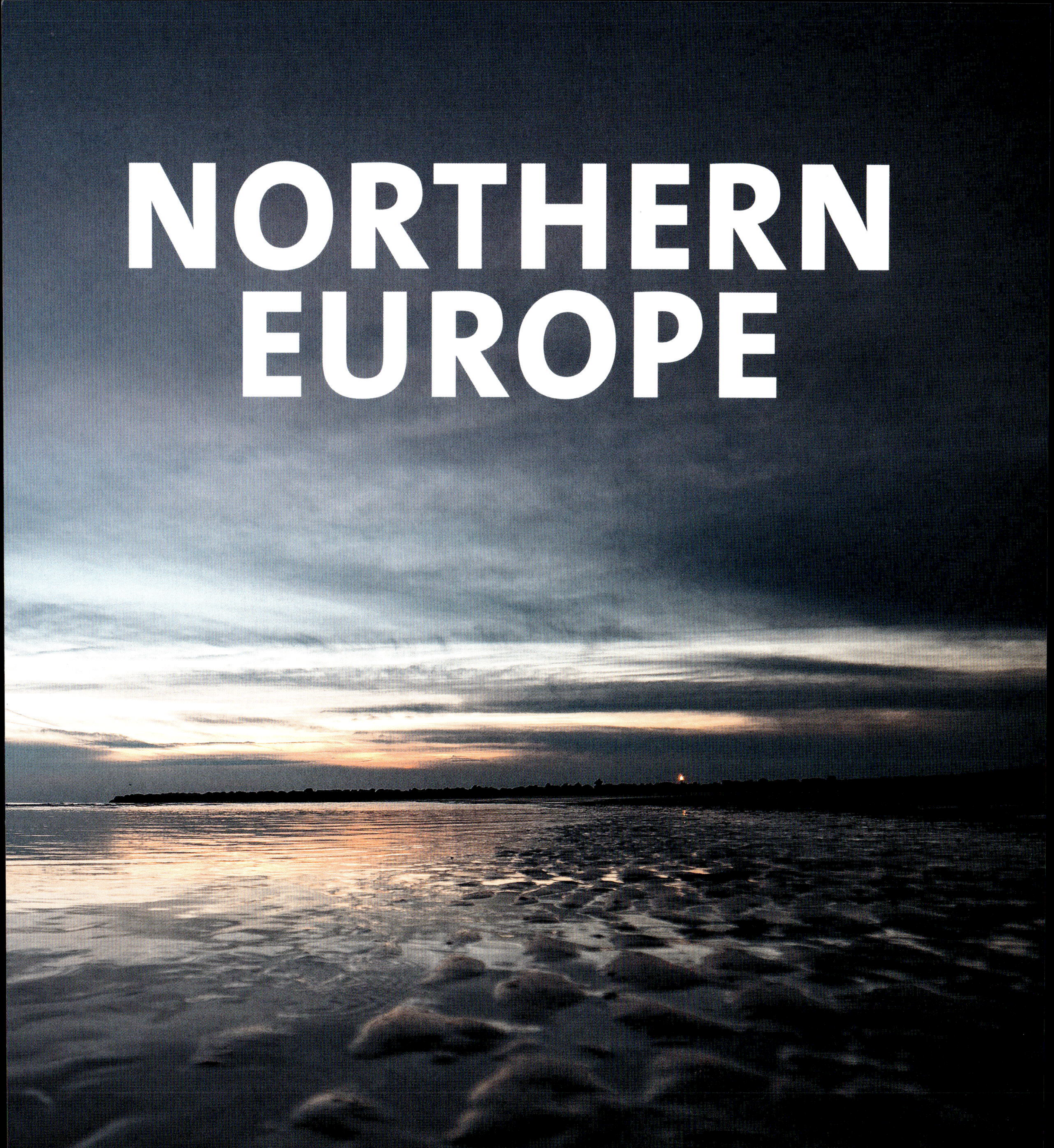
NORTHERN
EUROPE

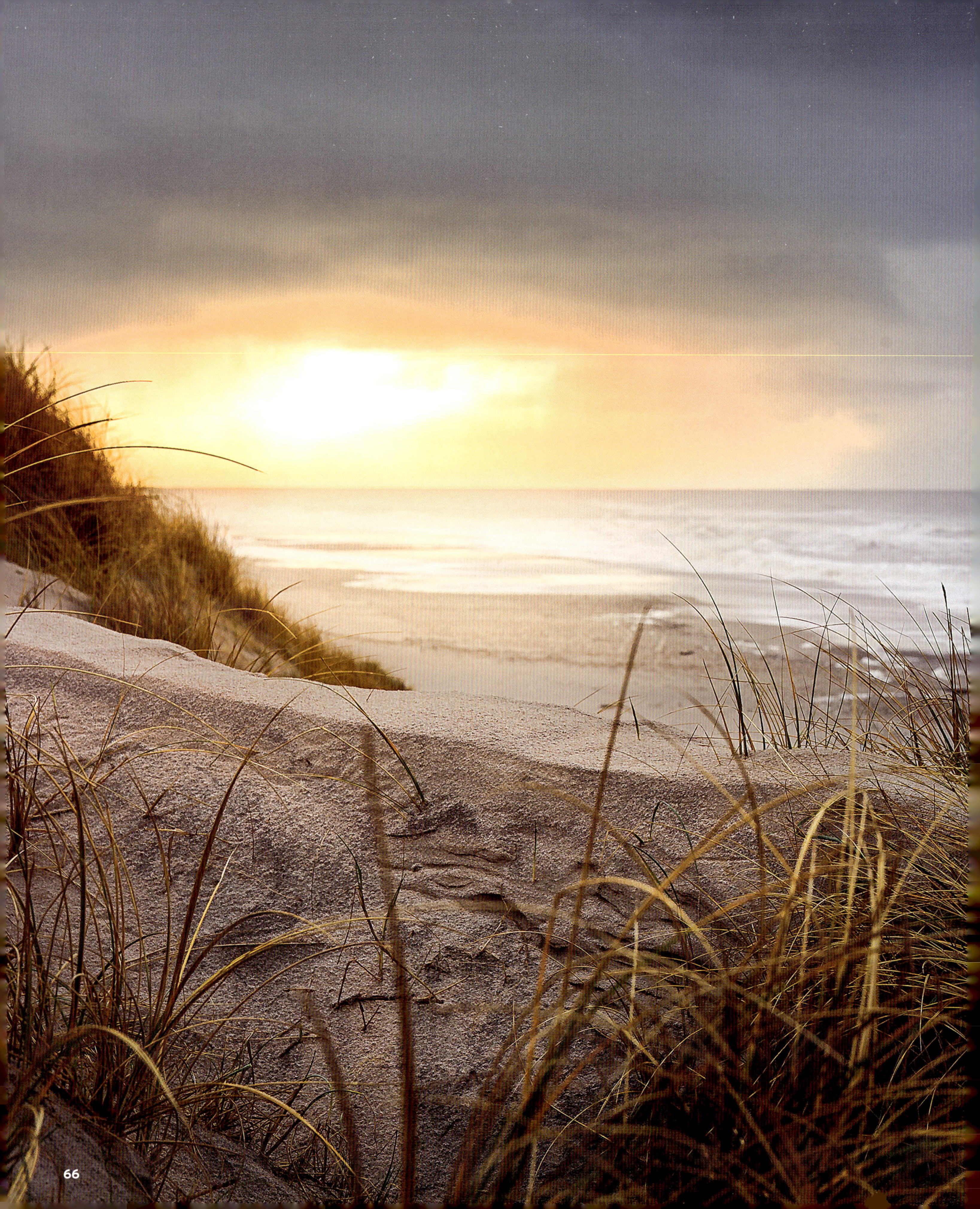

101. KITESURFING ON RINGKØBING FJORD

Sand dunes, beaches, solitude, quiet, and serenity: Ringkøbing Fjord, around 186 sq. mi. (300 km²), is extremely shallow and is a kiting area with some of the steadiest winds in Europe. A good 50 percent of this huge inland body of water is waist deep and thus offers perfect flat-water conditions for kitesurfing. A narrow headland, in some places only a few hundred yards wide, delimits it from the open (and sometimes rough) North Sea. This is what makes Ringkøbing Fjord so versatile and why it is unique for kitesurfers, attracting beginners as well as the skilled: from the sheltered smooth-water piste you can go straight into the thundering waves, from freestyling over the dunes to wave surfing in the open sea—all in one spot.

Kiting area: shallow flat water in the fjord, waves in the North Sea
Wind: 4–6 on the Beaufort scale; mostly westerly; onshore, side shore, or offshore depending on the spot
Best time of year: July–September.
Suitable for: everyone, including foil kites on the North Sea
Starting point: Hvide Sande
Information: www.kite-college.com

102. HIKING THE ØHAVSSTIEN ARCHIPELAGO LOOP TRAIL

One of the pearls of the South Funen Archipelago is this scenic loop hike that begins in Svendborg and ultimately ends up there again. The Øhavsstien Archipelago Trail on the South Funen Archipelago was opened in 2007, and at 137 mi. (220 km) it is one of the longest hiking trails in Denmark. It leads through the idyllic island landscape of Funen, which the Danish mainland can barely surpass: it runs along the coast with marvelous panoramic views over the South Funen Archipelago through colorful, blooming meadows and deep-green forests, past romantic castles and manor houses, and through charming, tranquil villages. You hike across the islands of Tåsinge and Langeland and then take the ferry to Ærø before closing the loop in Svendborg.

Length: 137 mi. (220 km)
Starting point/trail's end: Svendborg
Information: naturturisme.dk

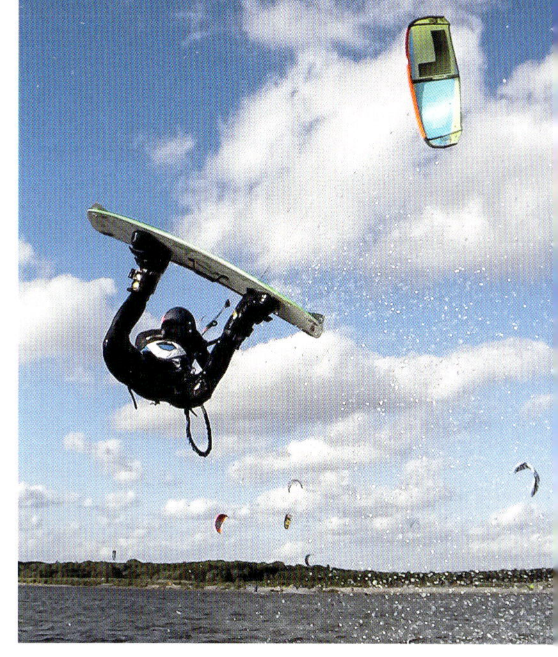

103. AN OYSTER SAFARI ON THE WADDEN SEA

A culinary mudflats hike—discover the secrets of the Wadden Sea when they are exposed to the tidal current. Your boots make a smacking sound as they sink a few inches into the wet sand and make the same noise as you pull your footwear—attached to a shapeless wader—out again to take the next step. You keep walking steadily toward the horizon, which shimmers and reflects the North Sea surf at a seemingly infinite distance. After a long march through this equally unprepossessing and unbelievably silted-up world of lugworms and cockles, you reach the mussel banks. For some years now, American oysters have been attaching themselves here on the ancestral places of the mollusks. Armed with oyster knives, you can both counteract this threat and feel like an outdoor gourmet as you shuck the oysters professionally and slurp up the contents in style under the sun.

Length / time it takes: 5 mi. (8 km) / 4–6 hours
Highest elevation: 33 ft. (10 m)
Starting point: Vadehavscentret, Vester Vedsted
Information: www.vadehavscentret.dk

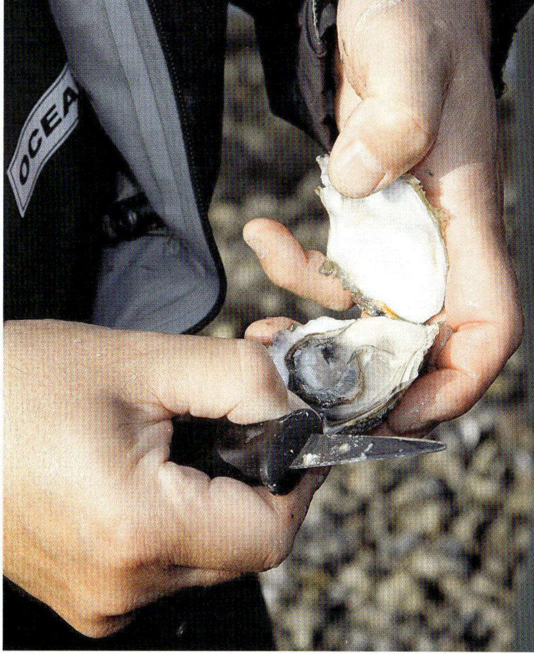

104. FOOTBALL GOLF AT NØRRE ASMINDRUP

Putting with a pickax—the cultivated-green-lawn sport conquers a new dimension. Golf is definitely considered a popular sport in Denmark, yet it remains an expensive and time-consuming leisure activity. This is how the "Danish Dynamite 2.0" generation invented football golf. With well-aimed kicks, players drive an ordinary leather ball over the fairways, which are somewhat more rustic than for the traditional game. Then they sink it on the green, with plenty of feeling, and into the hole under a lively, waving flag. Obstacles, trees and knolls, winding lines, small ponds, and sand traps make up the typical challenges for this young, highly entertaining sport. Playing the eighteen holes is a very entertaining pleasure, combined with great nature experiences and encounters with deer, hare, and fox, as well as requiring very few technical skills.

Length / time it takes: 1.86 mi. (3 km) / 2 hours
Requirements: par 72, 18 holes
Starting point: Kollekollevej 32, Svenstrup Hestehave, DK-4572 Nørre Asmindrup, Denmark
Information: www.odsherred-fodboldgolf.dk

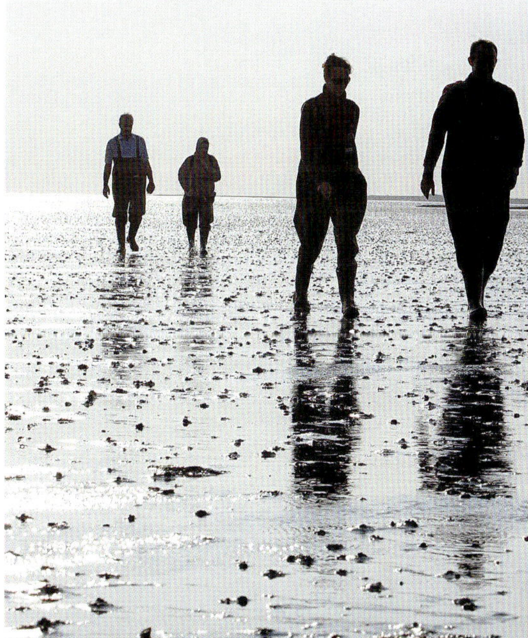

105. CONSTRUCTING A RAFT . . .

Ancient craftsmanship—the industrial necessities of past centuries become a haptic experience at close quarters. The Swedish forestry industry ran rafting operations on the Klarälven River until 1991. Lumberjacks bound the tree trunks, cut to a uniform length, together into blocks and transported these by barge on Lake Vänern to Karlstad. In a manner reminiscent of this time, you can construct your own raft—under expert guidance. This requires team spirit, strength, patience, and some skill to make sure that the process does not turn into a construction fiasco and, ultimately, a lot of dugouts end up drifting down the river. On the level riverbank, the 10 ft. long (3 m) trunks are first carefully sorted and then firmly tied together, using ropes as thick as a thumb and a special knotting technique. The finished vessel will later weigh as much as a full-scale Volvo.

Time it takes: 1–2 hours
Highest elevation: 1,247 ft. (380 m)
Starting point: Klarälven Camping, Stöllet, Värmland County
Information: www.vildmark.se/timmerfottspaket

106. . . . AND A RAFTING TRIP ON THE KLARÄLVEN RIVER

Let yourself drift—rediscover the leisure of slowness on board your homemade raft. The long tradition of commercial raft construction was maintained on the Klarälven River until 1991. Today you can make your own raft under expert guidance and experience an adventurous form of river cruising on the round, well-tensioned logs. The generally leisurely flow of the river is the only thing that propels the raft forward. You should always keep an eye out so as not to get too close to the bank and risk any unplanned stops. However, there is plenty of time for idleness and observing nature, because the wildlife along the riverbank is barely disturbed by your completely silent vessel. You experience particularly romantic moments in the evening when you cook and dine on deck. You spend the night in a tent under the stars.

Length / time it takes: 62 mi. (100 km) / 8 days
Highest elevation: 1,650 ft. (503 m)
Starting point: Branäs, Värmland County
Elevation difference: 1,191 vertical ft. (363 m)
Information: www.vildmark.se/de

107. HIGH ROPE ZIPLINE IN KLAVRESTRÖM

A special thrill when traversing Swedish forests at lofty heights. Just ascending to the impressive zipline alone commands a lot of respect, because after all, you have to climb a 66 ft. high (20 m) wooden tower. The adrenalin rush kicks in at the latest as you cross the 82 ft. long (25 m) and perilously swaying suspension bridge to the actual starting point at a height of 59 ft. (18 m). The bridge is distributed over eight ropes, and you cover a distance of more than 1.6 mi.(2.5 km) among treetops and over lake surfaces, well secured and powered only by your own gravity. The first rope in particular packs a punch: at just under 1,411 ft. (430 m) long, you can whiz along at a speed of up to 47 mph (75 km/hour). The airstream almost takes your breath away and brings tears to your eyes, so that the great view of nature almost gets hazy before your eyes. But that is precisely the attraction of a kamikaze zipline.

Length / time it takes: 1.6 mi. (2.5 km) / 3–4 hours
Highest elevation: 1,020 ft. (311 m)
Location: Klavreström-Norrhult, Småland Province
Elevation difference: 171 vertical ft. (52 m)
Information: www.swedenzipline.com

108. RAILBIKE TRIP THROUGH DALSLAND

A bike on rails—pedaling slowly through almost pristine nature—that's the credo of a relaxed railbike (draisine) ride. You can sometimes even see the king of the Swedish forests, the elk, from aboard your rustic steel companion. But that happens only if the drive chain is well lubricated and makes little noise, and the wheels roll gently over the rusty track. This long-disused railroad line is a true El Dorado, and not only for cyclists who want to ride on a predetermined course and be as relaxed as possible. Wherever you like, you can take a break, enjoy the wonderful air, and listen to the diverse soundscape of the forest, or look pensively across some lovely, shining lakes. You can pick flowers during the trip, just like you can gather and taste berries along the way.

Length / time it takes: 31 mi. (50 km) / 6–8 hours
Highest elevation: 427 ft. (130 m)
Starting point: Bengtsfors train station, Dalsland Province
Elevation difference: 131 vertical ft. (40 m)
Information: www.dvvj.se/de/draisine/

109. HIKING ON THE SKÅNELEDEN TRAIL

As a hiking region, southern Sweden stands in the shadow of Lapland. You will not find the lonely, wild expanse of the northern part of the country on the Skåneleden Trail. Yet, its five different sections are definitely an alternative. The Kust till Kustleden Trail, which leads from Sölvesborg on the Baltic Sea coast to Ängelholm on the North Sea coast, is particularly exciting and can also be easily divided into individual stages. The trails through Söderåsen National Park are idyllic, with forested hills and the spectacular Skäralid Gorge. The sections along the coast present a completely different appearance. They sometimes run along dunes but also inspire hikers with the rocky stretches such as on the Kullen Peninsula. This, however, is already part of the Skåneleden as the Öresundsleden Trail.

Length: a total of 777 mi. (1,250 km); 105 stages, 5 trails
Length of Kust till Kustleden: 230 mi. (370 km); 26 stages
Starting point of Kust till Kustleden: Sölvesborg or Ängelholm
Information: skaneleden.se/de

110. CLIMBING ON SKULEBERGET MOUNTAIN

The Skuleberget rises almost 984 ft. (300 m) on the western Swedish "Höga Kusten" ("High Coast") above the sea, which is only a few hundred yards away. The special thing about this towering granite rock is that due to its steep eastern flank, it features four via ferratas of varying degrees of difficulty, ranging from medium to very difficult. The view of the strip of coast is fantastic; you can enjoy it either while climbing or from the staffed "Toppstuga" Top Cabin cafe. Interesting here: the via ferrata routes lead up a mountain that a few thousand years ago was still below sea level. Only the highest peak of the Skuleberget showed out of the water; this is shown by a marker on the summit. Climbing equipment can be rented on-site.

Time it takes: 1.5 to 2 hours per route
Highest elevation: 968 ft. (295 m)
Starting point: Skuleberget (Naturum visitor center), S-870 33 Docksta
Elevation difference: 820 vertical ft. (250 m)
Information: www.viaferrata.se

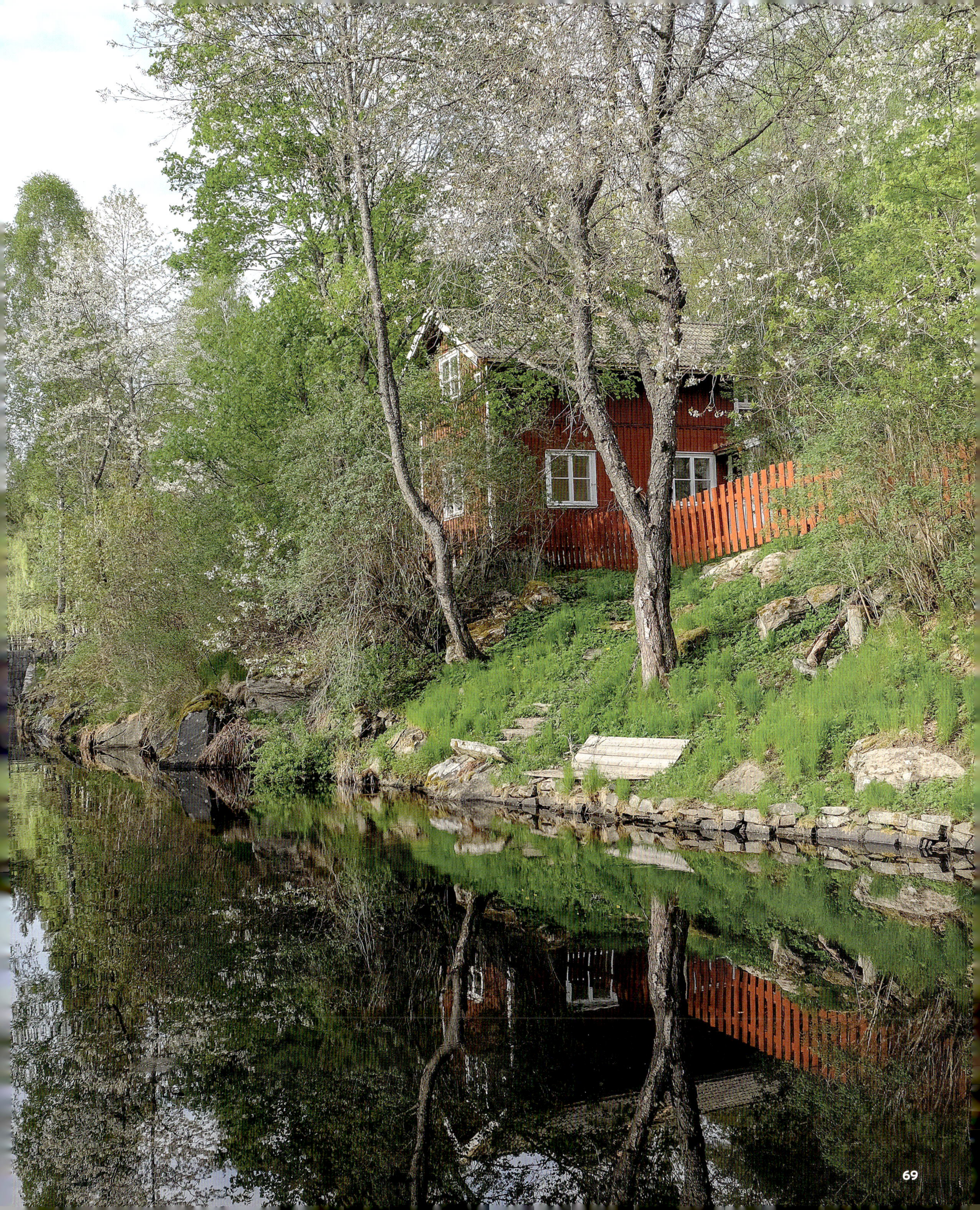

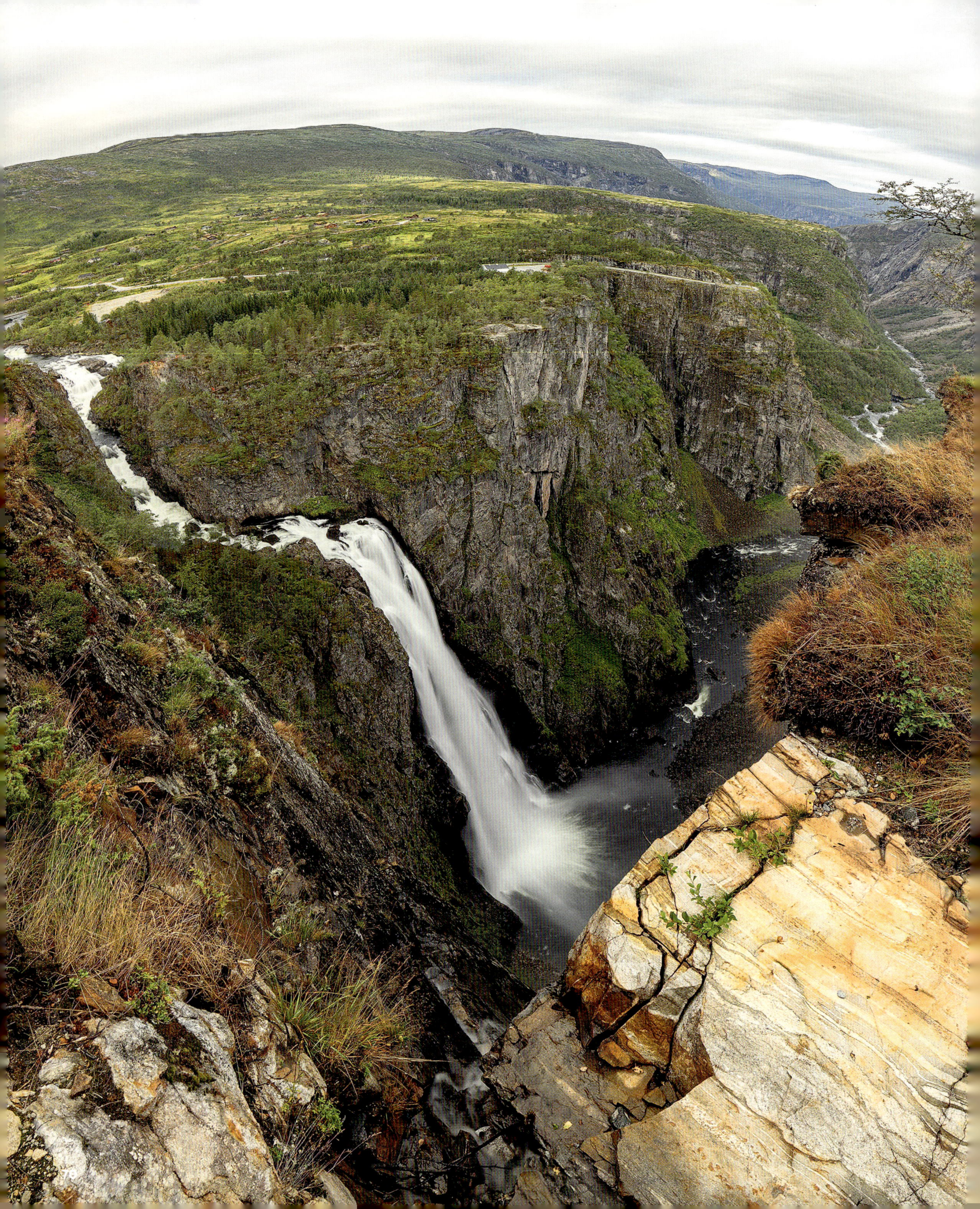

111. HIKE AROUND AND THROUGH TORGHATTEN MOUNTAIN

Mount Torghatten, on the island of Torget, is one of the most striking landmarks on the Norwegian coast. Only 853 ft. (260 m) high, halfway up there is a hole that is 525 ft. (160 m) long, 115 ft. (35 m) high, and 49 ft. (15 m) wide. According to legend, the arrow of a king's son pierced the mountain, but the truth is more prosaic. After the ice sheet that covered northern Europe during the most recent ice age melted, the continental plate, which had previously been pressed deep into the earth's mantle by the weight of the ice, rose again. The hole is nothing more than a cave dug into the rock by the ocean surf. Meanwhile, the cave has been lifted out of the surf by the land rising to 413 ft. (126 m) above sea level. One hiking trail leads around the mountain, while another one runs steeply uphill to the hole; you can climb through it and then descend again on the other side.

Time it takes: 3 hours for hiking around the mountain and the ascent
Highest elevation: 4,134 ft. (1,260 m)
Starting point: Torghatten Camping
Elevation difference: 394 vertical ft. (120 m)
Information: www.visitnorway.de

112. RAFTING ON THE BJOREIO RIVER

Hardly any other river in Norway offers such exciting rafting adventures as the Bjoreio, which rises in the heart of the Hardangervidda plateau. The trip from its source, which is at around 3,937 ft. (1,200 m) in elevation, to where it flows into the Eidfjord is 45 mi. (72 km) long. More than 883 cu. ft. (25 m^3) of water rush down into the valley every second, flooding over rocks that have been washed smooth for thousands of years or breaking against the steep rock faces. Whirling whitewater and narrow, fast passages in the stream alternate with quieter, yet nevertheless exciting and still-fast-flowing sections that are selected for rafting by the tour operator. Then there is the fascinating river landscape framed by dense forests. Swimwear and a towel—that's all you need to bring along. The tour operator will provide neoprene wetsuits, helmets, and other safety equipment. There is room for two to four people in the boats.

Time it takes: 3 hours
Highest elevation: 295 ft. (90 m)
Starting point: Øvre Eidford
Elevation difference: 262 vertical ft. (80 m)
Information: fatearth.no/en/

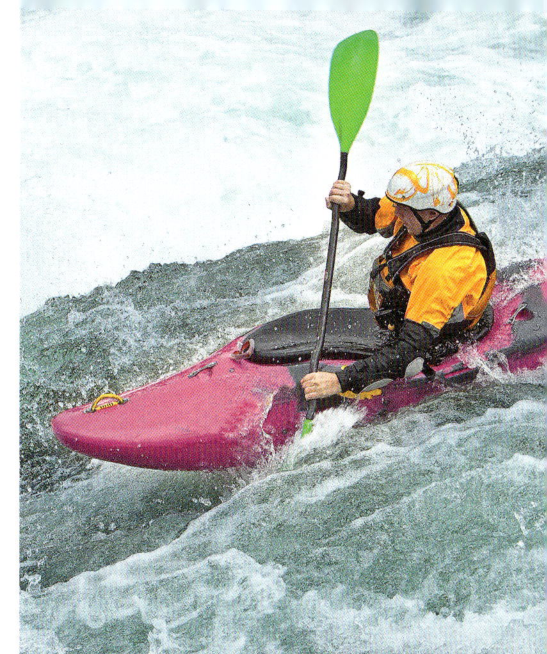

113. HIKING THROUGH THE HARDANGERVIDDA PLATEAU TO RAUHELLEREN

The Hardangervidda mountain plateau is a unique landscape—open, wild, and largely unspoiled, and more than 3,281 ft. (1,000 m) above sea level. Countless streams and rivers run through the country. They are extremely rich in fish, as are the many lakes, ponds, and sloughs that are embedded in the tundra landscape. Some 15,000 reindeer, Norway's largest wild population, still live today on the Hardangervidda, along with twenty-three other mammal species, including lynx, arctic fox, wolverine, alpine hare, and mountain lemming. A 746 mi. (1,200 km) network of hiking trails managed by DNT (Den Norske Turist Forening) runs throughout the plateau. The route from Ustaoset to Rauhelleren leads into the heart of Hardangervidda National Park. The Tuva and Heinseter mountain huts, where you can spend the night, are along the way. You are more flexible with your own tent.

Length / time it takes: 45 mi. (72 km) / 4 days
Highest elevation: 4,012 ft. (1,223 m)
Starting point/trail's end: Ustaoset
Elevation difference: 656 vertical ft. (200 m)
Information: www.dnt.no

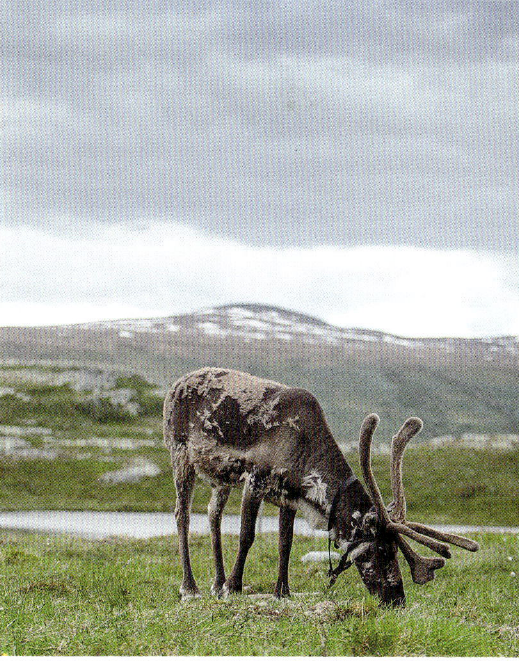

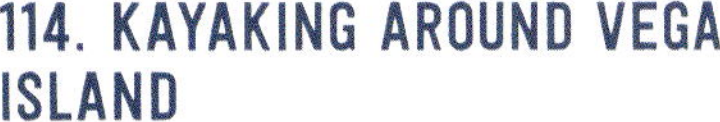

114. KAYAKING AROUND VEGA ISLAND

The Vega Archipelago, with its more than 6,500 islands, islets, and skerries, is the ideal place for kayaking. Beginners as well as advanced kayakers can pursue their passion here. Quiet coves, narrow channels, and of course the open ocean are often only a few minutes' paddling apart. For beginners, it is a good idea to take the kayaking course that is offered in the village of Nes on the main island. Even experienced kayakers can avail themselves of professional help. Local tour guides offer trips of various degrees of difficulty and lengths. If you have a lot of time and the necessary fitness, you can follow in the footsteps of the Vega Challenge, an annual international kayak race around the main island. To do this, you have to cover 28.5 mi. (46 km), a challenge even for experienced paddlers.

Time it takes: as you wish: a few hours to several days
Information: www.vegaopplevelsesferie.no

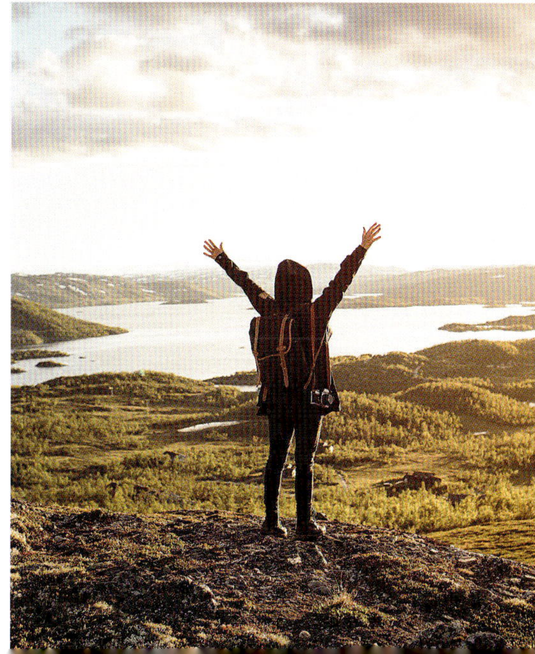

115. DOGSLEDDING TRIP IN NORTHERN NORWAY

The Sami people of Finnmark did not originally use sled dogs; they had reindeer pull their sleds. Today, sled dogs are used in Norway almost exclusively for tourists and for sporting purposes. When the days start getting longer again in April, the adventure begins in Tromsø. Part of the adventure is that the exact route is determined shortly before the start; it is based on the weather and snow conditions. Everyone can be their own musher with their own team of sled dogs. No previous knowledge is required, only that participants should be in good-enough physical condition to be able to stand on the sled for several hours over seven days and spend the nights in a tent. The tour operators provide not only information about the correct way to handle the dogs and sleds, but also provide warm outerwear, tents, and food.

Time it takes: depends on the weather (about 7 days)
Highest elevation: depends on the route
Starting point/trail's end: Tromsø
Elevation difference: depends on the route
Information: www.arcticadventuretours.no/dog-sleddine expedition-in-northern-norway/

116. NORTHERN LIGHTS SAFARI AT TROMSØ

Tromsø is one of the best places in the world to experience the fascinating northern lights, the aurora borealis. The city lies exactly on the auroral oval, where there is the highest probability of observing the aurora borealis. Here, far north of the Arctic Circle at 69°39'N, the natural spectacle can be seen during the long nights from late September to early April. Tromsø is easy to reach and offers hotels in all price ranges, as well as campsites and holiday cabins. Once you get there, you can simply wait until the lights appear. However, it is recommended that you link up with a service offering so-called northern lights safaris. Minibuses drive to where the all-around view is the greatest and no extraneous light disturbs observation. The prerequisite is a clear sky, in any case. That cannot be guaranteed.

Time it takes: 5–6 hours
Highest elevation: depends on visibility
Elevation difference: depends on visibility
Starting point/trail's end: Tromsø

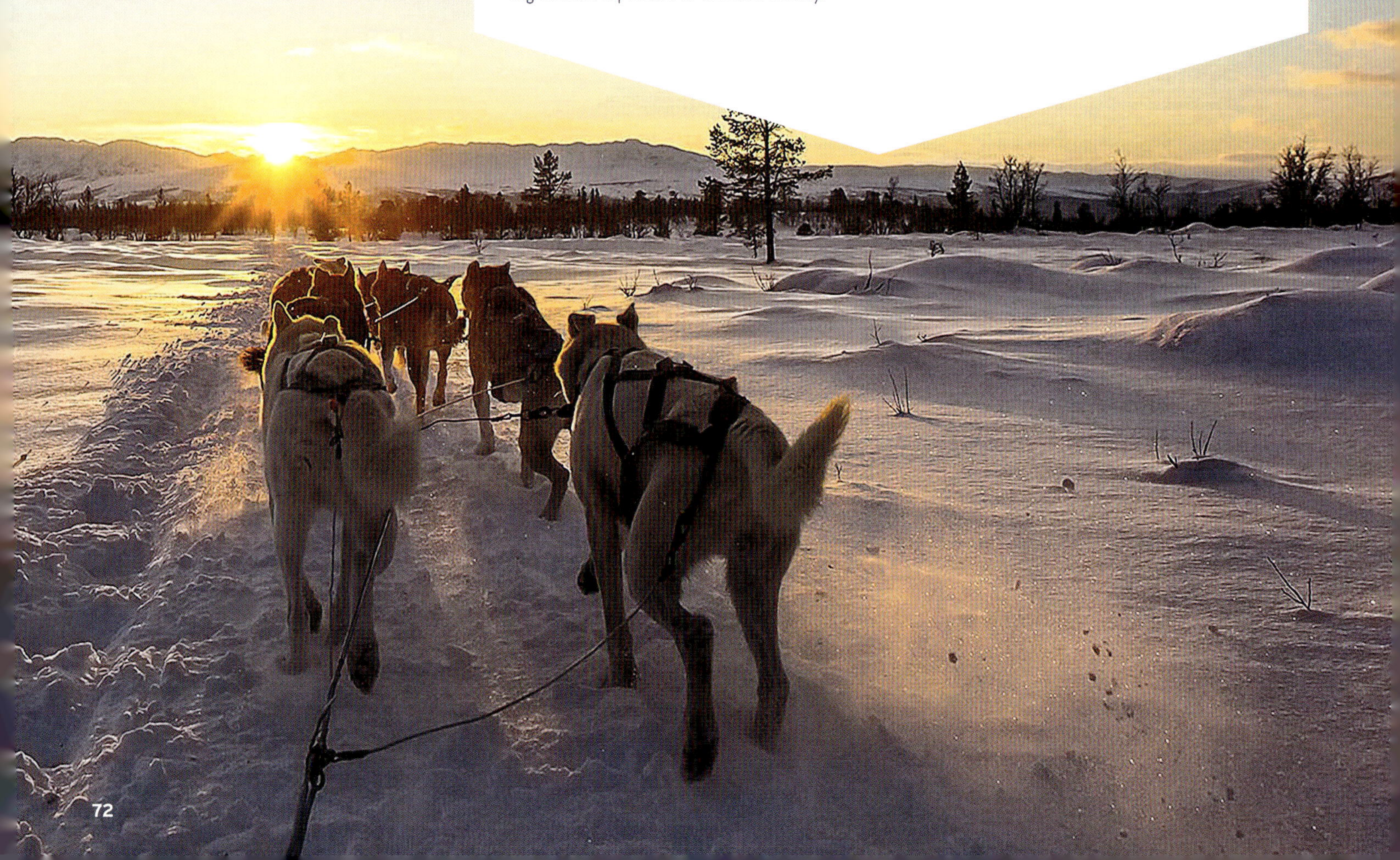

117. KAYAKING AND CLIMBING ON SPITSBERGEN ISLAND

The Hiorthfjellet is a prominent mountain on the northeast side of the Adventfjord. Its plateau, which has earned it its name as a "plateau mountain," is situated at 3,035 ft. (925 m) above sea level. From there, your view glides over the entire Adventfjord, far into the Isfjord with its glacier-covered, rugged mountains more than 3,281 ft. (1,000 m) high, with the bird cliffs and the beach plains in front of it—a fantastic panorama that is second to none. However, the goal is not easy to attain. First of all, you have to cross the Adventfjord by kayak. From the starting point at Longyearbyen to the foot of the mountain in Hiorthhamn, the Hiorth harbor, you have to cover 1.86 mi. (3 km) in a kayak. Then the steep ascent to the summit plateau begins. At 1,800 ft. (550 m) in elevation, the trail leads past an old coal mine. Because of the danger of polar bears, the local guide is armed.

Length / time it takes: 6.2 mi. (10 km) / 1 day
Highest elevation: 3,035 ft. (925 m)
Starting point/trail's end: Longyearbyen
Elevation difference: 3,035 vertical ft. (925 m)
Information: www.fremdenverkehrsamt.com/reisefuehrer/reiseziel/longyearbyen/index.htm

118. SNOWMOBILING TO BARENTSBURG

An unforgettable trip by time machine into the socialist past. The Russian mining settlement of Barentsburg is an enclave in Norwegian territory, and with around 500 inhabitants, it is the second-largest settlement after Longyearbyen, the main town. Most recently, 100 people have moved here and are now making a living in research rather than in coal mining. Several outfitters in Longyearbyen rent out colorful snowmobiles for tours of the polar bears' turf outside the settlement. Even more exciting is the fast ride on the slightly hilly and at first seemingly endless ice, until after 34 mi. (55 km) and the endless white and gray, surreal prefabricated buildings and a statue of Lenin suddenly appear. The local Krasniy Medved brewery serves hearty Russian cuisine with plenty of alcohol.

Length / time it takes: 68 mi. (110 km) / 1 day
Highest elevation: 295 ft. (90 m) above sea level
Starting point: Longyearbyen
Information: visitsvalbard.com

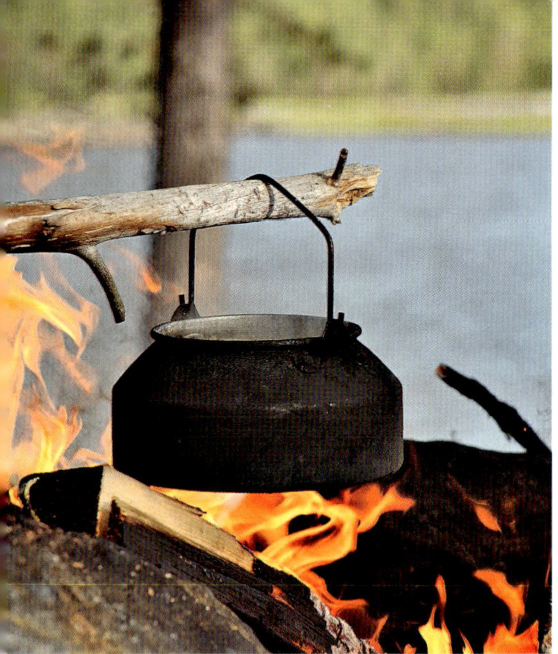

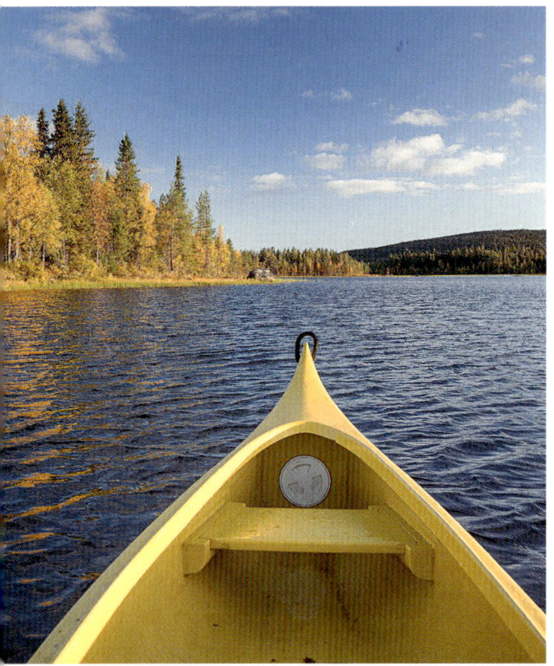

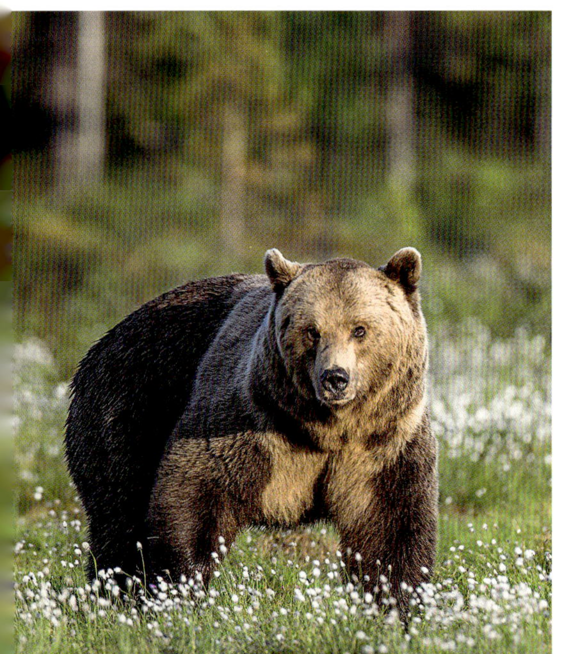

119. HIKING IN THE KOLI MOUNTAINS

Some 50 mi. (80 km) of hiking trails lead through Koli National Park. Most of the area is in a pristine state of nature, and it has already received several awards for the sustainable tourism practiced here. Even if the highest point, the Ukko-Koli, rises to just 1,138 ft. (347 m) in elevation, it offers an impressive view of the surrounding forests and the huge Pielinen Lake, on the shores of which people first lived and hunted as early as the Stone Age. The trails meander over the soft forest floor to the rocky peaks and plateaus, over which the traces of the glaciers of the most recent ice age stretch like scars. While the trees bend under the weight of the snow in winter, the landscape features blooming willows in summer.

Highest elevation: 1,138 ft. (347 m)
Starting point: Ukko Visitor Center
Information: www.nationalparks.f/en/kolinp

120. HIKE TO PIELPAJÄRVI WILDERNESS CHURCH

The starting point for this hike is the Sami People Cultural Center on the northern edge of the small village of Inari. From there, an asphalt trail leads for the first few miles along the shore of Inarijärvi Lake, until you reach a fork where you keep to the left and continue along narrow trails. Passing by small lakes and through lonely forests, there is a high probability that you will see reindeer during the hike. After about 6 mi. (10 km), the wilderness church of Pielpajärvi comes into view. The dark wood—looking almost as if it is charred—of which it was built forms a wonderful contrast to the light birch forests. In the eighteenth century a Sami winter village stood here; the seminomads visited it during the cold season. Midsummer services are still celebrated today in this small church in the midst of the wilderness.

Length / time it takes: 13 mi. (20 km) / 5.5 hours
Starting point: Cultural center and museum of the Sami people
Information: www.nationalparks.f/en/kolinp Pielpajarviwildernesschurch

121. CANOEING ALONG THE SQUIRREL ROUTE

The starting point for this canoe trip is the campsite in the small Finnish municipality of Juva. Here you can rent everything that you might need for the trip, and you also get a map with a description of the route. Depending on the time of year and weather conditions, the water level can vary and make it necessary to portage the canoe along some passages. The canoer can expect a varied route through lonely nature. The squirrel signs lead along narrow river courses, sometimes with protruding tree trunks, and then again across lakes—mirror clear—so that you can watch the fish from your canoe. Rapids, which will speed up the canoe, provide variety. There are always small resting places along the banks where you can spend the night and where there is even ready-cut wood for a cozy campfire.

Length / time it takes: 28 mi. (45 km) / 2 days
Starting point: Juva

122. HIKING ON THE KARHUNKIERROS TRAIL (BEAR'S LOOP)

One of Finland's most popular hiking routes is in Oulanka National Park, near the Arctic Circle. The area was shaped by the Oulankajoki River, which created a spectacular landscape of rocky gorges and waterfalls Brown bears have found ideal living conditions here, especially in the northern part of this protected nature area, and they gave the route its name. There are small trail huts, shelters, and tent sites along the well-signposted hiking trail. There are also places where you can stock up on provisions. Those who do not have the time or fitness for the entire loop can also go on equally beautiful day trip sections. The Bear's Loop can be done at any time of the year, but in winter the temperatures can drop to -40 degrees, and the masses of snow lengthen the hiking time accordingly.

Length / time it takes: 51 mi. (82 km) / 5 days
Starting point: Hautajärvi

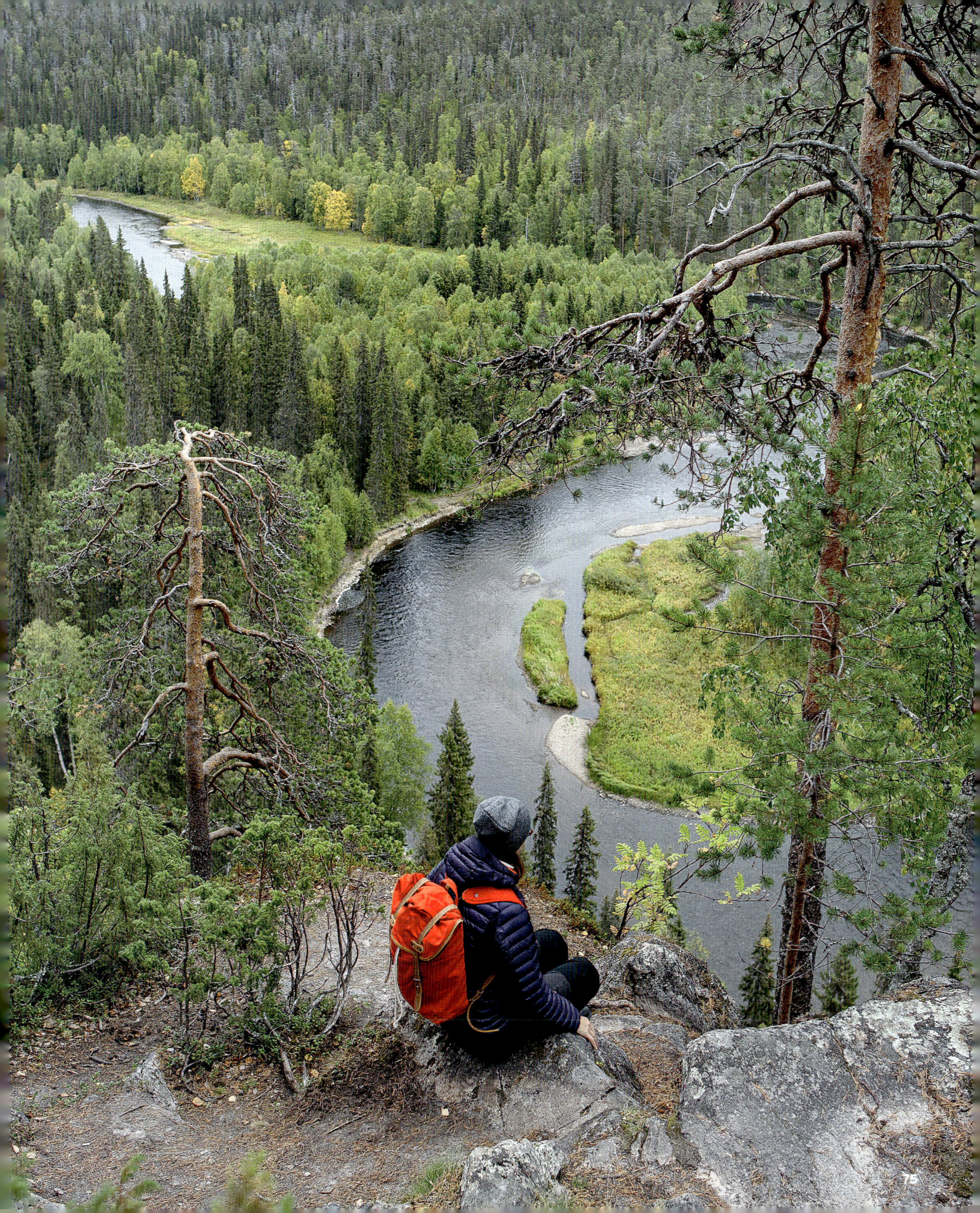

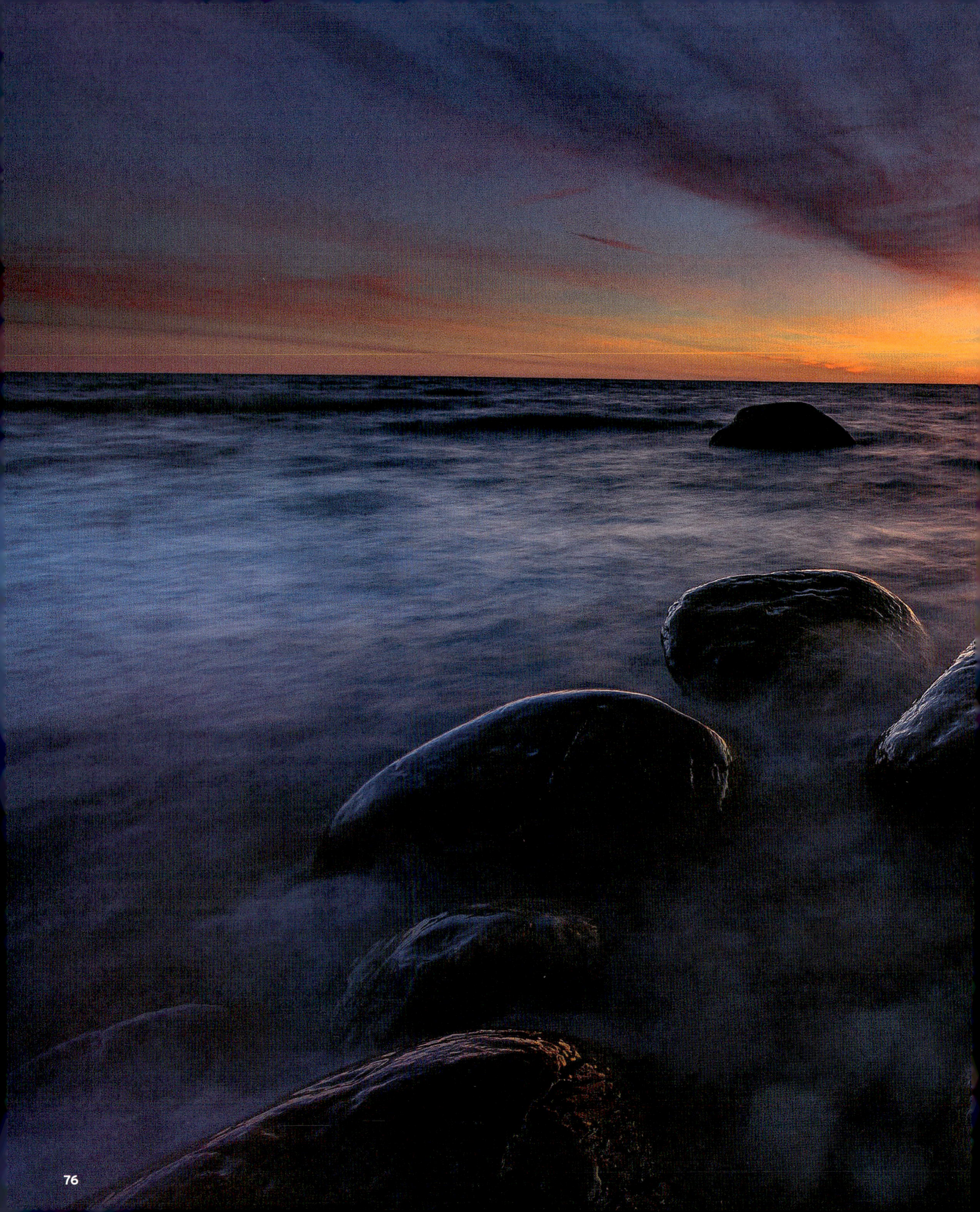

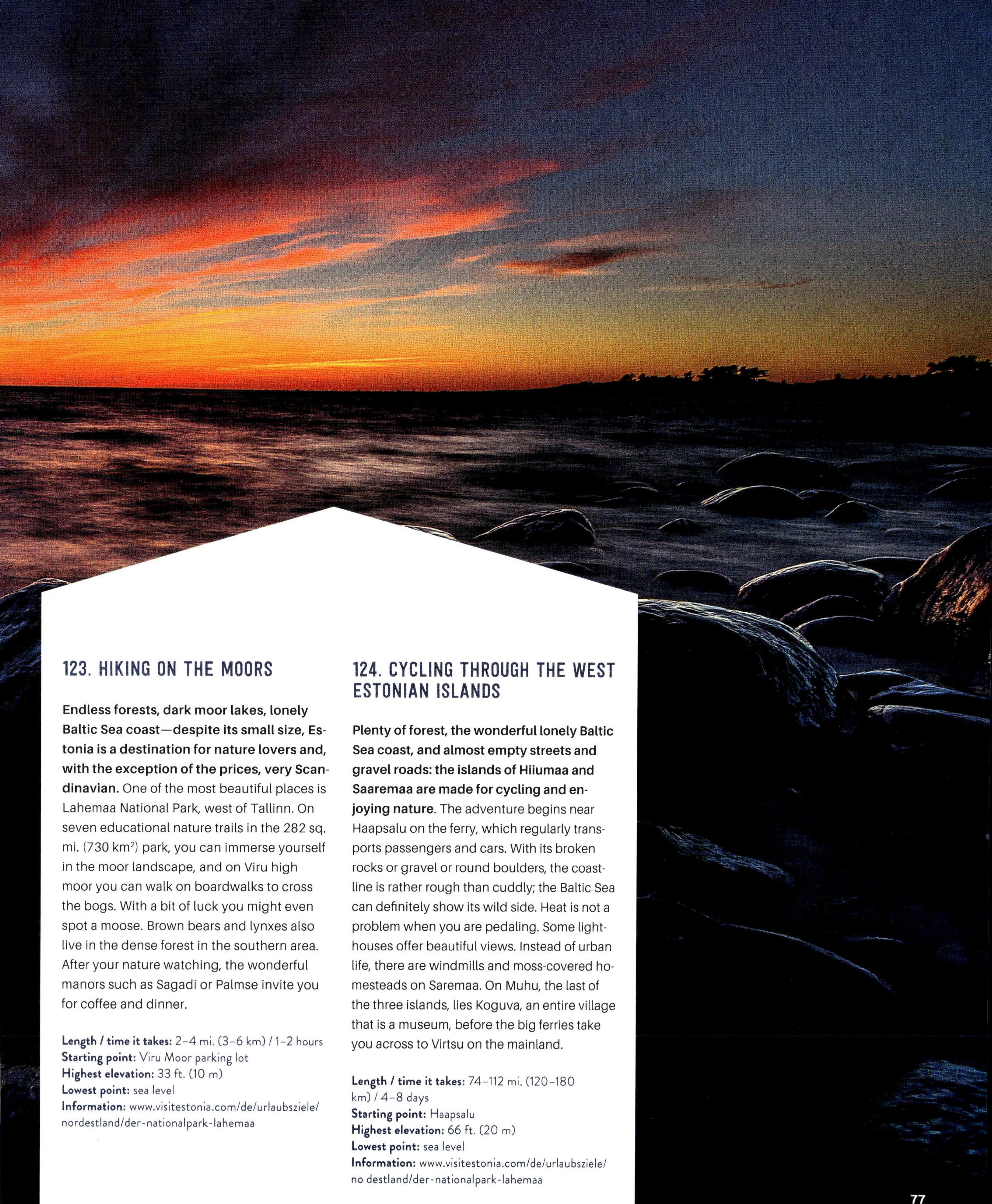

123. HIKING ON THE MOORS

Endless forests, dark moor lakes, lonely Baltic Sea coast—despite its small size, Estonia is a destination for nature lovers and, with the exception of the prices, very Scandinavian. One of the most beautiful places is Lahemaa National Park, west of Tallinn. On seven educational nature trails in the 282 sq. mi. (730 km^2) park, you can immerse yourself in the moor landscape, and on Viru high moor you can walk on boardwalks to cross the bogs. With a bit of luck you might even spot a moose. Brown bears and lynxes also live in the dense forest in the southern area. After your nature watching, the wonderful manors such as Sagadi or Palmse invite you for coffee and dinner.

Length / time it takes: 2–4 mi. (3–6 km) / 1–2 hours
Starting point: Viru Moor parking lot
Highest elevation: 33 ft. (10 m)
Lowest point: sea level
Information: www.visitestonia.com/de/urlaubsziele/nordestland/der-nationalpark-lahemaa

124. CYCLING THROUGH THE WEST ESTONIAN ISLANDS

Plenty of forest, the wonderful lonely Baltic Sea coast, and almost empty streets and gravel roads: the islands of Hiiumaa and Saaremaa are made for cycling and enjoying nature. The adventure begins near Haapsalu on the ferry, which regularly transports passengers and cars. With its broken rocks or gravel or round boulders, the coastline is rather rough than cuddly; the Baltic Sea can definitely show its wild side. Heat is not a problem when you are pedaling. Some lighthouses offer beautiful views. Instead of urban life, there are windmills and moss-covered homesteads on Saremaa. On Muhu, the last of the three islands, lies Koguva, an entire village that is a museum, before the big ferries take you across to Virtsu on the mainland.

Length / time it takes: 74–112 mi. (120–180 km) / 4–8 days
Starting point: Haapsalu
Highest elevation: 66 ft. (20 m)
Lowest point: sea level
Information: www.visitestonia.com/de/urlaubsziele/no destland/der-nationalpark-lahemaa

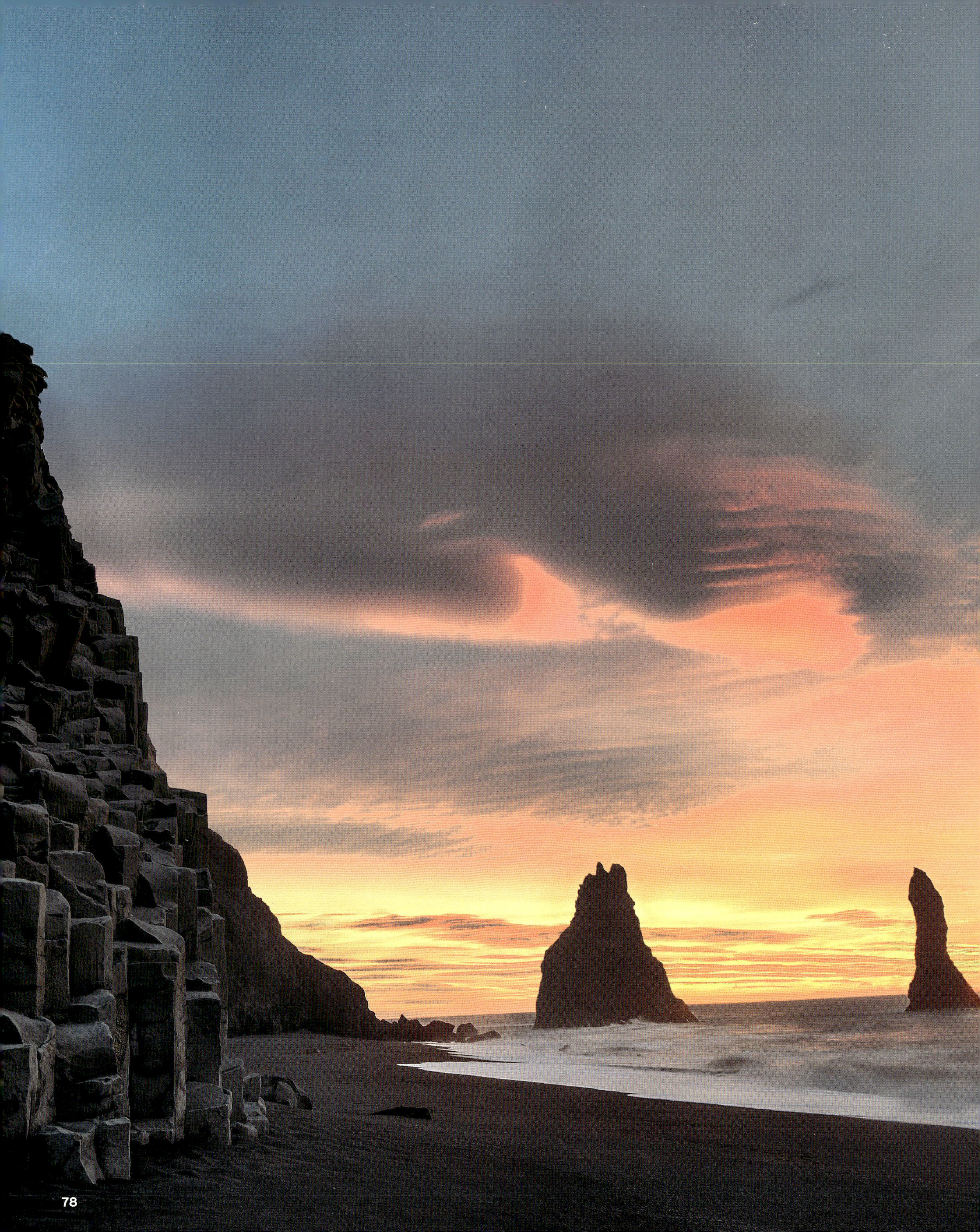

125. DESCENT INTO THRIHNUKAGIGUR VOLCANO

A volcano like this is a hot business. Fine and good, Thrihnukagigur Volcano near Reykjavik has now been quiet for 4,000 years. But it cannot hide its fiery past. The one-hour hike through the bizarre landscape is already interesting. But what is really breathtaking is to hover in a gondola, going vertically down about 492 ft. (150 m) into the depths of the magma chamber. Once you reach the bottom, you climb over blocks and look with fascination at the walls of the cooled chimney, formed of colored rock. The daylight shimmers only faintly through a narrow crevice in the ceiling of the huge hall. This is the only way to get out again, after a unique but—at around $390, not exactly inexpensive—three-quarters of an hour inside a volcano.

Time it takes: 6 hours
Starting point: Reykjavik (hotel pickup) or Breiðabliksskáli ski hut in Bláfjöll
Elevation difference: −492 ft. (−150 m)
Information: https://insidethevolcano.com

126. VOLCANO TREKKING ON NAMAFJÁLL LOOP TRAIL

Northern Iceland is already an extremely attractive travel destination in summer. The contrasts between fire and ice become even clearer in winter, when snow covers the land. In addition to the Dimmuborgir lava field, where you can also hike, the Hverir geothermal area is particularly interesting. Here the water steams in hot thermal springs, mud pools and mud pots boil, and foul-smelling steam rises and shoots from the fumaroles and solfataras—the breath of the underworld. The central volcano Krafa, around 5 mi. (8 km) north, is responsible for all this. A paved trail that is almost a mile (or perhaps more) long leads through the area (you should not leave the trail, because the bedrock is fragile in some places). In good conditions, the ascent to nearby Mount Námafjall is worthwhile; with its elevation of 1,581 ft. (482 m), it allows a great all-around view of the Mývatn area.

Length / time it takes: 2.2 mi. (3.5 km) / 1.5 hours
Starting point: Hverir parking lot (ring road)

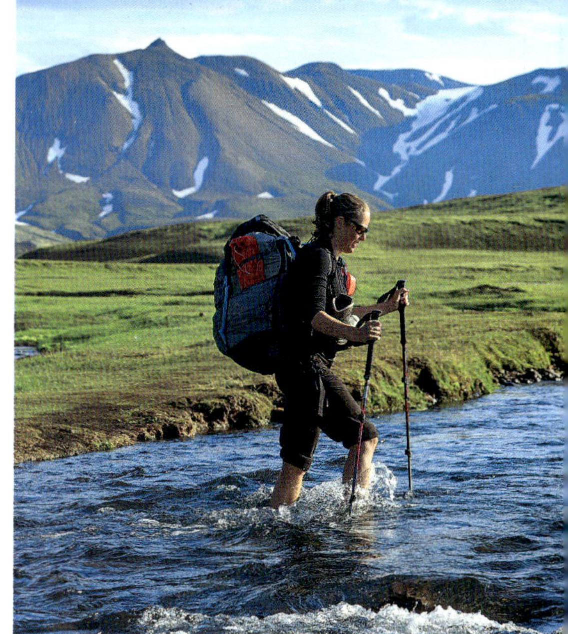

127. ON THE LAUGAVEGUR TREKKING TRAIL

This spectacular hiking route displays the entire spectrum of Icelandic landscapes; on the trail, you cross the cooled lava of Eyjafjallajökull Volcano, which erupted in 2010. The popular long-distance hiking trail and the classic of all Icelandic trips leads from Landmannalaugar ("way of the hot springs"), over 3,510 ft. high (1,070 m) Hrafntinnusker Mountain, and through deep valleys to the rugged Þórsmörk mountain ridge, whose name goes back to the Nordic god Thor. Along the way you can spend the night in well-equipped huts or in your own tent. The last part of the demanding hike over Fimmvörðuháls Pass is considered the most difficult part of the route. The Laugavegur Trail gives you the opportunity for intimate contact with a mighty natural spectacle, especially when unexpectedly wading through rivers.

Length / time it takes: 34 mi. (54 km) / 4–5 days
Highest elevation: 3,510 ft. (1,070 m)
Starting point: Landmannalaugar
Elevation difference: 1,768 vertical ft. (539 m)

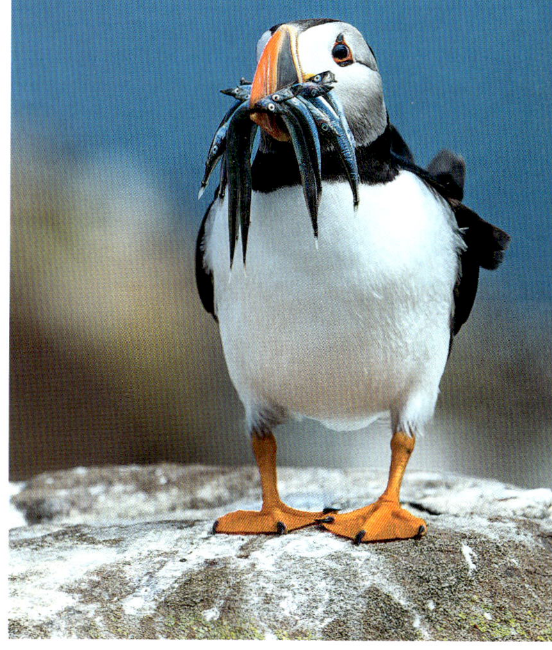

128. HIKING ON VÍK BLACK SAND BEACH

On the trail of *Star Trek* and *Game of Thrones*: this beach, made famous by several science fiction and fantasy movies, is on the south coast of Iceland. At the foot of Katla Volcano lies Iceland's southernmost village of Vík, with its famous church with its red roof. A walk to the west of the village leads to Reynisfjara Beach. Here lies the country's most famous beach, but the sand is black! Created by the erosion of volcanic rock, the grains of sand got their color as they hardened. A visit to the large cave in the rock shows the dramatic side of the Atlantic with its powerful, potentially dangerous waves. The 217 ft. high (66 m) Reynisdranger rocks, in the middle of the ocean, are visible from afar.

Length / time it takes: 6 mi. (10 km) / one day
Highest elevation: 33 ft. (10 m)
Starting point: Vík
Elevation difference: 778 vertical ft. (237 m)
Information: extremeiceland.is

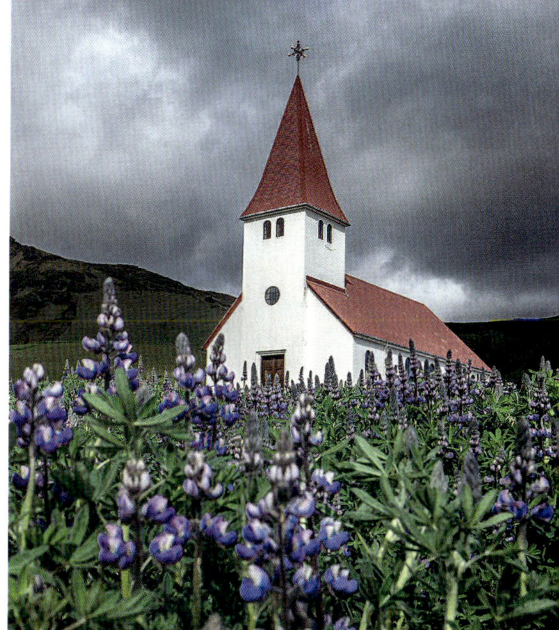

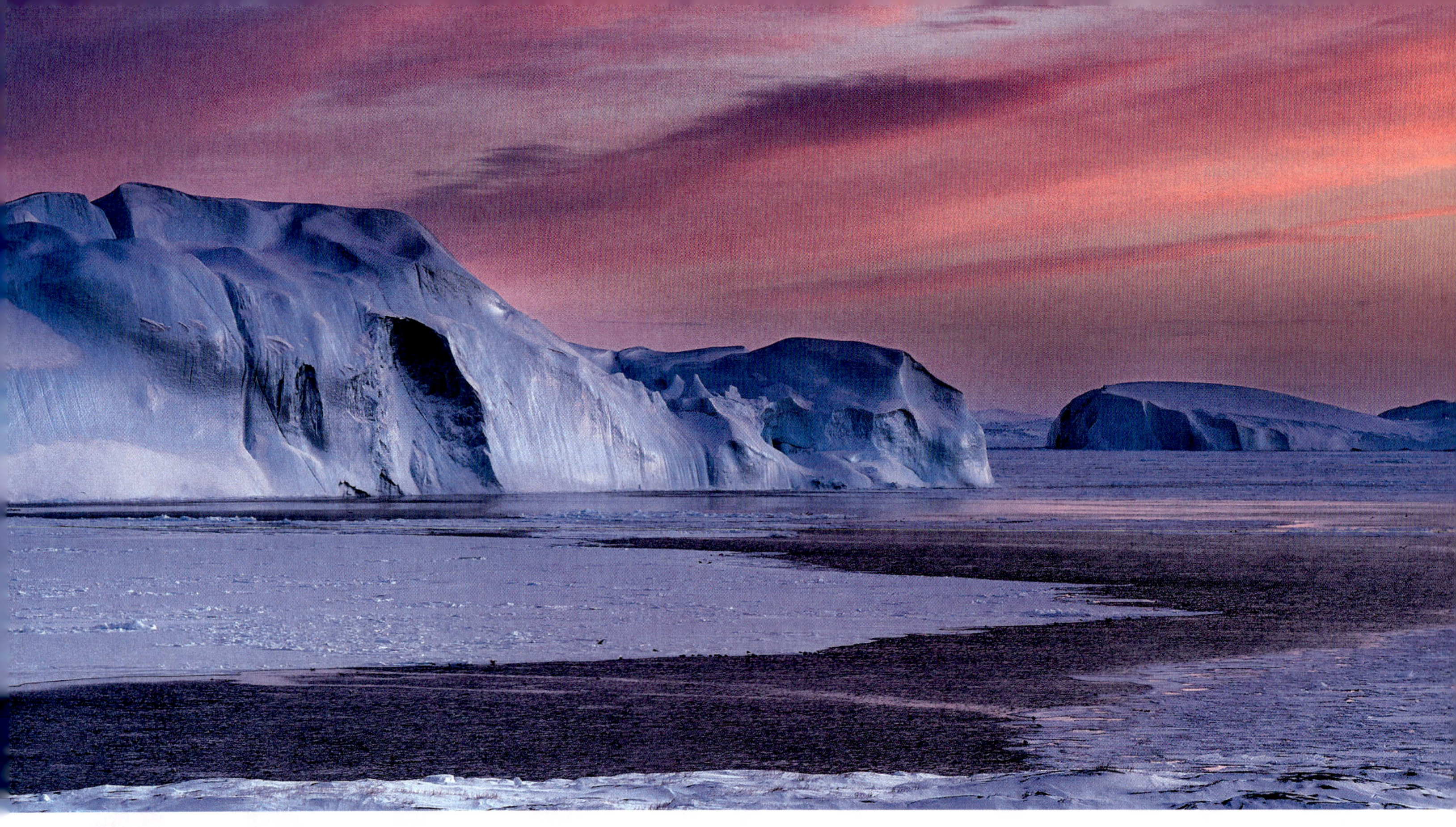

129. SNOWSHOEING AT KANGIA ICEFJORD

The netting in wide snowshoes glides gently over the covered ice, and you can make good progress with a breathtaking view of the 25 mi. long (40 km) and 4 mi. wide (7 km) ice fjord. It is the extension of Sermeq Kujalleq, one of the most active glaciers in the world, with a flow rate of 131 ft. (40 m) per day. It is from here that most of the Northern Hemisphere ice masses fall into the ocean; it is also verifiable that this was the source of the iceberg that later struck the *Titanic*. The glacier calving is worth hearing in summer, when huge pieces break off from the glacier with a loud roar. A day trip takes you on snowshoes from the town of Ilulissat in western Greenland to the UNESCO World Heritage Site of the Ilulissat Icefjord, then to the historic site of Holmens Bakke with its panoramic view.

Length / time it takes: 2.5 mi. (4 km) / 1 day
Starting point: Ilulissat
Information: pgigreenland.com

130. TREK AROUND EQI, THE CALVING GLACIER

Icebergs rise out of the sea like huge skyscrapers in a symphony of white, gray, and blue. As soon as the boat switches off the engine, there is silence until suddenly there is a crash along the 3 mi. wide (5 km) front. With a deafening noise the calving suddenly begins, and the first masses fall downward while the spectacle is reflected in the water. From Ilulissat, it is only 50 mi. (80 km) by cutter to this active glacier. A four-hour hiking trip extends from the spacious Glacier Lodge Eqi to the edge of the moraine. With a little more activity, after ten hours you can reach the inland ice, 3,281 ft. (1,000 m) in elevation, and even spend the night there. Incidentally, this was the region where the movie *Miss Smilla's Feeling for Snow* was filmed.

Length / time it takes: 100 mi. (160 km) / 1–3 days
Starting point: Ilulissat
Elevation difference: none
Information: greenlandtours.gl

131. HIKE THROUGH THE LOVOZERO TUNDRA, KOLA PENINSULA

The mountain range, shaped like a horseshoe, surrounds 5 mi. long (8 km) Seidosero Lake on the Russian Kola Peninsula. Thanks to the arrangement of the mountains, which protect the area around the lake from the icy north winds, a microclimate has been created that distinguishes the nature here from the otherwise usual polar flora. The uninviting town of Revda can serve as a starting point. From here you can take hikes of several days into the seclusion of the Russian tundra. The special atmosphere and the beauty of the landscape make some people believe that it is a special place of power; UFO sightings are not unheard of. The Sami people call the area around the lake "Luyavtzor," which can be translated as "hill near the lake of power."

Time it takes: several days
Starting point: Revda

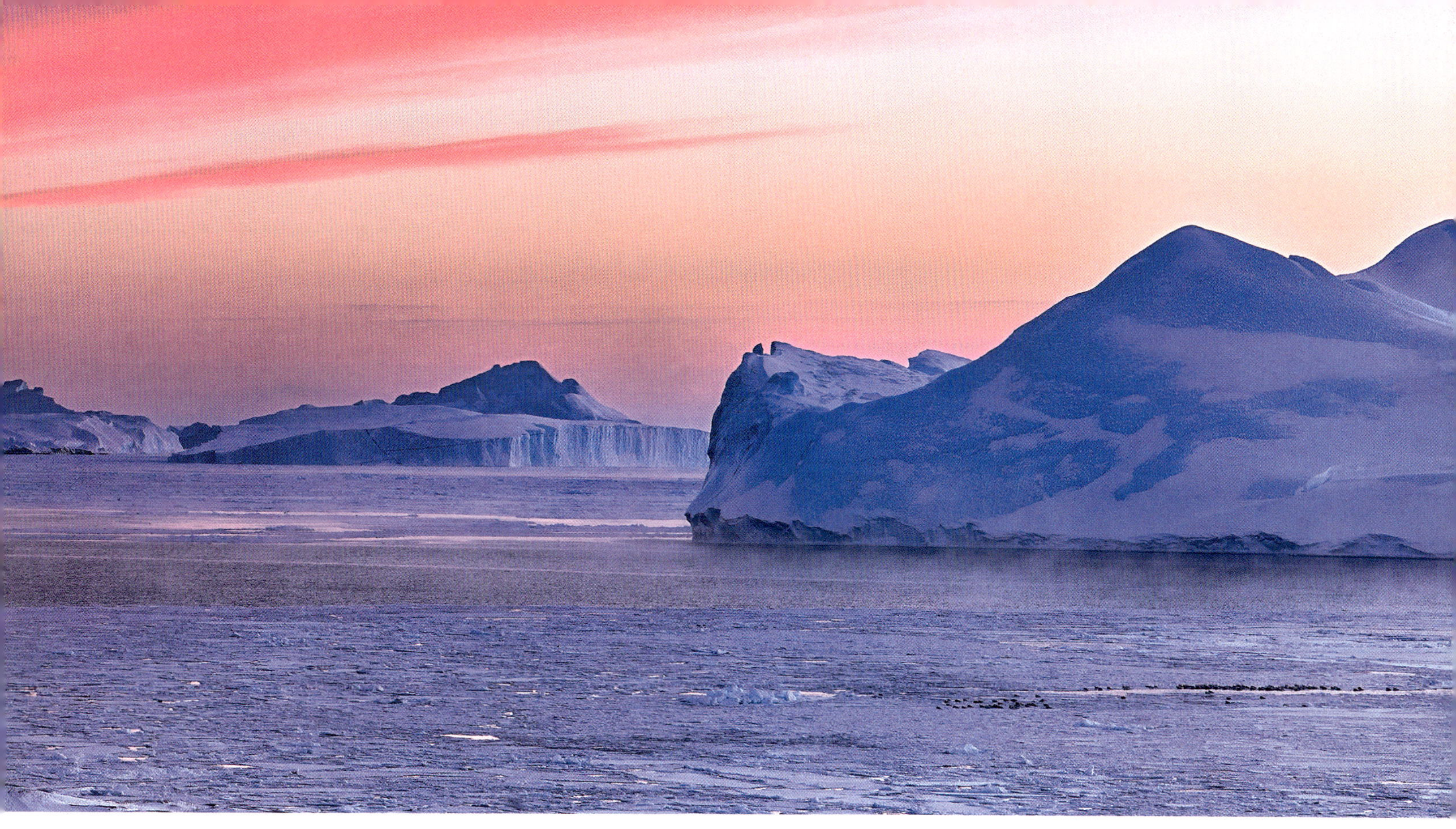

132. HIKING IN THE HEART OF THE KHIBINY MOUNTAINS

The dreary sight of the mining town of Kirovsk is quickly forgotten once you have set foot in the almost circular Khibiny Mountains. With a diameter of 28 mi. (45 km), the mountain range is the world's largest pluton; the magma piled up to a height of more than 3,281 ft. (1,000 m) some 360 million years ago. Chasnachorr Mountain in the Khibiny range is the highest peak on the Kola Peninsula. Winter lasts a good seven months here, and even in summer temperatures rarely rise above 46°F (8°C). Over the debris in an almost dry river and the lichens typical of the Russian tundra, the trail leads through the remoteness of the Far North, deeper and deeper into the mountains. Right in the middle is a small mountain rescue station, the only human settlement far and wide.

Length / time it takes: 12 mi. (20 km) / 2 days
Highest elevation: 3,940 ft. (1,201 m)
Starting point: Kirovsk
Elevation difference: 8,202 vertical ft. (2,500 m)

133. HIKE THROUGH FRANZ JOSEF LAND TO THE NORTHERNMOST ARCHIPELAGO ON EARTH

This inhospitable region on the Barents Sea, with its 191 islands, reaches to just before the North Pole and is considered one of the last adventures for individual travelers. The only way to explore the Russian part of the Arctic Ocean, the territory of the former polar explorers, is by an expedition ship of the icebreaker class. Because of the constantly changing ice conditions and Russia's renewed military interest in the Arctic, it is necessary to be extremely flexible. In the land of polar bears and walruses, there are a range of possible activities on land: you can visit the British station established on Bell Island in 1880 or the former Soviet-era meteorological station on Calm Bay. You can take pretty pictures of nature, such as of sea birds, on Cape Flora.

Time it takes: 14–21 days
Highest elevation: 59 ft. (18 m)
Starting point: Murmansk
Information: auroraexpeditions.com.au

134. DOGSLEDDING TO THE KOLA PENINSULA

Already on the third day a well-forged team: the dogs and the traveler on their way as a musher on this sledding adventure trip through the vastness of northern Russia. The Kola Peninsula is not far from the Finnish border between the White Sea, a marginal sea of the Arctic Ocean, and the Barents Sea. From Murmansk, with its international connections, the first stop is the base station at the mining town of Kirovsk, where the five-day guided tour starts. The peninsula is crisscrossed by long rivers and several lakes for unforgettable walks. Here you can also go ice fishing or warm yourself up in the sauna. Accommodation in simple hotels in a sparsely populated region lets you forget civilization; this journey feels like taking a time machine into a forgotten era.

Length / time it takes: 106 mi. (170 km) / 5 days
Highest elevation: 1,247 ft. (380 m)
Starting point: Murmansk
Information: paneurasia.de

EASTERN EUROPE

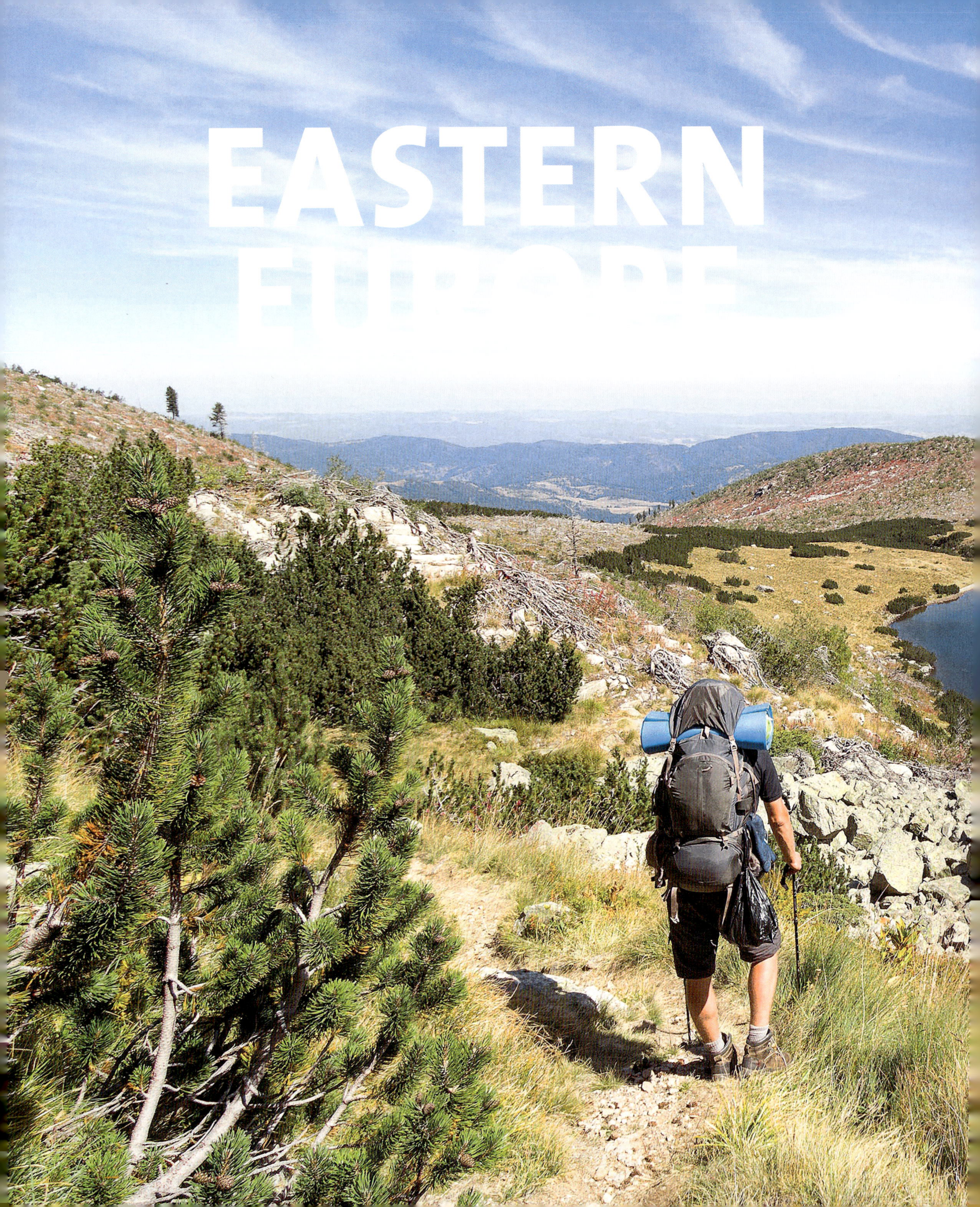

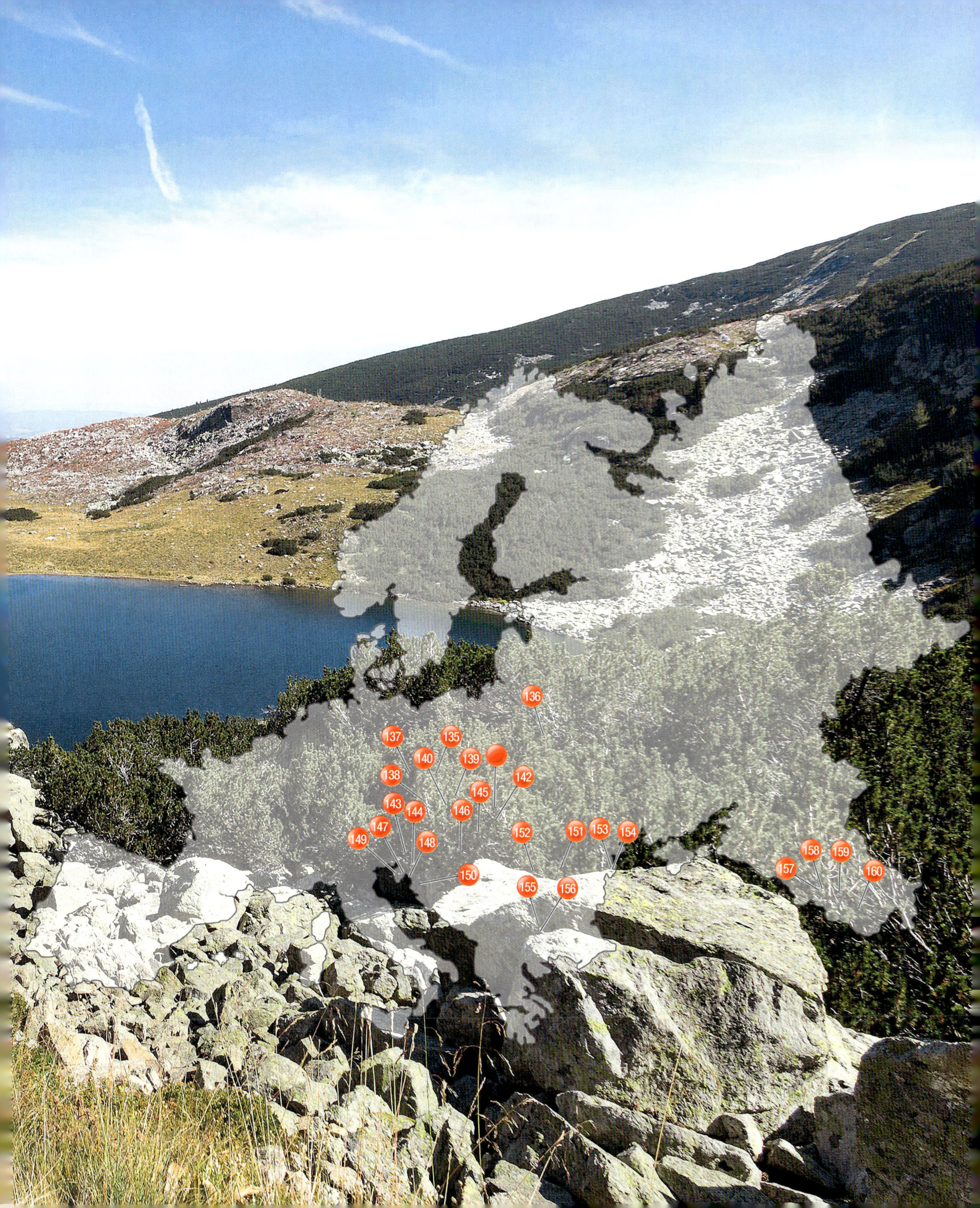
136
137
135
140
139
138
142
145
143
144
146
147
152
151
153
154
149
148
158
159
157
160
150
155
156

135. MOUNT SNĚŽKA/ŚNIEŻKA

The Giant Mountains and their highest peak, the 5,259 ft. high (1,603 m) Mount Sněžka—the Czech name; in Polish it is Mount Śnieżka—still have a magical appeal. The first documented ascents to its summit were made in the fifteenth century, and today a mountain hike on Mount Sněžkae is still a dream come true for many people. Once at the top, the panorama in all directions proves them right. The border between Poland and the Czech Republic runs right across the summit. On the summit, which is well prepared for the many hikers and climbers in summer, there is more to be found than the Laurentius Chapel: on the Czech side there is a post office, and on the Polish side a UFO-like mountain hut, with weather station and observatory, which was built between 1969 and 1974.

Length / time it takes: 3 mi. (5 km) / 1 day, there and back
Highest elevation: 5,249 ft. (1,600 m)
Starting point: Karpacz, Poland
Elevation difference: 3,675 vertical ft. (1,120 m)
Information: www.karpacz.pl/bergwanderungen

136. BIAŁOWIZA NATIONAL PARK

In the land of the bison: Podlaskie Voivodeship (province) lies in far northeastern Poland and, with its total of four national parks, is among the country's natural paradises. You can find evidence of its multicultural past and present everywhere throughout the Voivodeship. Historically the region of Podlasie, it covers a total area of 7,792 sq. mi. (20,180 km^2), has 1.2 million inhabitants, and has a population density of sixty inhabitants per square mile. Białowieża National Park is the oldest nature reserve in Poland and home to the largest wild herd of bison in the country. Białowieża, the primeval forest village not far from the Belorussian border, has around 3,000 inhabitants. You can roam through the strictly protected Hwoźna Protected Area without a guide, but you can visit the Orłówka Protective Unit on the "To the Oak of Jagiełło" path only if you are accompanied by a ranger and have registered beforehand.

Length / time it takes: 9 mi. (15 km) / 1 day, there and back
Highest elevation: 820 ft. (250 m)
Starting point: Białowieża
Elevation difference: 213 vertical ft. (65 m)
Information: bpn.com.pl

137. THE ORE MOUNTAINS BEER TRAIL

Because of Pilsner Urquell and Budweiser: the traditions of growing hops and brewing beer in the Ore Mountains looks back more than 800 years. It was not long ago, in May 2018, that this cross-border route was created. It offers a boozy highlight: anyone who collects stamps along the way and can show a stamp from ten of the participating Czech or German breweries will receive a special beer mug. The new association of seventeen mainly small breweries supports regional diversity and promotes truth that transcends borders when enjoying the nectar of the barley. You can often look right over the brewmaster's shoulder. However, caution is advised: alcohol consumption is forbidden on Czech roads, so it is better to cover the hilly route by bike.

Length / time it takes: 256 mi. (412 km) / 1–3 days
Highest elevation: 2,989 ft. (911 m)
Starting point: Teplice and Ústí nad Labem in the Czech Republic, Freiberg in Germany, and other cities
Elevation difference: 14,341 vertical ft. (4,371 m)
Info: https://www.erzgebirge-tourismus.de/bier-route

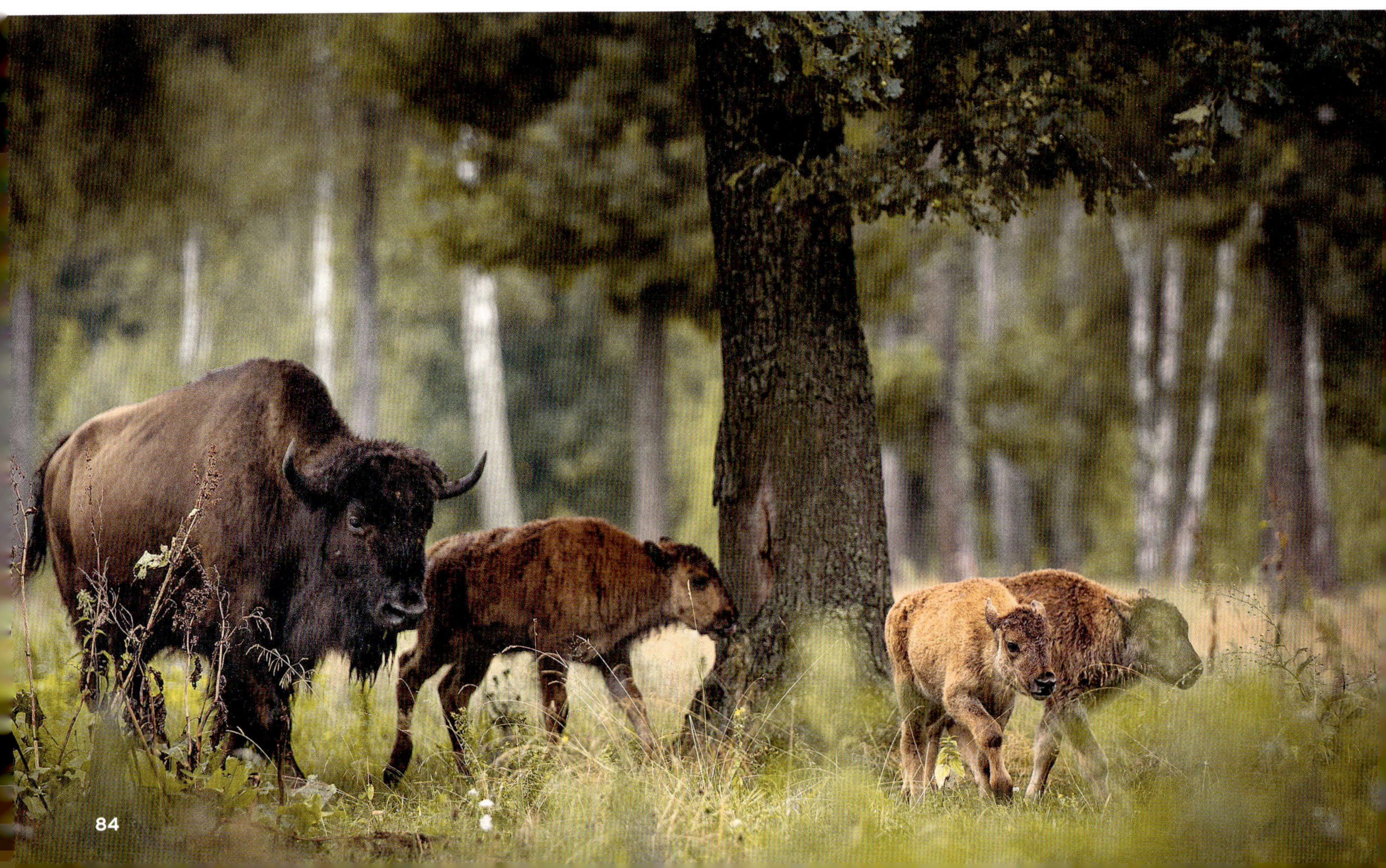

138. BIKE TRIP THROUGH THE TŘ BOŇ CARP PONDS

Carp ponds, peat bogs, and water meadows—the well-developed network of bike paths views the sparsely populated area of South Bohemia around the Czech town of Třeboň from its natural side. Two dozen multilingual information boards along the central educational trail provide information about the 500 or so pond systems and the flora and fauna of the region, which is known in the Czech Republic primarily because of the carp eaten at Christmas. The Renaissance town has also made a name for itself as a health spa. Rožmberk fish pond, an artificial pond east of the town, is actually the largest pond in the country. Anyone wanting to explore the region more extensively on a bike can continue cycling to the beer paradise of České Budějovice (home of Budweiser) or to the picturesque town of Český Krumlov.

Length / time it takes: 24–43 mi. (39–70 km) / 1–2 days
Highest elevation: 1,424 ft. (434 m)
Starting point: Třeboň
Elevation difference: at ground level

139. BOBSLEDDING AND DOWNHILL SKIING

Racing down the snowy slopes and riding a Hummer off-road vehicle over hill and dale: the Fun Park has it all year round. The village of Hlubočky is about 12 mi. (20 km) east of Olomouc, capital city of the Czech Olomouc administrative region, and is a big attraction there thanks to its small but fine ski resort. The six ski slopes on 2,297 ft. high (700 m) Odra Mountain's foothills ensure some real fun on the slopes. The area is also popular in summer, when the bobsled runs are used as a fun park. The only bobsled run in Moravia is 5,039 ft. (1,536 m) long and features seventeen curves, two spiral turns, a 148 ft. long (45 m) bridge, and a tunnel. Those who feel like even more adrenaline can drive over inhospitable terrain in a Hummer.

Length / time it takes: 1 mi. (1.6 km) / 1 day
Highest elevation: 1,247 ft. (380 m)
Starting point: Olomouc
Information: skiarealhlubocky.cz

140. NORDIC WALKING NEAR BRNO

The Moravian Karst north of Brno, the second-largest city in the Czech Republic after Prague, with its canyon-like valleys, deep depressions, and rushing streams, gives the impression it is a wild region. Appropriately, the half-marathon-distance Nordic walking tour over the slopes here is considered a difficult one. Yet, it is also extremely lovely thanks to the views the walk provides, both of the city and amazing natural beauties, such as the Adamov rock sphinx. The route, which is not always ideally signposted, leads from Obřany, across Výškovnice peak, and past sparse forests to the town of Adamov. Once you are already here: in the Moravian Karst region you should make time for 453 ft. deep (138 m) Macocha Gorge and the Punkva stalactite cave, both of which are high on the list of the top sights near Brno.

Length / time it takes: 13.3 mi. (21.4 km) / half a day
Starting point: Brno

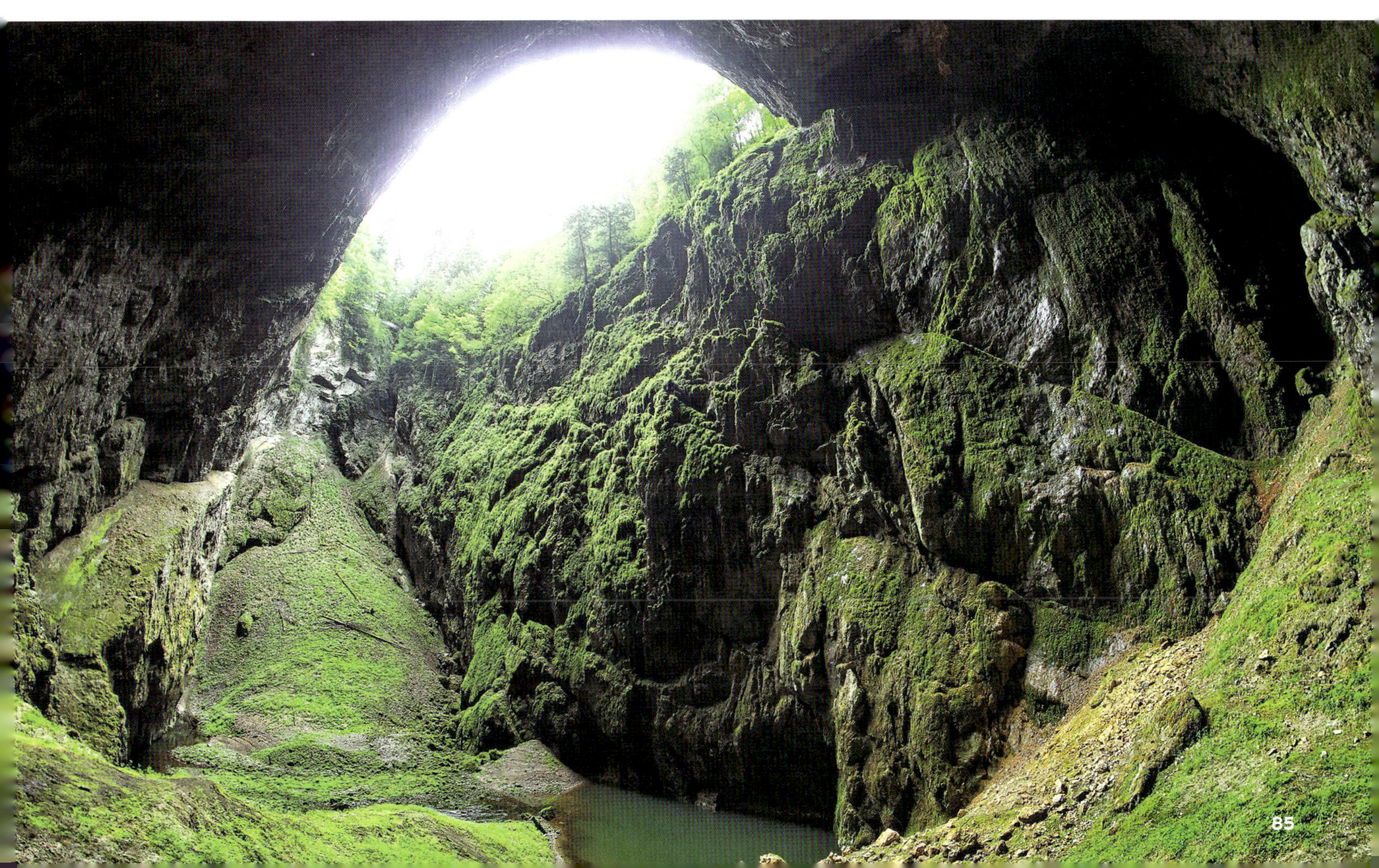

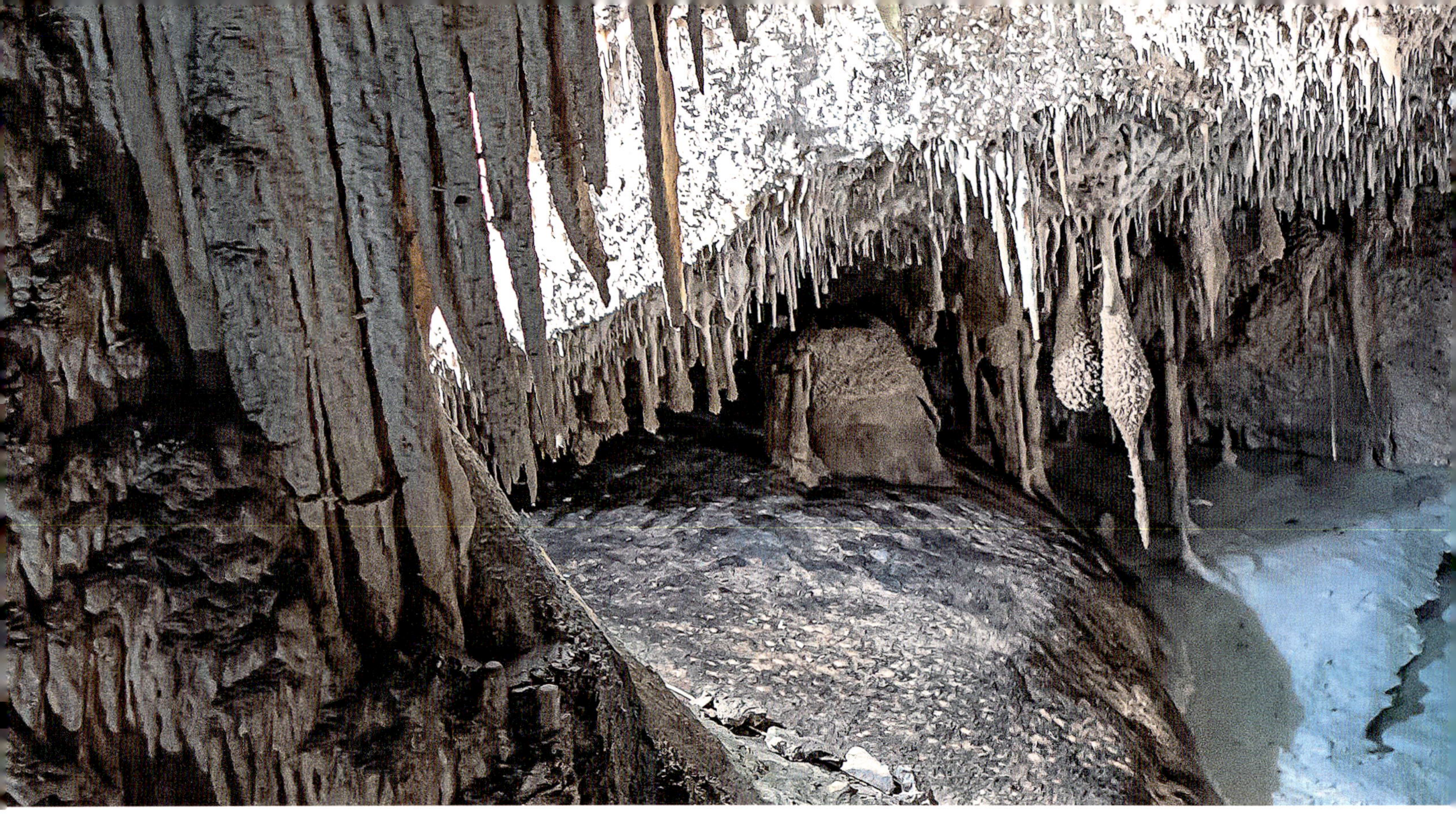

141. SLOVAK KARST NATIONAL PARK

Wild gorges carved by deep-lying rivers cut through the largest karst system in central Europe. The region has a diversity of natural lakes. UNESCO declared the first biosphere reserve in Slovakia a World Heritage Site thanks to its extensive underground cave systems. From the town of Rožňava, you can take a wide range of hikes through the dense forests and wide steppe landscapes of the enchanting Národný Park Slovenský kras (Slovak Karst National Park). The educational trail through 1.37 mi. long (2.2 km) Zádielska tiesňava Gorge is one of the most popular attractions. Other highlights include Jelení vrch Mountain, the highest peak at 3,107 ft. (947 m), and the thundering waterfall near the village of Háj. Cross-country skiing on the region's seven plateaus is also a possibility during the winter.

Time it takes: 1–2 days
Highest elevation: 1,112 ft. (339 m)
Starting point: Rožňava
Elevation difference: 2,008 vertical ft. (612 m)
Information: tikroznava.sk

142. DOMICA CAVE

A descent into prehistoric times: Domica Cave, in the Slovak Karst, southwest of Košice on the Hungarian border, has been a UNESCO World Heritage Site since 1995. Together with Baradla Cave, which is in Hungarian territory, it forms an underground system that is 16 mi. (25 km) long. It was discovered by Ján Majko in 1926. The cave's outstanding features include the large number of pagoda-shaped stalagmites and unique shields and drums that are not to be found in this form in any other cave in the world. In addition, there are sixteen different species of bats alone. If you walk among the Mesozoic Wetterstein limestone, from the Middle Triassic, you will be able to admire the 350,000-year-old "Samson's Columns" or the almost 328 ft. high (100 m) "Majkos Cathedral." Taking an underground boat trip can also make your visit a special experience.

Length / time it takes: .8 mi. (1.3 km) / 1–2 hours
Highest elevation: 1,115 ft. (340 m)
Starting point: Plešivec
Elevation difference: 98 vertical ft. (30 m)
Information: www.ssj.sk

143. HIKING IN TRIGLAV NATIONAL PARK

From rock faces to waterfalls, there is an amazing variety for unlimited adventure here. The only national park in Slovenia is named after 9,396 ft. high (2,864 m) Mount Triglav and is in the Julian Alps. Experienced mountaineers can complete the 9 mi. (14 km) ascent of Triglav from Aljažev dome within eleven hours. Hikers will find seemingly endless trails and interesting educational nature trails amid almost unspoiled nature. The twenty-five climbing walls ensure pure adrenaline; the 3,937 ft. high (1,200 m) north face of Mount Triglav is the third-highest north face in Europe. Thundering masses of water rush down a drop of 167 ft. (51 m) over Savica waterfall and into a reservoir. From there the water flows on into clear Bohinj Lake, a paradise for lovers of rowing, canoeing, kayaking, and SUPing.

Length / time it takes: 8.7 mi. (14 km) / 1–2 days
Highest elevation: 9,396 ft. (2,864 m)
Starting point: Bohinjska Bistrica
Elevation difference: 6,233 vertical ft. (1,900 m)
Information: tnp.si

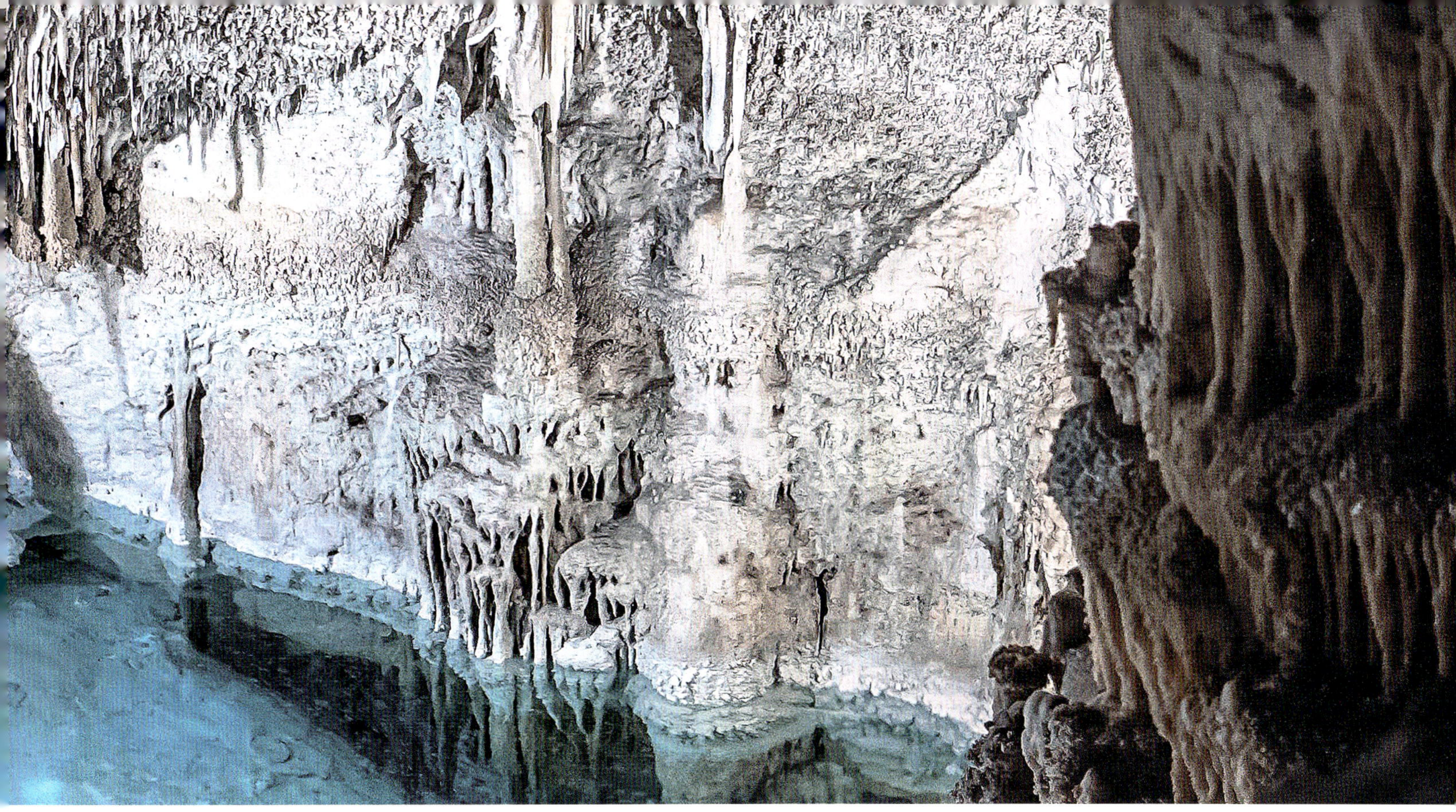

144. REGIONAL PARK–POSTOJNA STALACTITE CAVE

In the karst region of southwestern Slovenia you can take a train to an imposing underground paradise: since 1872, visitors have been traveling the first 2.3 mi. (3.7 km) of the 3 mi. (5 km) route in Postojna Cave by rail. The second-longest cave system in the world, discovered about 800 years ago, has a total of 15 mi. (24 km) of underground passages and halls. One of them unexpectedly boasts a crystal chandelier, which reminds us that the former Austrian imperial couple were among the guests 200 years ago. The cave post office, which opened in 1899, still accepts postcards today, despite smartphones. Incidentally, Postojna Park and the impressive medieval Predjama cave castle, which is carved into the rock, are just over 31 mi. (50 km) from Ljubljana, making them an ideal day trip destination.

Length / time it takes: 3 mi. (5 km) / 1 day
Highest elevation: 1,817 ft. (554 m)
Starting point: Postojna
Information: postojnska-jama.eu

145. FROM SZENTENDRE TO VISEGRÁD

A day trip from Budapest takes you from the small baroque town of Szentendre through Duna (Danube)-Ipoly National Park to Visegrád and back again. Just as the Danube divides Budapest into a hilly and a flat half, Szentendre is on the hilly western bank of the Danube, while 24 mi. long (38 km) Szentendre Island belongs to the lowlands opposite. This moderately difficult route, with a broad view of the Danube Bend, first leads through the national park, with its karst rocks, to the town of Visegrád. International summits were held at the castle of Visegrád during the Middle Ages. The way back is on the slightly lower west side. Instead of the set loop trip, you can also return from Visegrád along the Danube.

Starting point: Szentendre
Highest elevation: 1,821 ft. (555 m)
Elevation difference: 2,986 vertical ft. (910 m)

146. SAILING ON LAKE BALATON

Thanks to its warm water and exciting attractions, Lake Balaton is a paradise for water sports enthusiasts. The lake's twenty marinas and seven sailing schools provide a wide range of options, along with its generally shallow water and water temperatures of up to 82°F (28°C) in summer. In addition to the main town of Siófok in the east, the glamorous health resort of Balatonfüred in the north and the resorts of Balatonföldvar and Fonyód in the south are among the preferred vacation destinations. It is not uncommon for locals to bring their own boats along to ride on the waves during a storm. Organized sailing tours and regattas are held on Balaton several times a year, which let experienced sports sailors prove their skills. If sailing is not enough for you, there are also opportunities for SUP paddling, waterskiing, wakeboarding, and kitesurfing on Lake Balaton.

Length / time it takes: 49 mi. (79 km) / 2–7 days
Highest elevation: 341 ft. (104 m)
Starting point: Siófok
Information: balatonihajozas.hu

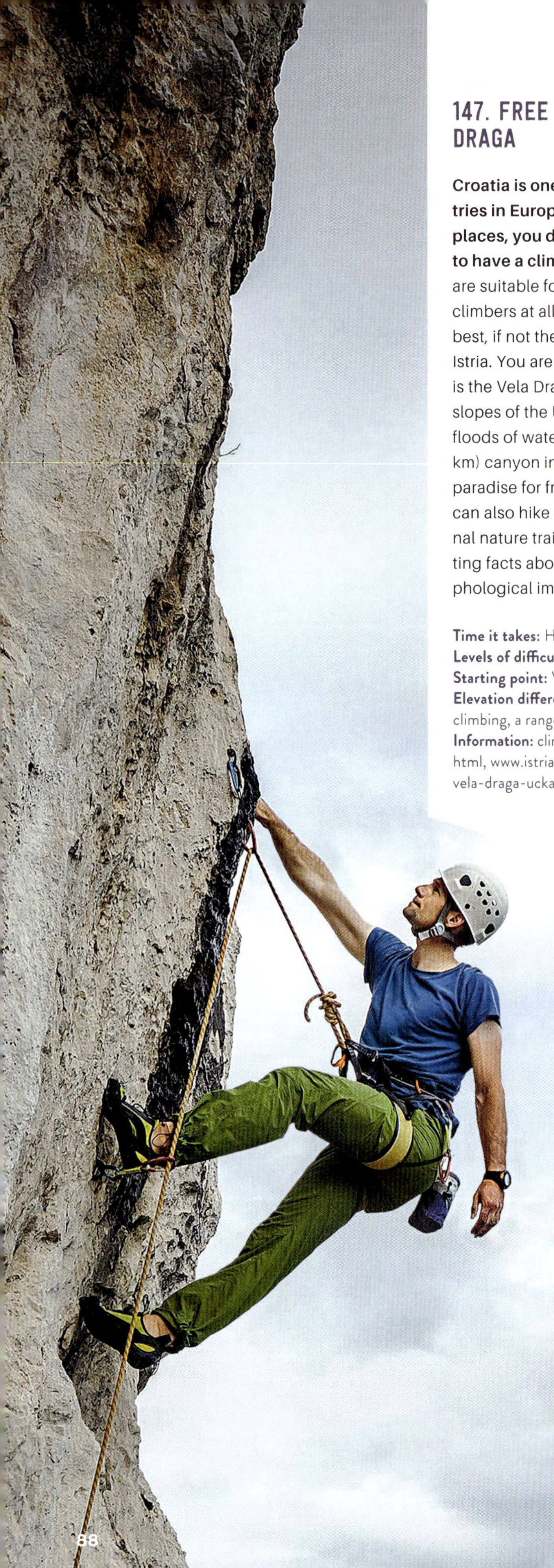

147. FREE CLIMBING IN VELA DRAGA

Croatia is one of the most interesting countries in Europe for climbers. And in many places, you don't have to be a professional to have a climbing experience. Most areas are suitable for beginners and for advanced climbers at all levels of difficulty. One of the best, if not the best, regions for climbers is Istria. You are spoiled for choice. But one tip is the Vela Draga, a canyon on the western slopes of the Učka mountain range. Here, floods of water have dug a 2 mi. long (3.5 km) canyon into the limestone and created a paradise for free climbers. Alternatively, you can also hike along the Vela Draga educational nature trail and learn all kinds of interesting facts about the extraordinary geomorphological importance of this region.

Time it takes: Hike 1.3 mil. (2.1 km) / 30 minutes
Levels of difficulty: 4b to 6a
Starting point: Vela Draga parking lot near Zrinšćak
Elevation difference: hiking, 561 vertical ft. (171 m); climbing, a range of elevations
Information: climb-europe.com/De/KletternKroatien.html, www.istria-culture.com/de/vela-draga-ucka-i130

148. BY SEA KAYAK AROUND RAB ISLAND

Right, left, right, left—the paddles pull evenly through the water. A white, salty layer builds up on your arms. You reach the small lighthouse at Donja Punta, the northwesternmost point on Rab Island. This is where you picnic: a spread of tomatoes, olives, flatbread, and yogurt. Then, strengthened, you go on: 1.25 mi. (2 km) across Kamporska Draga, an open bay. On the other side of the bay, the Dumici Peninsula is ideal for taking a coffee break at a beach bar. It is the second day of the trip circumnavigating the island. The only day when the bora, the strong Adriatic Sea fall wind, can bother the paddlers. Further stages along this fascinating circumnavigation of the island include St. Gregory and Goli, Medova Buža cave in the Geopark, and Pudarica beach, where the journey ends.

Time it takes: 6 days, 5 nights
Starting point: The town of Rab

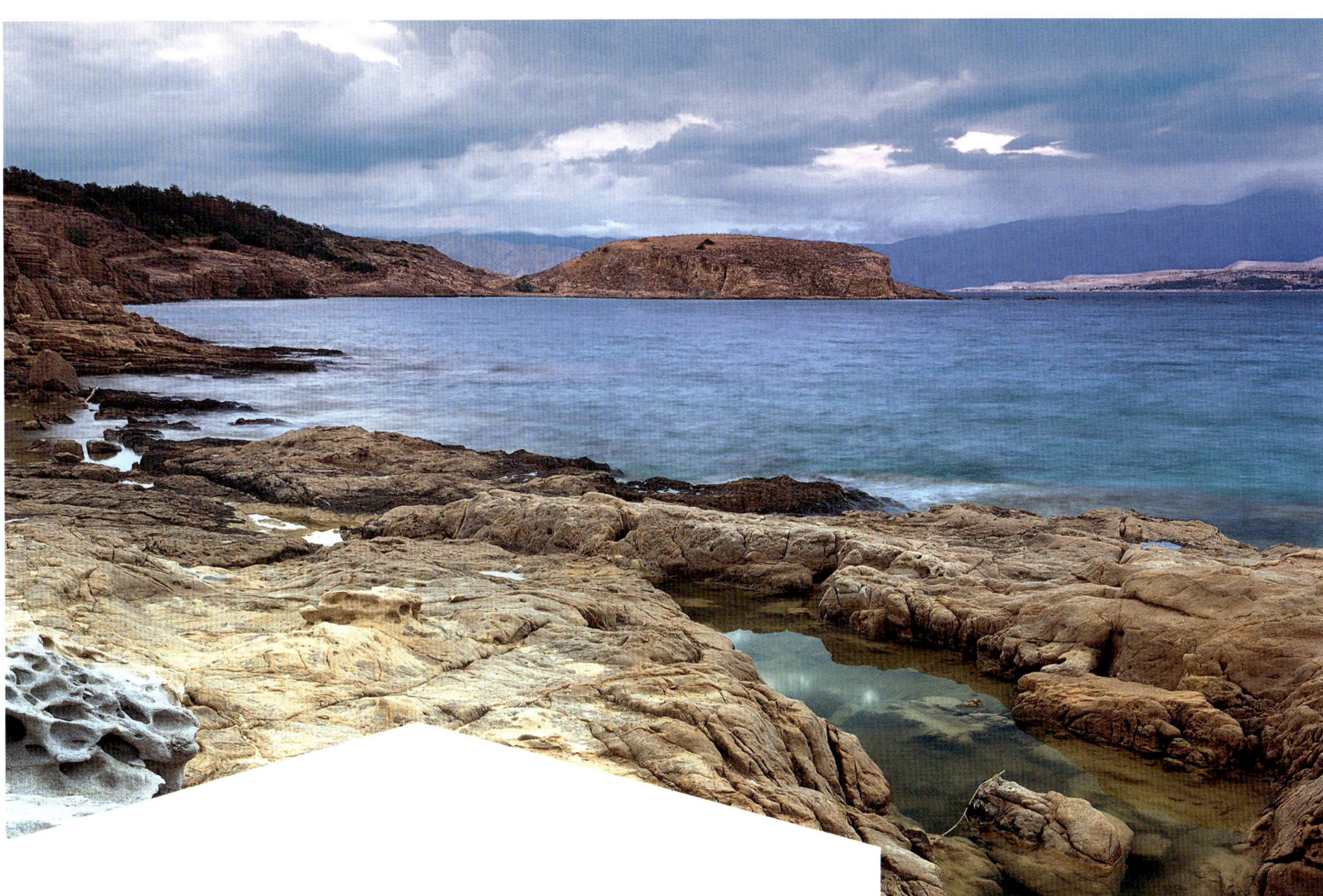

149. CYCLING ALONG THE WINE ROUTE AROUND POREČ

Poreč is the wine capital of the Istria Peninsula. Stone monuments from when it was ruled by the Roman Empire attest to viticulture as early as Roman times. The route is signposted in red. Always along the coast, it first goes north to the village of Červat Porat. You cycle slowly through rolling vineyards toward Tar, in the interior, and then in an eastern direction toward Višnjan. From now on, there will be plenty of good red-wine vineyards, and it is time to finally try one of the wines at one of the many wineries or at a konoba, the typical wine cellars of Istria, along the way. But be careful: Malvasia and other varieties are full-bodied and strong wines, and your legs can become unsteady. Always keep going a southerly direction, now through the hill country, again by the coast, and back.

Starting point: Poreč

150. THE PREMUŽIĆEVA STAZA TRAIL

In the hinterlands of the Croatian Adriatic region. The Premužic educational trail runs through Croatia's Northern Velebit National Park and is named after forestry engineer Ante Premužic. It was created under his technical direction during the 1930s. The most frequented and most spectacular part leads from the area of Zavižan, where you will find a botanical garden in addition to a mountain hut and a chapel, to Veliki Alan, a pass above the village of Jablanac, where there is another mountain hut, and past this to the craggy cliffs and peaks of the Rožanski kukovi. There are many beautiful views along the entire route, and the numerous, well-maintained drystone walls are impressive. From Veliki Alan the trail continues to the village of Oštarije.

Length: 31 mi. (50 km)
Starting point: Zavižan
Trail's end: Oštarije

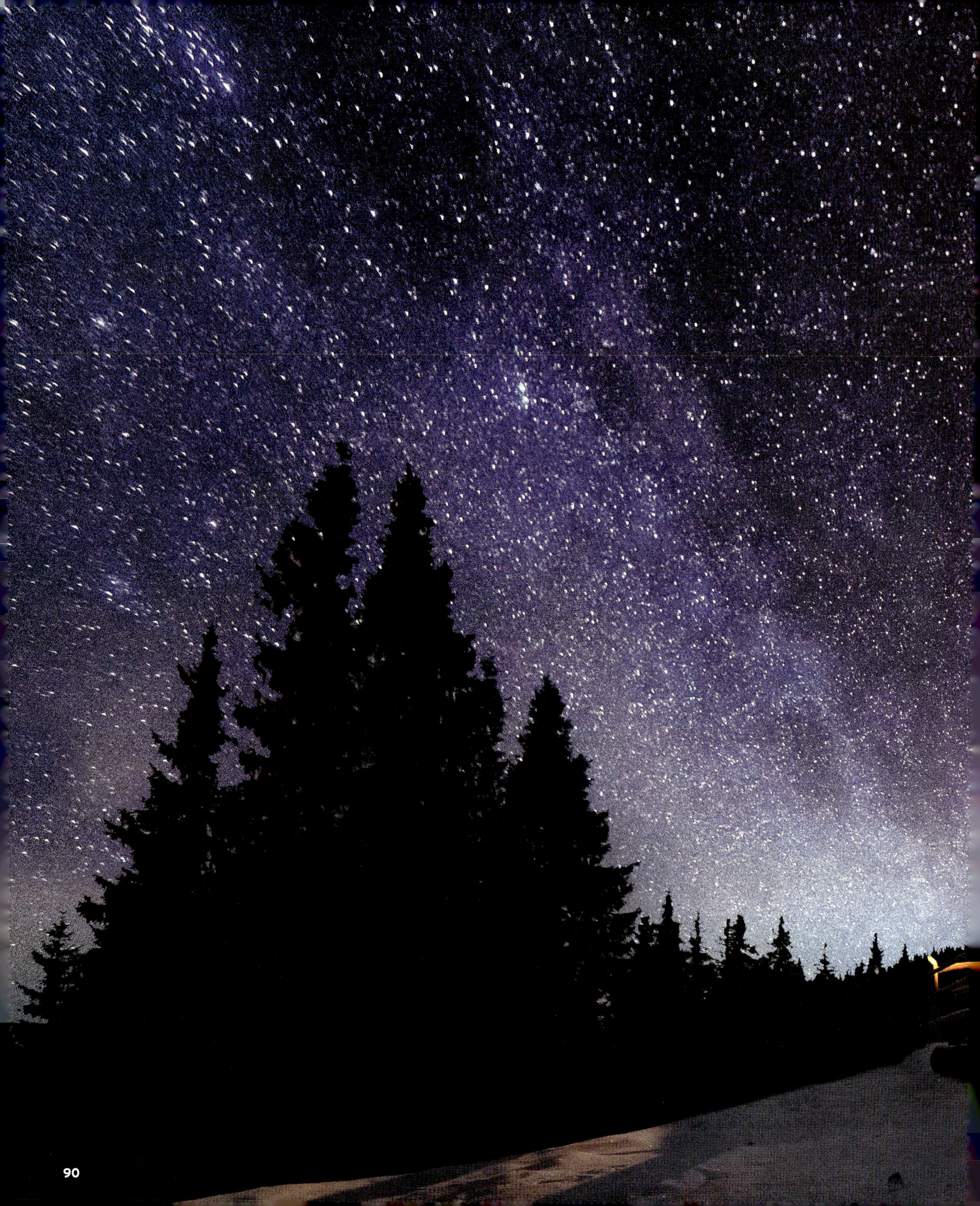

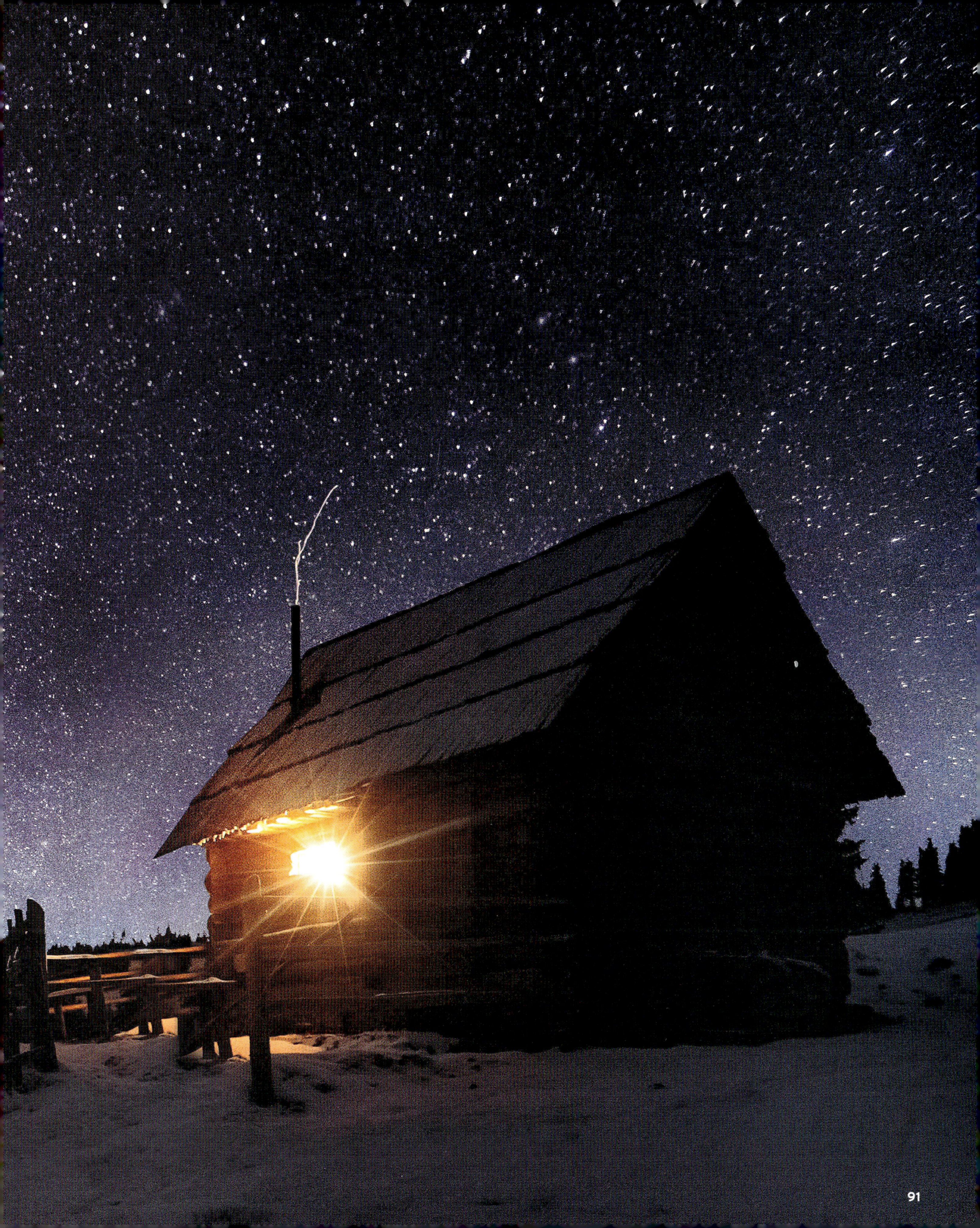

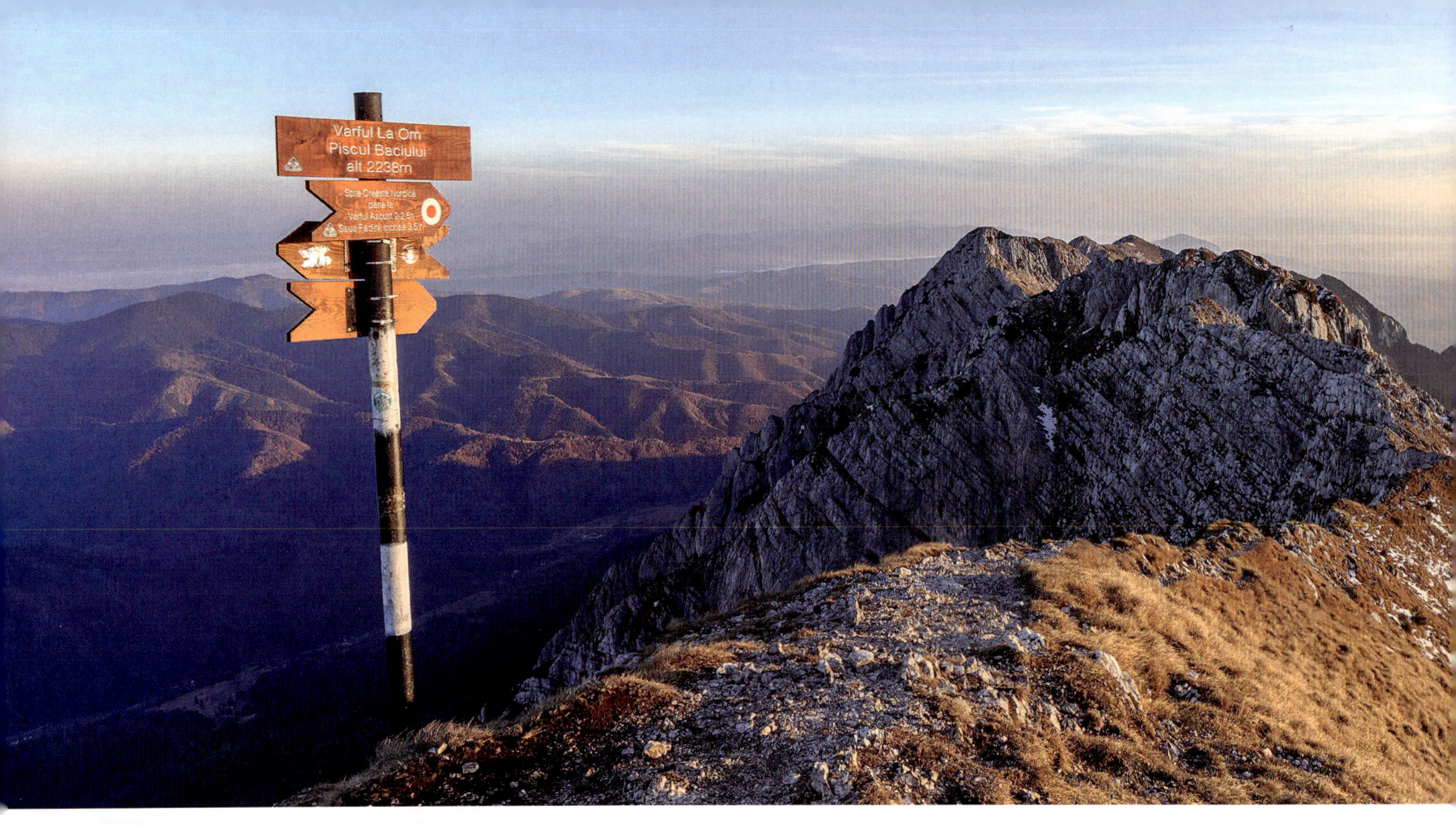

151. RIDGE HIKE IN THE PIATRA CRAIULUI MOUNTAINS–CARPATHIANS

This rocky limestone ridge, which runs 16 mi. (25 km) in a north-south direction, has a good network of hiking trails. You can take an impressive ridge hike over many peaks, including the highest, La Om, at 7,342 ft. (2,238 m) in elevation. In the vicinity of Măgura, trail 22 climbs steadily uphill from the banks of Cheia Brook to the Crăpăturii saddle, from where you can reach Turnu peak. This is where the number 26 ridge trail starts. This is one you won't forget, thanks to its magnificent views of the surrounding landscape and sometimes difficult climbing passages. This hike is recommended only for experienced, sure-footed mountaineers and should not be tackled in bad weather. Trail 26 ends at Funduri saddle. An interesting alternative is descending via trail 8 to Plaiul Foii mountain hut. This trail leads past the La Zazplaz rock formations over some rope-secured passages.

Length / time it takes: 16 mi. (25 km) / 10 hours
Highest elevation: 7,342 ft. (2,238 m)
Starting point: Măgura
Elevation difference: 2,920 vertical ft. (890 m)

152. THE RETEZAT MOUNTAINS–CARPATHIANS

Into the heart of the Carpathian massif in southwest Romania. A blue-and-white signposted trail leads steadily uphill from Staţiunea Cârnic, past beautiful Lolaia waterfall. The trail climbs higher and higher through dense forests and along a mountain stream to Pietrele mountain hut. It is about 656 vertical ft. (200 m) from there to Cabana Genţiana, where the mountain hut host has a warm cup of tea ready for hikers. Always going uphill and above the treeline, the trail leads to Curmătura Bucurei, a pass at more than 7,218 ft. (2,200 m) above sea level, from where you have a magnificent view of the surrounding peaks and the valley where Bucura glacier lake sparkles. Here you can take a break or spend the night in a tent. From the top of the pass, you can climb to the summits of Peleaga or Pietrele Mountains. There is also a trail over Poiana Pelegii down to the village of Buta, on the other side of the Retezat massif.

Length / time it takes: 9 mi. (15 km) from Pietrele to Buta / 7–8 hours
Highest elevation: 7,238 ft. (2,206 m) (without ascent to the summit)
Starting point: Staţiunea Carnic
Trail's end: Buta
Elevation difference: 3,363 vertical ft. (1,025 m)

153. BY CANOE–THE DANUBE DELTA

Probably the most pleasant way to travel in the Danube delta. Mila 23, a quaint little village in the middle of the delta, can be reached from the city of Tulcea by water taxi, for example. Here is where you transfer to canoes/kayaks. Head northeast along old arms of the Danube River or through a branching labyrinth of lakes. In the delta, navigation can be very difficult even with a map, since the smaller channels are often difficult to find or can even be blocked by floating reed islands; a local guide is therefore a must, especially for paddlers with little experience. The trip eventually takes you to Letea. This beautiful fishing village is on a sandbank that the Danube has piled up here over many centuries. A mysterious primeval forest and wild horses can be discovered nearby. If you haven't had enough yet, you can paddle on to the town of Sulina from here.

Information: www.romaniaforall.com/uber-das-donau-delta/

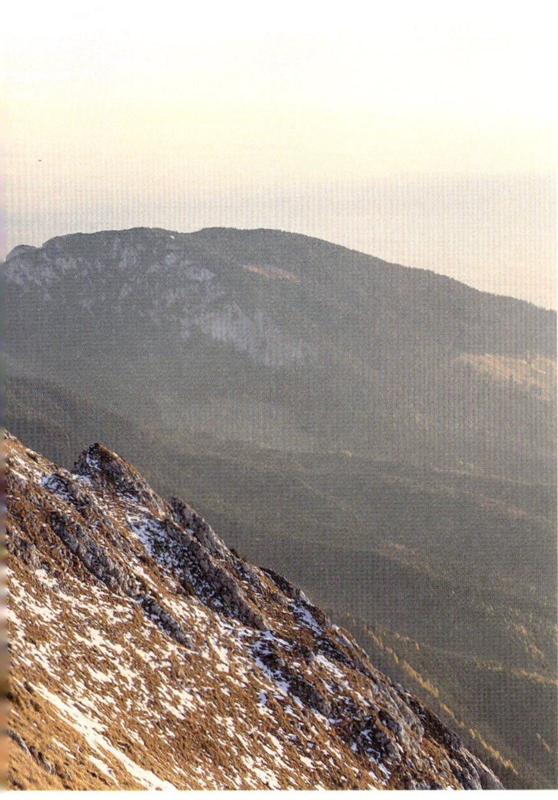

155. THE RILA MOUNTAINS SEVEN-LAKE LOOP TRAIL

A real highlight among Bulgaria's highest mountains—the seven lakes, an increasingly popular destination on the Balkan Peninsula. Rila National Park covers 80,000 hectares. Rila means something like "mountains rich in water." And that they are. From here the trail goes uphill in stages, along the ice-age lakes to the highest one at 8,202 ft. (2,500 m), and then back down again. The lakes, fed by rain and meltwater, are named according to their appearance, such as "the Tear" lake. But there is also a Lake Eye and a Lake Kidney, and they all radiate incredible natural energy and power! And all of them are fantastically beautiful. This moderately difficult hike thus guarantees you some impressive photos. If you want to go higher, you can climb 9,596 ft. high (2,925 m) Musala Peak. Best time to travel is July and August, since there may still be snow in early June.

Time it takes: 4–5 hours
Highest elevation: 8,202 ft. (2,500 m)
Starting point: Rilski Ezera mountain hut
Elevation difference: 1,312 vertical ft. (400 m)
Information: www.bulgariatravel.org

156. RILA MONASTERY: HIKE TO MALJOVIZA HUT

A spiritual center meets trekking at the Rila Monastery on the European E4 long-distance hiking trail. Attention, pilgrims on the Camino de Santiago! Here you make a pilgrimage to Rila Monastery—a UNESCO World Heritage Site, national shrine, and pilgrimage site with a history of more than 1,000 years. Spend the night in the monastery, with its beautiful facades and frescoes and the magnificent library, and feel as if you are in the movie *The Name of the Rose* with Sean Connery and Christian Slater, which was shot here. Tip: visiting in the evening, especially during the summer, guarantees less of a crush—the monastery is open until 9 p.m. From the monastery, set in a valley on the Rilska River, we recommend taking a three-hour hike to the fish lakes or a day trip to the Malyovitza mountain hut. The trail that continues from the hut to 8,953 ft. high (2,729 m) Malyovitza summit is part of the European E4 long-distance hiking trail.

Length / time it takes: 12 mi. (19 km) / 11.5 hours
Highest elevation: 7,205 ft. (2,196 m)
Starting point: Malyovitza mountain hut
Elevation difference: 6,086 vertical ft. (1,855 m)
Information: www.rilamonastery.info, www.rilskimanastir.org

154. BY BIKE—THE DANUBE DELTA

The Danube delta is barely accessible by car, motor boats are loud and far too fast, and so a bike tour is ideal. One longer loop trip starts in Tulcea, which can be considered the gateway to the delta. Along village streets and through small fishing settlements, it follows the northern Chilia distributary channel via Pardina to the district of Chilia Veche and finally leads, at times by boat, to Sulina, on the middle arm of the Danube, a dreamy settlement with an eventful history. A lonely, stony road takes the cyclist farther south to the city of Sfântu Gheorghe, on the wildest arm of the Danube. The route then becomes sandy on the way to Gura Portiței; this small village lies on a spectacular, narrow land bridge between Lake Razim and the Black Sea. Bird watching and wild camping round off this delta adventure, which finally takes you back to the starting point via Jurilovca.

Length / time it takes: about 217 mi. (350 km) on a bike / 9–10 days
Starting point: Tulcea
Information: www.romaniaforall.com/uber-das-donaudelta/ 93

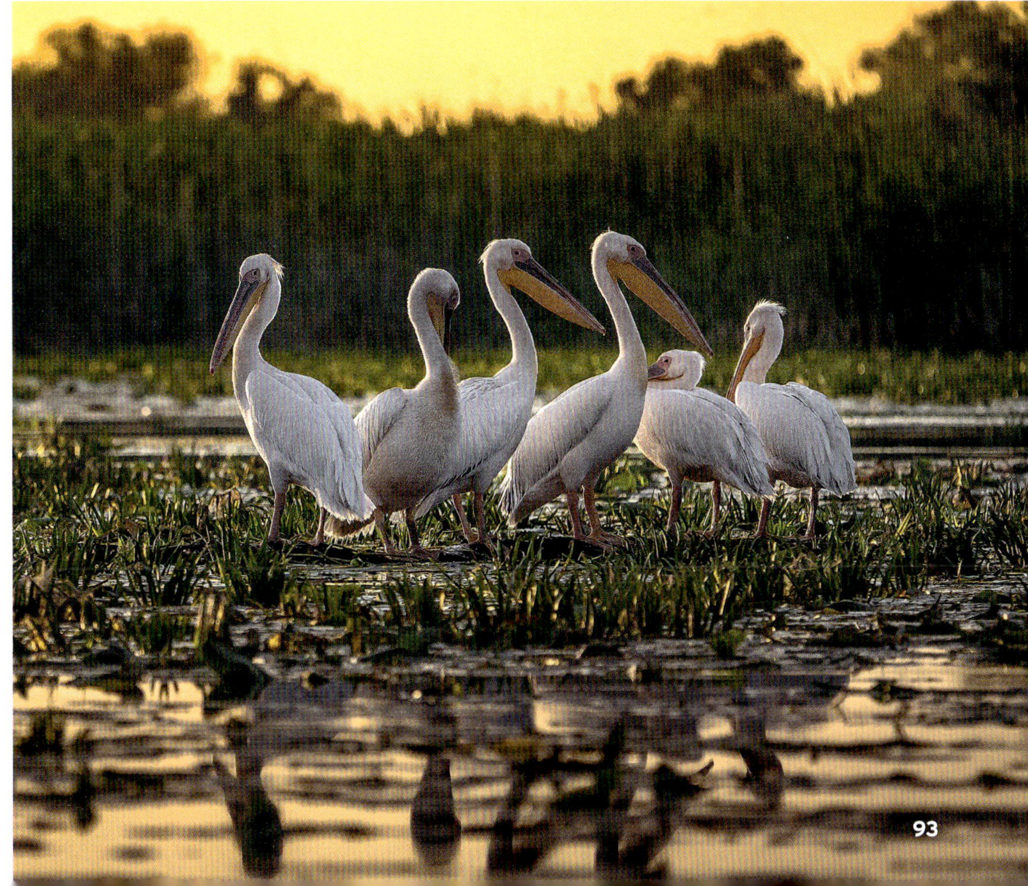

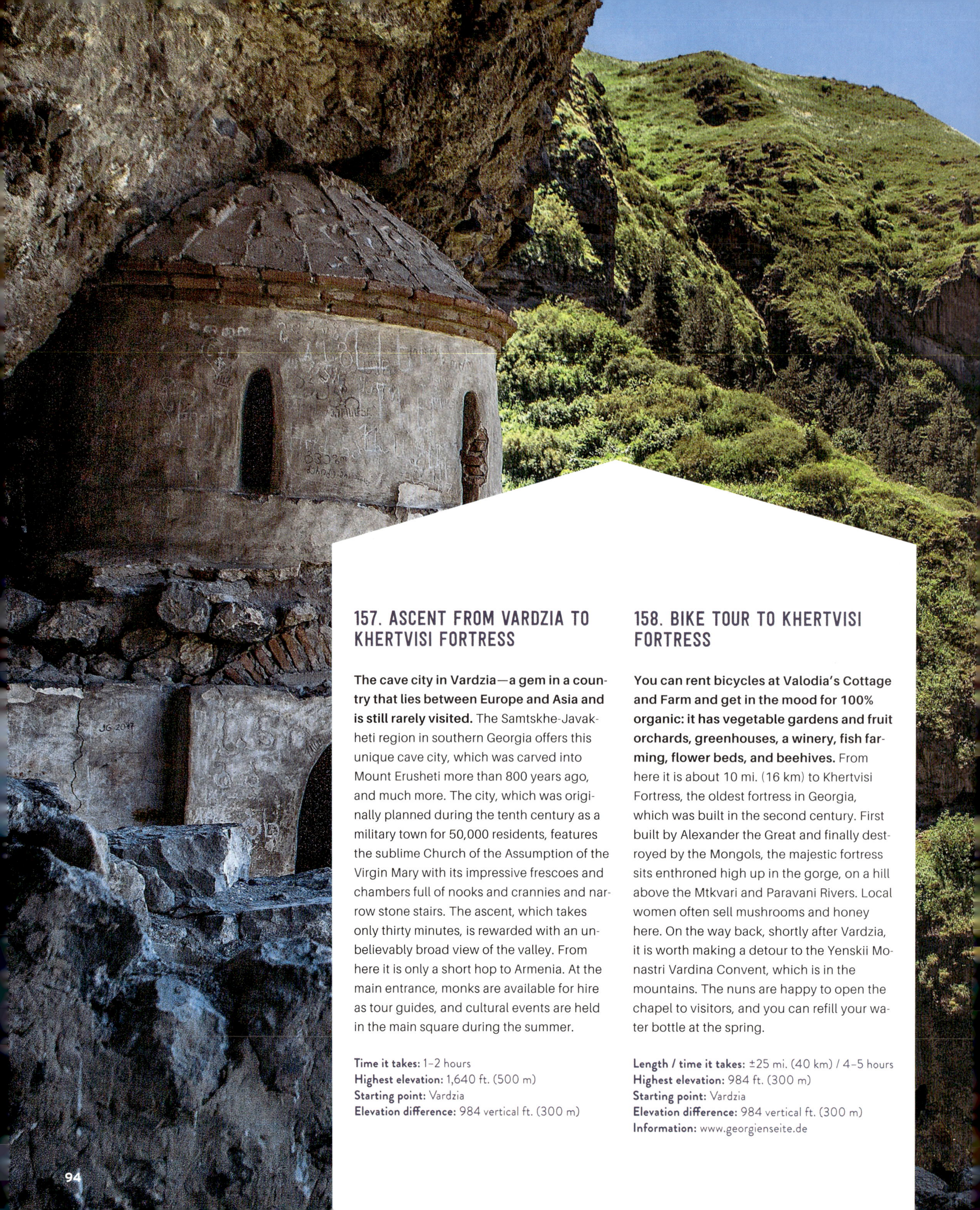

157. ASCENT FROM VARDZIA TO KHERTVISI FORTRESS

The cave city in Vardzia—a gem in a country that lies between Europe and Asia and is still rarely visited. The Samtskhe-Javakheti region in southern Georgia offers this unique cave city, which was carved into Mount Erusheti more than 800 years ago, and much more. The city, which was originally planned during the tenth century as a military town for 50,000 residents, features the sublime Church of the Assumption of the Virgin Mary with its impressive frescoes and chambers full of nooks and crannies and narrow stone stairs. The ascent, which takes only thirty minutes, is rewarded with an unbelievably broad view of the valley. From here it is only a short hop to Armenia. At the main entrance, monks are available for hire as tour guides, and cultural events are held in the main square during the summer.

Time it takes: 1–2 hours
Highest elevation: 1,640 ft. (500 m)
Starting point: Vardzia
Elevation difference: 984 vertical ft. (300 m)

158. BIKE TOUR TO KHERTVISI FORTRESS

You can rent bicycles at Valodia's Cottage and Farm and get in the mood for 100% organic: it has vegetable gardens and fruit orchards, greenhouses, a winery, fish farming, flower beds, and beehives. From here it is about 10 mi. (16 km) to Khertvisi Fortress, the oldest fortress in Georgia, which was built in the second century. First built by Alexander the Great and finally destroyed by the Mongols, the majestic fortress sits enthroned high up in the gorge, on a hill above the Mtkvari and Paravani Rivers. Local women often sell mushrooms and honey here. On the way back, shortly after Vardzia, it is worth making a detour to the Yenskii Monastri Vardina Convent, which is in the mountains. The nuns are happy to open the chapel to visitors, and you can refill your water bottle at the spring.

Length / time it takes: ±25 mi. (40 km) / 4–5 hours
Highest elevation: 984 ft. (300 m)
Starting point: Vardzia
Elevation difference: 984 vertical ft. (300 m)
Information: www.georgienseite.de

159. THE AZAT GORGE HIKE

Nestled in the wooded mountain slopes of the Azat Gorge, the Geghard rock monastery is one of the most beautiful monasteries in Armenia. There are some impressive churches along the way there; some of them are carved into the volcanic rocks at lofty heights and were built in the twelfth and thirteenth centuries. When descending into the valley, the hiking trail meets the Azat River, where you can admire the tallest basalt columns in Armenia. Above is the village of Garni, the former seat of the Armenian kings, with its Hellenistic-period pagan Temple of the Sun, dating from the first century.

Time it takes: 1 day
Elevation difference: 820 vertical ft. (250 m)
Starting point: Yerevan

160. HIKING THROUGH JERMUK GORGE

On the Arpa River, in a small forest area near Jermuk, there is a quiet, little-visited monastery surrounded by a beautiful natural landscape. Hikers reach Gndevank Monastery, which dates from the tenth century, through the grandiose gorge that the Arpa has dug through multiple levels of basalt columns. The defensive walls of the well-preserved and newly renovated core of the monastery rise magnificently from the river landscape. The hike, including entering Jermuk, can be accomplished in three to four leisurely hours.

Length / time it takes: 3–4 hours
Elevation difference: 1,640 vertical ft. (500 m)
Starting point: Jermuk

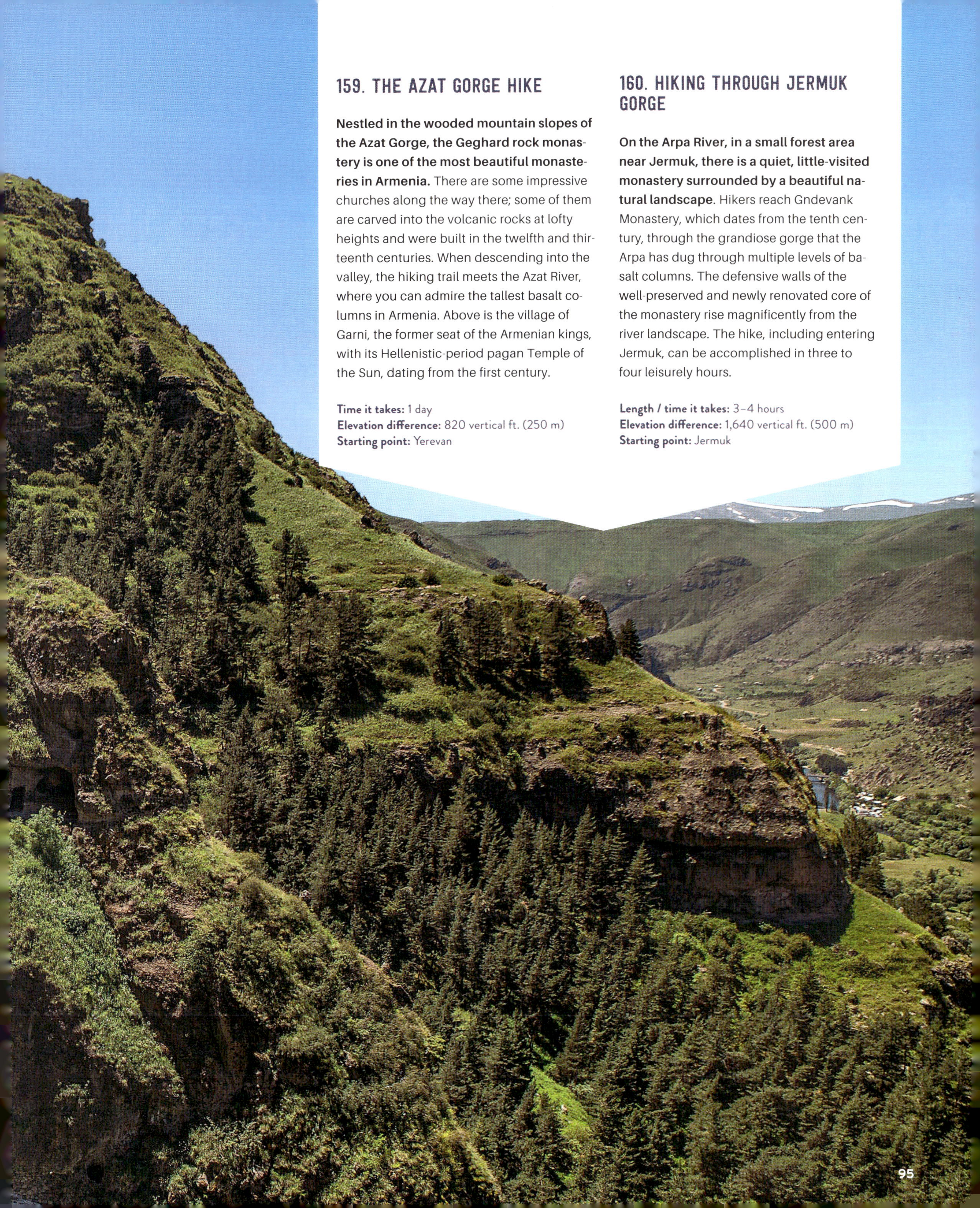

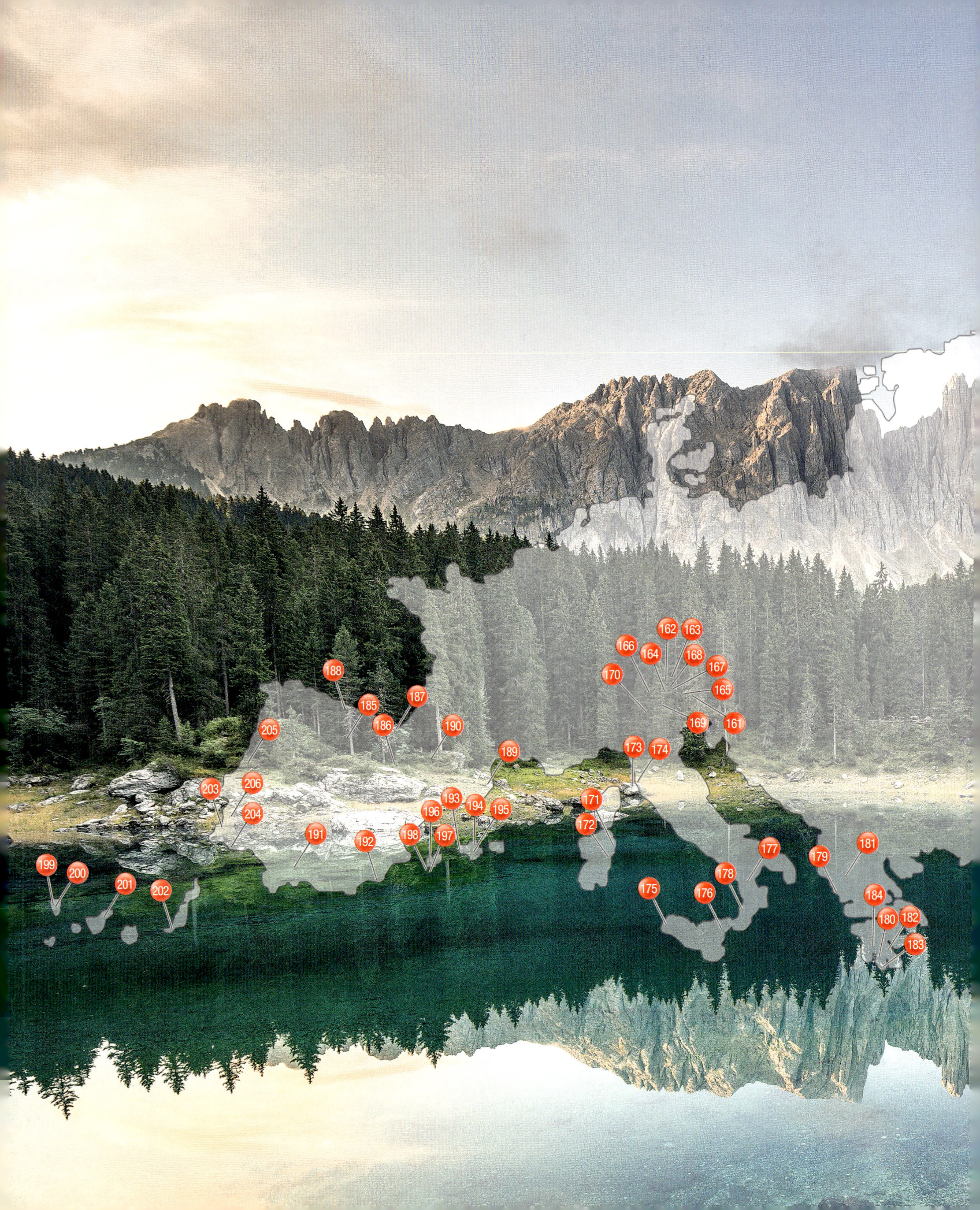
162
163
166
164
168
167
170
165
188
187
185
186
190
205
169
161
173
174
189
206
203
193
171
204
196
194
195
191
198
197
192
172
177
181
179
199
200
178
201
202
175
176
184
182
180
183

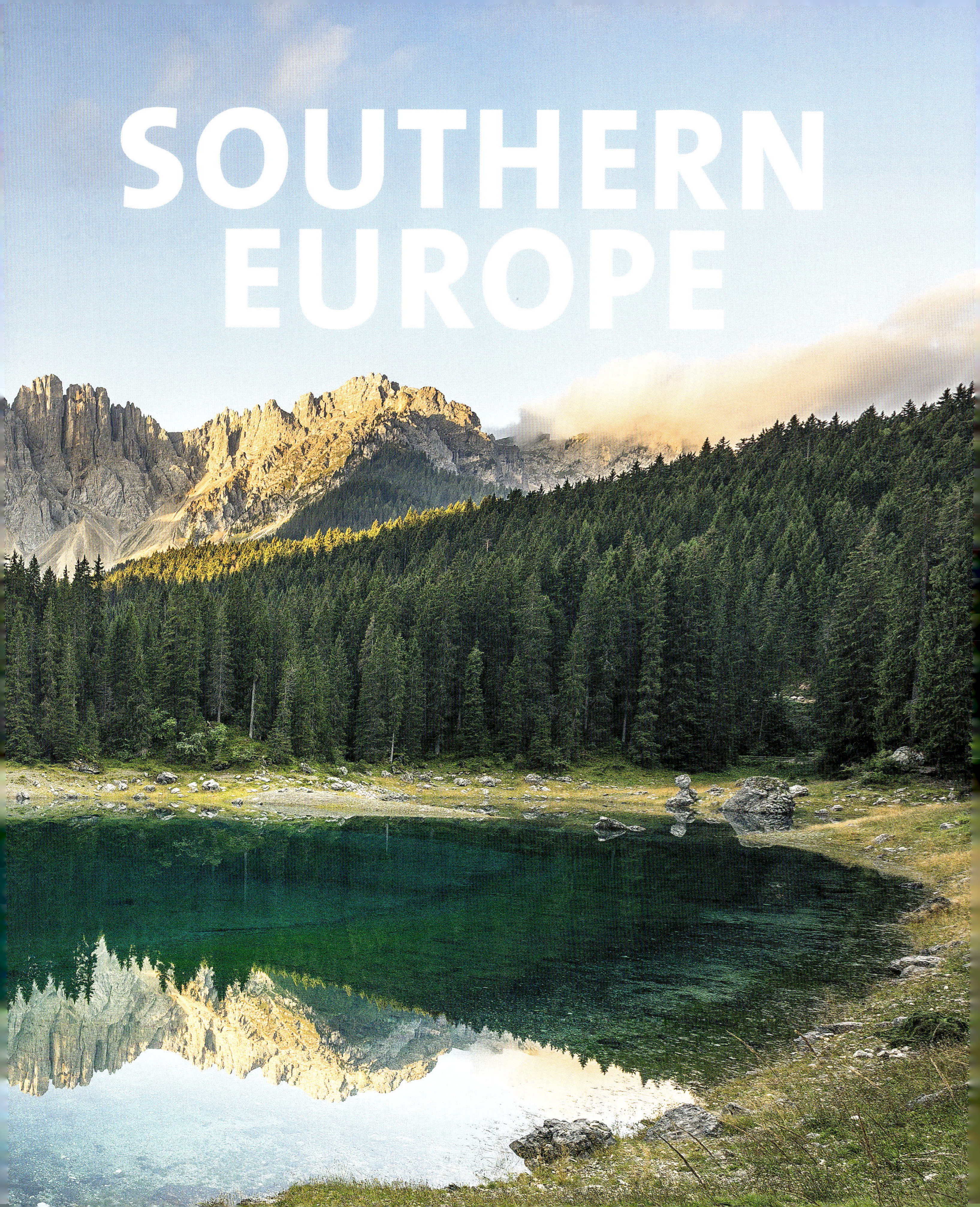

SOUTHERN EUROPE

161. HIKING ON THE ALPINI TRAIL

The Alpini Trail is an adventurous high-altitude trail running along vertical rock faces. The steep, well-secured trail runs along World War I front lines and is divided into two sections. The western part is reached from Zsigmondy mountain hut. It goes through rocky terrain to the beginning of the "Cengia della Salvezza" ledge, which runs across vertical cliffs, then the trail continues through a chimney-like gorge and over exposed ledges until finally the easier part ends. You can make your way back via the Elferscharte wind gap; experienced hikers continue on their way through the eastern section, which is dangerous or even impassable in bad weather. It crosses the northeast flank of the Elfer peak almost horizontally and climbs vertically on via ferrata ladders and rungs. The trail leads very steeply downhill via the Sentinellascharte wind gap.

Length / time it takes: 3 mi. (5 km) / 4 hours
Highest elevation: 8,694 ft. (2,650 m)
Starting point: Zsigmondy mountain hut
Requirements: a good head for heights; sure-footed for the eastern section

162. HIKING ACROSS SCHLERN MOUNTAIN AND THE ALPE DI SIUSI (SEISER ALM)

From the Hotel Bellavista, the route runs over the Alpe di Siusi, the largest high pasture in Europe, over wide trails up to 8,409 ft. high (2,563 m) Schlern Mountain. This is considered a landmark of South Tyrol, not least because of its striking appearance with its jagged Santner and Euringer peaks. From the large-staffed Schlern mountain huts, which were opened in 1885 or 1903, the path leads to the highest elevation, which bears the quaint name Monte Petz. The trail that continues to the Burgstall (8,251 ft. [2,515 m]), which forms the northern edge of the mountain and offers a wonderful view down into the Eisack River valley, is highly recommended. If you like, you can start the descent to Tierser Alpl and then on to the Rosengarten, or you can return to the Alpe di Siusi.

Time it takes: 7 hours
Highest elevation: 8,409 ft. (2,563 m)
Starting point: Hotel Bellavista
Information: www.bellavistaseiseralm.de

163. HIKE ON SECEDA MOUNTAIN

The 2.8 mi. long (4.5 km) cable car ride from Ortisei in Val Gardena surmounts almost 3,281 ft. (1,000 m) in elevation on its way up Seceda Mountain. Alternatively, you can tackle the ascent of about three hours on trail 2A, through steep forest slopes and mountain meadows. From the summit of Seceda there is a fantastic view of the Val Gardena mountain world. While Seceda Mountain gives the impression of an evenly rising Alpine pasture from the southeast, its western face especially falls off in steep and broken rock walls. The route runs through wide pastures on the steep trail 1, going down about 1,575 vertical ft. (480 m) to the Regensburger mountain hut, which is on beautiful Alpine slopes below the peaks of the Geisler mountain group. The hut is a good starting point for many hikes and rock climbs in the Geisler-Puez group. It continues on broad trails down the valley to St. Christina, from where it is not far to the starting point in Ortisei (St. Ulrich).

Time it takes: 7 hours
Highest elevation: 8,264 ft. (2,519 m)
Starting point: Ortisei (St. Ulrich) in Val Gardena

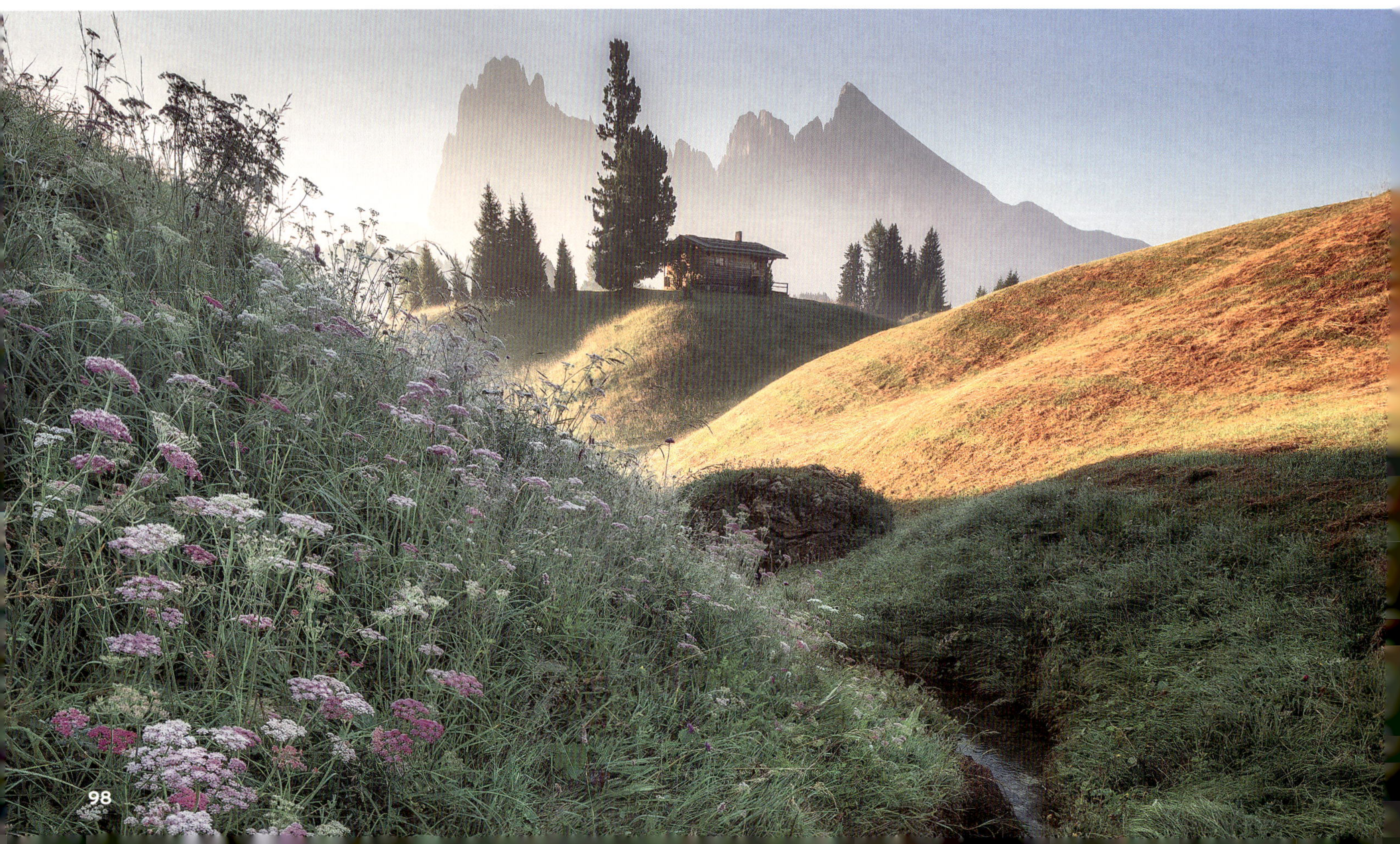

164. CROSSING LATEMAR MOUNTAIN

The Latemar is a lonely and little-developed mountain massif in the Dolomites. On the two-day trip across it, there is only one small-staffed overnight accommodation, the Rifugio Torre di Pisa mountain hut at the summit of Cima Valbona Mountain. It can be reached from Obereggen via trail 21A. Near the mountain hut you can see the eponymous rock formation, which is around 66 ft. (20 m) high and recalls the Leaning Tower of Pisa. Following trail 516, the terrain first drops off a bit before it rises going east to the Great Latemarscharte pass. From here the trail continues on a moderately difficult hike—you must be sure-footed—up to the 9,186 ft. high (2,800 m) and panoramic eastern Latemar mountain peak. The steep trail 18 runs through the Little Latemarscharte pass down to the Karersee (Lake Carezza), where such eminent persons as the last Austrian empress, Sisi; Karl May; Agatha Christie; and Sigmund Freud stayed at the Residence Grand Hotel Carezza.

Time it takes: 2 days
Highest elevation: 9,186 ft. (2,800 m)
Starting point: Obereggen

165. TREKKING ON THE THREE PEAKS LOOP TRAIL

Few sights make such an impression on your memory as that of the Three Peaks. A loop hike around these imposing mountain obelisks begins at the Auronzo mountain hut (7,612 ft. [2,320 m]). Trail 105 leads to the Three Peaks mountain hut, from where you can climb the via ferrata to the summit of Paternkofel (Paterno) Mountain. Here there are still tunnels from World War I, when the region was heavily fought over as part of the front line. Continue on trail 101 to the Lavaredo mountain hut, and from there back again to the starting point. On this loop hike you are always eye to eye with the stone triumvirate: the Cima Grande / Grosse Zinne (Big Peak) measures an impressive 8,839 ft. (2,999 m), and the smallest peak is at least 9,373 ft. (2,857 m) high. With their steep north walls, they are the symbol of the Dolomites and have been a UNESCO World Heritage Site since 2009.

Length / time it takes: 6.2 mi. (10 km) / 2.5 hours (without via ferrata ascent of the Paternkofel)
Highest elevation: 1,522 ft. (2,450 m)
Starting point: Auronzo mountain hut (7,612 ft. [2,320 m])
Elevation difference: 280 vertical ft. (450 m)

166. GOLDSEE TRAIL HIKE TO RIFUGIO FORCOLA MOUNTAIN HUT

Like a huge petrified giant, Ortles (Ortler) Mountain rises into the sky to an elevation of 2,426 ft. (3,905 m) in the southern Vinschgau Valley. The Ortles Group, with around a hundred glaciers and many 1,864 ft. (3,000 m) peaks, is a paradise for hikers and climbers. From Trafoi you can take a bus to the Stelvio Pass (Stilfser Joch), the second-highest Alpine pass. After a twenty-minute walk to the Piz da las Trais Linguas (Three-Language Peak), you are able to view the surrounding mountain landscape, including "König Ortler" ("King Ortler"), the highest peak in the eastern Alps. Before South Tyrol became part of Italy, this high mountain region, which was strongly contested during World War I, was situated in the border triangle among Italy, Switzerland, and Austria. Steep trail 20 leads from the Piz da las Trais Linguas, past Goldsee Lake, to the eastern ridge of the Tartscher Kopf; from there it goes through the Fassa (Furkel) Valley and finally on a beautiful forest path back down to Trafoi.

Length / time it takes: 7.7 mi. (12.4 km) / 4.5 hours
Highest elevation: 9,370 ft. (2,856 m)
Starting point: Trafoi/Stelvio Pass

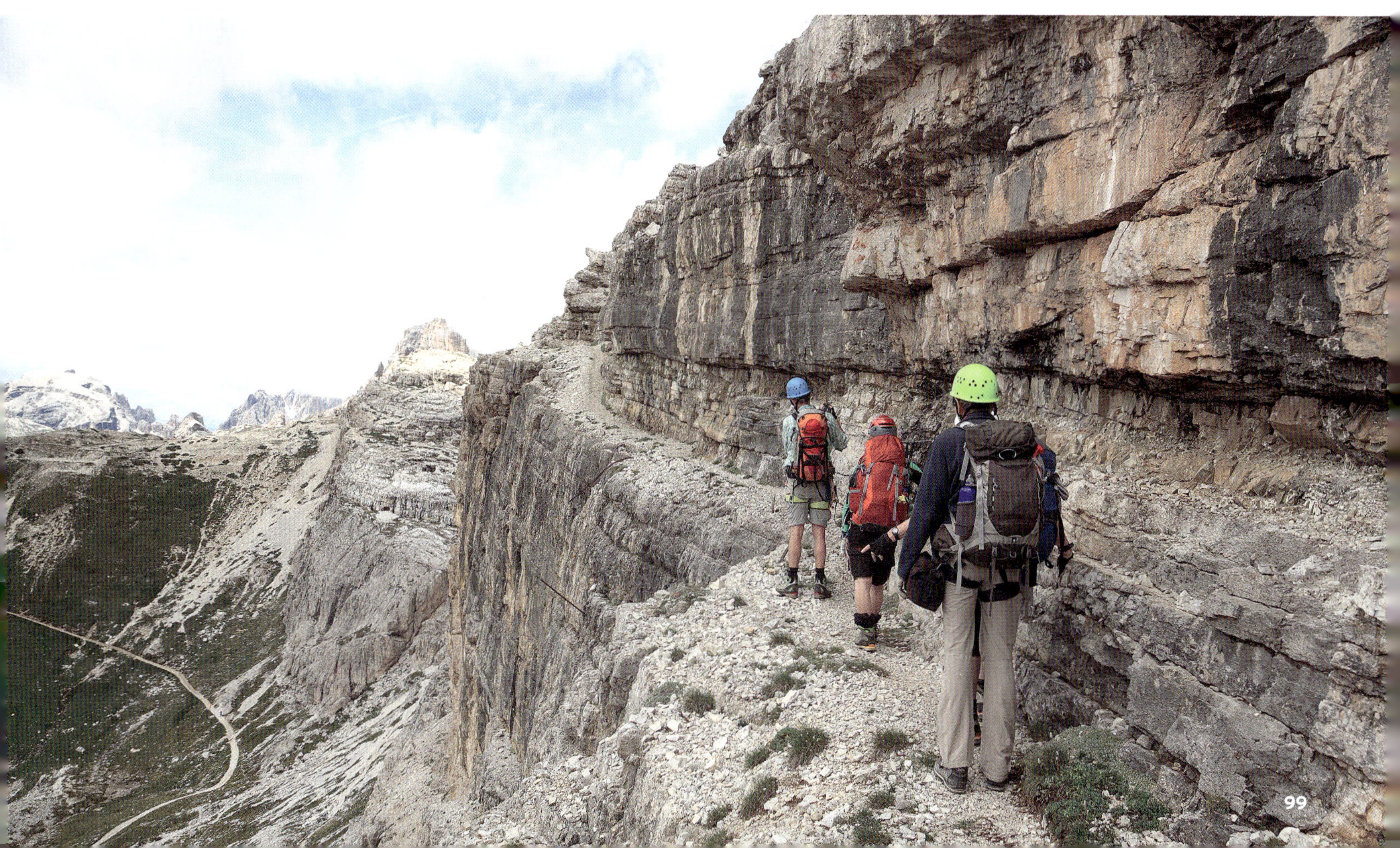

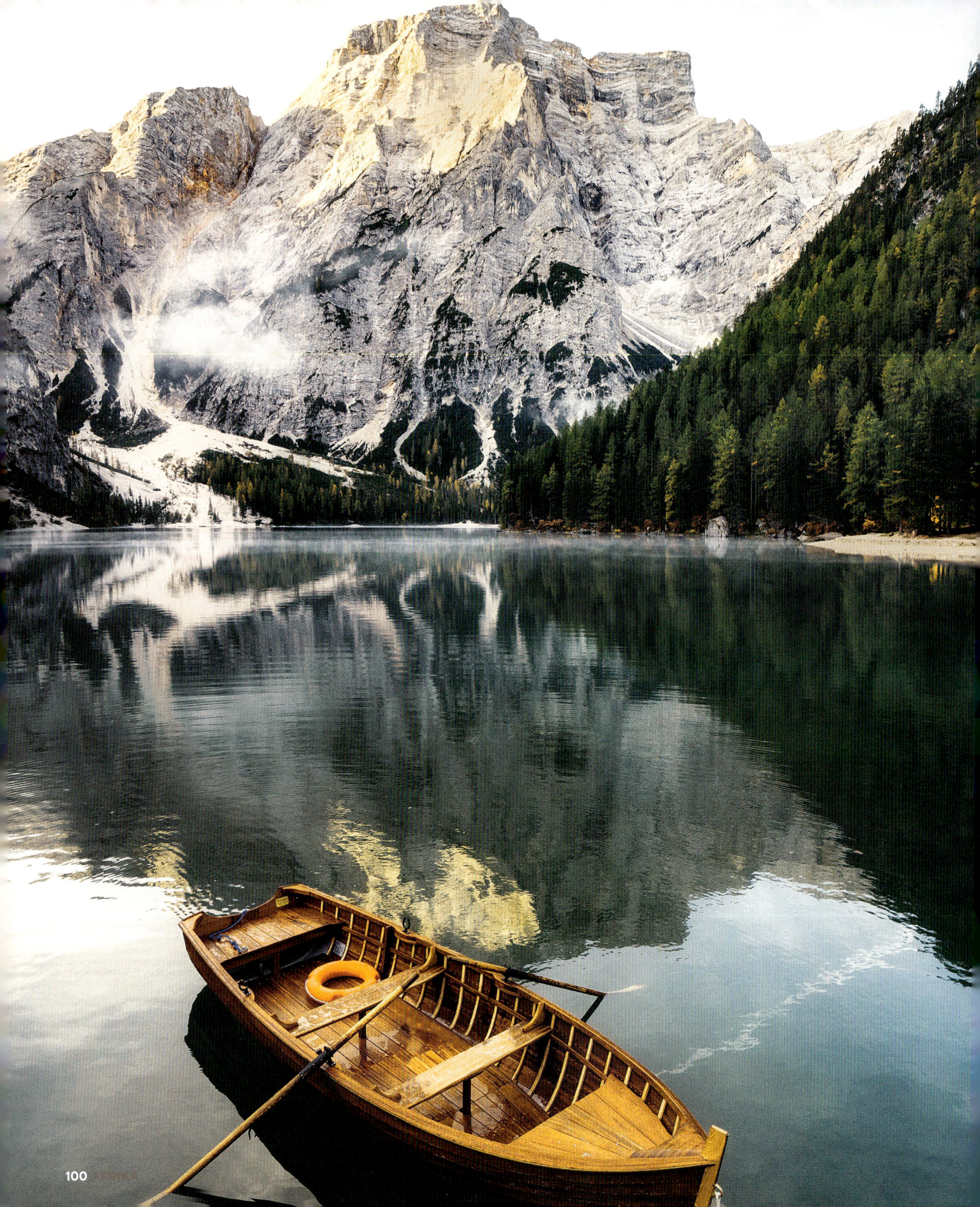

167. ASCENDING PIZ BOÈ MOUNTAIN

Although one of the most difficult via ferratas in the Alps leads to the only 1,864 ft. (3,000 m) summit in the Sella Group massif, there is also an easier way to reach the summit of Piz Boè Mountain. From Sass Pordoi, follow trail 627, which at some point becomes trail 638. From now on, the trail runs steeply uphill and through easy via ferrata passages, some of which are secured with wire cables, until you reach the 10,341 ft. high (3,152 m) summit. From here, there is a fantastic panorama and you can enjoy a unique view of Marmolada Mountain, the queen of the Dolomites. If you don't want to go back down the same way, you can cross the summit to continue the hike. The Dolomite High Route 1 leads through the Val Mezdì valley to the village of Calfosch/Kolfuschg. Alternatively, you can choose High Route 2 to the Gardena Pass.

Highest elevation: 10,341 ft. (3,152 m)
Starting point: Sass Pordoi

168. FREE RIDING IN THE DOLOMITES

After ascending with the Sass Pordoi cable car, the ambitious off-piste skier has two of the free-ride classics in the Dolomites to choose from: the Val Mezdì and the Passo Pordoi. You won't find any ski bars, piste markings, or safety fences here. Instead, narrow gullies with a steep gradient up to 45 degrees await, and not just expert free riders. Even skiers who are other than expert on the classic slope can easily get into the rush of gully skiing here. For inexperienced skiers, it is advisable to hire a professional mountain guide who can also provide the necessary avalanche safety equipment. This winter adventure should be tackled only in good conditions; in icy conditions, a fall in the gullies can be fatal.

Length: Val Mezdì about 2.5 mi. (4 km); Passo Pordoi about 1 mi. (1.5 km)
Time it takes: with arrival and departure via the Sella Ronda ski circuit: 1 day.
Highest elevation: 9,678 ft. (2,950 m)
Starting point: Pordoi Pass, Saas Pordoi cable car, Canazei
Elevation difference: Val Mezdì, 3,858 vertical ft. (1,176 m); Pordoi Pass, 2,297 vertical ft. (700 m)
Information: www.bergfuehrerdolomiten.it

169. HIKING ACROSS THE TRENTINO PLATEAUS

Fortresses, signal stations, command posts: there are few Alpine regions where the remains of World War I have such a presence as they do on the Folgaria, Luserna, and Lavarone plateaus. Today the "trail of fortresses" leads through this unique landscape with its rich cultural history. There are a number of routes for exploring it: via Fort Lusern and Passo Vezzena to Altopiano di Luserna (17 mi., 1,903 ft. ascent [27 km, 580 m ascent]); from Bertoldi via Passo Cost, Fort Belvedere, and Carbonare (13 mi., 1,181 ft. ascent [20 km, 360 m ascent]); or the big loop across the Altopiano di Folgaria plateau (about 22 mi., 3,281 ft. ascent [35 km, 1,000 m ascent]).

Length: 62 mi. (100 km)
Highest elevation: 3,281 ft. (1,000 m)

170. SUP TRIP ACROSS LAKE IDRO

Ponte Caffaro—in German that sounds like a "Kaff," or "hick town": compared to the nearby, usually very overcrowded Lake Garda, Ponte Caffaro, on Lake Idro (Italian: Lago d'Idro), is a rather quiet small town. This picturesque mountain lake is the right place for anyone looking for relaxing tranquility, heavenly nature, and almost complete seclusion. Those who are looking for some relaxation can paddle a stand-up paddleboard from the landing stage to the town of Vesta across wonderfully smooth water during a morning. There they treat themselves to a cappuccino, take a morning swim in the cool lake, and paddle back comfortably. Those who wait long enough will even have a tailwind, which usually starts at noon. The via ferrata along the lake or a round of kitesurfing can serve as a supplemental program!

Length / time it takes: 5 mi. (8 km) / 3 hours
Type of trip: easy
Starting point: Ponte Cafaro, landing stage
Information: www.mks-kite.com

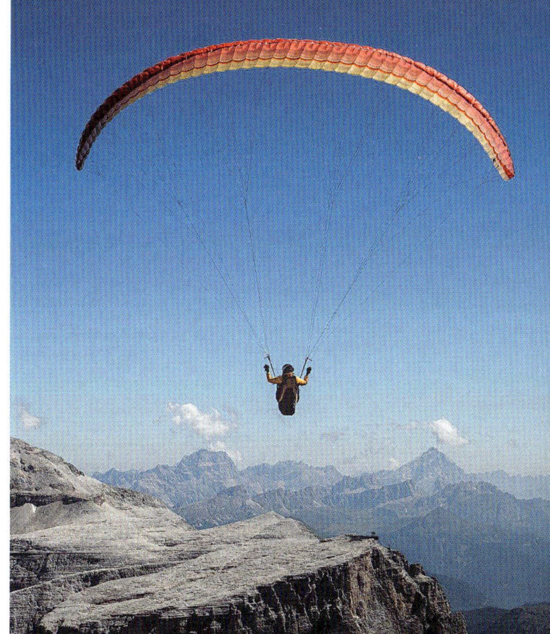

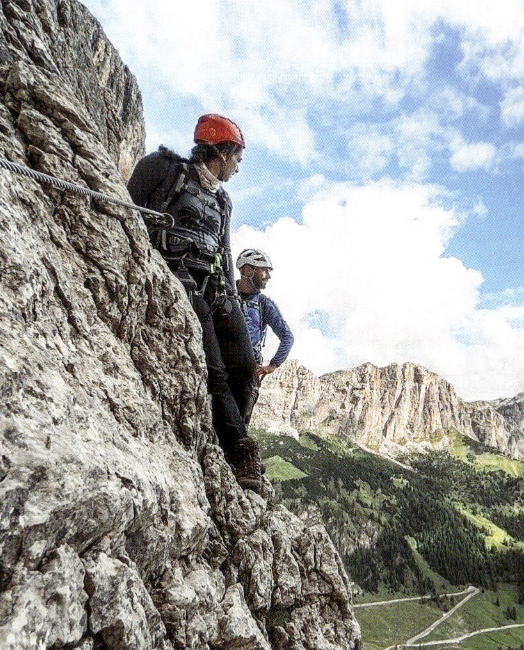

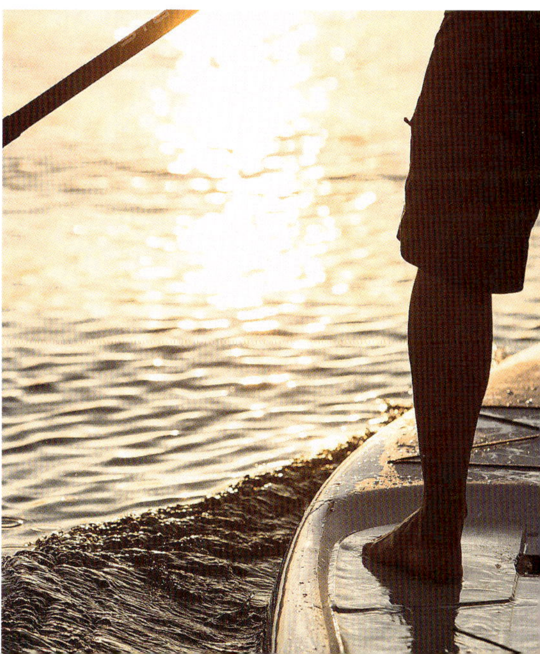

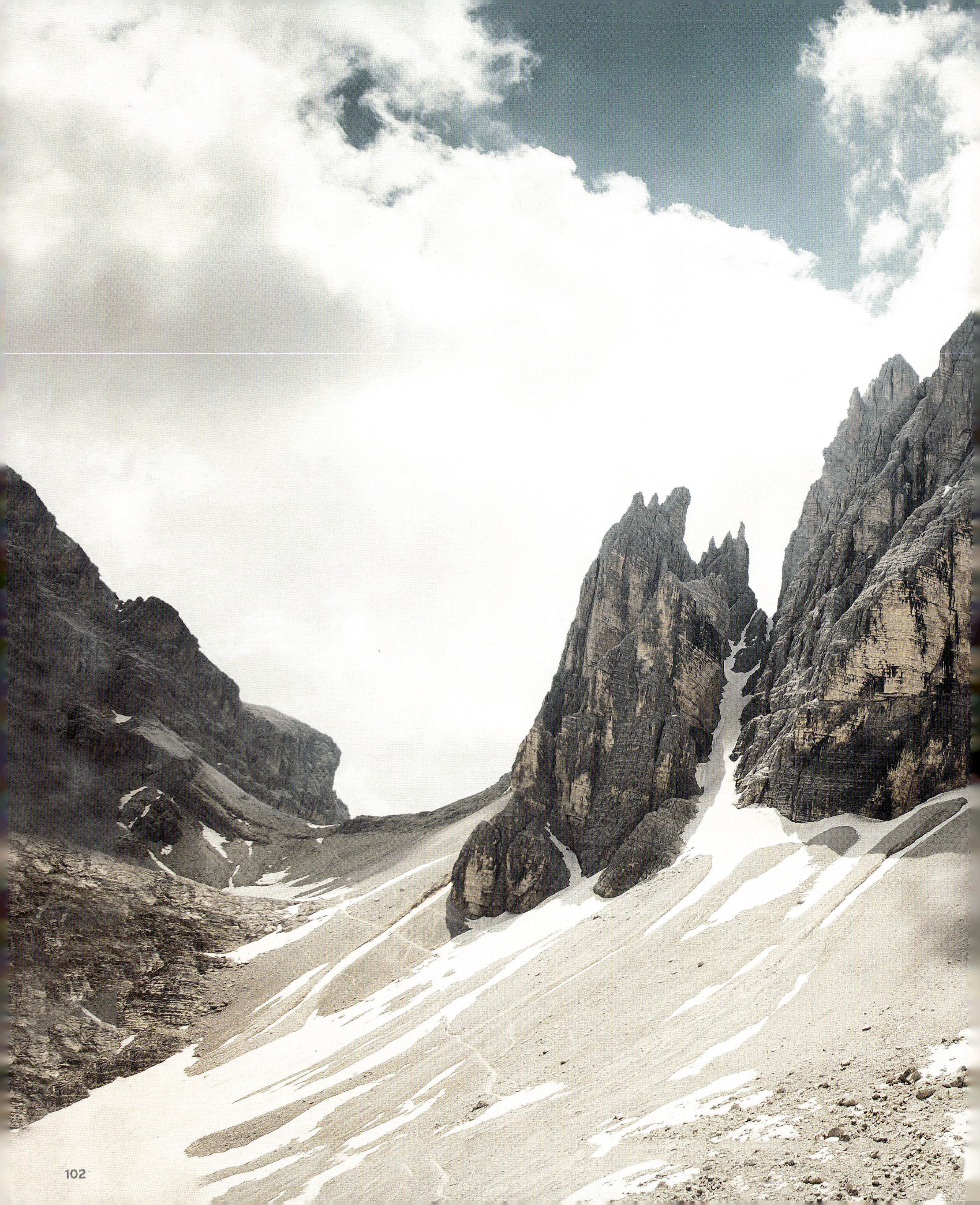

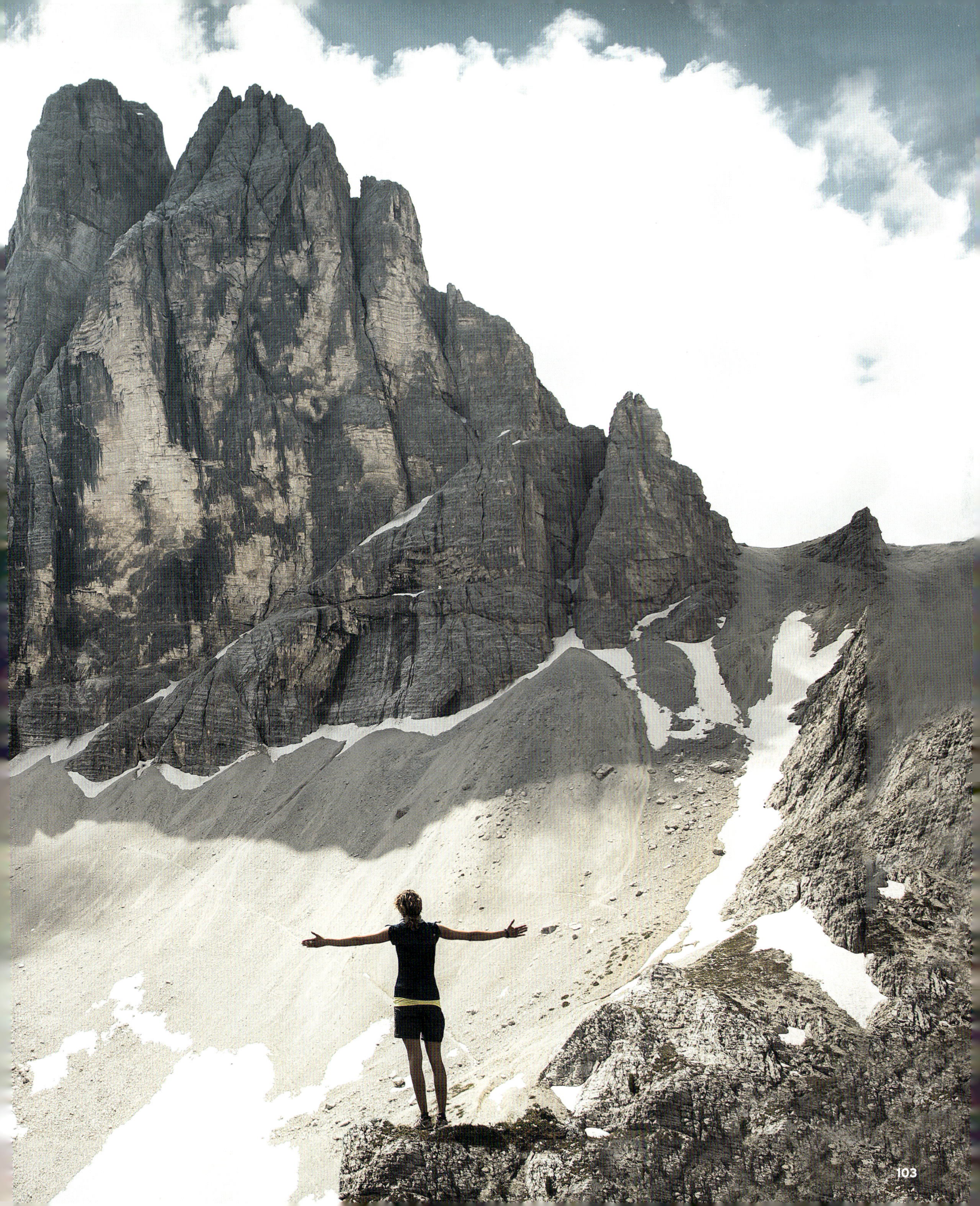

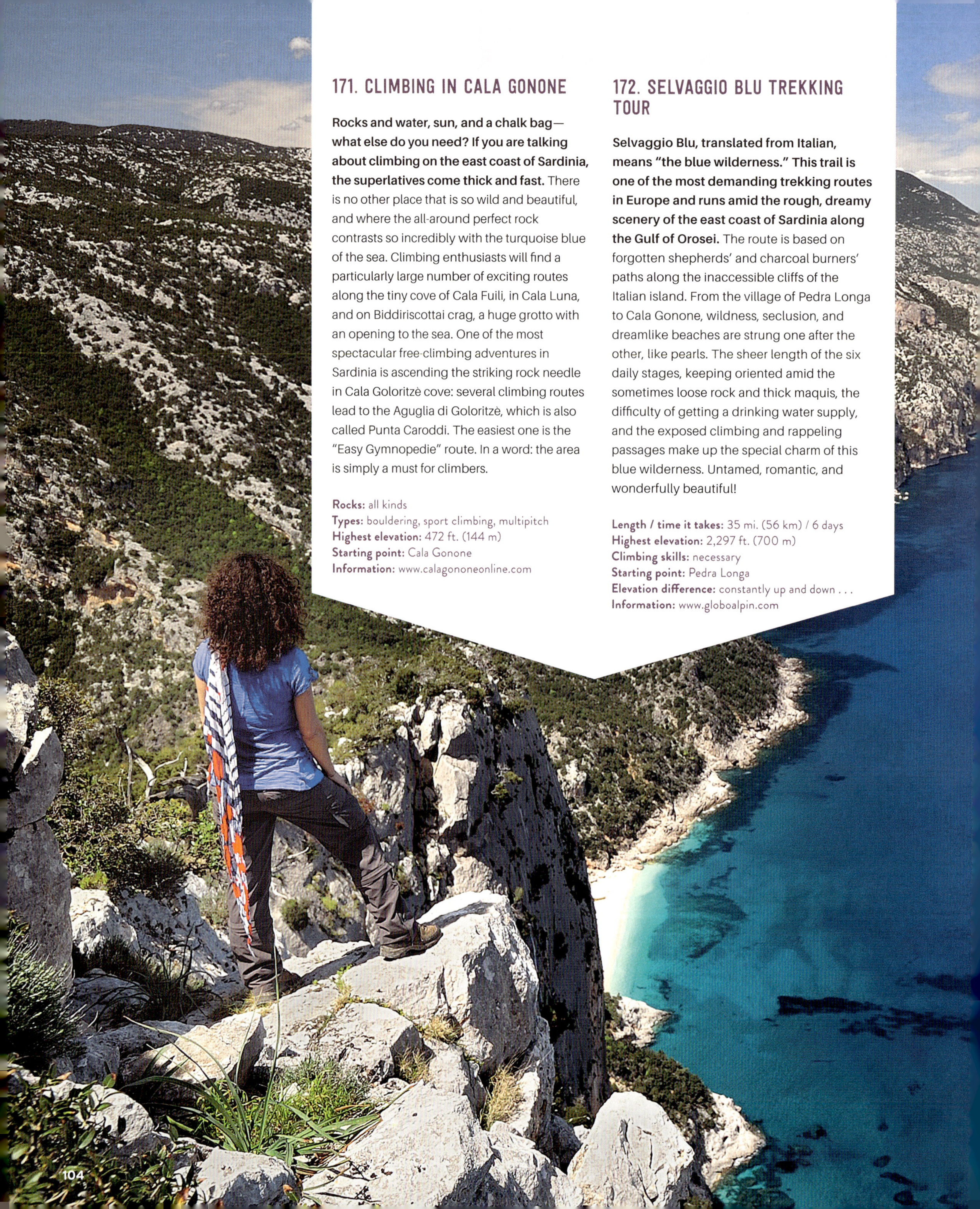

171. CLIMBING IN CALA GONONE

Rocks and water, sun, and a chalk bag—what else do you need? If you are talking about climbing on the east coast of Sardinia, the superlatives come thick and fast. There is no other place that is so wild and beautiful, and where the all-around perfect rock contrasts so incredibly with the turquoise blue of the sea. Climbing enthusiasts will find a particularly large number of exciting routes along the tiny cove of Cala Fuili, in Cala Luna, and on Biddiriscottai crag, a huge grotto with an opening to the sea. One of the most spectacular free-climbing adventures in Sardinia is ascending the striking rock needle in Cala Goloritzè cove: several climbing routes lead to the Aguglia di Goloritzè, which is also called Punta Caroddi. The easiest one is the "Easy Gymnopedie" route. In a word: the area is simply a must for climbers.

Rocks: all kinds
Types: bouldering, sport climbing, multipitch
Highest elevation: 472 ft. (144 m)
Starting point: Cala Gonone
Information: www.calagononeonline.com

172. SELVAGGIO BLU TREKKING TOUR

Selvaggio Blu, translated from Italian, means "the blue wilderness." This trail is one of the most demanding trekking routes in Europe and runs amid the rough, dreamy scenery of the east coast of Sardinia along the Gulf of Orosei. The route is based on forgotten shepherds' and charcoal burners' paths along the inaccessible cliffs of the Italian island. From the village of Pedra Longa to Cala Gonone, wildness, seclusion, and dreamlike beaches are strung one after the other, like pearls. The sheer length of the six daily stages, keeping oriented amid the sometimes loose rock and thick maquis, the difficulty of getting a drinking water supply, and the exposed climbing and rappeling passages make up the special charm of this blue wilderness. Untamed, romantic, and wonderfully beautiful!

Length / time it takes: 35 mi. (56 km) / 6 days
Highest elevation: 2,297 ft. (700 m)
Climbing skills: necessary
Starting point: Pedra Longa
Elevation difference: constantly up and down . . .
Information: www.globoalpin.com

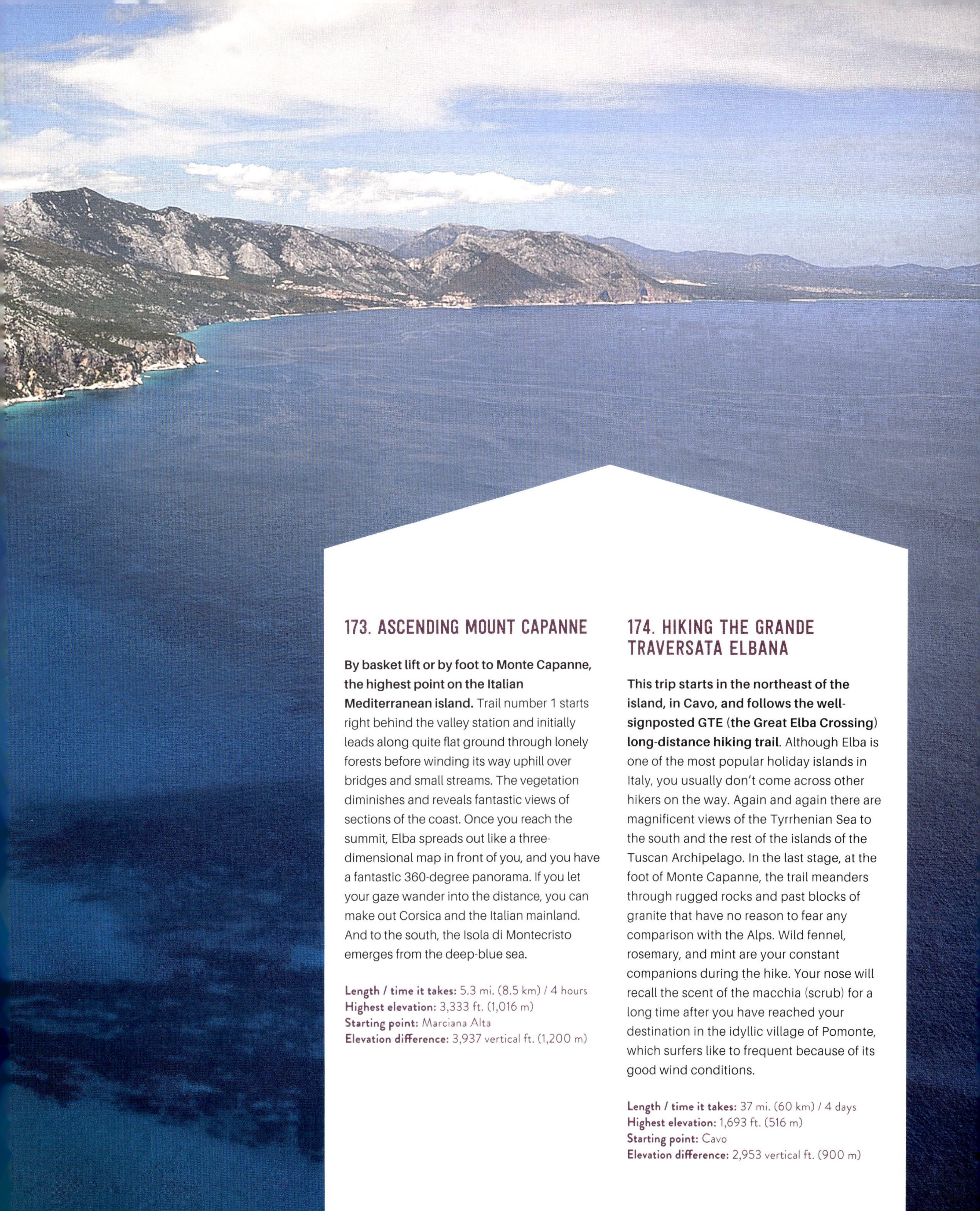

173. ASCENDING MOUNT CAPANNE

By basket lift or by foot to Monte Capanne, the highest point on the Italian Mediterranean island. Trail number 1 starts right behind the valley station and initially leads along quite flat ground through lonely forests before winding its way uphill over bridges and small streams. The vegetation diminishes and reveals fantastic views of sections of the coast. Once you reach the summit, Elba spreads out like a three-dimensional map in front of you, and you have a fantastic 360-degree panorama. If you let your gaze wander into the distance, you can make out Corsica and the Italian mainland. And to the south, the Isola di Montecristo emerges from the deep-blue sea.

Length / time it takes: 5.3 mi. (8.5 km) / 4 hours
Highest elevation: 3,333 ft. (1,016 m)
Starting point: Marciana Alta
Elevation difference: 3,937 vertical ft. (1,200 m)

174. HIKING THE GRANDE TRAVERSATA ELBANA

This trip starts in the northeast of the island, in Cavo, and follows the well-signposted GTE (the Great Elba Crossing) long-distance hiking trail. Although Elba is one of the most popular holiday islands in Italy, you usually don't come across other hikers on the way. Again and again there are magnificent views of the Tyrrhenian Sea to the south and the rest of the islands of the Tuscan Archipelago. In the last stage, at the foot of Monte Capanne, the trail meanders through rugged rocks and past blocks of granite that have no reason to fear any comparison with the Alps. Wild fennel, rosemary, and mint are your constant companions during the hike. Your nose will recall the scent of the macchia (scrub) for a long time after you have reached your destination in the idyllic village of Pomonte, which surfers like to frequent because of its good wind conditions.

Length / time it takes: 37 mi. (60 km) / 4 days
Highest elevation: 1,693 ft. (516 m)
Starting point: Cavo
Elevation difference: 2,953 vertical ft. (900 m)

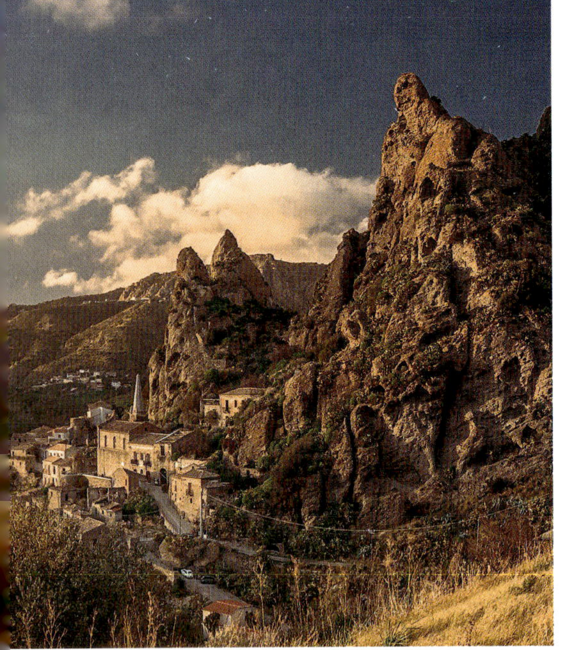

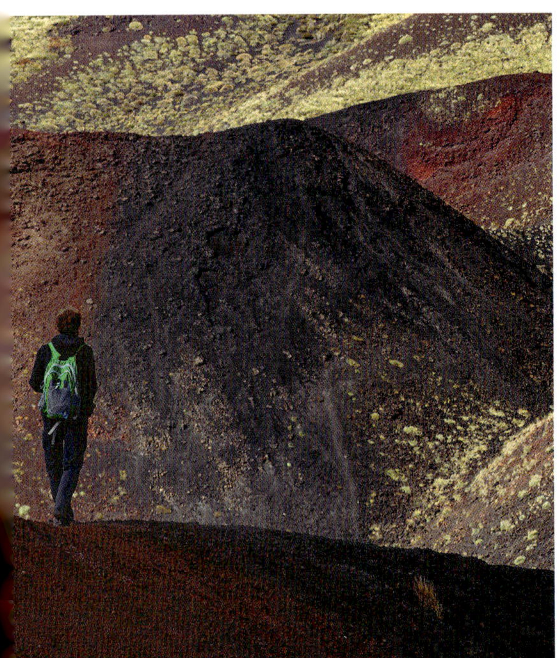

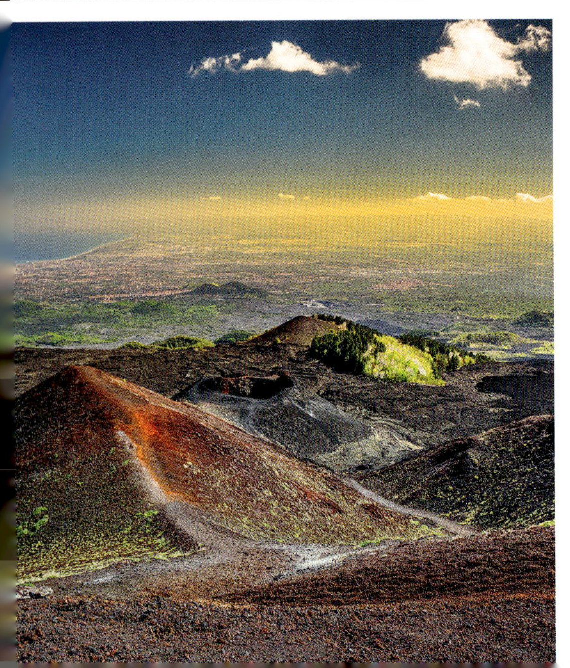

175. KITESURFING IN SICILY

The original windmills, old salt flats, pink flamingos, and former Mafia villages—kitesurfing at this hot spot on the Sicilian west coast is a spectacular experience overall. At first, Europe's largest shallow, flat-water lagoon was particularly popular with beginners and intermediates, but now more and more freestylers are discovering this brilliant area in the Stagnone Lagoon Natural Reserve in the Marsala Lagoon for themselves. You can discover the original backdrop of mountains of salt, windmills, and vines from the water while playing on your kite. The small island in front of the lagoon is quickly circumnavigated. In the hinterland, Mount Erice appears to rise straight out of the sea. Important: arrive at the kite station in time for sundowners to enjoy the magnificent sunsets with a spritzer.

Kitesurfing spot: Lagoon with shallow flat water
Wind: 3–5 Beaufort, as well as thermal, northwest/northeast and southeast winds
Best time of year: March to November
Suitable for: Beginners, advanced, free riding, freestyle
Starting point: Lo Stagnone
Information: www.kite-college.com

176. TREKKING ON MOUNT ETNA

Look into the throat of the earth: at 10,902 ft. (3,323 m) above sea level, on the "roof of the Mediterranean," you can look not only over almost all of Sicily, but also right into the boiling crater of the Etna volcano. A trip to one of the most active volcanoes on Earth is a challenge in several respects: sharp and angular terrain, snowfields, possible volcanic eruptions, escaping gases (sometimes toxic), the sheer elevation, etc. . . . the list is long. After a strenuous final ascent over scree fields, you reach the summit of Etna. The view inside the crater is an exciting natural spectacle. As a person, you feel small in the face of these forces of nature. There are several routes, including ones that shorten the trip by taking a gondola lift and a jeep.

Time it takes: ascent: 3.5 hours
Highest elevation: 10,902 ft. (3,323 m)
Starting point: Refugio Sapienza mountain hut (6,299 ft. [1,920 m])
Elevation difference: 4,692 vertical ft. (1,430 m)
Information: www.vulkankultour.de

177. WHITEWATER RAFTING IN POLLINO NATIONAL PARK

The Dolce Vita—the sweet life: after a delicious cappuccino (the Italians just know how to do it!), head out of the bar and into the wild gorges of the Lao River in Pollino National Park, between the southern Italian regions of Basilicata and Calabria. On the *fiume*—the river—kayakers and rafters immerse themselves in a bizarre world of rocks, light, and water. In the narrow canyon, the "Grande Gola del Lao," water adventurers have to take a good pull on their oars from time to time so that the raft can make its way along the narrow channel between the steep rock faces and the boulders in the river. While it is usually really hot above the gorge, those brave souls who splash about among the rocks below will be rewarded with cool, refreshing water. This is how to lead the Dolce Vita in whitewater!

Length / time it takes: 18 mi. (28 km) / 1 day
Level of difficulty: WW 2–3 (one section is 4), technical
Starting point: Laino Borgo
Best time of year: May to October
Information: www.global-kayak.com

178. KITESURFING IN CALABRIA

"Sun and wind in Kalavrika": in southern Italy, sun-hungry kiters can let it rip as early as spring without freezing. "Kalavrika," as the locals snippily call their Italian province of Calabria (not just because of its geographical proximity to Africa), offers a perfect kiting spot on Hangloose Beach—and Lamezia Terme airport ensures good accessibility, including for short breaks from everyday life. The varied coastline and beautiful mountain bike trips into the mountainous hinterland provide a guaranteed authentic outdoor experience. In summer, gigantic beach parties (until late in the morning) mar the otherwise rather calm atmosphere.

Kitesurfing spot: open sea, no standing area.
Wind: 4–5 Beaufort, mostly thermal winds. Side onshore
Best time of year: spring–summer
Suitable for: advanced, free riding, freestyle, foil
Starting point: Gizzeria Lido
Information: www.hangloosebeach.it

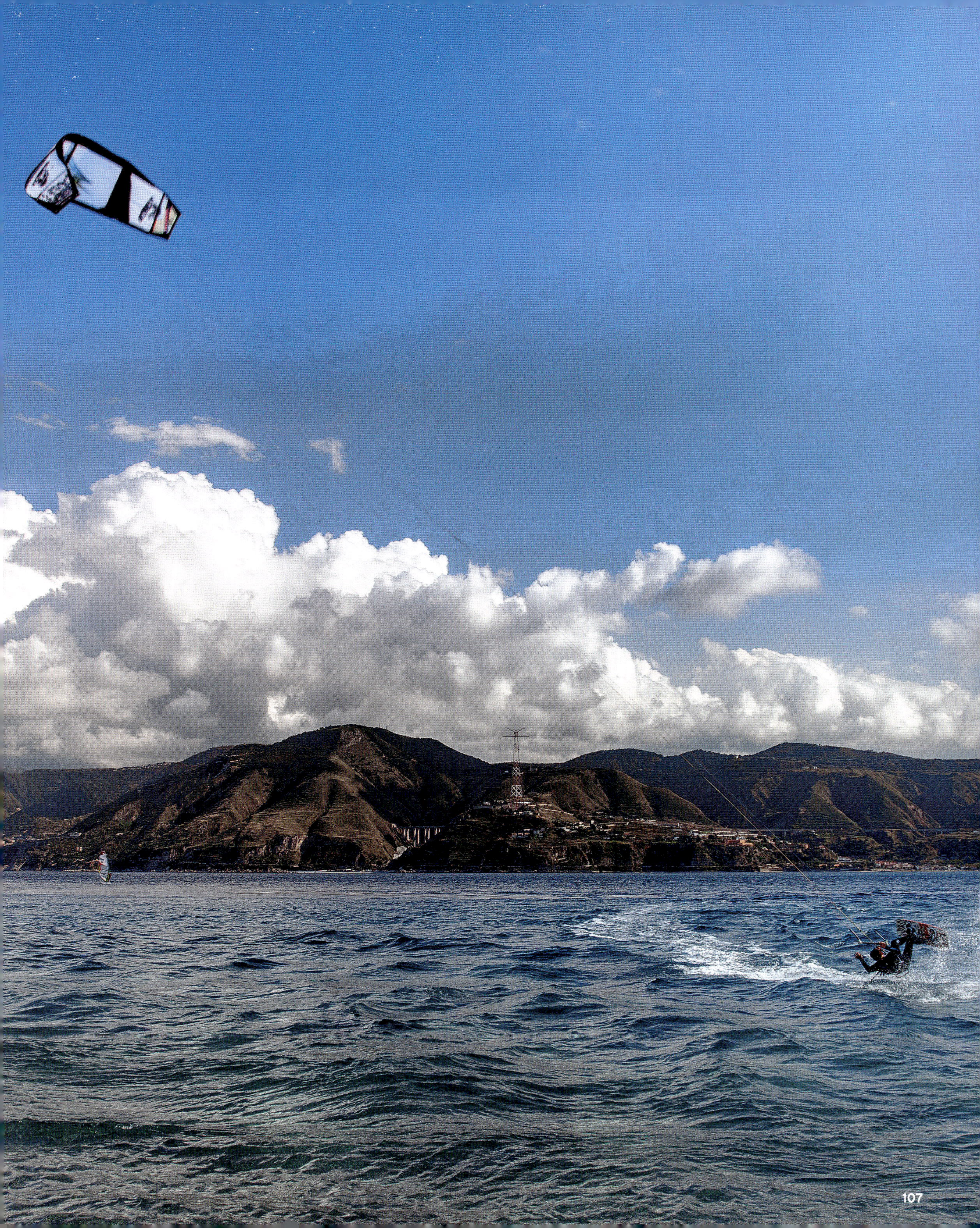

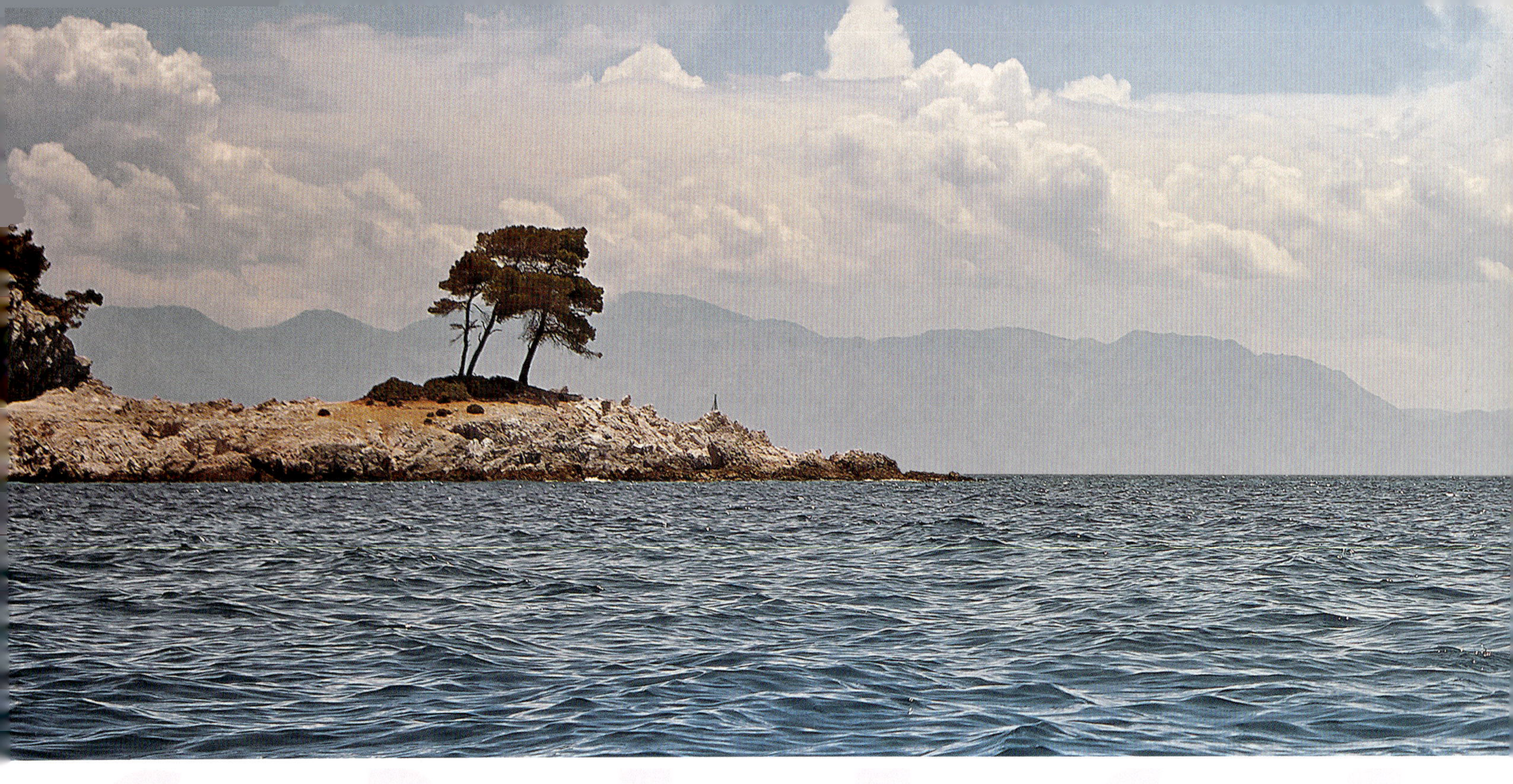

GREECE

179. TREKKING ON THE CORFU TRAIL

A few years ago a project was developed on the island of Corfu that is unique among the Greek islands: a long-distance hiking trail that stretches more than 124 mi. (200 km) across the island, from the flat south to the mountainous north. Yellow signs marked "CT" have been set up along the route to orient hikers. Away from touristy, developed areas, the trail integrates peasants' paths and mule tracks, besides paths across the fields and short sections of asphalt, as well as old paved footpaths and cart tracks. They all lead from one of the original villages to the next, over mountain passes, to remote monasteries and picturesque bays. The farther north you go, the more uphill the trail runs. The many forests and olive groves provide pleasant shade, and the small size of the island ensures constant views far out to sea.

Length / time it takes: 137 mi. (220 km) / 11 days
Highest elevation: 2,972 ft. (906 m)
Starting point: Arkoudillas
Trail's end: Cape Agía Ekateríni
Elevation difference: various, between 492 and 2,132 vertical ft. (150–650 m)
Information: www.corfutrail.gr

180. BY SEA KAYAK ALONG MANI PENINSULA

In a sea kayak along the wild "Mani" to the second-most-southern point in Europe. This paddling trip starts out nicely at Kalamata: we paddle south amid picturesque beaches and dreamy olive groves, past water caves and fishing villages. But little by little, the coast is getting rougher, more rocky and barren: this used to be the domain of pirates, smugglers, and, of course, the wild Maniots in their fortified stone houses. For us kayakers, the adventurous part also begins here, with the circumnavigation of the mighty "Cavo Grosso." On the last day of kayaking we reach our destination, Cape Tainaron. A moving moment for us kayakers, who have paddled along the full length of the Mani coast under our own power—with the certainty that we have come a little closer to Africa, which lies beyond the horizon.

Length / time it takes: 60 nautical mi. (70 mi. [112 km]) / 6 days
Starting point: Kalamata, Peloponnese Peninsula
Level of difficulty: intermediate
Best time of year: April to October
Information: www.global-kayak.com

181. WHITEWATER RAFTING AND KAYAKING IN THE PINDOS MOUNTAINS

Whitewater in the land of the gods: the landscape in the Pindos Mountains doesn't correspond to the classic island idyll of Greece. Instead, here there are rugged mountains and wild gorges. Right in the middle—in the valley of the Arachtos River—is the small area of Plaka: a northern Greek center for whitewater, rafting, and kayaking sports has developed here, which is not surprising if you look at the surroundings. The best way to explore the crystal-clear waters, great canyons, and enchanted mountain villages from here is on the foaming waterways. In addition to the Arachtos, runs on the Kallaritikos and Voidomaitis Rivers are a must for enthusiastic nature and water lovers. The absolute highlight—but only for kayakers: through the narrow rocky gateway on the Acheron River you can paddle straight into Hades, the Greek underworld.

Length / time it takes: from 6.2 mi. (10 km) / up to 7 days
Level of difficulty: WW 2–3
Starting point: Plaka
Best time of year: April to May
Information: www.global-kayak.com

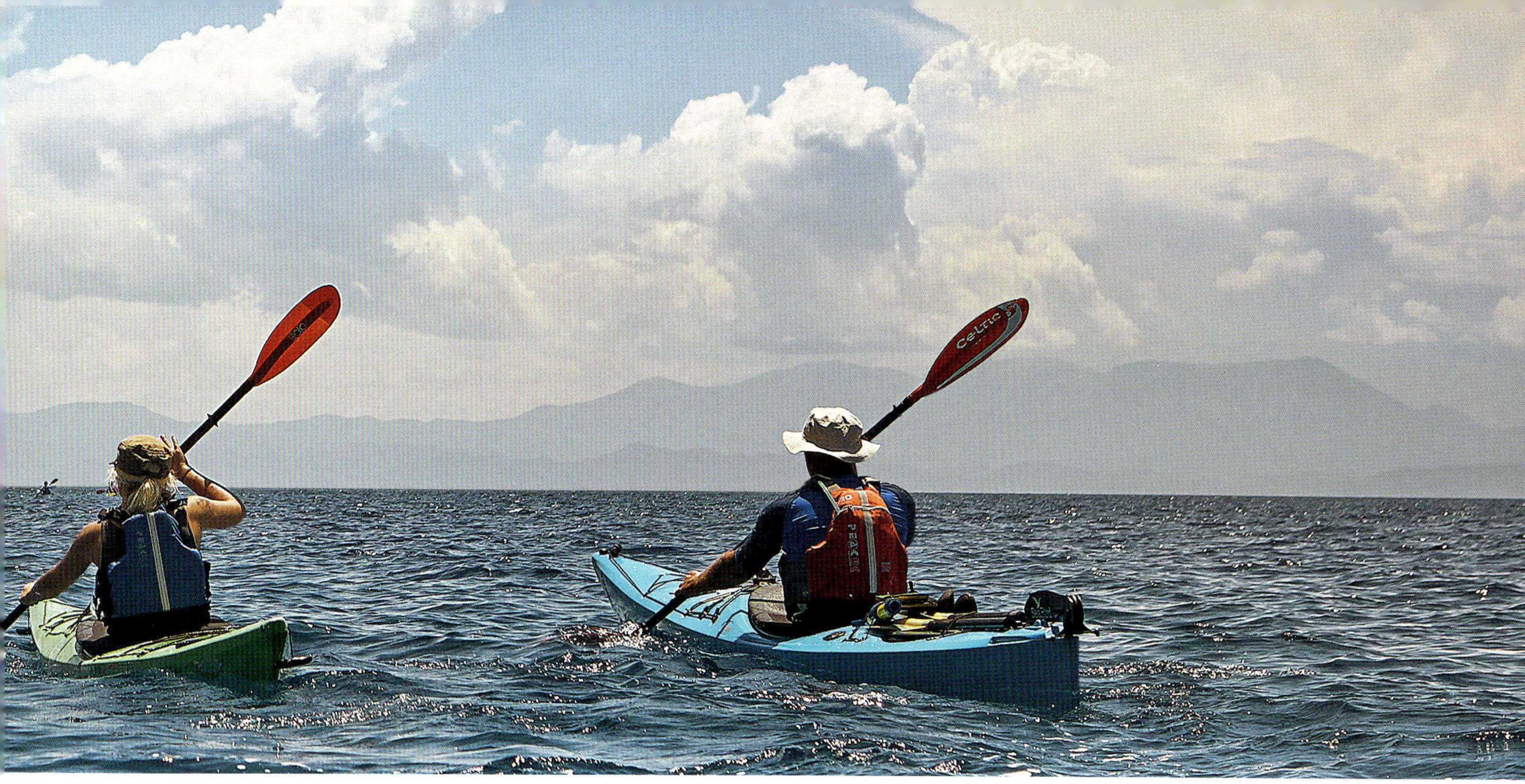

182. CLIMBING IN LEONIDIO

On a journey of discovery in one of the largest, still largely undeveloped climbing areas in Europe: climbing in Leonidio means vertical sports in their most beautiful form: graspable rock—namely, limestone with sintering and stalactites—great routes, lots of sun, and the wonderful deep-blue sea. So far, only a few insiders know about the climbers' El Dorado on the Greek Peloponnese Peninsula—and appreciate it because of the almost inexhaustible potential of the rocks. Leonidio will make a name for itself in the climbing scene: bouldering, sports climbing, multipitch—everything is possible. Many of the routes are well secured, some less so. Starting in May it can get really hot. That is to say, Leonidio is your winter climbing escape from cold central Europe—preferably from October to April!

Highest elevation: 33 to 656 ft. (10 to 200 m)
Starting point: Leonidio, Peloponnese Peninsula
Level of difficulty: 4b–9a (French scale)
Information: www.climbinleonidio.com

183. HIKE TO THE MANI LIGHTHOUSE AT CAPE TAINARON

On foot to the lowest point in Europe: before Cape Tainaron, at a depth of 16,801 ft. (5,121 m), lies the deepest point in the Mediterranean, the Calypso Deep. The nearby entrance into a large cave system is considered a gateway to Hades, the Greek underworld. And where such extremes of nature and mythology await, long-venerated sanctuaries are also not far away: the ancient Spartans built some temples here at Cape Tainaron to honor this mystical place appropriately. Today, you take a stony path to get to the prominent, exposed lighthouse on one of the southernmost headlands in Europe. A short, mystical, and often very windy hike through the rough beauty of nature to the deep-blue sea. It is a good idea to wear windproof clothing.

Length / time it takes: 4.35 mi. (7 km) / 2 hours
Highest elevation: 492 ft. (150 m)
Starting point: Kokkinogeia
Elevation difference: 459 vertical ft. (140 m)
Information: www.insidemani.gr

184. SEA KAYAKING IN MESSINIA

This heavenly sea-kayaking trip along the Messinian coast, in Greek Messinia, is like a bubble bath for the soul. It takes paddlers to Caribbean-like beaches, neat villages, and mighty castles. Quaint taverns and traditional Greek hospitality are of course also part of the package. You float your kayak, paddling casually and relaxed, over the deep-blue Greek sea from Marothopoli to Koroni, from hotel to hotel. This luxury version gently introduces less experienced kayaking enthusiasts and outdoor newbies to the experience of kayaking—nature and the sea. Real old outdoor campaigners take a tent along with them and camp in the wild by the beautiful bays. But beware: the beauty of the Greek coast and the fascination of sea kayaking are potentially addictive!

Length / time it takes: 50 nautical mi. (56 mi. [90 km]) / 6 days
Level of difficulty: easy
Starting point: Marothopoli, southwestern Peloponnese Peninsula
Best time of year: April to October
Information: www.global-kayak.com

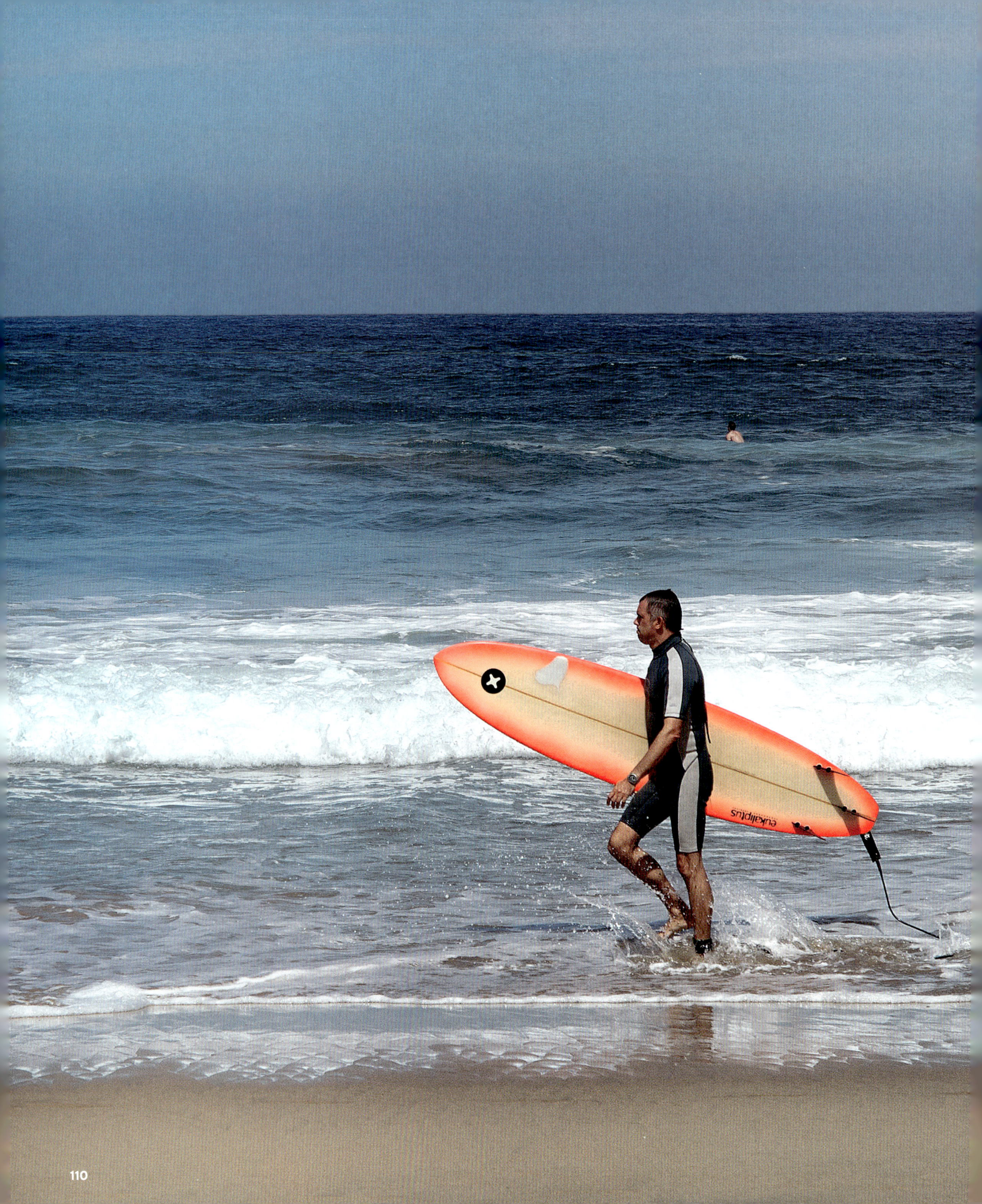
eukaliplus

185. PILGRIMAGE ON THE CAMINO DE SANTIAGO (WAY OF ST. JAMES)

The pathway climbs steeply through the forest. It smells of eucalyptus. Breathing becomes difficult. Your backpack presses on your shoulders. Sunbeams fight their way through the canopy of leaves and conjure up artistic patterns on tree trunks, roots, and ferns. After the end of the ascent, the situation eases. The pathway runs among meadows and cattle pastures. Not much longer, then that will be it for today as you reach the next pilgrims' hostel. More and more people in northern Spain want to experience this authentic sense of pilgrimage, which can be found here in the Galicia region between the towns of Melide and Boente. In total, there are almost 497 mi. (800 km) on the main Camino de Santiago route, from the Pyrenees to the longed-for destination of Santiago de Compost la, where the (reputed) tomb of the apostle James has been venerated since the Middle Ages. Many pilgrims also arrive by themselves via the Santiago de Compostela detour.

Length / time it takes: 497 mi. (800 km) / at least 5 weeks
Starting point: St-Jean-Pied-de-Port, France

186. BY BIKE ALONG A VÍA VERDE (GREEN WAY)

The wide, marked route, shared by cyclists and hikers, extends wonderfully into the distance. Here, on the "Vía Verde de Tarazonica" between Tarazona and Tudela, it is accompanied by the sound of silence. It is a prime example of the network of the Vía Verdes, the "Green Ways." In Spain they have been built on many former railroad lines. The rail lines had either been closed down or, although built, never put into operation. The fact that nature fans enjoy them today is very commendable. So far, there are more than 1,678 mi. (2,700 km) of Vía Verdes trails, of which the "Vía Verde de Tarazonica" accounts for a modest 14 mi. (22 km). The historic quarters of both towns, Tarazona in Aragon and Tudela in Navarra, are ideal for making a pleasant stop before or after—far away from the usual flows of tourists.

Length: 14 mi. (22 km)
Information: www.viasverdes.com

187. SURFING SAN SEBASTIÁN

Spain's best city beach for surfers? In response to this question, many people would choose Playa de la Zurriola in the Basque town of San Sebastián. "The beach is well protected and simply perfect. There are no rocks, just sand. And good waves 340 days a year," says resident surfing champ and instructor Oscar García, emphasizing the almost unbeatable advantages. There are several surfing schools, such as the Zurriola Surf Escola and Bera Bera, that offer courses and equipment rentals. The backdrop is the beautiful and elegant city of San Sebastián, with its promenades, its panoramic hills, the main bay, and a legendary density of pubs in the old town. There you can ultimately replenish the calorie levels that you have depleted on the Bay of Biscay.

Information: zurriolasurfeskola.com, www.beraberasurf.com

188. HIKING ON THE RUTA DEL CARES TRAIL

The wild gorge scenery of the Cares Gorge (Garganta del Cares) in the mountains and Picos de Europa National Park really has it all! A spectacular hiking trail leads right through the middle of this mountain-and-valley scenery: the Ruta del Cares. Here, 7.5 mi. (12 km) of hiking lie between the outskirts of Poncebos and the village of Caín, and sometimes the trail also runs along unprotected precipices toward the gorge. For this reason you should never start a trek in strong winds and rain, because the ground then becomes dangerously slippery and there is a risk of falling rocks. Along the way, the river sends its murmurs up as a greeting from the depths. The path widens, draws itself together, leads over stones, and squeezes along rock faces. Los Callaos is a classic panorama point along the way. Time for some photos and taking a breather.

Length / time it takes: 7.5 mi. (12 km) / 3 hours for a one-way route
Elevation difference: 1,509 vertical ft. (460 m)
Information: www.rutadelcares.org

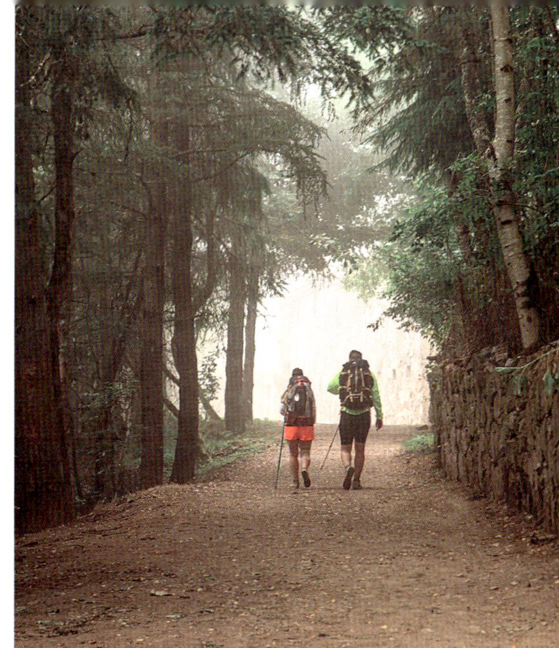

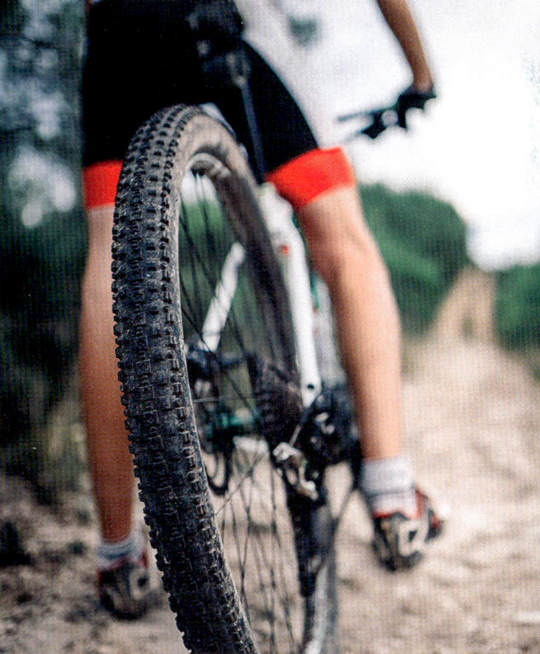

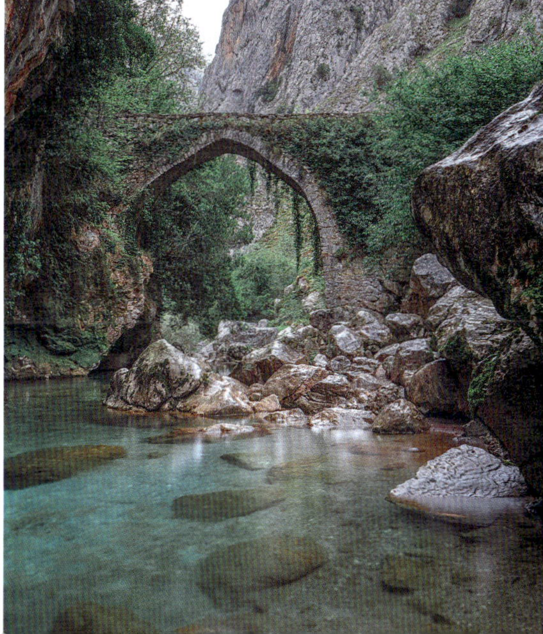

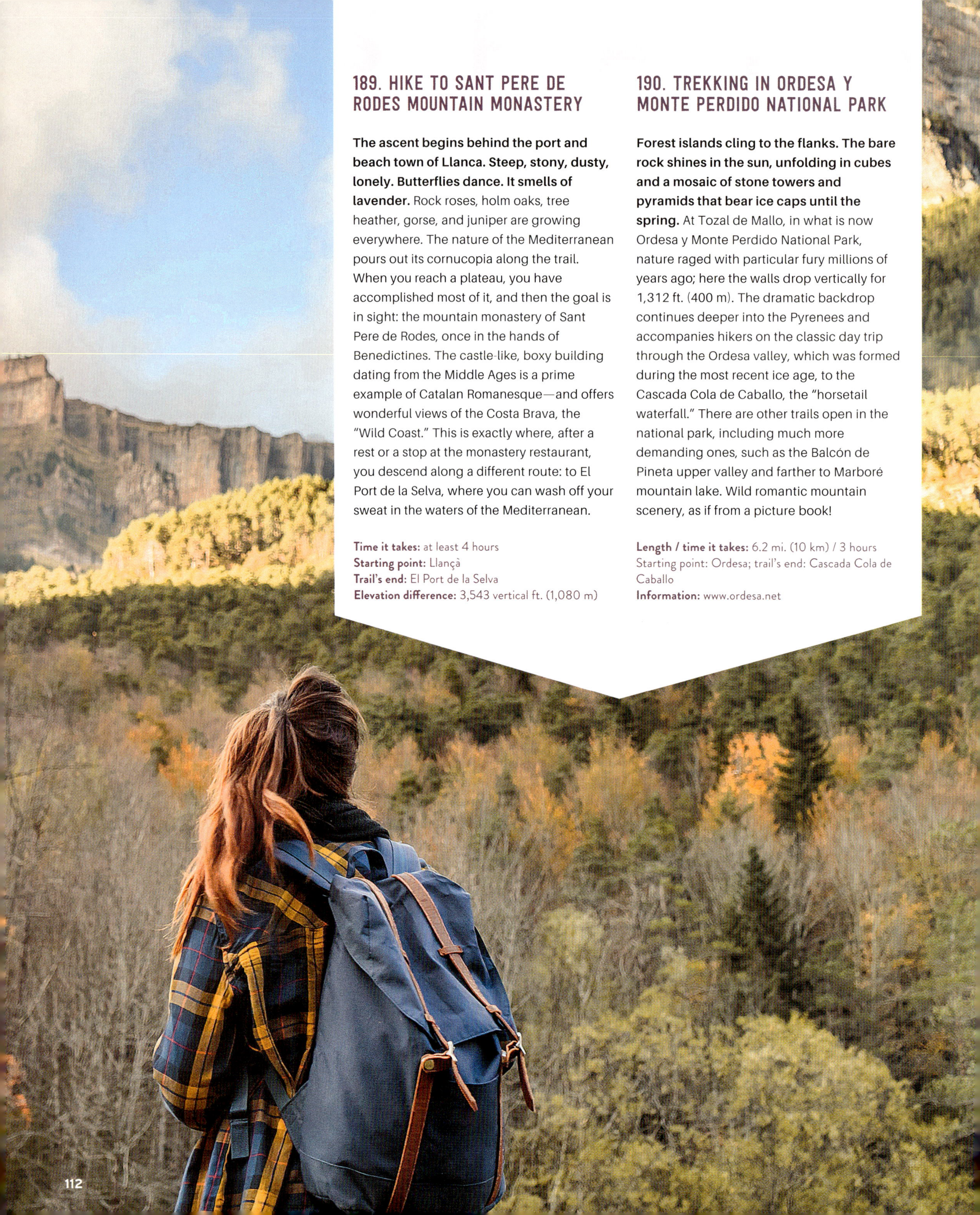

189. HIKE TO SANT PERE DE RODES MOUNTAIN MONASTERY

The ascent begins behind the port and beach town of Llanca. Steep, stony, dusty, lonely. Butterflies dance. It smells of lavender. Rock roses, holm oaks, tree heather, gorse, and juniper are growing everywhere. The nature of the Mediterranean pours out its cornucopia along the trail. When you reach a plateau, you have accomplished most of it, and then the goal is in sight: the mountain monastery of Sant Pere de Rodes, once in the hands of Benedictines. The castle-like, boxy building dating from the Middle Ages is a prime example of Catalan Romanesque—and offers wonderful views of the Costa Brava, the "Wild Coast." This is exactly where, after a rest or a stop at the monastery restaurant, you descend along a different route: to El Port de la Selva, where you can wash off your sweat in the waters of the Mediterranean.

Time it takes: at least 4 hours
Starting point: Llançà
Trail's end: El Port de la Selva
Elevation difference: 3,543 vertical ft. (1,080 m)

190. TREKKING IN ORDESA Y MONTE PERDIDO NATIONAL PARK

Forest islands cling to the flanks. The bare rock shines in the sun, unfolding in cubes and a mosaic of stone towers and pyramids that bear ice caps until the spring. At Tozal de Mallo, in what is now Ordesa y Monte Perdido National Park, nature raged with particular fury millions of years ago; here the walls drop vertically for 1,312 ft. (400 m). The dramatic backdrop continues deeper into the Pyrenees and accompanies hikers on the classic day trip through the Ordesa valley, which was formed during the most recent ice age, to the Cascada Cola de Caballo, the "horsetail waterfall." There are other trails open in the national park, including much more demanding ones, such as the Balcón de Pineta upper valley and farther to Marboré mountain lake. Wild romantic mountain scenery, as if from a picture book!

Length / time it takes: 6.2 mi. (10 km) / 3 hours
Starting point: Ordesa; trail's end: Cascada Cola de Caballo
Information: www.ordesa.net

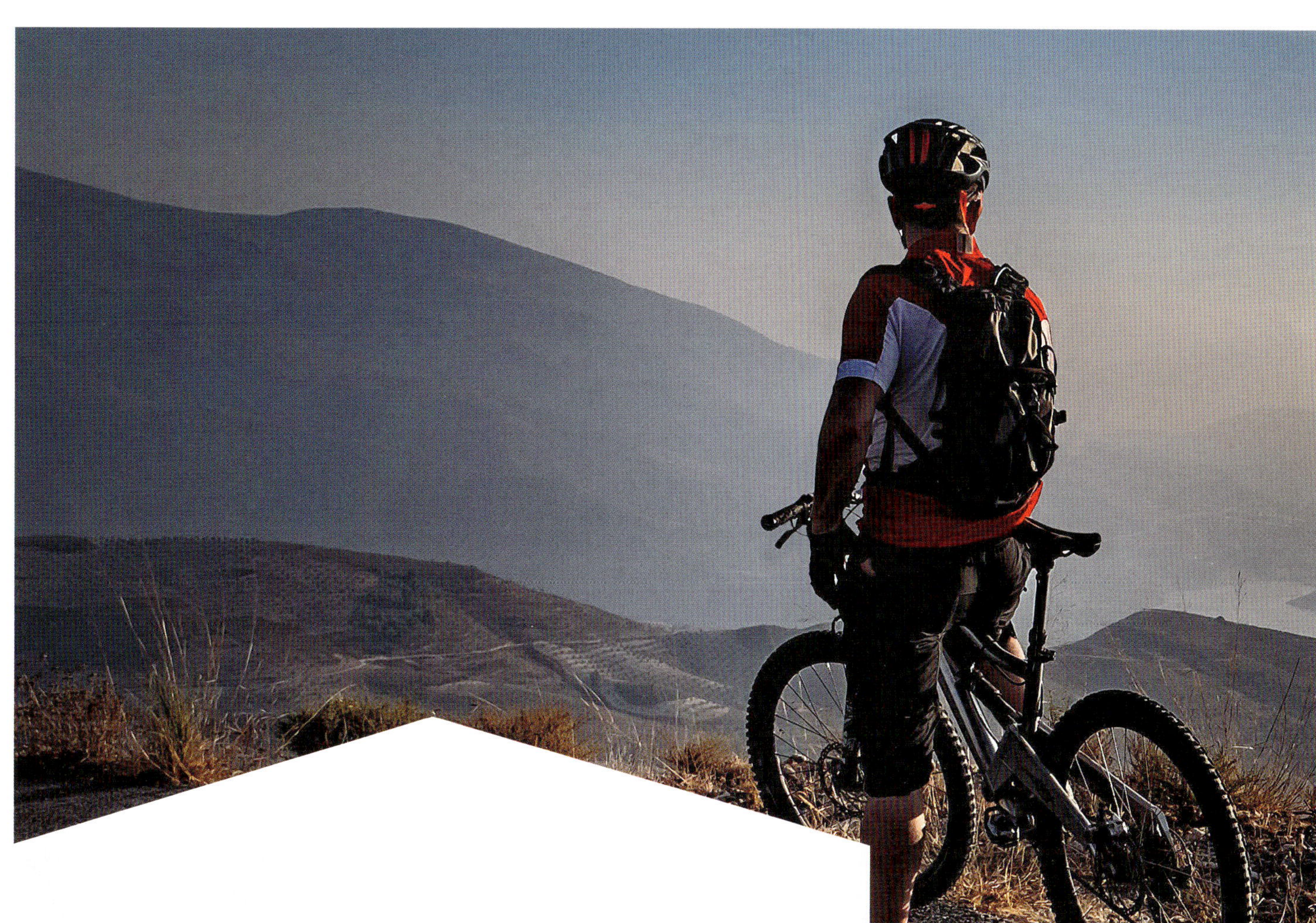

191. HIKING ON THE CAMINITO DEL REY CLIFF WALKWAY

The Caminito del Rey, the "Little Way of the King," in Andalusia is simply dizzying. It is named after Spanish ruler Alfonso XIII, who, in 1921, visited the site to see this breakneck path laid out at the beginning of the century. It was originally intended to give workers access to a hydropower plant and to facilitate the transport of materials. After being closed for a long time, the Caminito del Rey has celebrated a resurgence due to tourists; at the same time, security concerns are meticulously heeded. This is because the trail, which begins in the municipality of Ardales and ends in that of Álora, consists of one via ferrata passage and suspension bridge after another. The panoramas of the gorge en route are fantastic. Visitor access is limited.

Length: 4.8 mi. (7.7 km)
Information: www.caminitodelrey.info; use this official website to make reservations by date and time

192. CYCLING IN ANDALUSIA

Volcanic land, semidesert, bays, and mountains full of prickly pear cacti and agaves. The Cabo de Gata Nature Park is a very special one in southwest Europe and is a wonderful place to cycle through: either on your own or with bike guides, such as Francis Segura and Christel Steinhauser, who have the best insider knowledge. Almería Bike Tours offers a range of package tours. They know tracks hardly anyone else does: over gravel roads, small side streets, through the dry riverbeds that create furrows through the nature park. The wheels of the mountain bikes dig deep into the sand and leave traces of their rolling side to side. A stretch of the trail, such as that between the villages of El Pozo de los Frailes and Los Albaricoques, becomes a huge effort and a balancing act—and runs through completely unknown sides of Andalusia. By the way, the summer doesn't work as a cycling season because it gets too hot.

Information: for example, almeria-bike-tours.de
Best travel time: Spring or autumn

193. HIKING ON THE "DRAGON ISLAND," SA DRAGONERA

Don't worry, there are no fire-breathing creatures on Sa Dragonera, the "dragon island." The "dragons" are in small format: Lilford's wall lizard, endemic to the Balearic Islands (Latin: *Podarcis lilfordi giglioli*). They make sure that there are creepings and rustlings everywhere. How many are living here is not statistically recorded. It is estimated that there are two lizards per square yard.

Sa Dragonera is a protected nature conservancy. From Sant Elm, Majorca, you can reach the uninhabited island by passenger boat. Hiking trails start right from the pier. An easy one leads to the old (Far de) Tramuntana lighthouse, running by olives, pines, and rocks. The hikes to Far de Llebeig on the southwestern tip and to Far Vell on the summit of the island, at around 1,148 ft. (350 m) in elevation, are more strenuous. Adequate footwear is essential.

Length / time it takes: Hiking trails 0.68–2.8 mi. (1.1–4.5 km) / 30 minutes–2 hours
Information: de.balearsnatura.com

194. TORRENT DE PAREIS WALKING TOUR

Experience the wildest sides of the natural beauty of Majorca in the far north in the Torrent de Pareis. The "torrent" refers to the eroded valley of a flash flood stream that carries water only after a heavy rainfall. A grueling, demanding classic hike leads along the Torrent de Pareis amid the gorge scenery, where the rock walls rise 656 ft. (200 m) in some places. The gorge begins at s'Entreforc, where the two source streams, Lluc and Gorg Blau, meet. Accompanied by a hiking guide, you move between huge blocks and mysterious formations that stimulate the viewer's imagination. The outlet is in Sa Calobra Bay, which is ideal for a refreshing swim after all the exertion!

Time it takes: 6 hours; at the trail's end there is return transport by bus or boat to Port de Sóller
Elevation difference: 2,067 vertical ft. (630 m)
Information: for example, mondaventura.com

195. SAILING TRIPS IN THE BALEARIC ISLANDS

It doesn't matter whether Majorca mutates into what German tourists call "Malle" over the summer and Ibiza into a gigantic party platform—anyone launching a sailing trip around the Balearic Islands is well out of it. Or only mixes with all the people if they really want to when they go ashore. On board, you can keep your distance from overpriced hotel accommodations, tourist traps, overcrowded beaches, and beach bars with throbbing rhythms. It is all the more pleasant to let the silky air caress your face and take in the dreamy panoramas of quiet bays and spectacular rocky scenery. You change your perspective and experience the islands from the seaside. If you are traveling with a skipper, all you need to do is let yourself go in every respect—and if it is at an insider spot, in the crystal-clear water.

Information: for example, www.deltayachtcruisers.com (also for families with children; Spanish, English)

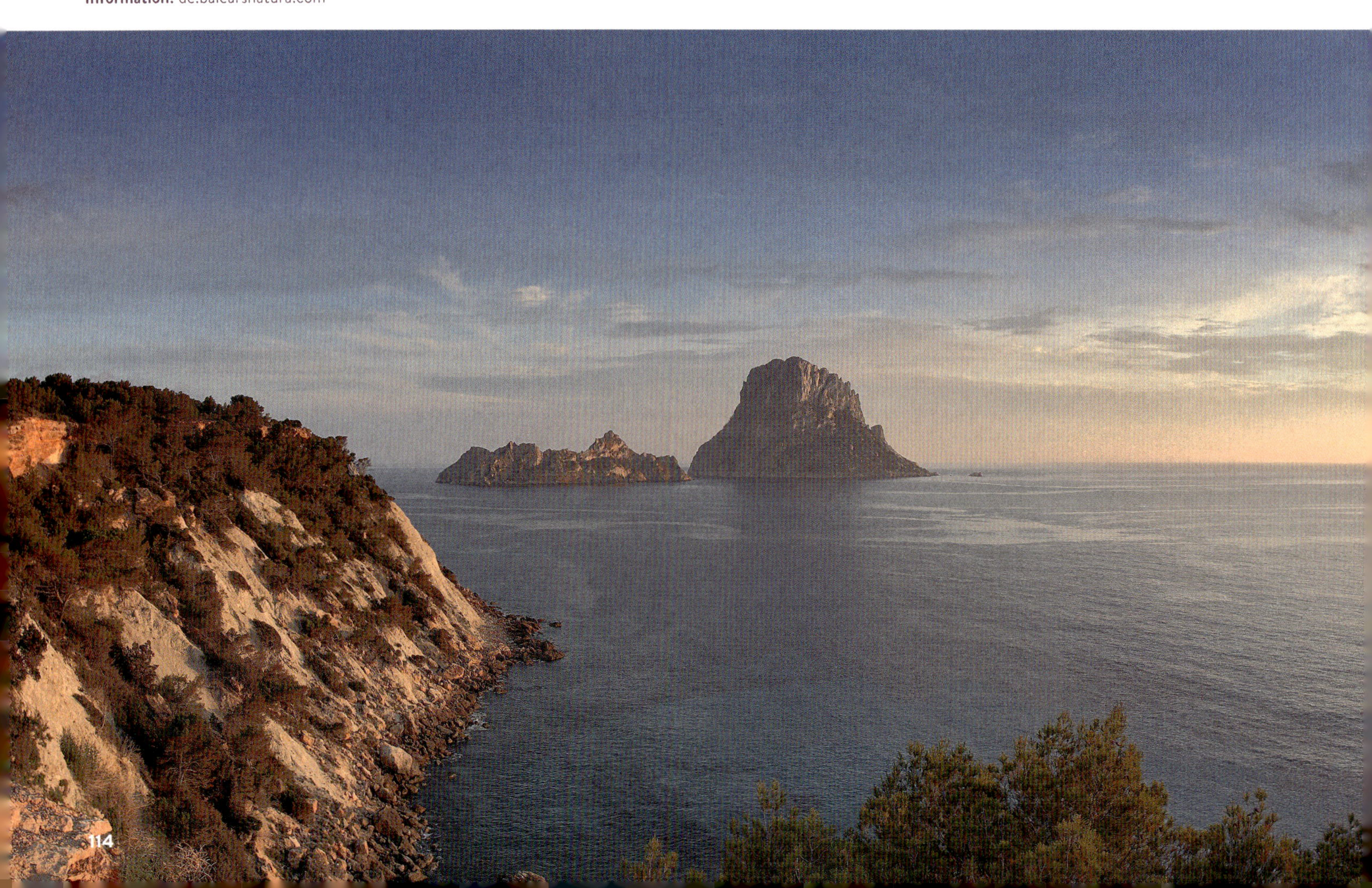

196. ASCENDING IBIZA'S HIGHEST MOUNTAIN

Sa Talaia Mountain is calling to you! At 1,558 ft. (475 m) in elevation, this is the highest peak on Ibiza. Sure, you could just drive up over a winding, bumpy route, but that's pointless. It is much more pleasant to rely on your leg muscle power to take you up from the inland town of Sant Josep de sa Talaia on a trail over steep and rocky stages and through magnificent pine forests, with the scent of herbs in your nose. If you should lose sight of the trail, you can orient yourself by the antenna masts on top of Sa Talaia. As the elevation increases, the view widens out, with a clear view of the east and west coasts of Ibiza. You feel rewarded for every drop of sweat and are comfortably removed from the summer hustle and bustle on the beaches, in Ibiza city, and in the town of Sant Antoni de Portmany. After taking a well-deserved break, go back down along the same route.

Time it takes: ascent: about 45 minutes
Highest elevation: 1,558 ft. (475 m)

197. FORMENTERA ISLAND BY SEA KAYAK

Caribbean feeling! The sky is cloudless, and the Mediterranean glitters turquoise to deep blue. With steady strokes, the sea kayak takes you along the north coast of Formentera. When paddling through the strait to the neighboring island of S'Espalmador, the crystal-clear water lets you get a glimpse of the bottom. Your intermediate goal is a shallow, wide-carved bay. You can't miss the seaweed that has washed up, forming huge underwater meadows. "Posidonia oceanica," explains tour guide Asier. "This is the lungs of the Mediterranean and a sign of healthy waters; very important for nature to cleanse itself." A visit to shore leads to a hill with an old watchtower and a view over to Ibiza. "You can really party over there; not with us," says Asier. Making it all the more pleasant to continue enjoying nature later on in the sea kayak.

Information: for example, www.wet4fun.com

198. HIKING ON FORMENTERA

"The last paradise in the Mediterranean" is what the island is sometimes exuberantly called. On Formentera, a 32 sq. mi. (82 km^2), small Balearic island, hikes bring a wonderful variety to vacation living. One runs inland along the south and west shores of the "Foul-Smelling Lake," the Estany Pudent, while another covers the northwestern part of the island. There, a nice little hike takes you from the Can Marroig picnic area to Torre Sa Gavina, 1.24 mi. (2 km) southwest. The bulging defense tower, dating from the eighteenth century, sits on the rocky promontory of the same name, from where you have beautiful views across to Ibiza. The shrubbery and wooded lands along the way are striking. It smells of lavender; lizards dart across the path. On the rocky slabs near the historical tower, the wash of the waves mingles with the sound of silence.

Time it takes: about 1 hour, there and back

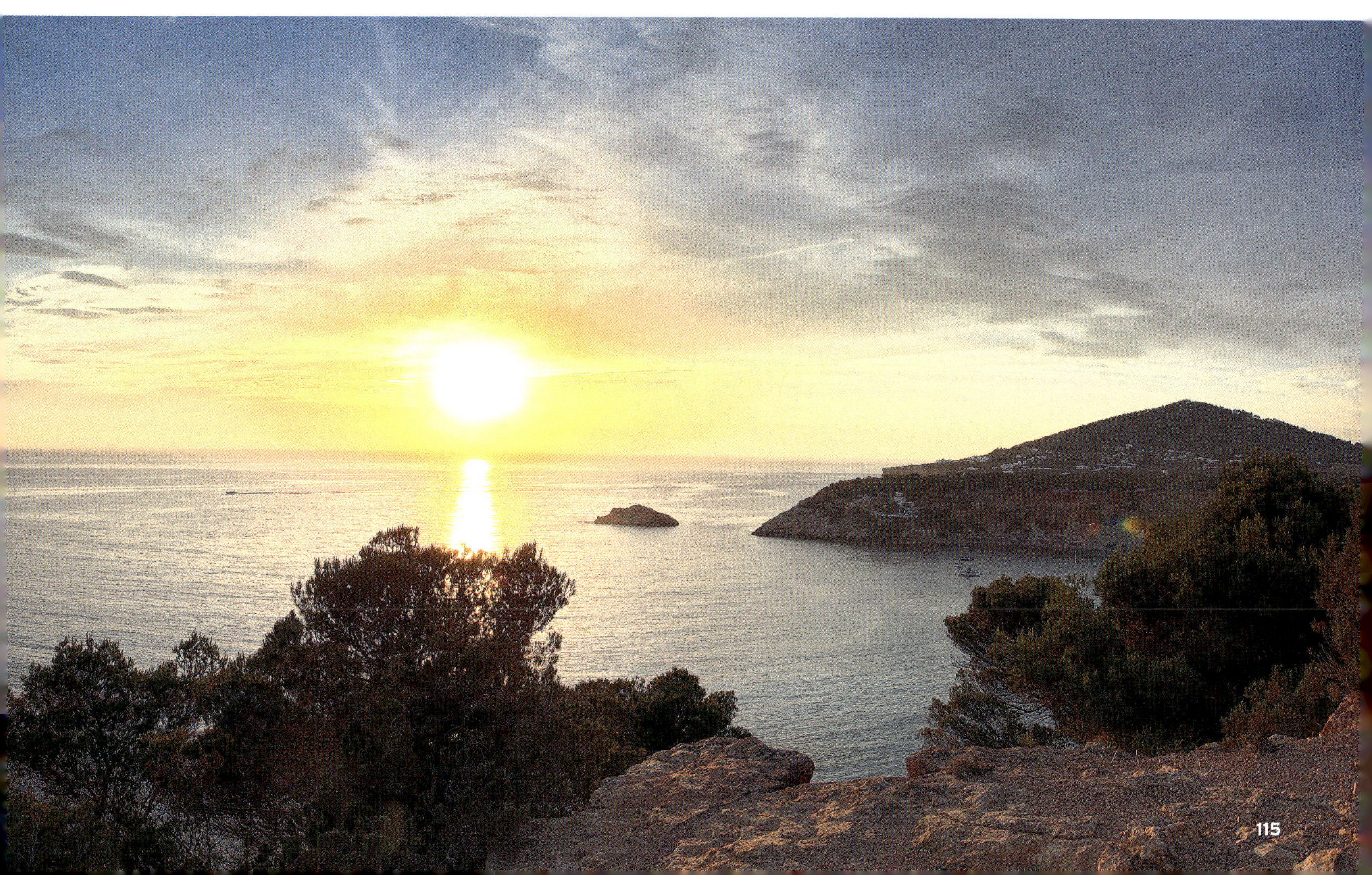

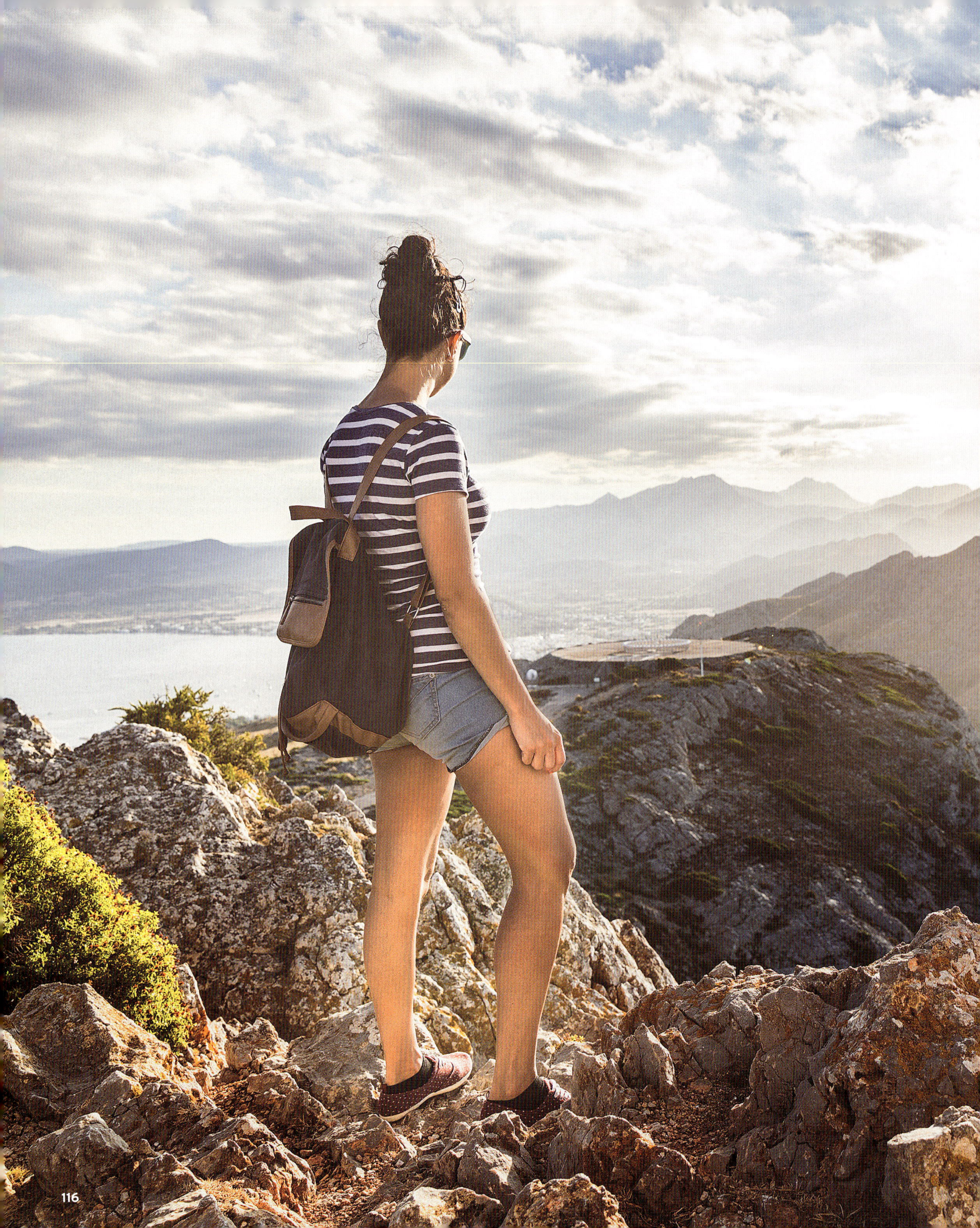

199. PARAGLIDING ON LA PALMA

La Palma has its charms, and not just from the perspectives of land and water. On a paraglider (Spanish: *parapente*) you can discover this Canary Island from a different perspective, preferably between September and the spring. In any case, you must be prepared at all times for the fact that the weather situation can change quickly. The west side is the most popular side for pilots; this is where international competitions are held. The landing site is in Puerto Naos. A tandem flight is a good way for a newcomer to paragliding to get started. To do this, all you need is good footwear and a windproof jacket and you're ready for adventure. A short flight from the cliff over Puerto Naos takes five to fifteen minutes, while a panoramic flight from a starting point at 3,117 ft. (950 m) takes up to half an hour.

Time it takes: 5–30 minutes
Information: paragliding school Palma Club Aventura (www.palmaclub.com)

200. TREKKING ON THE RUTA DE LOS VOLCANES TRAIL, LA PALMA

This demanding hiking trail is one of the most beautiful on La Palma, Canary Islands. You start from Refugio El Pilar, a barbecue area in the middle of the pine forest, and walk along the Cumbre Vieja mountain range, which separates the small island into a western and an eastern part. As you walk, the trail ascends to several peaks (Pico Birigoyo, 5,906 ft. [1,800 m]; Hoyo Negro volcano, 6,188 ft. [1,886 m]; and Duraznegro, 6,066 ft. [1,849 m]). The highest point is the Deseada II volcano at 6,339 ft. (1,932 m). Make sure you take enough water along with you. The exciting views of the Atlantic on both sides of the island and into various volcanic craters, and the hike through the pine forest and above the treeline, will compensate you for the blisters you may get on your feet. As a reward, the Bar Parada in the town of Fuencaliente has the best almond cookies in the Canaries!

Length / time it takes: 11 mi. (18 km) / 5–7 hours
Highest elevation: 6,339 ft. (1,932 m)
Starting point: Refugio El Pilar
Trail's end: Fuencaliente, 2,362 ft. (720 m); start at 4,774 ft. (1,455 m) above sea level
Requirements: good physical condition, mountaineering experience, and hiking footwear

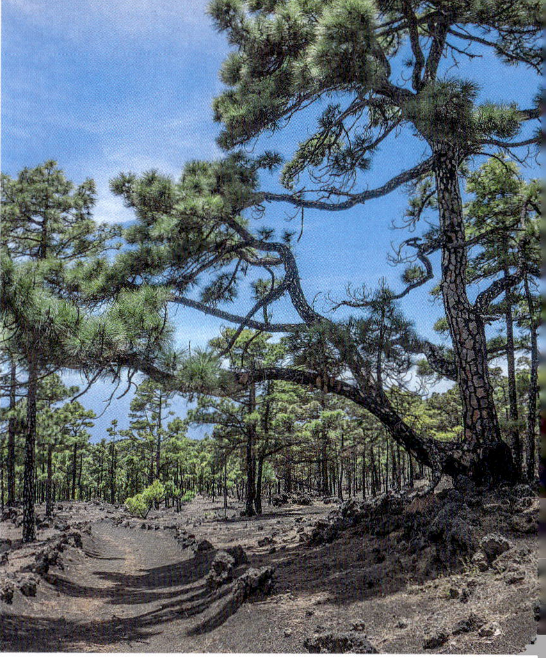

201. STARGAZING ON TENERIFE

Tenerife's mountains are considered a worldwide paradise for stargazers, as evidenced by the observatory. The reasons that this is an ideal spot are simple: its isolated island location in the Atlantic, the high elevations, the low light pollution, plus 300 days of a clear view each year. Certainly you could use standard binoculars in the mountains. But taking part in a guided tour ensures getting the best observation results for your money: with professional equipment and guides such as Miguel Ángel Pérez Hernández. The moon and a few planets are the appetizers, then you swivel to the Milky Way. Shooting stars provide special effects in between. By the way, Miguel Ángel used to be an administrative clerk until he no longer wanted to work in an office, but rather "under the stars."

Information: for example, discoverexperience.com, www.volcanoteide.com

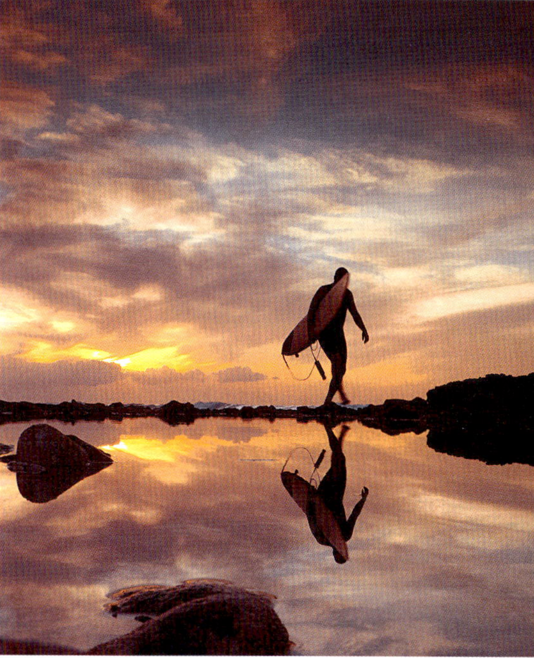

202. COASTAL AND BEACH TREKKING ON FUERTEVENTURA

Crystal-clear water. Sand dunes. Beaches stretching to the horizon. It is as if Fuerteventura was created for fans of the sea. And a dream terrain for beach and coastal hikers! In the far south, the island is building up its longest sandbanks, including along the eastern lee side of the coastline around Playa de Sotavento. A 12.5 mi. (20 km) hike from Morro Jable to Costa Calma provides a challenge along this eastern lee side for anyone who is physically fit. Passages over the sand and rocks alternate with dunes and wetlands. On the isolated pebble beaches, the pebbles roll noisily back and forth in the surf. The coast is anything but dead straight. You hike on headland after headland along its geographical profile—and you can cool off with a swim along the way. At some point the architectural wilderness of Costa Calma comes into view.

Length / time it takes: 12.5 mi. (20 km) / at least 4.5 hours, return by bus.

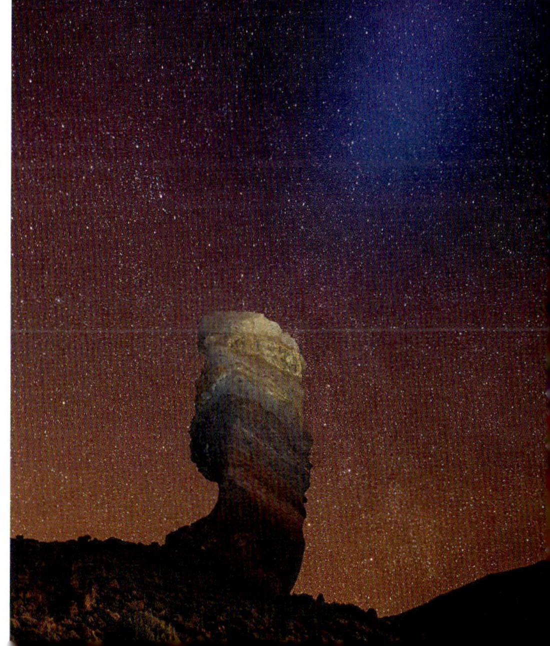

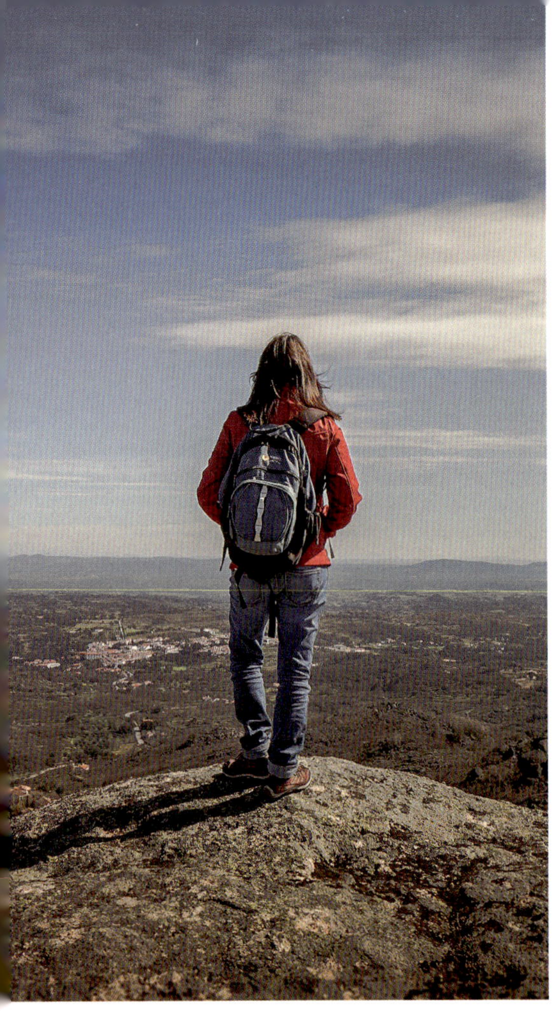

203. SURFING THE COSTA VICENTINA COASTLINE

Every beginning is difficult, including for surfing, but the Algarve is ideal terrain for practicing. Surfing instructor Ricardo Gonçalves lets his gaze wander over Praia do Amado, one of the top beaches on the wild and romantic Costa Vicentina, which is part of the Algarve. He encourages newcomers and says, "Things are ideal for beginners here, because the waves break early and roll out for a long time." Besides, between April and October you can keep working at ebb and flow, often under ideal conditions: clear and mild air and a restrained wind. Then comes lesson after lesson, until finally you can stand at the first attempt and no longer disappear into the water without a trace. For advanced surfers, Ricardo admits, Praia do Amado is not demanding enough. But don't worry: the Algarve has more choices available!

Information: for example, www.future-surf.com

204. LONG-DISTANCE HIKE ON THE VIA ALGARVIANA TRAIL

The Algarve, which is extremely popular as a sun-drenched vacation region, gets you going. Especially on the Via Algarviana, a long-distance hiking trail that stretches more than 186 mi. (300 km) from Alcoutim, on the border with Spain, to Cape São Vicente. This trail doesn't go from beach to beach; the Via Algarviana runs through the hinterland away from sand, channels, and concrete-covered coastlines. This gives you the opportunity to get to know the Algarve and its residents in a completely different, more original way. Alpine difficulty levels do not apply. The ideal time for hiking is the spring, when it is not that hot; everything is blooming and smells intensely of herbs. At the end of the trail, on the coastal cliff around the cape—the land's end of the continent of Europe—wild nature plays its soundtrack of wind and waves.

Length / time it takes: 186 mi. (300 km) / about 2 weeks
Information: www.viaalgarviana.org

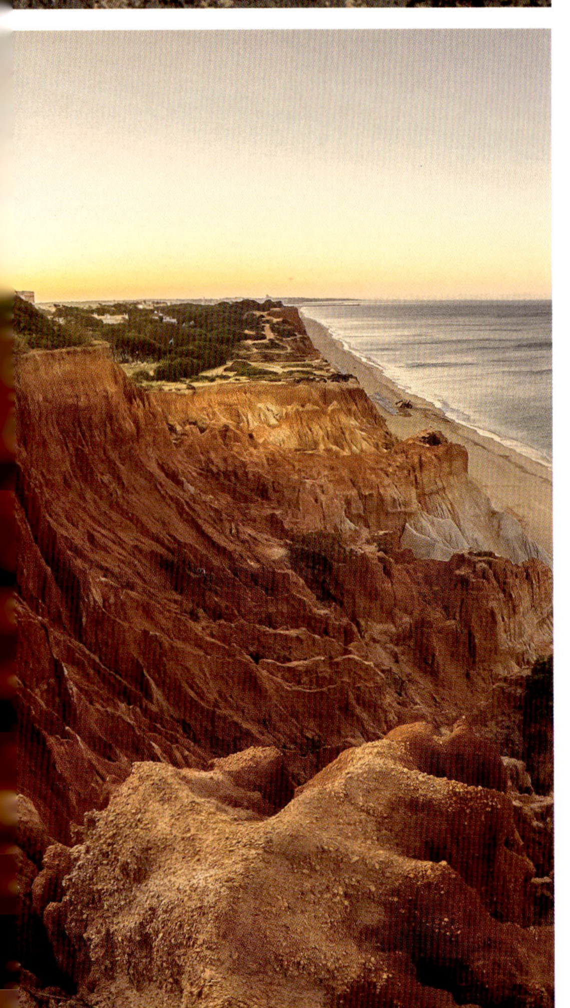

205. SKYDIVING OVER THE ALENTEJO REGION

The vast expanses of the Greater Alentejo region are rapidly approaching—at least for those who dare to make the tandem jump from 13,780 ft. (4,200 m) above sea level. The starting point is the skydiving school airfield about 25 mi. (40 km) west of the city of Beja, near Figueira dos Cavaleiros. High up in the sky, the roller door opens quickly. The wind hisses murderously. The sun stings your eyes. Companion João pushes his protege to the exit until his legs are flapping high above the distant patchwork of cork oak and olive groves, fields, and pastures. No time to turn back. João gently pushes his protege out. The kick is not long in coming; you lack the strength for making a planned primal scream. The folds of skin on your face are flapping at more than 124 mi. (200 km) per hour. The free fall takes just under a minute. Then João pulls the ripcord. The rest is just peacefully floating.

Time it takes: 1 minute
Information: skydiveeurope.com

206. LONG-DISTANCE HIKE ON THE ROTA VICENTINA TRAIL NETWORK

With the fishing port of Vila Nova de Milfontes left behind, a sandy passage awaits you above the sea on the Rota Vicentina long-distance hiking trail. You become absorbed; dune vegetation spreads on both sides of the path, and the Atlantic rumbles and magnetizes. But you shouldn't let yourself get too distracted. Sometimes the trail runs without guards along precipices that are 66 ft. (20 m) and higher along this section to Malhão Beach. The Rota Vicentina comprises a network of hiking trails in southwestern Portugal, a total of 280 mi. (450 km). A small, stubborn core of outdoor enthusiasts are responsible for designating this trail. The main routes are the coastal "fisherman's path" (the most beautiful!) and the interior "historical trail," which come together at certain places. Cape São Vicente is the southernmost trail's end of both variants.

Information: de.rotavicentina.com

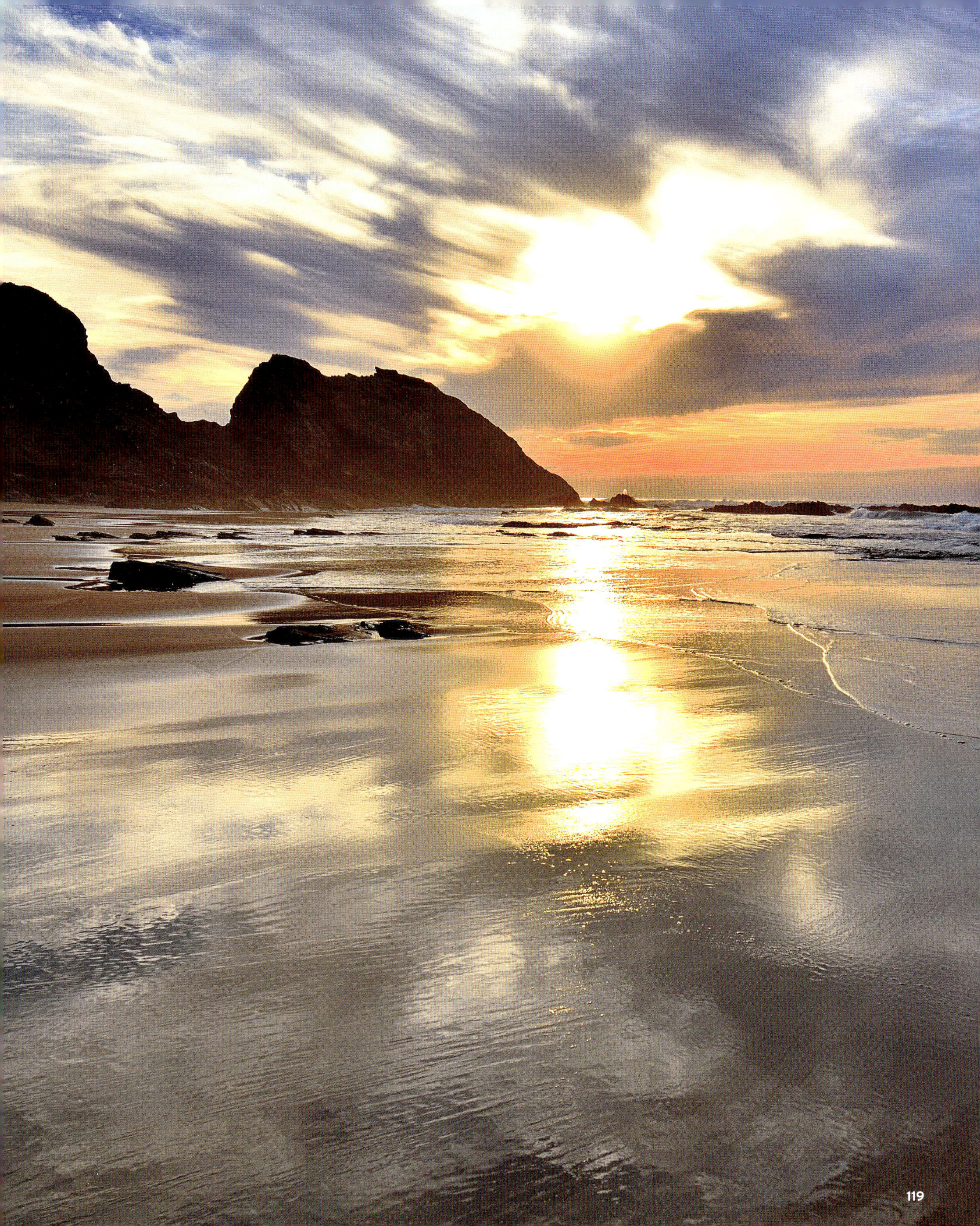

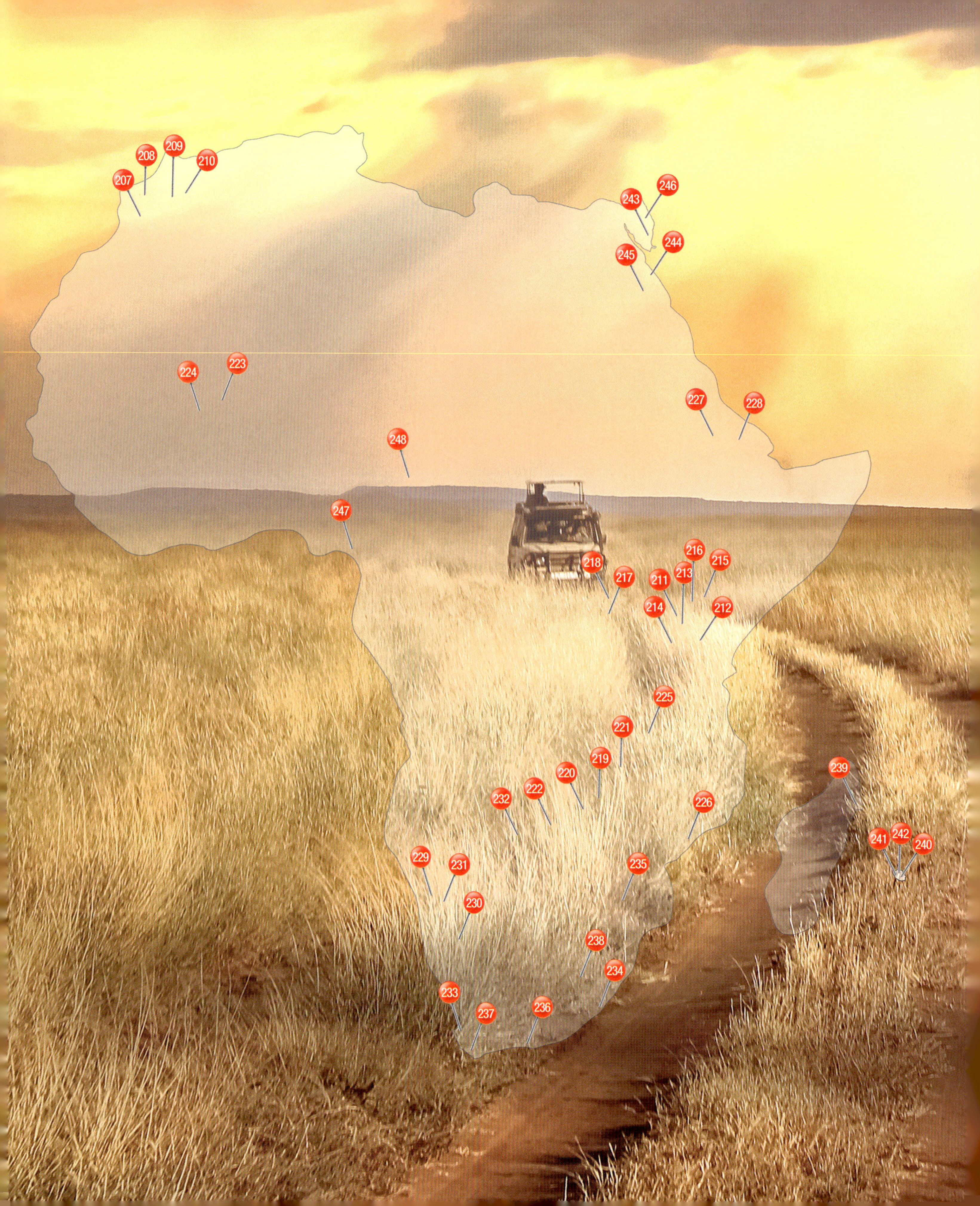
209
208
210
207
246
243
244
245
223
224
227
228
248
247
216
218
215
217
211
213
214
212
225
221
219
239
220
222
232
226
242
241
240
229
231
235
230
238
234
233
236
237

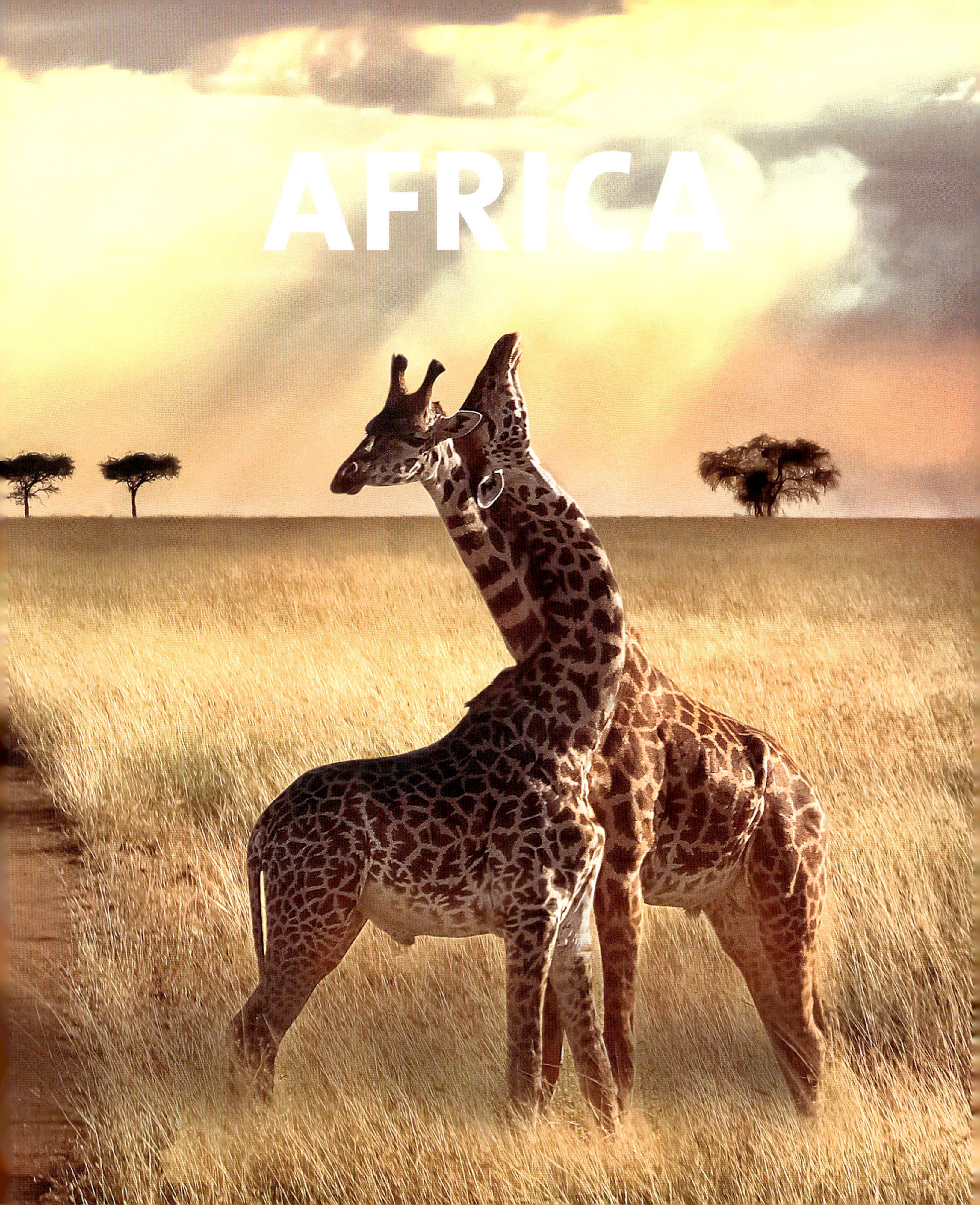
AFRICA

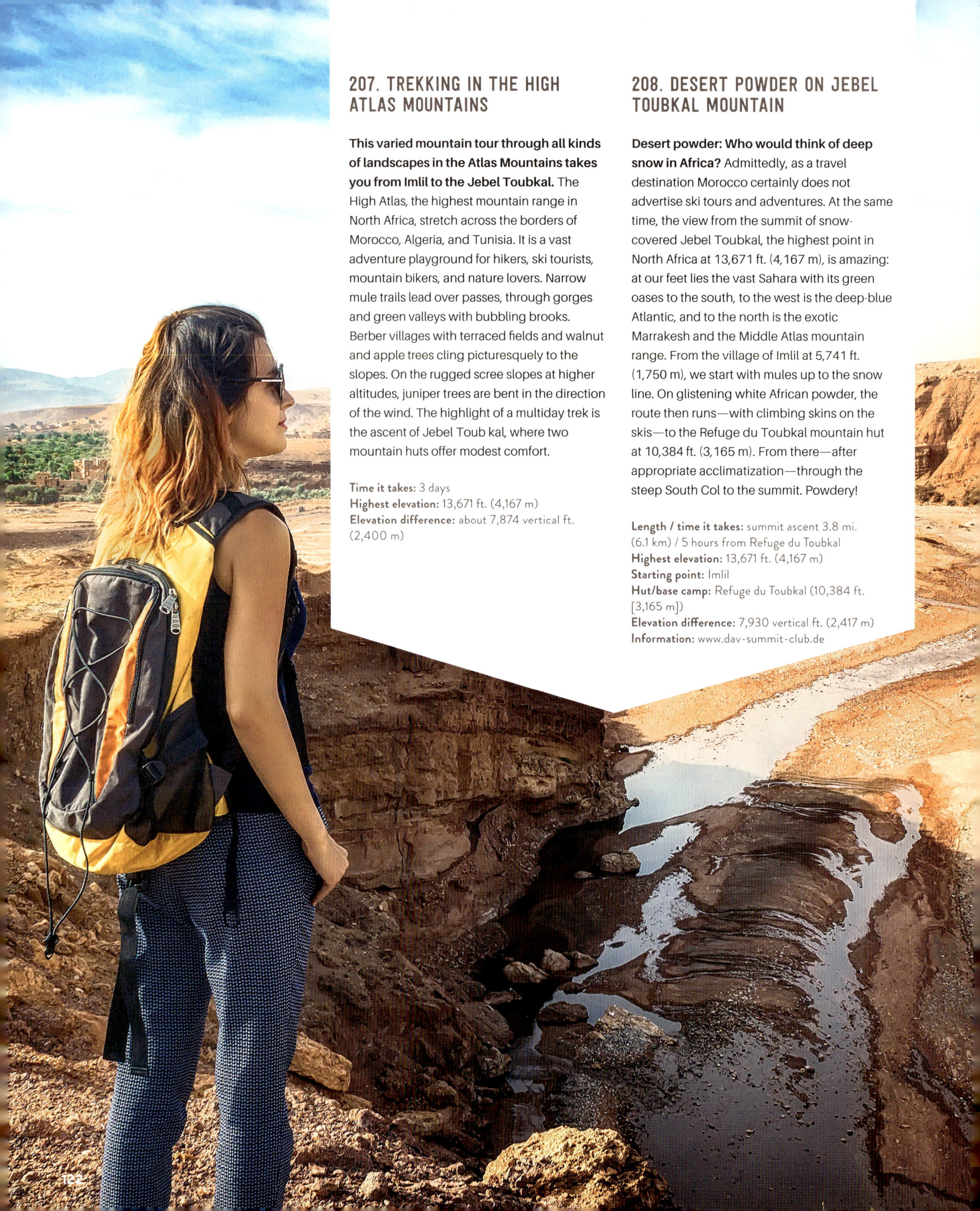

207. TREKKING IN THE HIGH ATLAS MOUNTAINS

This varied mountain tour through all kinds of landscapes in the Atlas Mountains takes you from Imlil to the Jebel Toubkal. The High Atlas, the highest mountain range in North Africa, stretch across the borders of Morocco, Algeria, and Tunisia. It is a vast adventure playground for hikers, ski tourists, mountain bikers, and nature lovers. Narrow mule trails lead over passes, through gorges and green valleys with bubbling brooks. Berber villages with terraced fields and walnut and apple trees cling picturesquely to the slopes. On the rugged scree slopes at higher altitudes, juniper trees are bent in the direction of the wind. The highlight of a multiday trek is the ascent of Jebel Toub kal, where two mountain huts offer modest comfort.

Time it takes: 3 days
Highest elevation: 13,671 ft. (4,167 m)
Elevation difference: about 7,874 vertical ft. (2,400 m)

208. DESERT POWDER ON JEBEL TOUBKAL MOUNTAIN

Desert powder: Who would think of deep snow in Africa? Admittedly, as a travel destination Morocco certainly does not advertise ski tours and adventures. At the same time, the view from the summit of snow-covered Jebel Toubkal, the highest point in North Africa at 13,671 ft. (4,167 m), is amazing: at our feet lies the vast Sahara with its green oases to the south, to the west is the deep-blue Atlantic, and to the north is the exotic Marrakesh and the Middle Atlas mountain range. From the village of Imlil at 5,741 ft. (1,750 m), we start with mules up to the snow line. On glistening white African powder, the route then runs—with climbing skins on the skis—to the Refuge du Toubkal mountain hut at 10,384 ft. (3,165 m). From there—after appropriate acclimatization—through the steep South Col to the summit. Powdery!

Length / time it takes: summit ascent 3.8 mi. (6.1 km) / 5 hours from Refuge du Toubkal
Highest elevation: 13,671 ft. (4,167 m)
Starting point: Imlil
Hut/base camp: Refuge du Toubkal (10,384 ft. [3,165 m])
Elevation difference: 7,930 vertical ft. (2,417 m)
Information: www.dav-summit-club.de

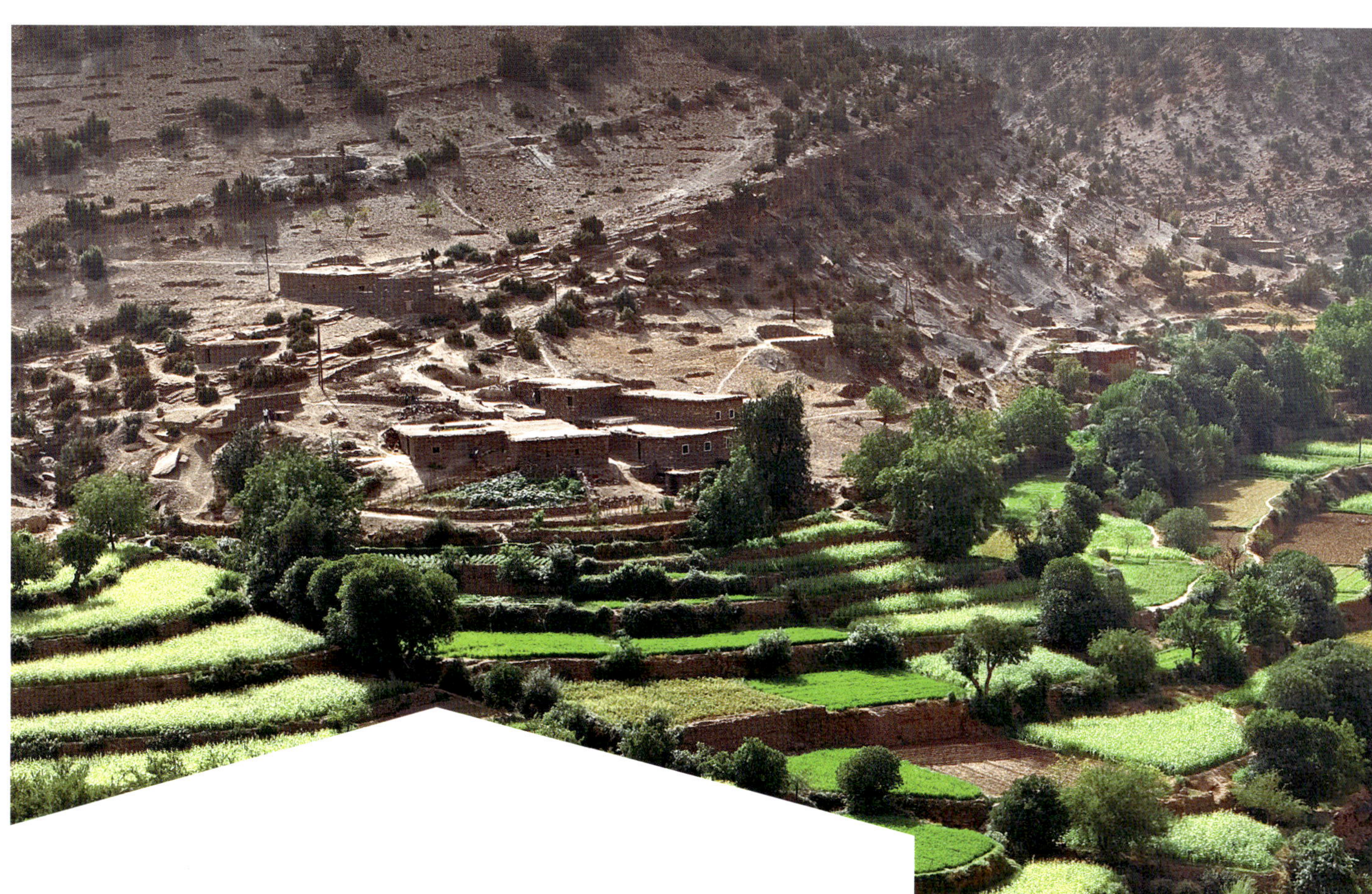

209. TREKKING THROUGH THE JEBEL-SAGHRO MASSIF

The Jebel Saghro massif, in southern Morocco on the edge of the Sahara, is fascinating with its clarity of shapes and colors. Bizarre rock towers reach into the crystal-blue sky; the black-brown volcanic stone shimmers in the midday heat. Just a few colorful dabs of flowers, which manage to get by in the barren conditions here, enliven the otherwise bare desert mountains. On a four-day trekking trip—five hours of walking per day—straight through from Boumalne Dades to Nkob, you go over some 6,562 ft. high (2,000 m) peaks and passes, then again through deep gorges, with a bit of civilization in the form of picturesque Berber villages. Encounters with Ait Atta nomads and spending the night in a tent under the starry sky ensure unforgettable experiences.

Time it takes: 4 days; daily hike, 5 hours
Starting point: Boumalne Dades; trail's end: Nkob
Elevation difference: 8,504 vertical ft. (2,592 m)

210. CLIMBING IN TODRA GORGE

Climbing in the Todra Gorge in Morocco means adventure, a foreign culture, and perfect, ocher-colored limestone. Sports climbing started to develop in the Todra Gorge when it was opened by French climbers in 1977. The conditions are simply ideal: the limestone is easy to grip, and there is all-around beautiful, strange nature—a fabulous spot for outdoor sports. Four decades later, there are more than 300 sports and alpine routes, and the number is rising. Add to that the exotic culture and the breathtaking landscape and it's clear that all this attracts countless sports climbers from all over the world. The routes are usually well secured; the Todra rock is very rough and aggressive. In any case, take enough tape along with you, or your fingers won't hold out very long.

Highest elevation: 82–820 ft. (25–250 m)
Starting point: Tinghir
Spots: more than 30 sectors and around 300 routes
Level of difficulty: 3 to 8b (French scale)
Information: www.oxfordalpineclub.co.uk

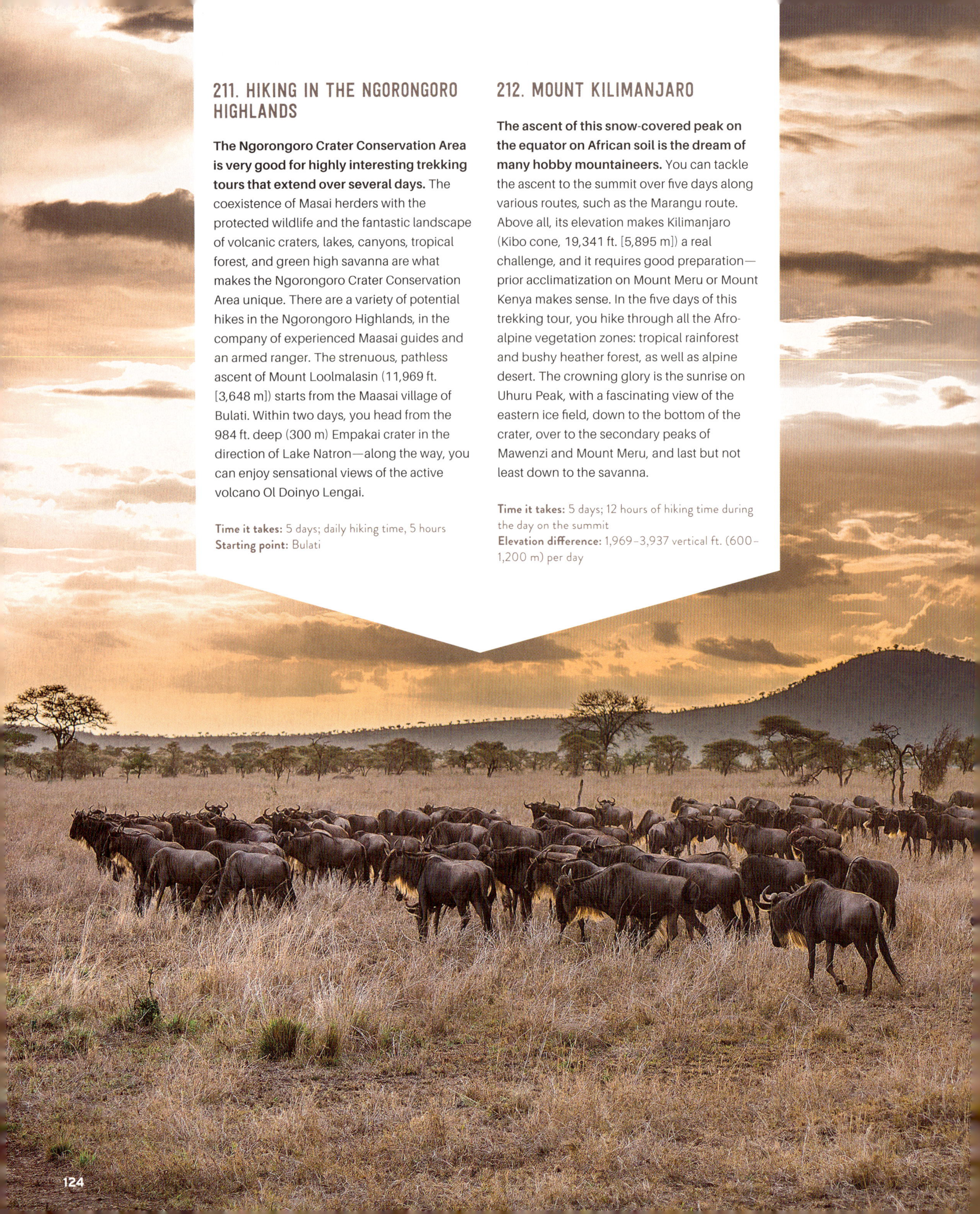

211. HIKING IN THE NGORONGORO HIGHLANDS

The Ngorongoro Crater Conservation Area is very good for highly interesting trekking tours that extend over several days. The coexistence of Masai herders with the protected wildlife and the fantastic landscape of volcanic craters, lakes, canyons, tropical forest, and green high savanna are what makes the Ngorongoro Crater Conservation Area unique. There are a variety of potential hikes in the Ngorongoro Highlands, in the company of experienced Maasai guides and an armed ranger. The strenuous, pathless ascent of Mount Loolmalasin (11,969 ft. [3,648 m]) starts from the Maasai village of Bulati. Within two days, you head from the 984 ft. deep (300 m) Empakai crater in the direction of Lake Natron—along the way, you can enjoy sensational views of the active volcano Ol Doinyo Lengai.

Time it takes: 5 days; daily hiking time, 5 hours
Starting point: Bulati

212. MOUNT KILIMANJARO

The ascent of this snow-covered peak on the equator on African soil is the dream of many hobby mountaineers. You can tackle the ascent to the summit over five days along various routes, such as the Marangu route. Above all, its elevation makes Kilimanjaro (Kibo cone, 19,341 ft. [5,895 m]) a real challenge, and it requires good preparation—prior acclimatization on Mount Meru or Mount Kenya makes sense. In the five days of this trekking tour, you hike through all the Afro-alpine vegetation zones: tropical rainforest and bushy heather forest, as well as alpine desert. The crowning glory is the sunrise on Uhuru Peak, with a fascinating view of the eastern ice field, down to the bottom of the crater, over to the secondary peaks of Mawenzi and Mount Meru, and last but not least down to the savanna.

Time it takes: 5 days; 12 hours of hiking time during the day on the summit
Elevation difference: 1,969–3,937 vertical ft. (600–1,200 m) per day

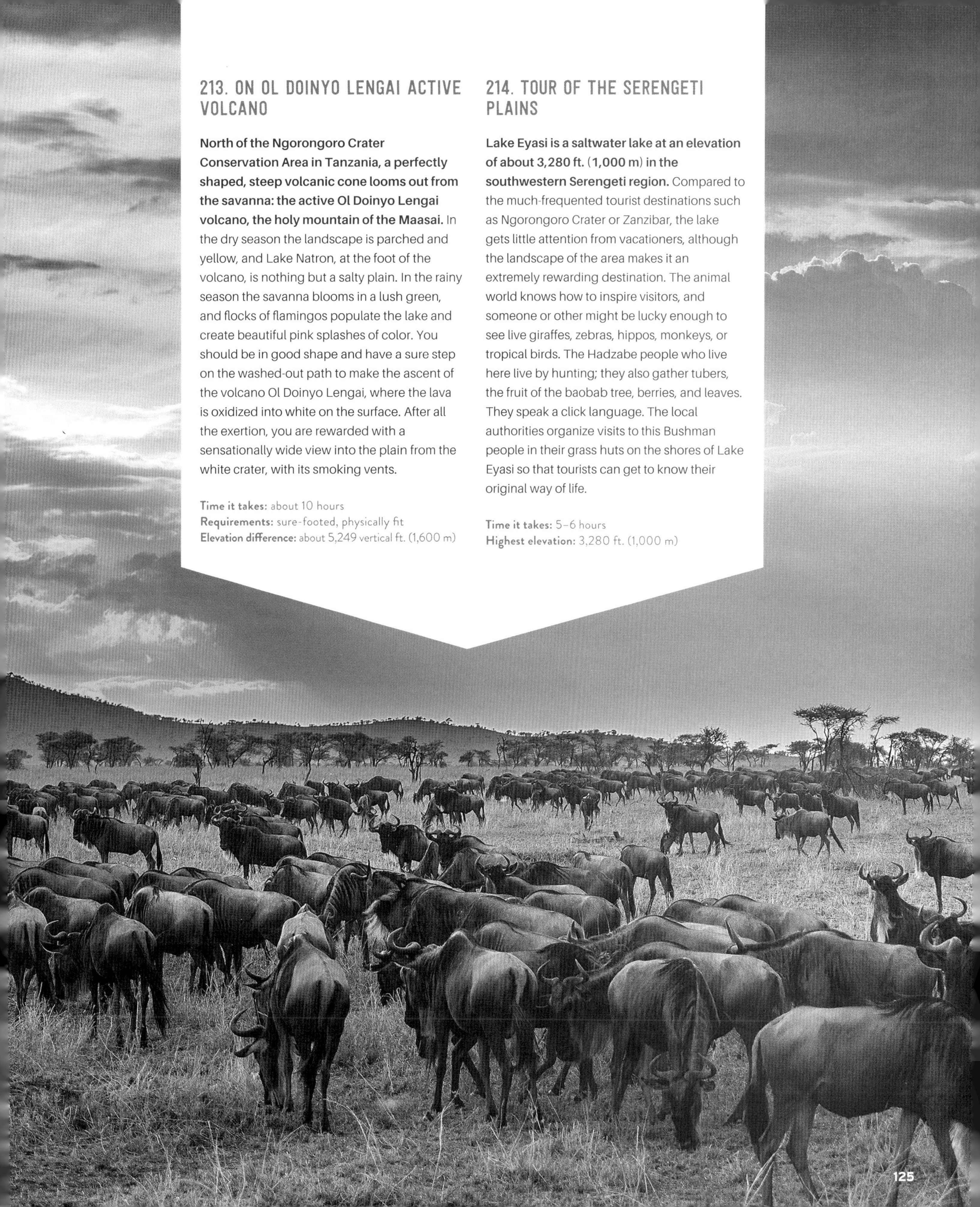

213. ON OL DOINYO LENGAI ACTIVE VOLCANO

North of the Ngorongoro Crater Conservation Area in Tanzania, a perfectly shaped, steep volcanic cone looms out from the savanna: the active Ol Doinyo Lengai volcano, the holy mountain of the Maasai. In the dry season the landscape is parched and yellow, and Lake Natron, at the foot of the volcano, is nothing but a salty plain. In the rainy season the savanna blooms in a lush green, and flocks of flamingos populate the lake and create beautiful pink splashes of color. You should be in good shape and have a sure step on the washed-out path to make the ascent of the volcano Ol Doinyo Lengai, where the lava is oxidized into white on the surface. After all the exertion, you are rewarded with a sensationally wide view into the plain from the white crater, with its smoking vents.

Time it takes: about 10 hours
Requirements: sure-footed, physically fit
Elevation difference: about 5,249 vertical ft. (1,600 m)

214. TOUR OF THE SERENGETI PLAINS

Lake Eyasi is a saltwater lake at an elevation of about 3,280 ft. (1,000 m) in the southwestern Serengeti region. Compared to the much-frequented tourist destinations such as Ngorongoro Crater or Zanzibar, the lake gets little attention from vacationers, although the landscape of the area makes it an extremely rewarding destination. The animal world knows how to inspire visitors, and someone or other might be lucky enough to see live giraffes, zebras, hippos, monkeys, or tropical birds. The Hadzabe people who live here live by hunting; they also gather tubers, the fruit of the baobab tree, berries, and leaves. They speak a click language. The local authorities organize visits to this Bushman people in their grass huts on the shores of Lake Eyasi so that tourists can get to know their original way of life.

Time it takes: 5–6 hours
Highest elevation: 3,280 ft. (1,000 m)

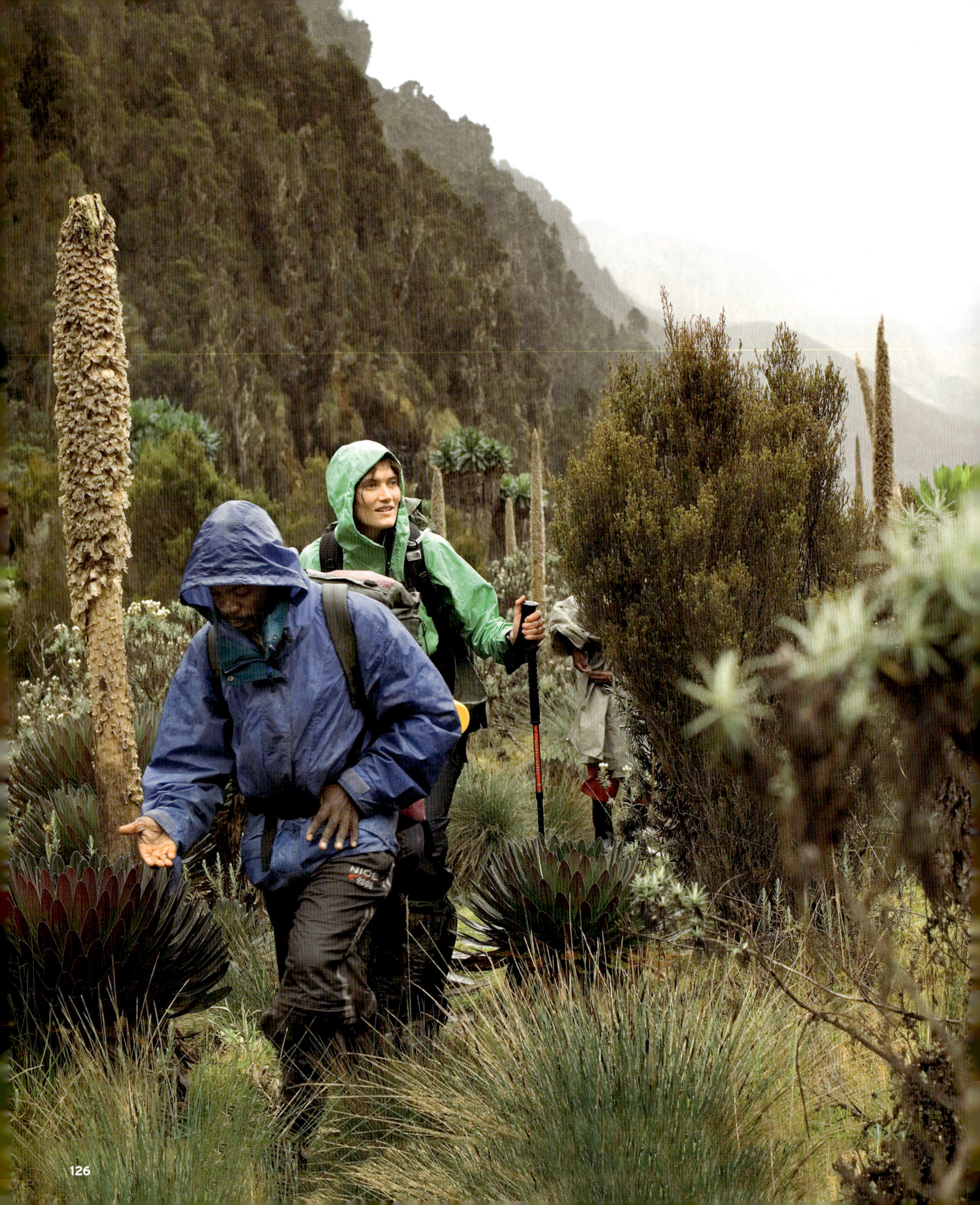

215. TREKKING IN MOUNT KENYA NATIONAL PARK

The glaciated summit of Mount Kenya, an extinct volcano, rises like an island from the sea of the East African savanna, right at the equator. Trekking tours in Mount Kenya National Park take you through dense heather and bamboo forest, and over flat moorland where tussock grass and giant groundsel grow. You keep on hiking, facing the mountain's jagged peaks Batian (17,057 ft. [5,199 m]) and Nelion (17,021 ft. [5,188 m]), pass by some mighty glaciers and crystal-clear crater lakes, and gather countless new impressions. Batian can be scaled only by rock climbing; however, the trekking summit Point Lenana (16,355 ft. [4,985 m]) also offers mountaineers plenty of unforgettable and unhindered views of the expanses of the flat equatorial landscape.

Time it takes: 5 days; daily hiking time, 5–8 hours
Trail's end: Point Lenana
Elevation difference: maximum 3,281 vertical ft. (1,000 m) per day

216. TREKKING IN ABERDARE NATIONAL PARK

An untouched and misty-humid mountain landscape, encounters with wild animals, and views of the Rift Valley and Mount Kenya; all this awaits hikers in the Aberdare mountain range in central Kenya. A multiday trek certainly does require some organizational effort, but you will experience such diverse landscapes as the impressive Karuru Falls (919 ft. [280 m]), the lush vegetation in the rainforest and bamboo forest, and the moor and tussock grass landscape on the high plateau at more than 9,843 ft. (3,000 m) in elevation. An armed ranger protects adventurous tourists from the buffalo, rhinos, and elephants. . . . The highest peak, Ol Doinyo Le Sattima (13,127 ft. [4,001 m]), can be climbed in about three days.

Time it takes: 5–7 days, south–north crossing on the Mulwa Trail; 2–3 days for ascent of Le Sattima

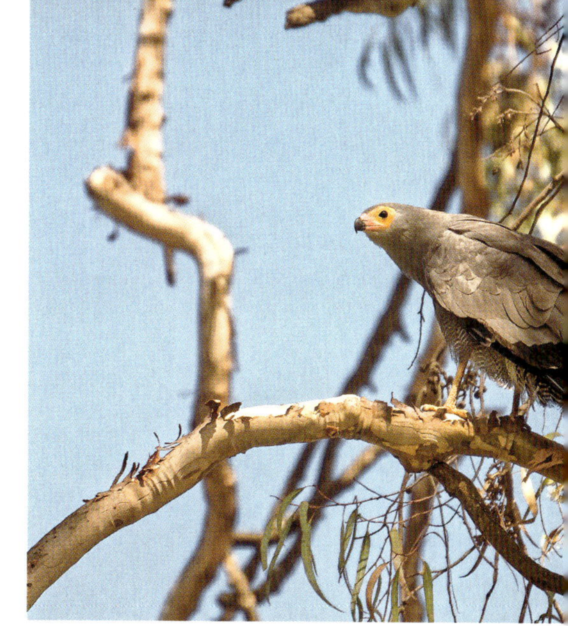

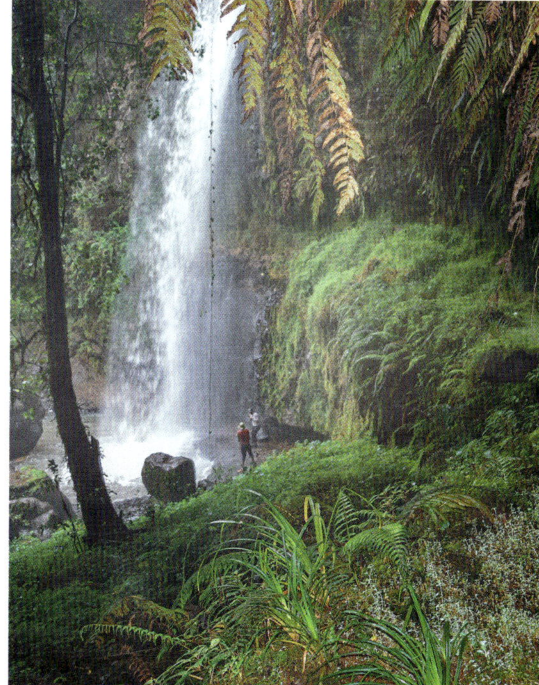

217. TRACKING IN BWINDI NATIONAL PARK

An absolute highlight for every animal lover: tracking down the traces of the last mountain gorillas in the African rainforest on foot and watching them in their natural habitat. The mountain forest in Bwindi Impenetrable Forest National Park in southwestern Uganda is the home of about 340 mountain gorillas. Accompanied by a ranger armed with a machete and a rifle, visitors track gorillas through the dense undergrowth. With a bit of luck you can track them down in about an hour, although it can sometimes take longer. You then spend an unforgettable hour watching a silverback and his family. On the way back, with a bit of luck, you can spot other representatives of the four-legged beings who make their home here, including chimpanzees, monkeys, and forest elephants.

Time it takes: 1–5 hours tracking, 1 hour watching
Information: www.bwindiforestnationalpark.com/

218. HIKING IN THE RUWENZOR MOUNTAINS

Rainproof footwear is an essential part of this trekking tour in East Africa between the Republic of the Congo and Uganda, because you are guaranteed to get wet feet in the rainy Ruwenzoris. There is hardly a more primeval and contrasting landscape in Africa: below is a lush rainforest with huge giant groundsel (*Dendrosenecio*), yard-long beard lichens, ferns, and mosses, while at the top on the summit there is eternal ice and snow. The "Mountains of the Moon" are a paradise for experienced mountaineers who are not afraid to jump from bog hole to bog hole for days on end over untracked terrain. Ascending Margherita peak at 16,762 ft. (5,109 m) requires experience in climbing glaciers.

Time it takes: 8 days; daily hiking time, 4–12 hours (on the day on the summit)
Requirements: Experience and being sure-footed
Elevation difference: up to 3,609 vertical ft. (1,100 m)

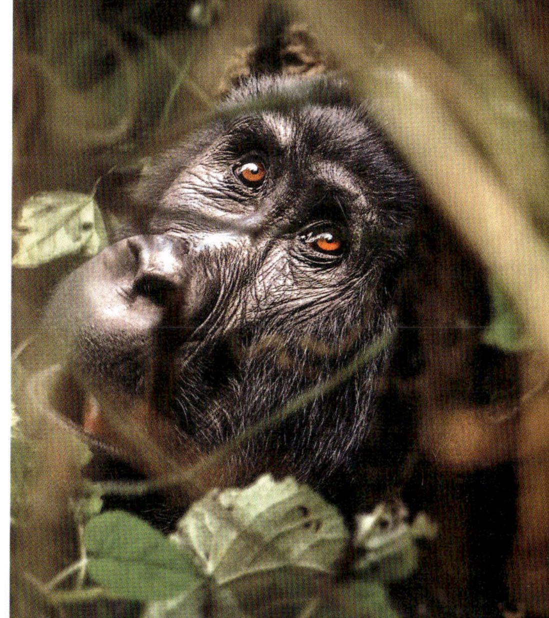

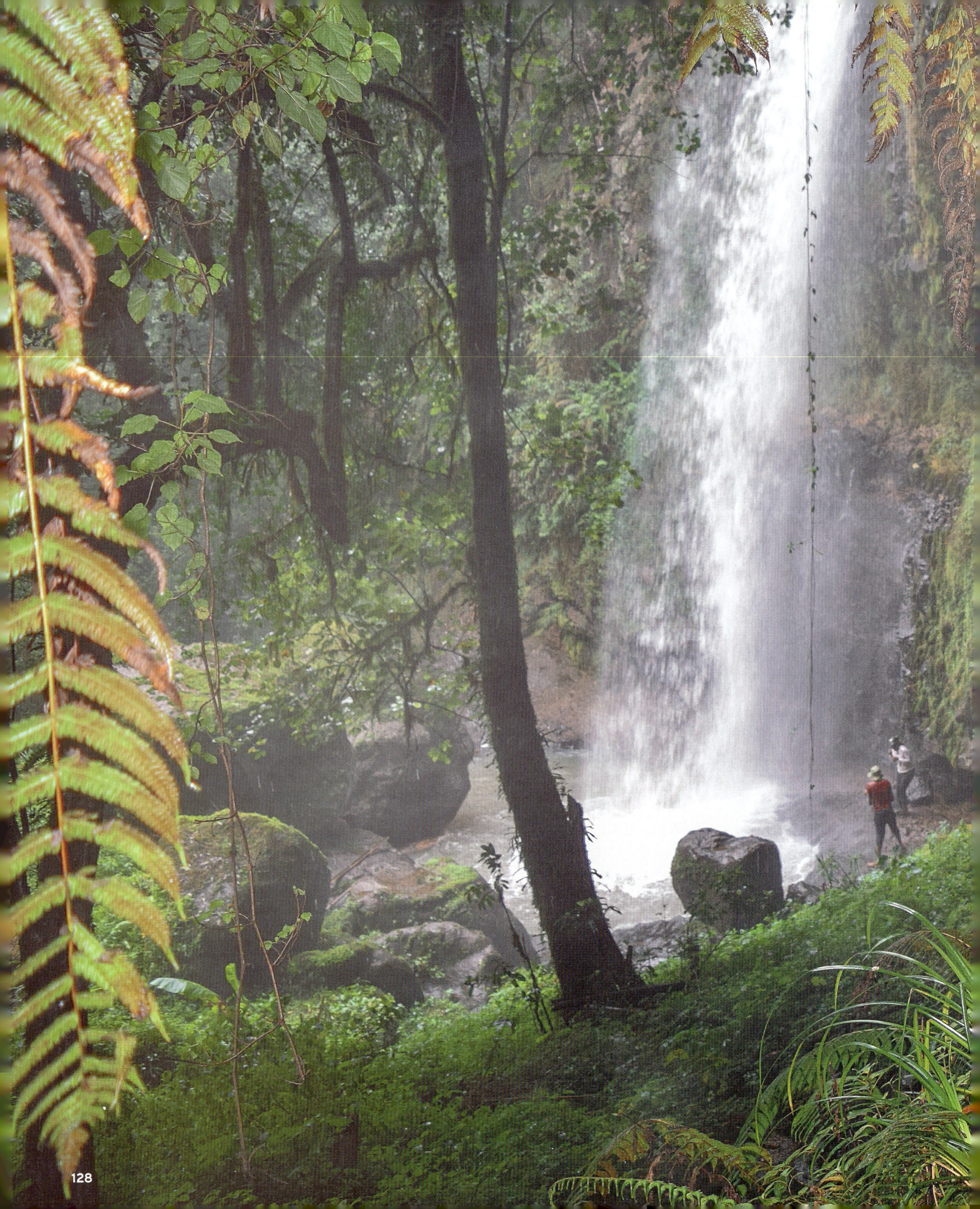

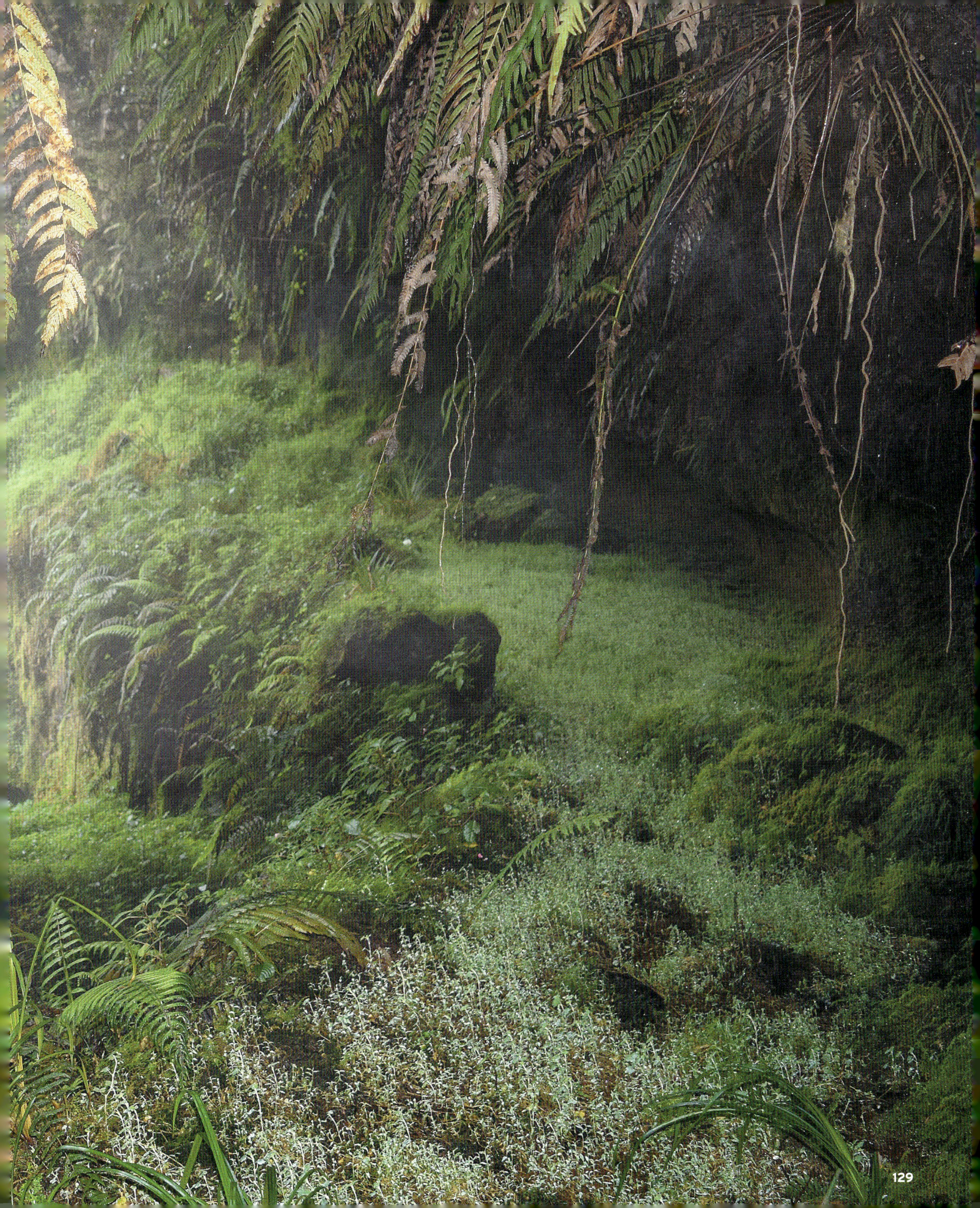

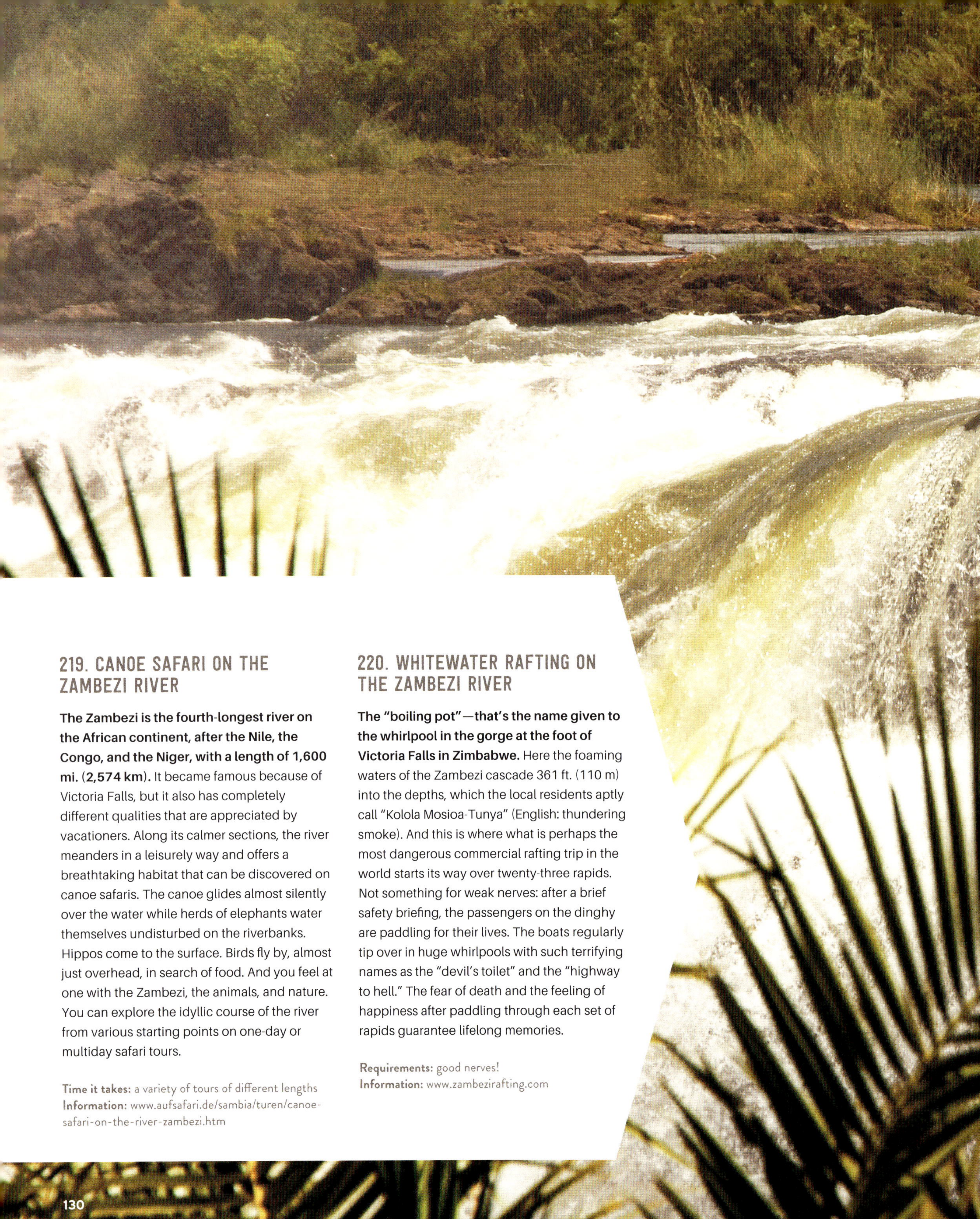

219. CANOE SAFARI ON THE ZAMBEZI RIVER

The Zambezi is the fourth-longest river on the African continent, after the Nile, the Congo, and the Niger, with a length of 1,600 mi. (2,574 km). It became famous because of Victoria Falls, but it also has completely different qualities that are appreciated by vacationers. Along its calmer sections, the river meanders in a leisurely way and offers a breathtaking habitat that can be discovered on canoe safaris. The canoe glides almost silently over the water while herds of elephants water themselves undisturbed on the riverbanks. Hippos come to the surface. Birds fly by, almost just overhead, in search of food. And you feel at one with the Zambezi, the animals, and nature. You can explore the idyllic course of the river from various starting points on one-day or multiday safari tours.

Time it takes: a variety of tours of different lengths
Information: www.aufsafari.de/sambia/turen/canoe-safari-on-the-river-zambezi.htm

220. WHITEWATER RAFTING ON THE ZAMBEZI RIVER

The "boiling pot"—that's the name given to the whirlpool in the gorge at the foot of Victoria Falls in Zimbabwe. Here the foaming waters of the Zambezi cascade 361 ft. (110 m) into the depths, which the local residents aptly call "Kolola Mosioa-Tunya" (English: thundering smoke). And this is where what is perhaps the most dangerous commercial rafting trip in the world starts its way over twenty-three rapids. Not something for weak nerves: after a brief safety briefing, the passengers on the dinghy are paddling for their lives. The boats regularly tip over in huge whirlpools with such terrifying names as the "devil's toilet" and the "highway to hell." The fear of death and the feeling of happiness after paddling through each set of rapids guarantee lifelong memories.

Requirements: good nerves!
Information: www.zambezirafting.com

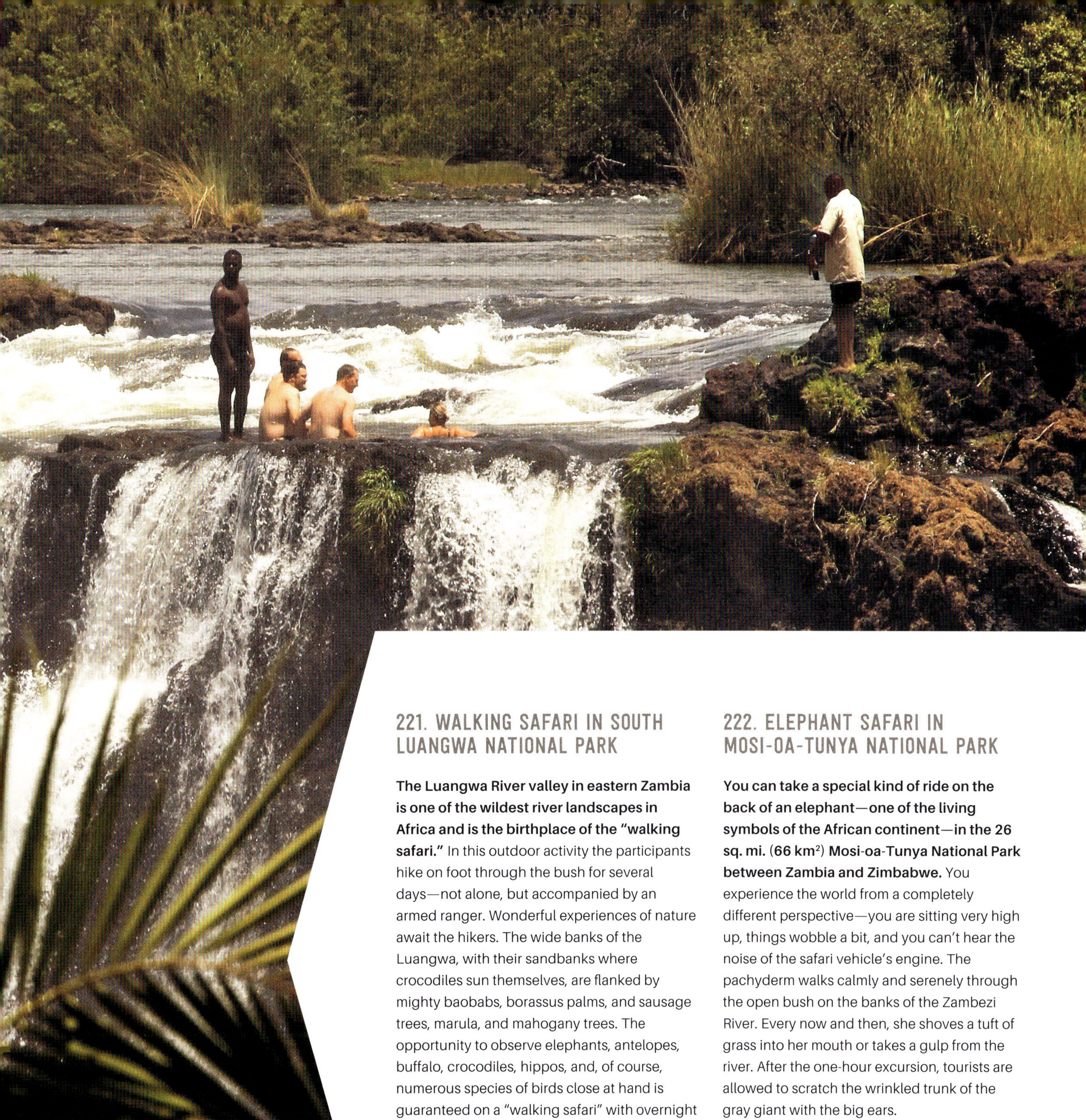

221. WALKING SAFARI IN SOUTH LUANGWA NATIONAL PARK

The Luangwa River valley in eastern Zambia is one of the wildest river landscapes in Africa and is the birthplace of the "walking safari." In this outdoor activity the participants hike on foot through the bush for several days—not alone, but accompanied by an armed ranger. Wonderful experiences of nature await the hikers. The wide banks of the Luangwa, with their sandbanks where crocodiles sun themselves, are flanked by mighty baobabs, borassus palms, and sausage trees, marula, and mahogany trees. The opportunity to observe elephants, antelopes, buffalo, crocodiles, hippos, and, of course, numerous species of birds close at hand is guaranteed on a "walking safari" with overnight stays in rustic bush camps.

Time it takes: several days; daily hiking time, 3 hours
Starting point: Kapani Lodge

222. ELEPHANT SAFARI IN MOSI-OA-TUNYA NATIONAL PARK

You can take a special kind of ride on the back of an elephant—one of the living symbols of the African continent—in the 26 sq. mi. (66 km²) Mosi-oa-Tunya National Park between Zambia and Zimbabwe. You experience the world from a completely different perspective—you are sitting very high up, things wobble a bit, and you can't hear the noise of the safari vehicle's engine. The pachyderm walks calmly and serenely through the open bush on the banks of the Zambezi River. Every now and then, she shoves a tuft of grass into her mouth or takes a gulp from the river. After the one-hour excursion, tourists are allowed to scratch the wrinkled trunk of the gray giant with the big ears.

Time it takes: 1 hour
Information: www.zambiatourism.com/destinations/national-parks/mosi-oa-tunya/

223. TRIPS IN THE HOMBORI MOUNTAINS

Summer is the ideal time to visit eastern Mali. Then the otherwise arid Sahel landscape comes to life: the river beds, which have been completely dry for a long time, carry water, and the barren thornbushes turn green. The Hombori table mountains rise proudly into the sky all year round. Gushing waterfalls pour down from the steep rock faces of the massif. The most striking mountain is the "Hand of Fatima," with its rock towers, which you can climb on a gravel path. The trip gives you experiences of nature that are sure to be remembered. With a little climbing experience, you can also ascend the steep slopes of Hombori Tondo (3,789 ft. [1,155 m]), Mali's highest mountain—from above, there is an unforgettable panorama of one of the most beautiful landscapes in West Africa.

Time it takes: a variety of day hikes of different lengths

224. ALONG THE FALAISE DE BANDIAGARA CLIFFS

Bandiagara, in southeastern Mali, is the starting point for excursions into the Dogon country that are rich in cultural history. These black African farming people have a great cultural tradition and are known for their idiosyncratic architecture and complicated mythology. Most of their villages, built on the slopes of a cliff (falaise), can be reached only on foot. With a local guide you get an insight into the fascinating world of Dogon beliefs. You will spend the night in villages, built in the customary local architecture with its traditional thatched mud huts. You can admire the local rock paintings there, visit traditional markets, and, with a lot of luck, attend a masked dance based on popular folklore.

Time it takes: a variety of hikes of different lengths
Starting point: Bandiagara

225. NYIKA NATIONAL PARK

Huge granite blocks overgrown with moss and lichens interrupt the image of the gently undulating grassy landscape in Nyika National Park, which extends across both Malawi and Zambia for more than 3,827 sq. yd. (3,200 m^2). Zebras, jackals, kudus, elands, and the largest roan antelope population in Africa live on the rough high plateau. There is an opportunity to take a variety of day hikes that lead from Chelinda Lodge through pine and juniper forest: you can aim for the summit Nganda Peak (8,553 ft. [2,607 m]), Chisanga Falls, or the Chipome Valley. On a three-day trip with an armed ranger, you descend over the eastern escarpment down to Livingstonia.

Time it takes: a variety of hikes of different lengths
Starting point: Chelinda Lodge

226. SNORKELING WITH WHALE SHARKS IN MOZAMBIQUE

"Whale sharks! Jump in!" When this call comes from the guide on a snorkeling tour off the coast of Mozambique, snorkel and diving goggles are in position within a minute. The tourists frantically jump into the Indian Ocean waves so that they don't miss anything. In the blink of an eye, the gigantic mouth of a whale shark opens right in front of the divers. Quite impressive and commanding respect, but fortunately the biggest fish in the world is interested only in plankton and passes by silently. During the season from November to March, the diving schools in Ponto do Ouro, Tofo, and Vilanculos, along the coast of Mozambique, offer boat trips to see these giants of the seas. Yet, even if you do not encounter a whale shark, the underwater world of the reef, with its exotic fish and other creatures, is a revelation for diving fans.

Information: for example, www.scubaadventures.co.za, www.tofoscuba.com, www.odysseadive.com

227. SIMIEN NATIONAL PARK

The highlands of northern Ethiopia are crowned by 15,157 ft. high (4,620 m) Ras Dashen Mountain in Simien National Park. The wild and unspoiled mountain world on the "roof of Africa" is an El Dorado for nature lovers and trekking tourists alike: you hike over broad, high plateaus and through spectacular gorges, cross 13,123 ft. high (4,000 m) peaks, and gather first-class experiences of nature. With a bit of luck you can also experience unforgettable animal watching. On a several-day mountain trip especially, you can observe Abyssinian ibexes, Semien foxes, klipspringers, baboons, hyenas, and bearded vultures.

Time it takes: at least 3 days
Highest elevation: 15,157 ft. (4,620 m)

228. ON ERTA ALE ACTIVE VOLCANO

Because of its extremely remote location, difficult terrain, and inhospitable climatic conditions, few expedition travelers visit Erta Ale in the world's hottest desert, the Danakil in northeastern Ethiopia—and if they do, only when accompanied by an Afar guide. This shield volcano rises from the Danakil Depression and is almost permanently active. Anyone who climbs to the 2,011 ft. high (613 m) crater rim is faced with a downright hellish spectacle: glowing fountains of lava keep breaking through the lumpy, solid magma crust in the approximately 279 ft. deep (85 m) caldera—this phenomenon is particularly impressive under a black night sky. Of course, this is possible only if you are daring enough to spend the night up there in a bivouac!

Time it takes: 4–5 hours
Requirements: good nerves, experience in rough terrain
Elevation difference: 1,312 vertical ft. (400 m)

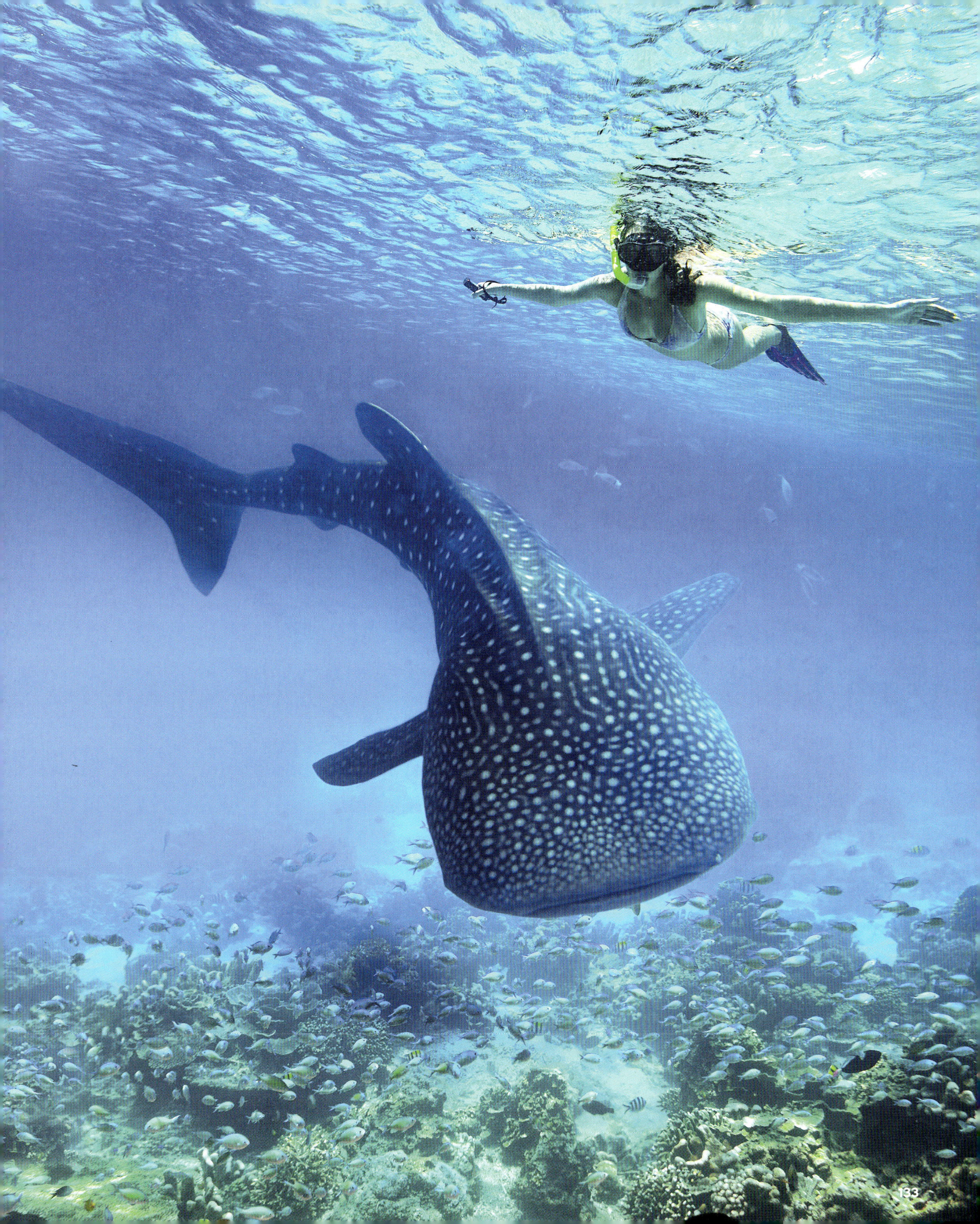

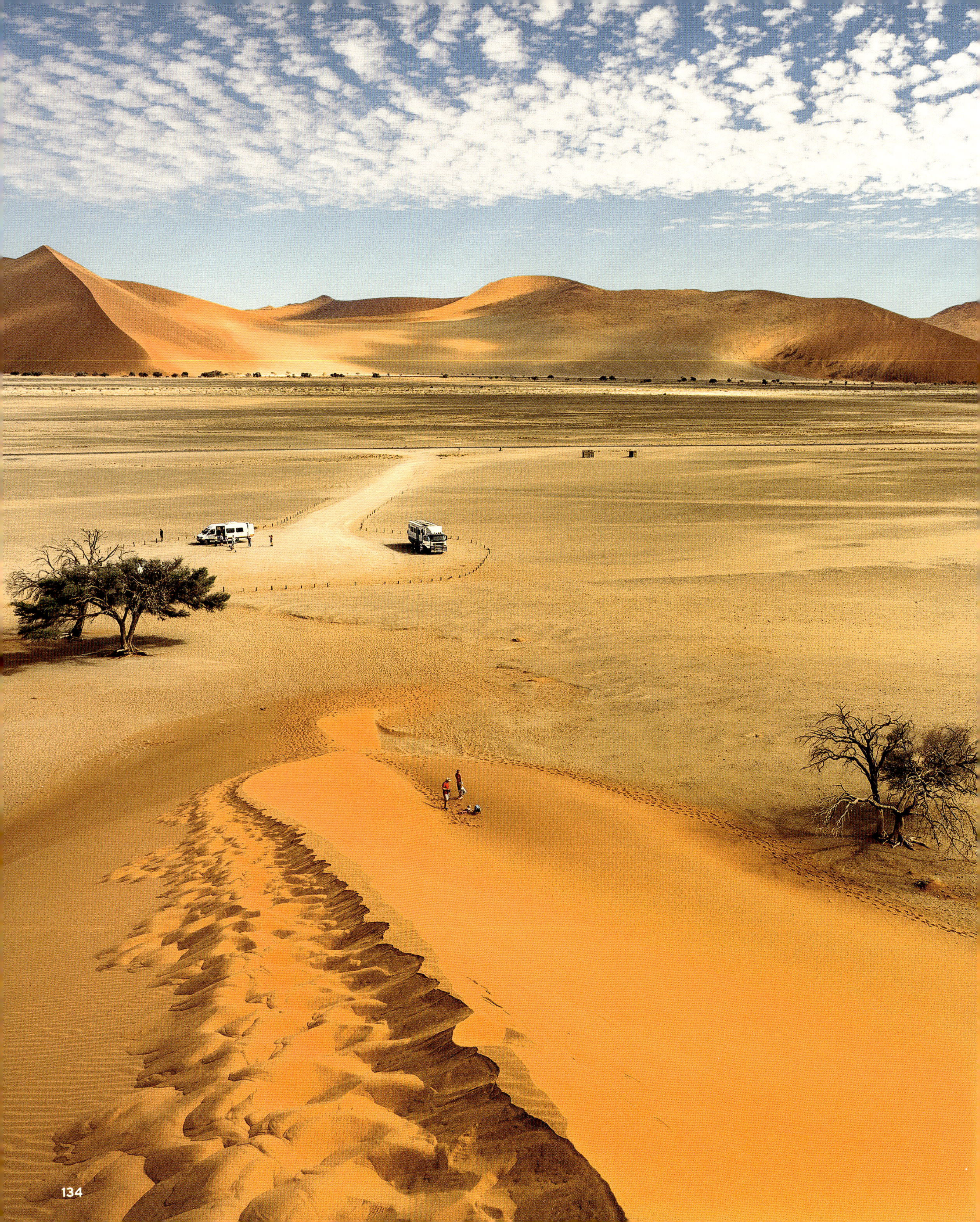

229. TO THE SOSSUSVLEI AND DEADVLEI STAR DUNES

Who hasn't already seen them, the atmospheric images of these huge reddish dunes ribbed behind a black tree, its withered branches towering into the steel-blue sky? The Deadvlei, a clay plain surrounded by sand dunes in Namib-Naukluft National Park, is certainly one of the most fascinating desert landscapes in the world. A tarred road leads from Sesriem into the park, and Dune 45 presides right along the way, not to be missed. The ascent of this sand mountain at dawn is an absolute must to see the sunrise from the top. The highest Sossusvlei dunes are more than 656 ft. (200 m) high; the climb in the soft sand is exhausting, but the view definitely makes up for it!

Time it takes: as desired; short hikes
Elevation difference: 656 vertical ft. (200 m)

230. TREKKING THROUGH FISH RIVER CANYON

This canyon, one of the largest in the world, is in the |Ai-|Ais / Richtersveld Transfrontier Park, a national park between Namibia and South Africa, and is 100 mi. (160 km) long. You will be able to hike more than half of the Fish River Canyon Trail in five days, but only during their wintertime (August or September), because of the scorching heat that prevails in the Namib Desert. The steep descent over scree—1,640 vertical ft. (500 m) down into the canyon—is the first hardship test. The hike continues without a trail, through a rugged gorge landscape, along the riverbank, and over difficult ground made up of sand and stones. River crossings are also part of the trip. This adventure trip ends with a swim in the |Ai-|Ais thermal hot springs.

Length / time it takes: 56 mi. (90 km) / 5 days, daily hiking time about 8 hours

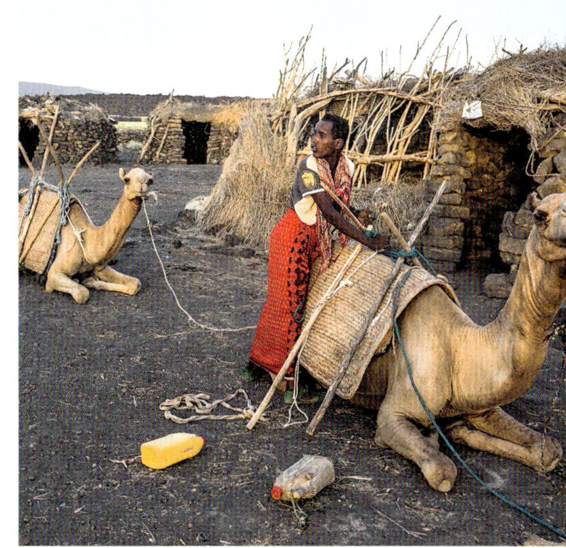

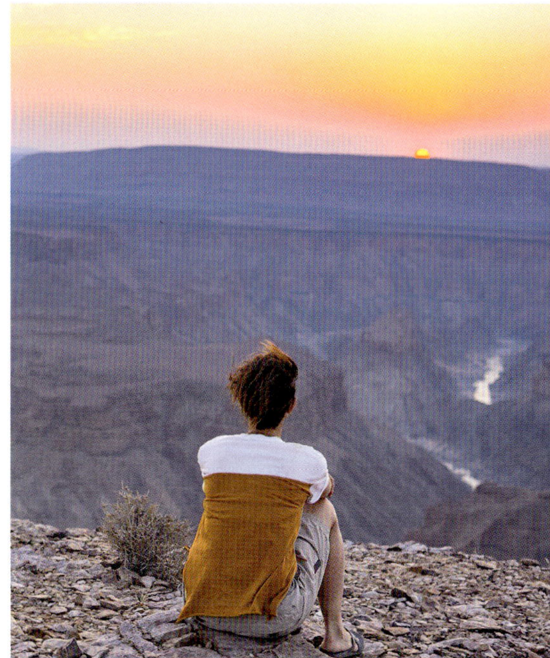

231. TREKKING ON THE TOK TOKKIE TRAIL

The Namib Desert measures 36,680 sq. mi. (95,000 km²) and is on the southwest side of the African coast. UNESCO has classified part of the desert, the "Namib Sand Sea," as a World Heritage Site. Hikers experience the fascinating landscape of the Namib in all its diversity and beauty along the three-day Tok Tokkie Trail, which leads through the southern NamibRand nature reserve. Red sand dunes, wide gravel plains, and desert mountains alternate with each other as oryx antelopes and springboks cross your way. Accompanied by a guide, you will learn the tricks that plants and animals use to survive in the extremely dry conditions. In the evening at the latest, you will get a real desert feeling in your camp bed: nowhere does the starry sky shine more clearly and brighter than in the desert.

Length / time it takes: 19 mi. (30 km) / 3 days, a daily maximum of 6.2 mi. (10 km) and 5 hours of hiking time

232. A MOKORO SAFARI IN THE OKAVANGO DELTA

The mokoro, a centuries-old form of dugout boat, is the classic means of transportation in the Okavango delta. A one-to-several-day guided tour in these traditional boats starts from the city of Maun, or from the Mboma Boat Station in Moremi Game Reserve in northern Botswana. The mokoro glides almost silently through the water, and you can watch many birds, such as sea eagles and kingfishers. With a bit of luck, there will be an elephant or buffalo standing on the reedy bank. . . . Overnight stays are either in the vicinity of Maun (such as the Audi Camp), in (very expensive) lodges in the central Okavango delta, or in so-called tented camps (with safari tents) in the Moremi Reserve.

Information: www.audisafaris.com/audi-safaris/#nav_mokoro

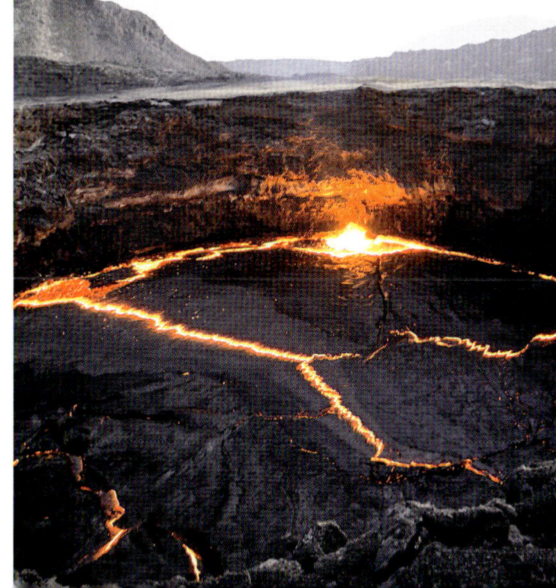

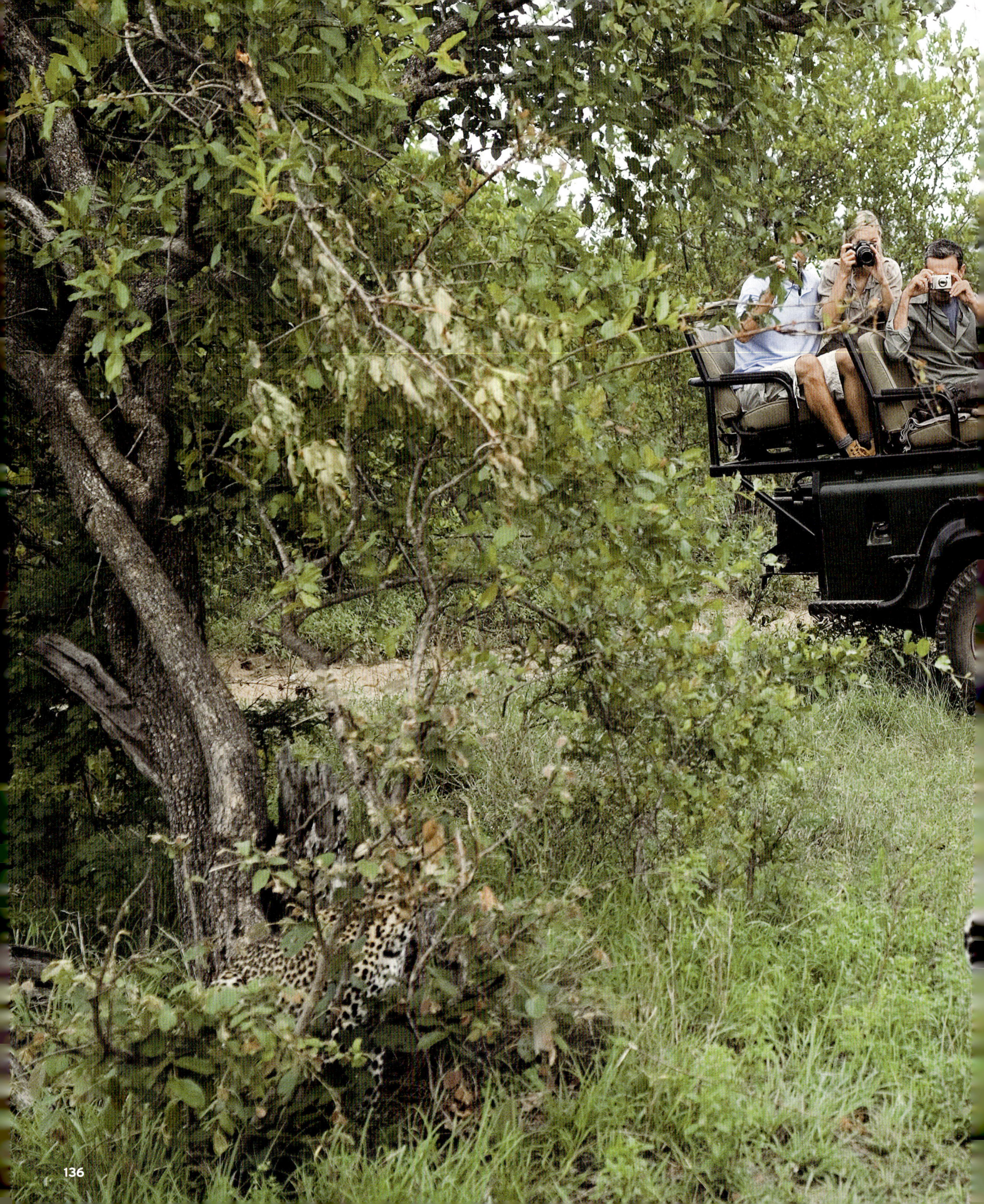

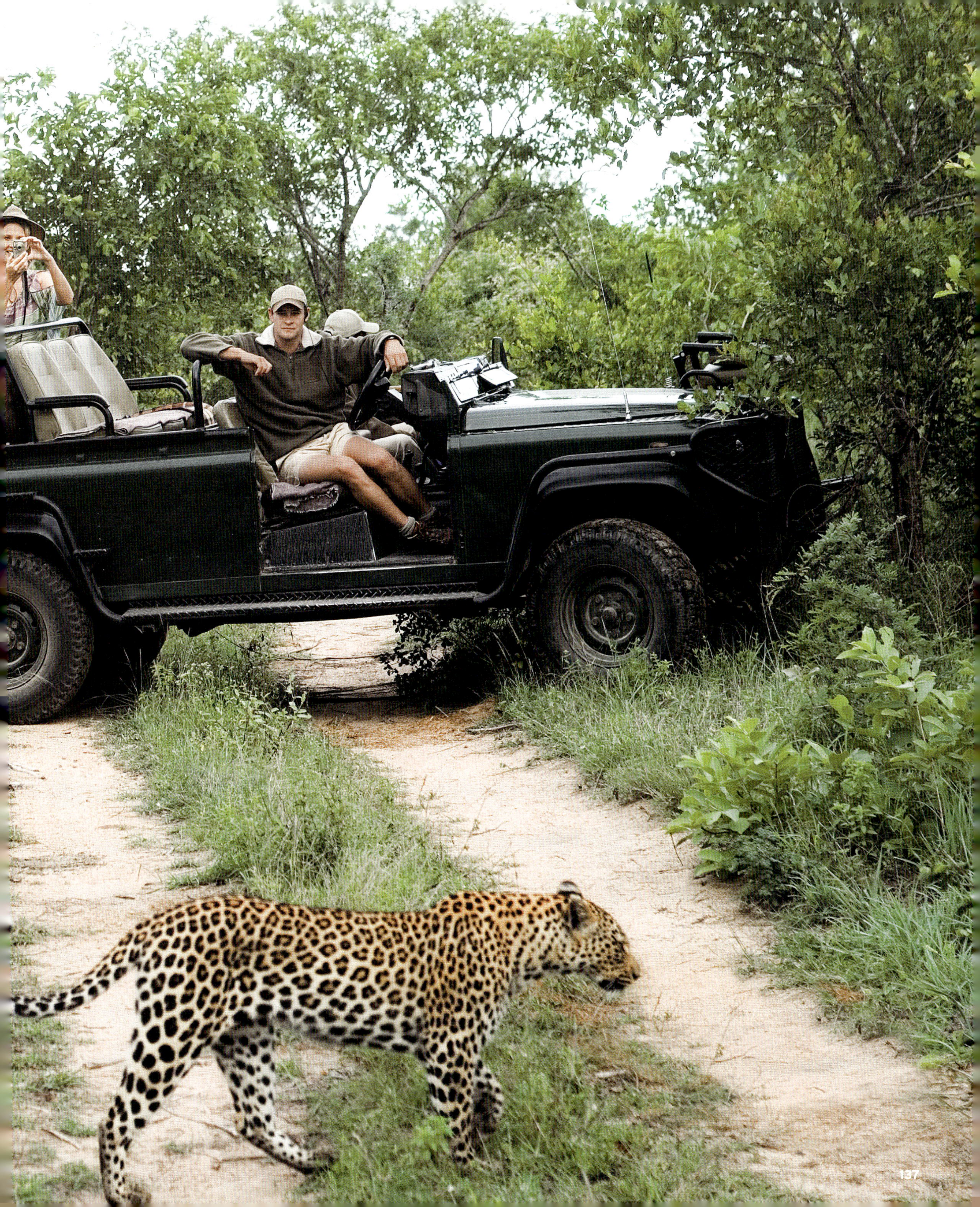

233. HIKING IN TABLE MOUNTAIN NATIONAL PARK

Table Mountain (3,563 ft. [1,086 m]) is part of a mountain range that stretches across the Cape Peninsula; it is about 32 mi. (52 km) long and 10 mi. (16 km) wide at its widest point. The Cape of Good Hope lies at its southern end. Cape Town tourists should definitely explore Table Mountain, with its unique fynbos vegetation, on foot—assuming you are sufficiently fit to take on the approximately 2,953 vertical ft. (900 m). Various ascent routes of varying levels of difficulty guarantee special experiences of nature and grandiose views of nearby Cape Town and the sea. The best-known routes are Platteklip Gorge (in the north), Nursery Ravine, and Skeleton Gorge (from Kirstenbosch). The two-day Cape of Good Hope Trail starts at the entrance to the Cape of Good Hope conservation area and runs along some 21 mi. (34 km) of lonely beaches and wild coastline.

Length / time it takes: 21 mi. (34 km) / 2 days
Highest elevation: 3,563 ft. (1,086 m)

234. ON THE WILD COAST TRAIL

This long-distance hiking trail runs between Durban and Port Elizabeth, on the east coast of South Africa, and leads from Port St. Johns to Coffee Bay. The trail is certainly the most strenuous trekking path in South Africa, but also one of the most beautiful! If you are put off by the great distance, you can simply explore just one of the five sections. From the sixteenth to the nineteenth centuries, many English and Portuguese ships became stranded on the dangerous coastline; hence the name Wild Coast. The trail keeps leading south along the rugged coast, past lonely pebble and sand beaches, rocky cliffs, lagoons, and mangrove swamps. In addition, you pass by some Xhosa villages and gain insight into the culture and way of life of the Xhosa people. The most famous attraction is the Hole-in-the-Wall: the rumbling surf beats its way through a mighty hole in a rocky island.

Length / time it takes: 155 mi. (250 km) / 5 days
Starting point: Port St. Johns
Trail's end: Coffee Bay
Information: for example, wildcoasthikes.com

235. WALKING SAFARI IN KRUGER NATIONAL PARK

Have you ever looked an elephant or lion right in the eye? In Kruger National Park, which is often incorrectly referred to in German as Kruger Park, a walking safari offers the opportunity to meet one or the other representative of the "big five." In the park's bushland, one of the largest wildlife sanctuaries in the world, you can get really close to African wild animals—accompanied by an armed ranger, of course. Several wilderness trails (such as the Wolhuter, Olifants, or Bushman Trails) take you to bush camps in the still-almost-pristine wilderness in this nearly 2,000,000-hectare national park over four days' time. Along the way you learn to identify the most-important plants of the South African Lowveld, to interpret animal tracks and scat, and to face a buffalo more calmly the next time . . .

Time it takes: about 4 days
Information: for example, www.krugerpark.co.za/

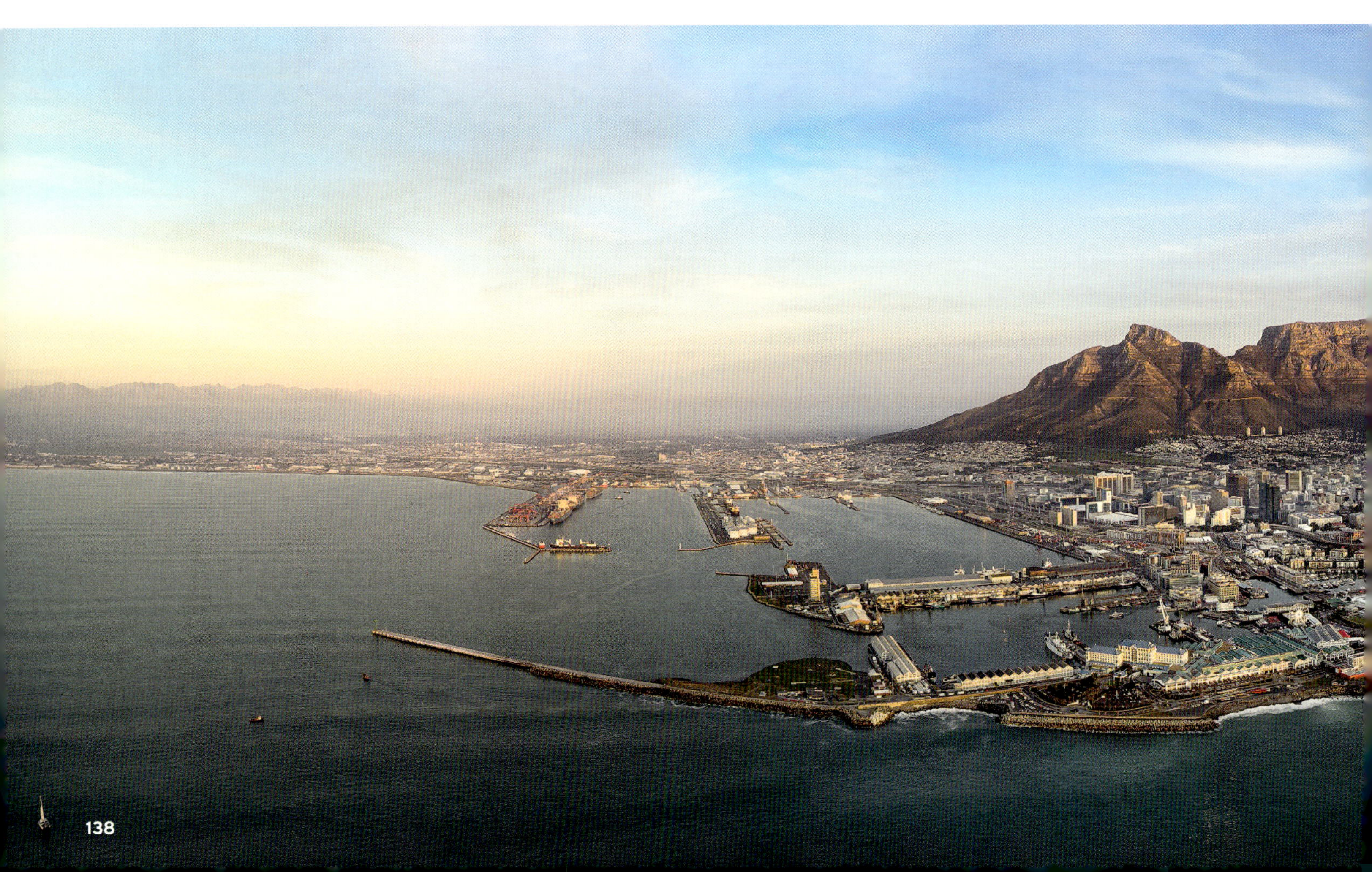

236. THE OTTER TRAIL

One of the most famous and popular South African trekking trails begins at Storms River Mouth Camp, just under 124 mi. (200 km) west of Port Elizabeth on the Indian Ocean. The moderately difficult hike runs through Tsitsikamma National Park. For five days you walk southward through the wilderness, always with a great view of the rugged coastal landscape on the left and dense primeval forest on the right. Tsitsikamma Forest, with its exotic-looking fynbos vegetation consisting of proteas, ferns, and orchids, forms the largest contiguous rainforest region in South Africa. The estuaries along the way often can be crossed only at low tide. You will spend the night in huts on the way and must be well prepared for the trip: you need to book up to twelve months in advance!

Length / time it takes: 26 mi. (42 km) / 5 days, at a daily hiking time of 5–14 hours
Information: www.sanparks.org/parks/garden_route/camps/storms_river/tourism/otter.php

237. SHARK DIVING IN GANSBAAI

The great white shark is the king of the seas. Along the coasts of South Africa, you have a particularly good chance of seeing these 13 to 20 ft. long (4–6 m) saltwater giants up close and to snorkel or dive with them. The boat takes you out to the sea from Gansbaai, on South Africa's southwest coast. Bait is cast at suitable places, and it usually takes only a few minutes for majestic bull, tiger, or great white sharks to come circling the boat. Now the word is "Get into the water!" When so-called cage diving, you climb into a cage, which comes up to your neck and is attached to the boat, and you can look the sharks straight in the eye. Experienced divers encounter sharks in a more natural way, without a cage; for example, in the Aliwal Shoal and Protea Banks diving areas. Don't worry; the sharks won't see the divers in their equipment as prey . . .

Information: http://aliwalshoal.co.za/baited-shark-dive/ www.sharkcagediving.co.za

238. ON LESOTHO'S HIGH PLATEAU

The rough, lush green high plateau of Lesotho towers like an island in the middle of South Africa. A bumpy road leads over Sani Pass (9,429 ft. [2,874 m]) into the kingdom of Lesotho, where civilization—at least so far—does not seem to have arrived yet. Small paths lead up and steeply down across the unspoiled, unforested mountainous area, past babbling waterfalls and picturesque Basotho round huts—the only evidence of human settlement in this remote area. The greatest challenge is certainly the ascent of the highest mountain in southern Africa, 11,424 ft. high (3,482 m) Thabana Ntlenyana. But there are also other trekking routes: from a day hike from the Sani Pass to multiday trekking accompanied by pack ponies and a local guide, all options are open for interested tourists.

Time it takes: one or several days, as desired
Highest elevation: 11,424 ft. (3,482 m)
Information: seelesotho.com/hiking-in-lesotho

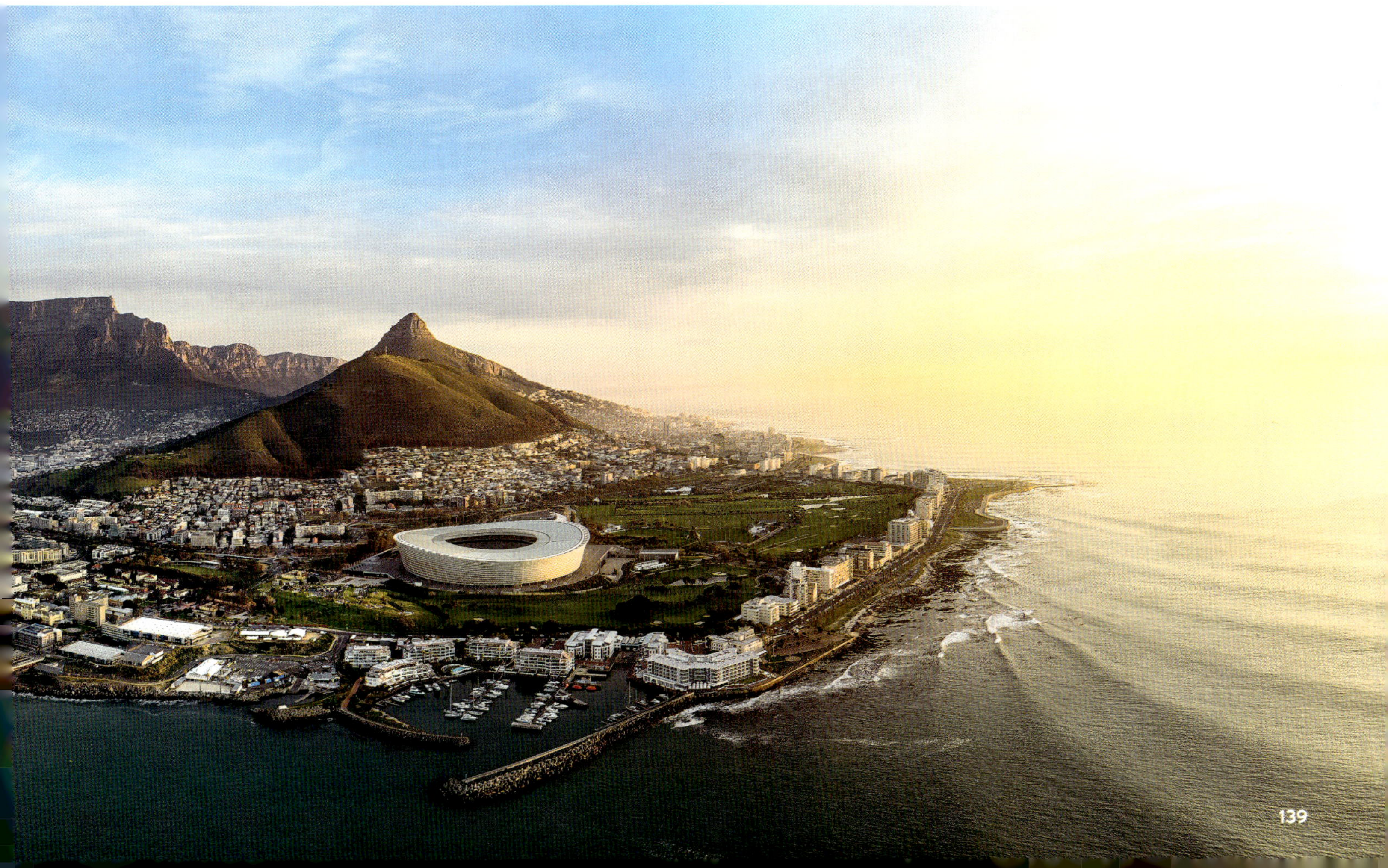

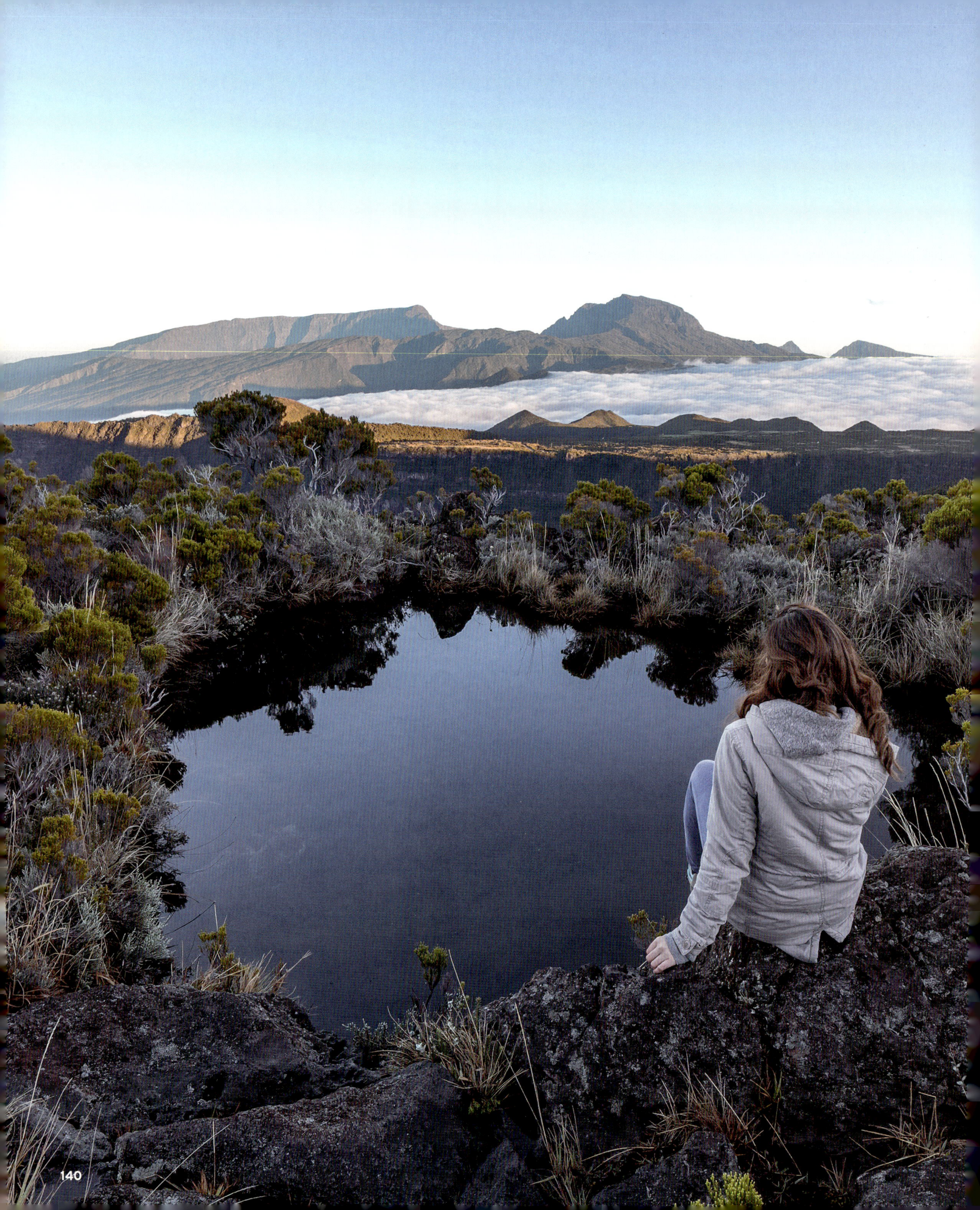

239. TREKKING ON THE MASOALA PENINSULA

Hikers discover wonderfully beautiful beaches, pristine rainforest, cascading waterfalls, and traditional villages on the Masoala Peninsula in northern Madagascar. A strenuous multiday trip for adventurous trekkers leads from Antalaha, the capital of vanilla, to Maroantsetra, at the end of Antongil Bay. You walk along muddy trails across rice terraces and vanilla and coffee plantations and come through small villages. At Ambatoledama you should definitely explore the jungle, where you have the chance to get to know the native fauna: here, various species of lemurs bravely move hand over hand from branch to branch. In Ampokafo you can extend the trek to Cap Est and Masoala National Park (a World Heritage Natural Site).

Length / time it takes: 68 mi. (110 km) /5 days
Starting point: Antalaha
Trail's end: Maroantsetra

240. VOLCANIC RÉUNION ISLAND

Although this fascinating island is part of France, it has absolutely nothing in common with that country, at least in terms of landscape. No wonder, since the volcanic island lies east of Madagascar and southwest of Mauritius, far out in the Indian Ocean. It is a truly heavenly spot, and trekking travelers will feel that they are in good hands here: well-marked hiking trails, a dozen mountain huts, and the diverse landscape between the mountains and the sea make it possible to undertake many day hikes and multiday trips. In the interior of the country, trails run through rugged mountain landscapes with active volcanoes, rainforests, gorges, and valley basins, and past solidified lava flows and rushing waterfalls. The absolute highlights are the ascent of Piton des Neiges (10,069 ft. [3,069 m]), the highest peak on the island, and Piton de la Fournaise (8,635 ft. [2,632 m]), one of the most active volcanoes in the world.

Time it takes: as desired, one day or several days
Highest elevation: 10,069 ft. (3,069 m)
Information: www.insel-la-reunion.com/

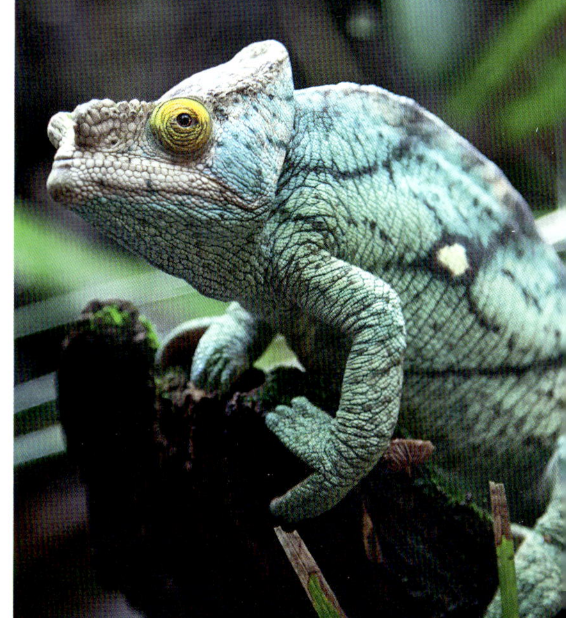

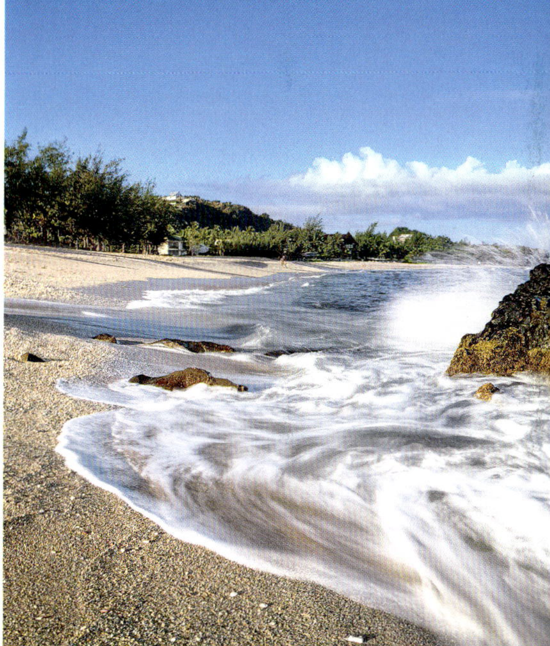

241. TANDEM PARAGLIDING AT SAINT LEU

Goosebumps guaranteed! On tandem paragliding flights, the nonpilots also have the opportunity to soar through the air like a bird. Paragliders can go high in the air on Réunion—with a breathtaking view down over this island of sports, nature, and "laissez-faire," and of letting yourself go. Gentle thermals (updraft winds) over the tropical mountains invite you to explore the island from above and land on the beach. Réunion is west of the well-known, larger island of Mauritius and is a French national département, despite its location in the Indian Ocean. The safety precautions for paragliding meet European standards, and the passenger can lean back and relax in the air. The island also has a lot to offer in culinary terms—its exotic French-Creole cuisine is known worldwide.

Time it takes: 1 hour
Starting altitude: 4,754 ft. (1,449 m)
Starting point: Saint Leu, on the west coast
Information: www.airreunion.re

242. TREKKING IN THE CIRQUE DE MAFATE CALDERA

This small island in the Indian Ocean is one of the last paradises on Earth. Its omnipresent exotic quality and the grandiose landscape blend together with reliable travel safety to make Réunion one of the top destinations worldwide. Réunion is only around 43.5 mi. (70 km) in diameter and rises from the sea to an elevation of 10,072 ft. (3,070 m) atop the extinct volcano Piton des Neiges. Its northwestern caldera is the Cirque de Mafate, a difficult-to-access valley basin with steep rock faces and deep canyons. Even the mailman comes here only on foot or by helicopter. Anyone willing to take on this steep, arduous trail will find a spectacular trekking area amid the volcano, the wilderness, and only nine tiny inhabited hamlets. The GR R3 long-distance hiking trail leads once around the Cirque de Mafate.

Length / time it takes: 25.5 mi. (41 km) / 5 days
Starting point: The Col des Boeufs parking lot
Elevation difference: about 15,289 vertical ft. (4,660 m)
Requirements: sure-footed
Information: www.outdooractive.com

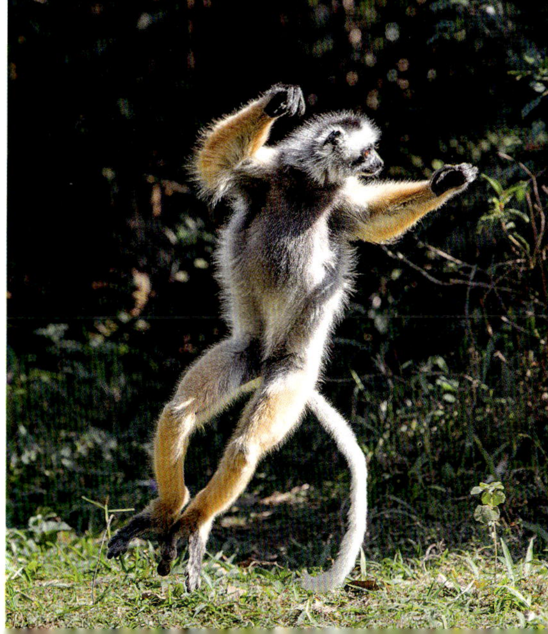

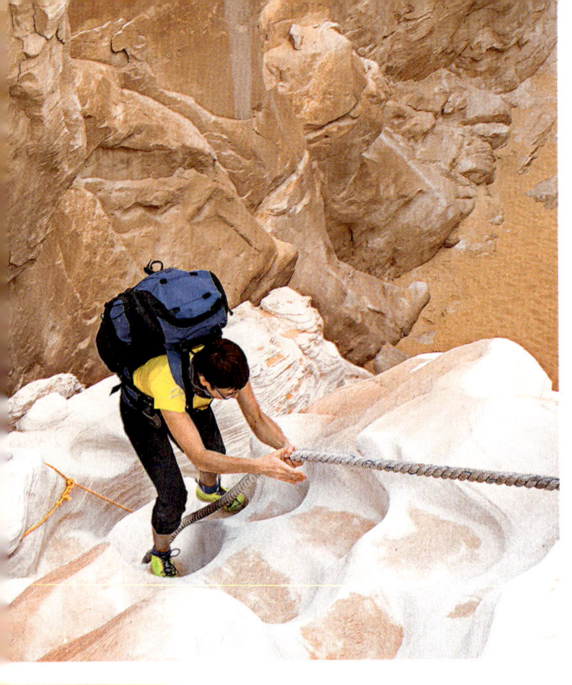

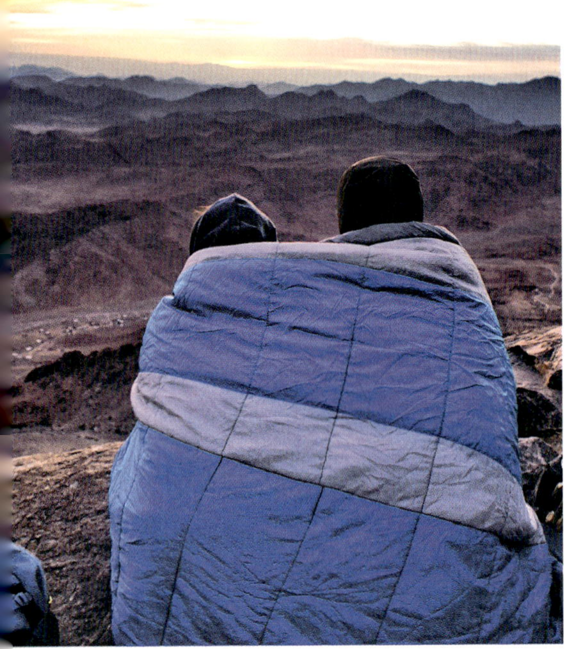

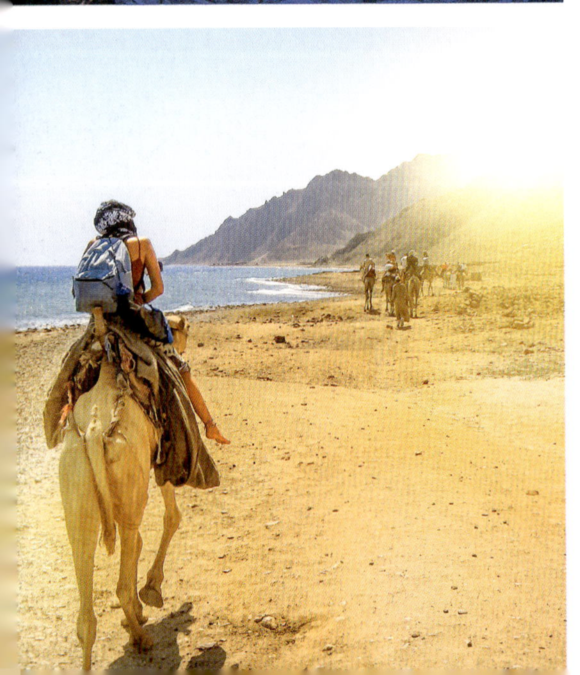

243. MOUNTAINEERING IN THE SINAI PENINSULA

You get an especially intense experience of the granite mountains and desert landscape of the Sinai Peninsula if you are in the company of Bedouins who know their way around or on the back of a dromedary. A multiday trek takes you through barren stone and sand deserts, through wadis—valleys or riverbeds where often, water runs only after rainfall—as well as deep gorges. The highlight is the ascent of the holy Jebel Musa (Mount Sinai, 7,497 ft. [2,285 m]), which you tackle before dawn; you will be rewarded for the strenuous ascent with the sunrise at the summit. No fewer than 3,750 steps lead steeply downhill again to the Monastery of St. Catherine. Right next door rises Jebel Katharina (Mount Katharina), which is the highest peak in Egypt at 8,668 ft. (2,642 m).

Time it takes: Mount Sinai hike: 3 hours
Elevation difference: 2,297 vertical ft. (700 m)
Time it takes: Mount Katharina hike: 5 hours
Elevation difference: 3,609 vertical ft. (1,100 m)

244. KITESURFING AT EL GOUNA

Turquoise-blue water, sandy beaches as far as the eye can see, and idyllic lagoons lined with chic hotel complexes alongside luxury yachts: this is El Gouna, the artificial lagoon city between desert and sea, perfectly thought out for tourism. This stretch of Egypt along the Red Sea is pampered by having the best wind conditions and pleasant water temperatures almost all year round. There is hardly a water sports enthusiast who can avoid this place, because El Gouna is just a stone's throw from Hurghada Airport, around thirty minutes' drive. There is enough space on the kitesurfing beaches even during peak times to let you enjoy the usually smooth water with the best winds. An offshore coral reef protects the standing areas. The food and ambience at the smart hotels and resorts are as legendary as the top service at the kiting stations.

Kiting area: Standing area, flat water, open sea
Wind: 4–6 Beaufort, side shore
Best time of year: all year round, very hot in summer
Suitable for: everyone
Starting point: El Gouna
Information: www.kite-college.com

245. OVER THEBAN MOUNTAIN INTO THE VALLEY OF THE KINGS

The rock face rises steeply behind the large mortuary temple of Hatshepsut, which is open to the Nile valley. To the right of the temple, a narrow trail leads steadily uphill and soon gives you a wonderful view and extraordinary perspective of the temple complex and the significantly older temple of Mentuhotep II to its southwest. With a view of the pyramid-shaped mountain El Qurn, the highest point of the Theban Mountains, you pass an old settlement where the workers from the Valley of the Kings once lived. From here the trail meanders down and leads in places over fields of scree, so it is advisable to wear good footwear. Again and again there is a breathtaking view of the famous Valley of the Kings. On the mostly lonely stretch there is almost no shade, so you should think about taking along enough water and wearing sun protection.

Length / time it takes: 1.6 mi. (2.5 km) / 1.5 hours
Starting point: Mortuary Temple of Hatshepsut
Trail's end: Valley of the Kings

246. CAMEL TREKKING THROUGH THE SINAI DESERT

The Sinai Peninsula, which juts out into the Red Sea, features deserts and rugged mountains, especially in the south. What could be a better idea than traveling through this landscape on the back of a camel? Bedouin camps in Dahab offer the opportunity to rent two-humped camels for several-day trips. The trail runs through granite and basalt gorges and valleys lined with acacia trees, sometimes going through steep passes at an elevation of more than 2,625 ft. (800 m). Beyond the passes you reach the plateaus, where the landscape changes; the ground becomes more sandy, and in the distance the first sandstone mountains stand out sharply against the sky. Unique rock formations and sand dunes towering up to 591 ft. (180 m) will ensure some impressive moments.

Length / time it takes: 44 mi. (70 km) / 4 days
Highest elevation: 3,281 ft. (1,000 m)
Starting point: Dahab
Elevation difference: 4,921 vertical ft. (1,500 m)

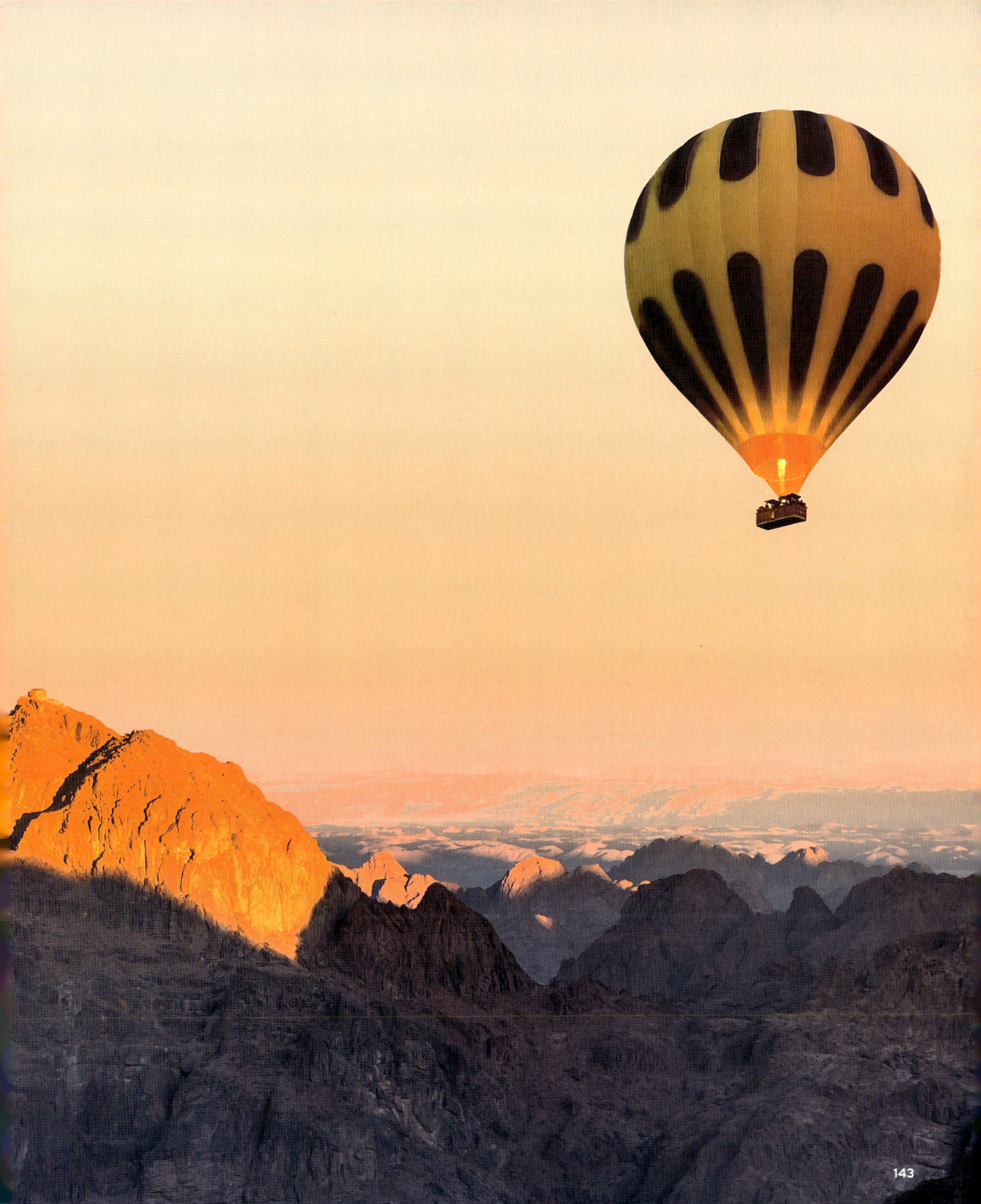

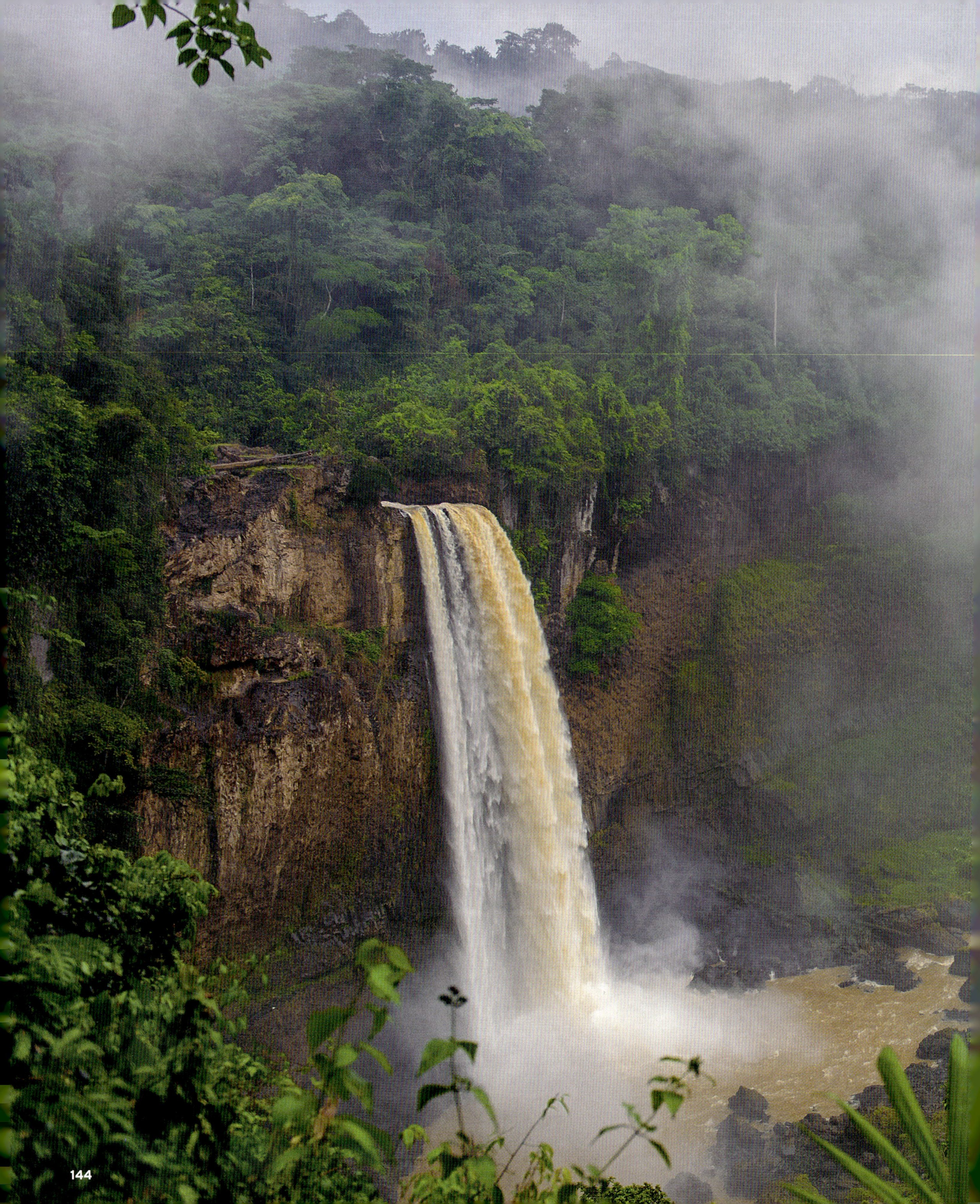

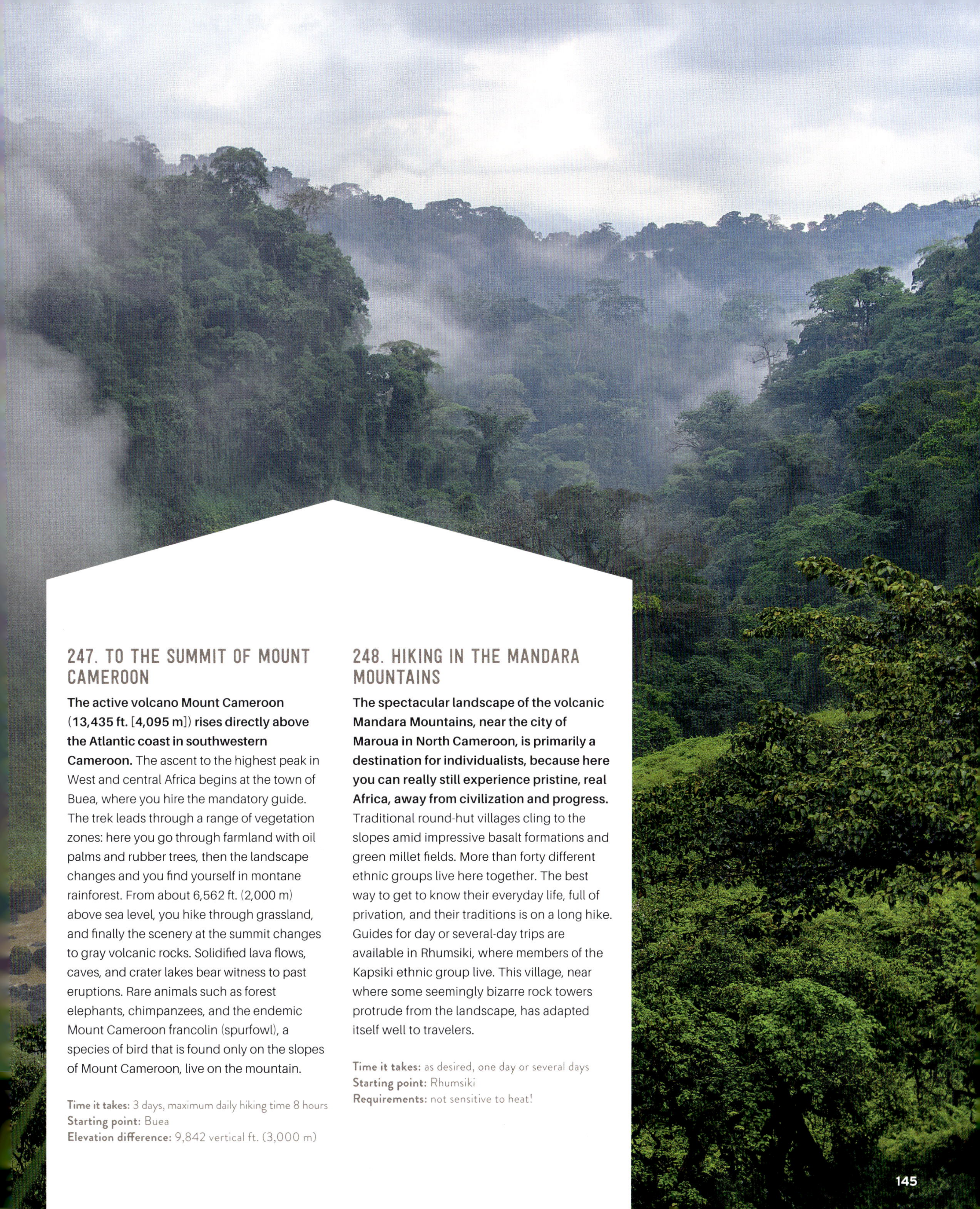

247. TO THE SUMMIT OF MOUNT CAMEROON

The active volcano Mount Cameroon (13,435 ft. [4,095 m]) rises directly above the Atlantic coast in southwestern Cameroon. The ascent to the highest peak in West and central Africa begins at the town of Buea, where you hire the mandatory guide. The trek leads through a range of vegetation zones: here you go through farmland with oil palms and rubber trees, then the landscape changes and you find yourself in montane rainforest. From about 6,562 ft. (2,000 m) above sea level, you hike through grassland, and finally the scenery at the summit changes to gray volcanic rocks. Solidified lava flows, caves, and crater lakes bear witness to past eruptions. Rare animals such as forest elephants, chimpanzees, and the endemic Mount Cameroon francolin (spurfowl), a species of bird that is found only on the slopes of Mount Cameroon, live on the mountain.

Time it takes: 3 days, maximum daily hiking time 8 hours
Starting point: Buea
Elevation difference: 9,842 vertical ft. (3,000 m)

248. HIKING IN THE MANDARA MOUNTAINS

The spectacular landscape of the volcanic Mandara Mountains, near the city of Maroua in North Cameroon, is primarily a destination for individualists, because here you can really still experience pristine, real Africa, away from civilization and progress. Traditional round-hut villages cling to the slopes amid impressive basalt formations and green millet fields. More than forty different ethnic groups live here together. The best way to get to know their everyday life, full of privation, and their traditions is on a long hike. Guides for day or several-day trips are available in Rhumsiki, where members of the Kapsiki ethnic group live. This village, near where some seemingly bizarre rock towers protrude from the landscape, has adapted itself well to travelers.

Time it takes: as desired, one day or several days
Starting point: Rhumsiki
Requirements: not sensitive to heat!

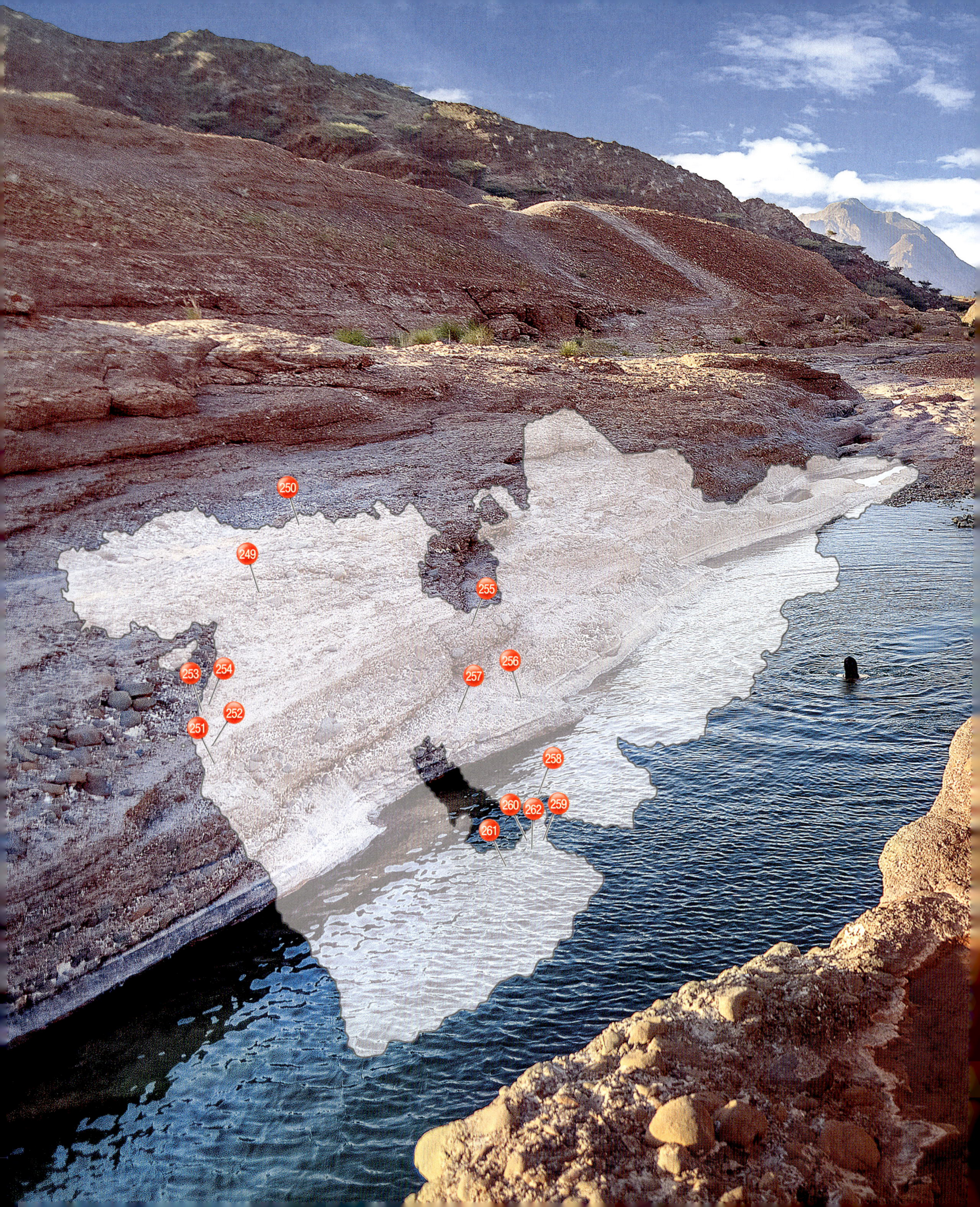
250
249
255
253
254
256
257
252
251
258
260
262
259
261

ASIA MINOR/ ARABIA

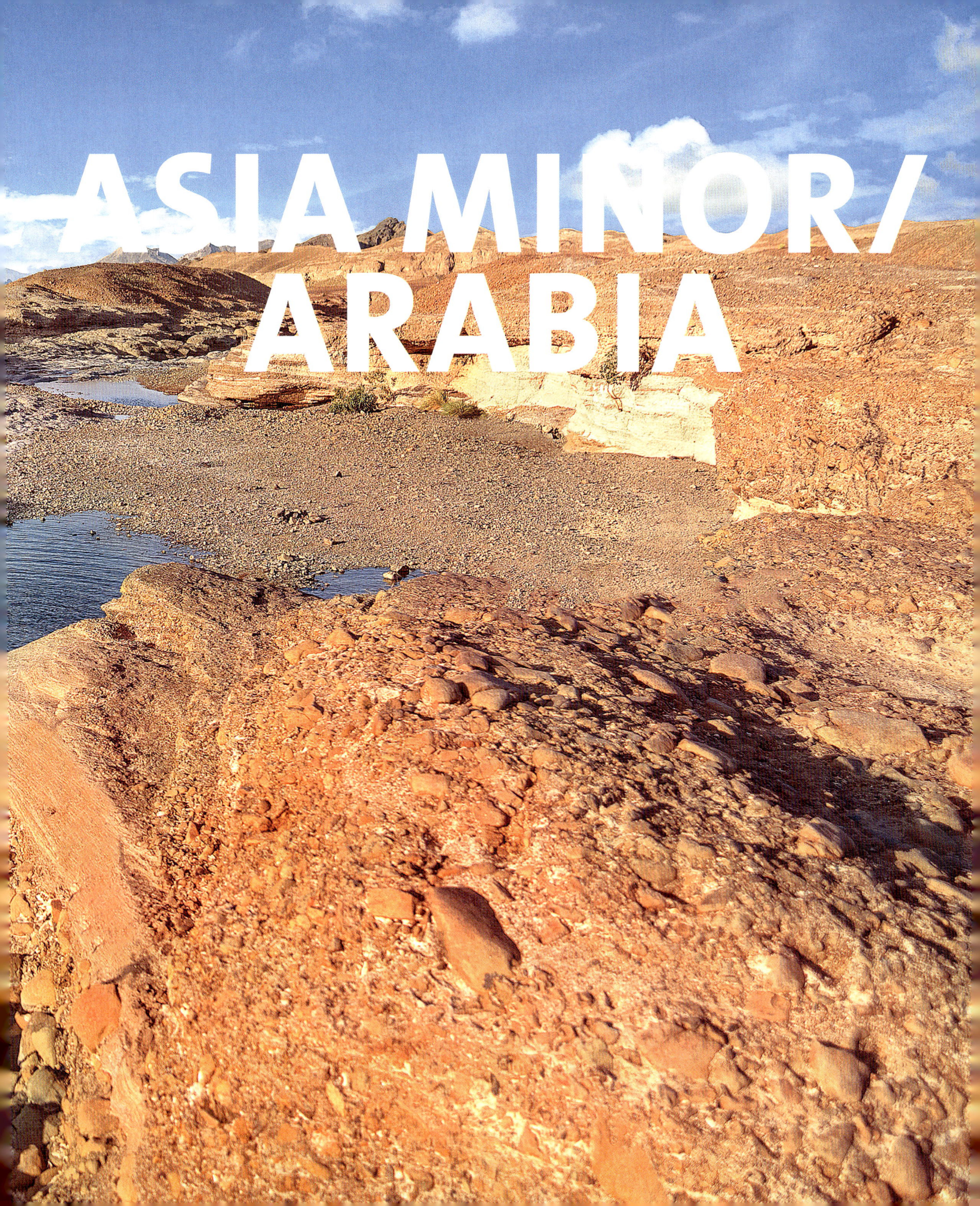

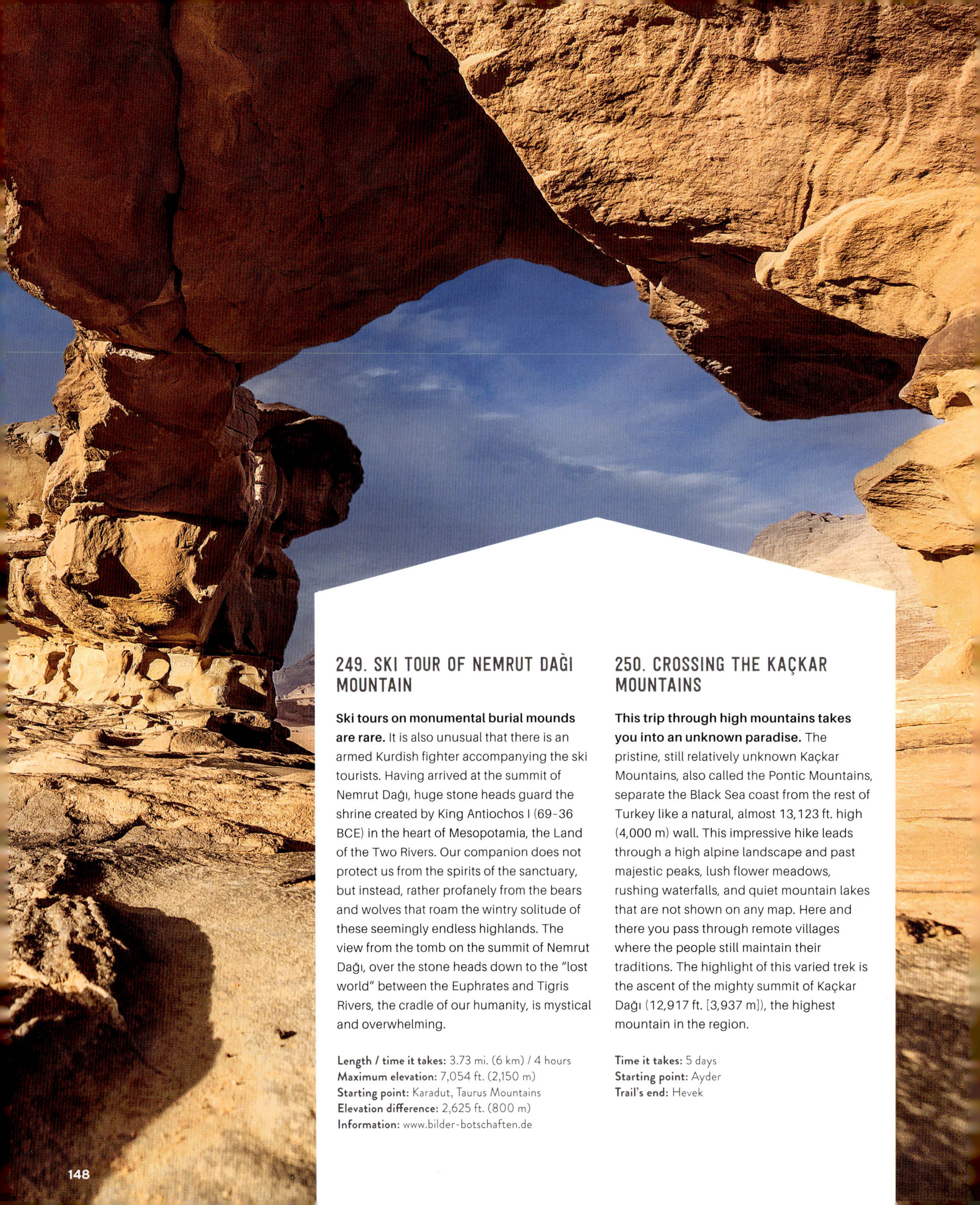

249. SKI TOUR OF NEMRUT DAĞI MOUNTAIN

Ski tours on monumental burial mounds are rare. It is also unusual that there is an armed Kurdish fighter accompanying the ski tourists. Having arrived at the summit of Nemrut Dağı, huge stone heads guard the shrine created by King Antiochos I (69–36 BCE) in the heart of Mesopotamia, the Land of the Two Rivers. Our companion does not protect us from the spirits of the sanctuary, but instead, rather profanely from the bears and wolves that roam the wintry solitude of these seemingly endless highlands. The view from the tomb on the summit of Nemrut Dağı, over the stone heads down to the "lost world" between the Euphrates and Tigris Rivers, the cradle of our humanity, is mystical and overwhelming.

Length / time it takes: 3.73 mi. (6 km) / 4 hours
Maximum elevation: 7,054 ft. (2,150 m)
Starting point: Karadut, Taurus Mountains
Elevation difference: 2,625 ft. (800 m)
Information: www.bilder-botschaften.de

250. CROSSING THE KAÇKAR MOUNTAINS

This trip through high mountains takes you into an unknown paradise. The pristine, still relatively unknown Kaçkar Mountains, also called the Pontic Mountains, separate the Black Sea coast from the rest of Turkey like a natural, almost 13,123 ft. high (4,000 m) wall. This impressive hike leads through a high alpine landscape and past majestic peaks, lush flower meadows, rushing waterfalls, and quiet mountain lakes that are not shown on any map. Here and there you pass through remote villages where the people still maintain their traditions. The highlight of this varied trek is the ascent of the mighty summit of Kaçkar Dağı (12,917 ft. [3,937 m]), the highest mountain in the region.

Time it takes: 5 days
Starting point: Ayder
Trail's end: Hevek

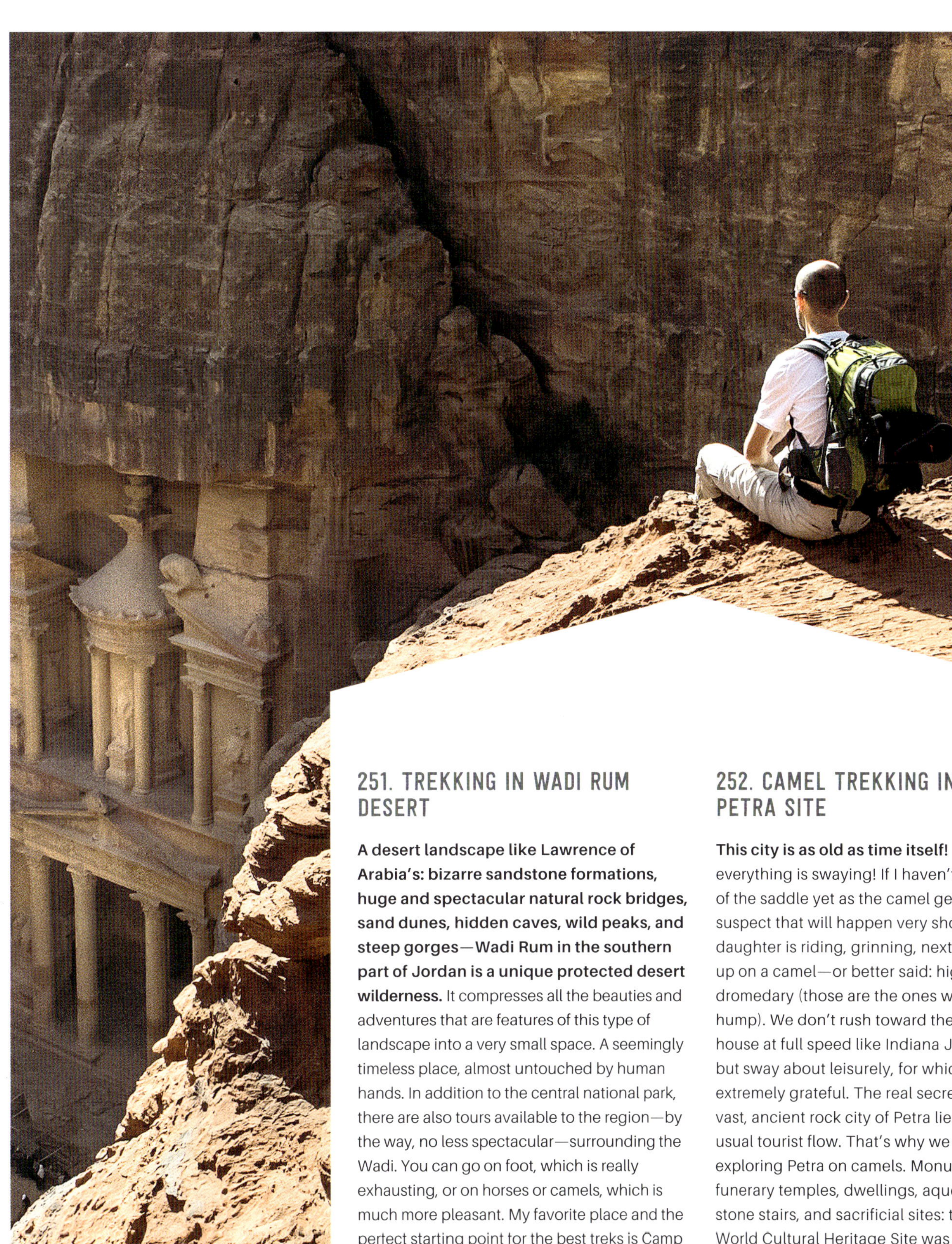

251. TREKKING IN WADI RUM DESERT

A desert landscape like Lawrence of Arabia's: bizarre sandstone formations, huge and spectacular natural rock bridges, sand dunes, hidden caves, wild peaks, and steep gorges—Wadi Rum in the southern part of Jordan is a unique protected desert wilderness. It compresses all the beauties and adventures that are features of this type of landscape into a very small space. A seemingly timeless place, almost untouched by human hands. In addition to the central national park, there are also tours available to the region—by the way, no less spectacular—surrounding the Wadi. You can go on foot, which is really exhausting, or on horses or camels, which is much more pleasant. My favorite place and the perfect starting point for the best treks is Camp Bait Ali: a magical place.

Time it takes: 1 day
Starting point: Camp Bait Ali, Shakaria
Information: www.baitali.com

252. CAMEL TREKKING IN THE PETRA SITE

This city is as old as time itself! And how everything is swaying! If I haven't flown out of the saddle yet as the camel gets up, I suspect that will happen very shortly. My daughter is riding, grinning, next to me, high up on a camel—or better said: high up on a dromedary (those are the ones with only one hump). We don't rush toward the treasure house at full speed like Indiana Jones did, but sway about leisurely, for which I am extremely grateful. The real secrets of the vast, ancient rock city of Petra lie beyond the usual tourist flow. That's why we are exploring Petra on camels. Monumental funerary temples, dwellings, aqueducts, stone stairs, and sacrificial sites: this huge World Cultural Heritage Site was carved directly out of the pink rock by the Nabataeans. Breathtaking.

Time it takes: 1 day
Starting point: Petra Visitor Center
Information: www.visitjordan.com

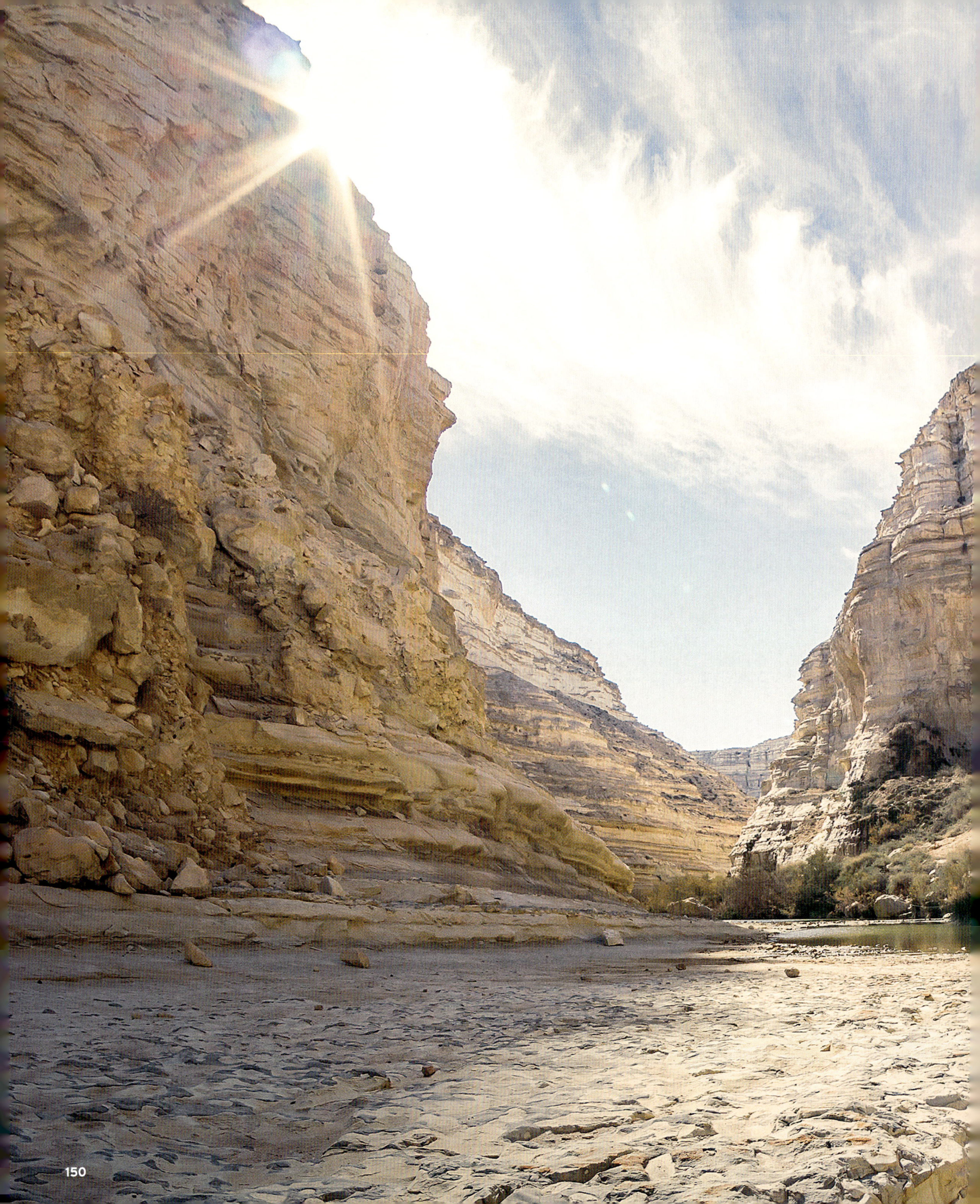

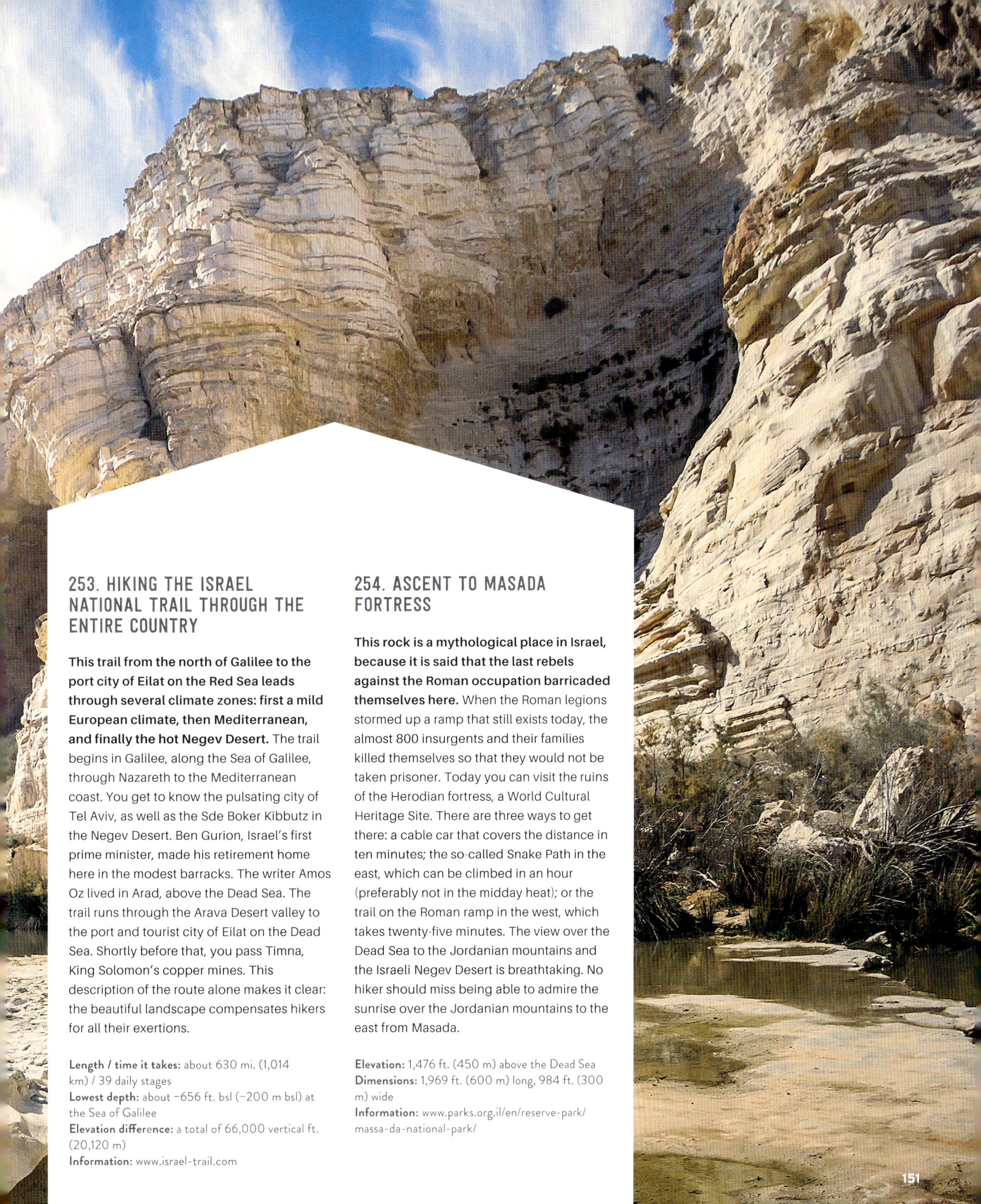

253. HIKING THE ISRAEL NATIONAL TRAIL THROUGH THE ENTIRE COUNTRY

This trail from the north of Galilee to the port city of Eilat on the Red Sea leads through several climate zones: first a mild European climate, then Mediterranean, and finally the hot Negev Desert. The trail begins in Galilee, along the Sea of Galilee, through Nazareth to the Mediterranean coast. You get to know the pulsating city of Tel Aviv, as well as the Sde Boker Kibbutz in the Negev Desert. Ben Gurion, Israel's first prime minister, made his retirement home here in the modest barracks. The writer Amos Oz lived in Arad, above the Dead Sea. The trail runs through the Arava Desert valley to the port and tourist city of Eilat on the Dead Sea. Shortly before that, you pass Timna, King Solomon's copper mines. This description of the route alone makes it clear: the beautiful landscape compensates hikers for all their exertions.

Length / time it takes: about 630 mi. (1,014 km) / 39 daily stages
Lowest depth: about −656 ft. bsl (−200 m bsl) at the Sea of Galilee
Elevation difference: a total of 66,000 vertical ft. (20,120 m)
Information: www.israel-trail.com

254. ASCENT TO MASADA FORTRESS

This rock is a mythological place in Israel, because it is said that the last rebels against the Roman occupation barricaded themselves here. When the Roman legions stormed up a ramp that still exists today, the almost 800 insurgents and their families killed themselves so that they would not be taken prisoner. Today you can visit the ruins of the Herodian fortress, a World Cultural Heritage Site. There are three ways to get there: a cable car that covers the distance in ten minutes; the so-called Snake Path in the east, which can be climbed in an hour (preferably not in the midday heat); or the trail on the Roman ramp in the west, which takes twenty-five minutes. The view over the Dead Sea to the Jordanian mountains and the Israeli Negev Desert is breathtaking. No hiker should miss being able to admire the sunrise over the Jordanian mountains to the east from Masada.

Elevation: 1,476 ft. (450 m) above the Dead Sea
Dimensions: 1,969 ft. (600 m) long, 984 ft. (300 m) wide
Information: www.parks.org.il/en/reserve-park/massa-da-national-park/

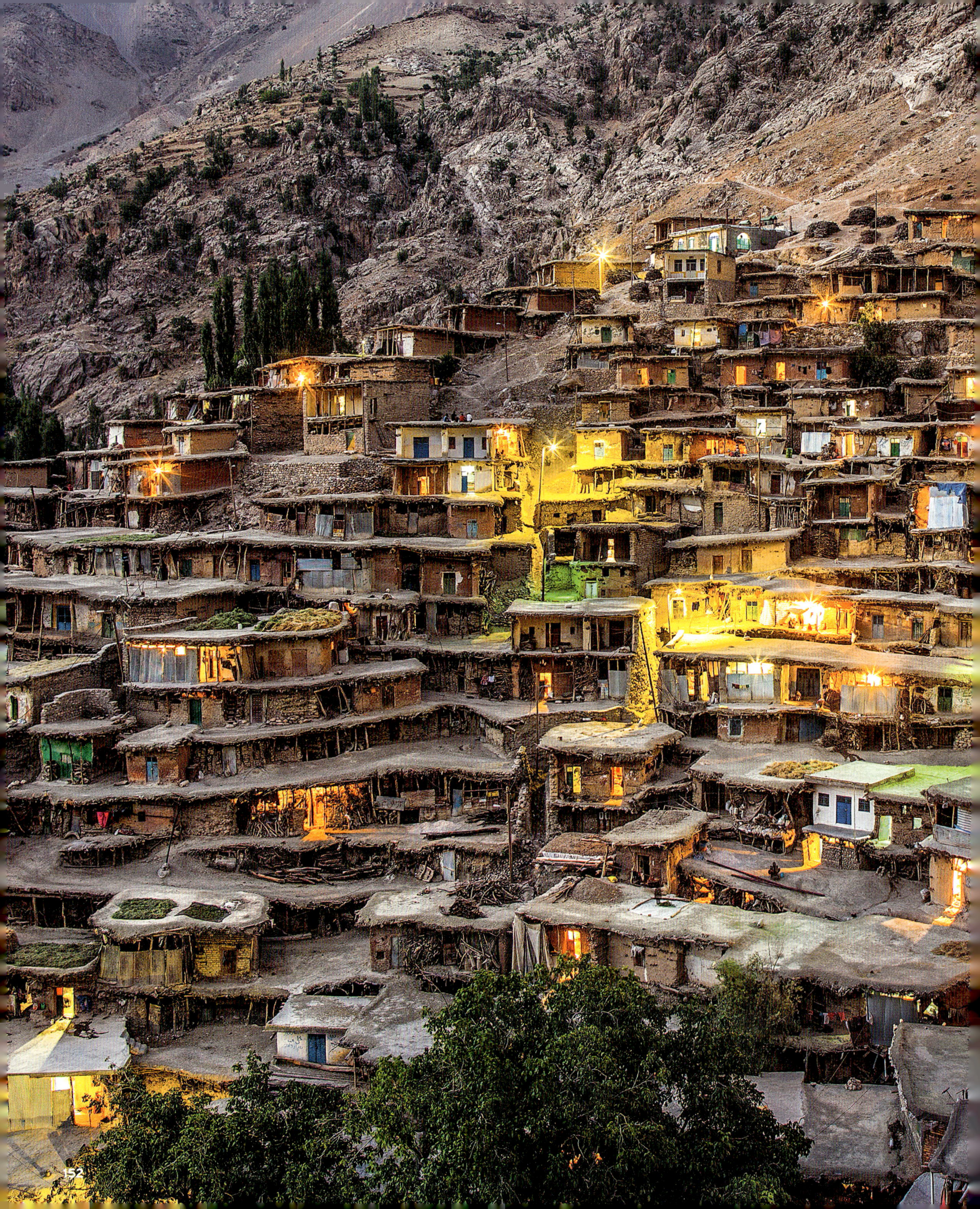

255. ASCENDING MOUNT DAMAVAND

Usually covered with a thick blanket of snow, the "Smoking Mountain" has always had a magical attraction for mountaineers. The base station is only two hours away from Tehran; thus, 18,606 ft. high (5,671 m) Mount Damavand is one of the most accessible peaks in the world and the highest point in Iran. The volcano, which has remained silent for more than 7,000 years, only occasionally releases clouds of smoke. The easiest way to get to the top is via the southern access, going through three successively staggered camps, ideally with a local guide. Don't be surprised: shortly before the summit, the earth spits out quite a lot of sulfur. The poetic Persian people have dedicated a variety of poetic legends and poems to this majestic mountain on the south coast of the Caspian Sea.

Length / time it takes: 32 mi. (52 km) / 4–7 days
Highest elevation: 18,606 ft. (5,671 m)
Starting point: Polour
Elevation difference: 11,155 vertical ft. (3,400 m)
Information: damavand.de

256. TREKKING TO THE DASHT-E KAWIR AND DASHT-E LUT DESERT OASES

Life in the void: on the fringes of the Iranian Dasht-e Kawir and Dasht-e Lut Deserts, there are many picturesque oases that owe their existence to the qanats. These are horizontal wells that draw groundwater from the mountains. The technique of building these underground irrigation channels spread from Iran along the Silk Road to the entire ancient world. Thousands of these wells are still working today and transform the emptiness of the Iranian sand and salt deserts into blooming green oases. When exploring these channels, you dive deep into the ancient oasis culture—and are completely on your own. Because of the persistently difficult political situation, there are no organized treks, but the incredibly friendly, mostly open-minded and courteous Iranians will help, always and everywhere.

Time it takes: requires a lot of time
Organization: explore on your own
Starting point: Kerman
Information: www.iran.de

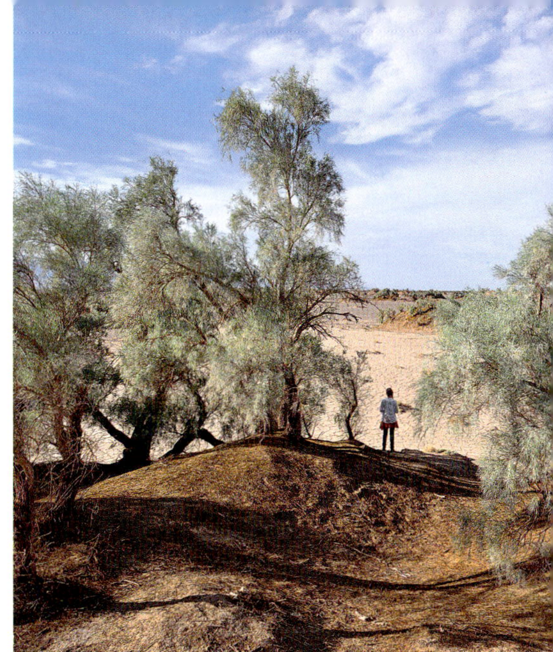

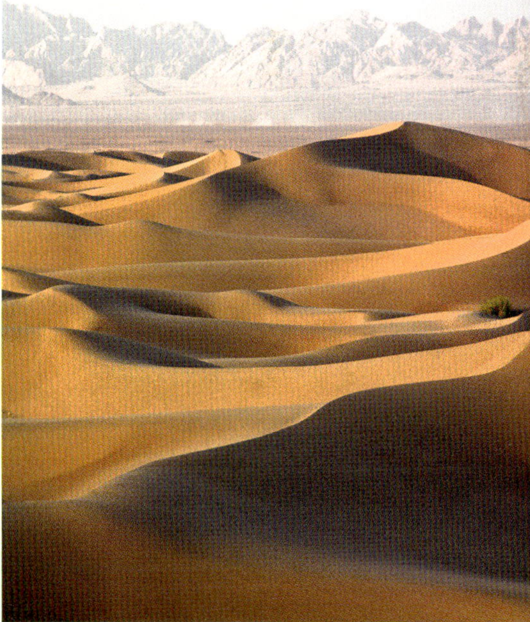

257. HIKES THROUGH THE ZAGROS MOUNTAINS

In the midst of the majestic mountainous landscape of the Zagros range and the fertile Koohrang River valley, you can experience the Bakhtiari people's still-living nomadic traditions. The starting point for every trip is the town of Chelgerd, around 118 mi. (190 km) and three hours' drive from Isfahan. Adventurers can join the nomadic life for a few days or reach the charming village of Sar Agha Seyed in three hours by jeep from Chelgerd over an eventful gravel road. The migration to the Zagros summer mountain pastures takes place from April to May, and the one to the winter pastures in Khuzestan Province from September to October. The gravel road crosses the breakneck trails of the old migration route so that you encounter migrating families again and again.

Length / time it takes: 29 mi. (46 km) / 1–5 days
Starting point: Chelgerd; time to travel: May to October
Information: nomad.tours

258. AMONG MANGROVES AND CANYONS IN QESHM GEOPARK

On the largest island in the Persian Gulf, you hike through breathtaking gorges and pass through lush mangrove forests, with watercourses that are planted with junglelike greenery. Qeshm, an island, is one of the global geoparks officially cited by UNESCO. The Hara mangrove area is home to more than 150 different migratory birds and is best explored by speedboat. You don't need to be an expert at canyoning to experience the bizarre rock formations of Chahkooh Canyon, but you do need sturdy footwear. The wide hiking trail leads into ever-narrower branches of this primary attraction. The easiest way to organize a loop tour is through one of the many guesthouses on the island. Great service: they also pick up guests from the ferry docks in Bandar e Laft or Qeshm City.

Length / time it takes: 37 mi. (60 km), or 56 mi. (90 km) from Qeshm City / 1 day
Starting point: Qeshm City or Bandar e Laft ferry terminals
Boat docks: near the villages of Tabl, Soheil, and Gavarzin
Accommodations / tour guides: Assad Homestay, email: assad2426@yahoo.com
Information: www.qeshmgeopark.ir

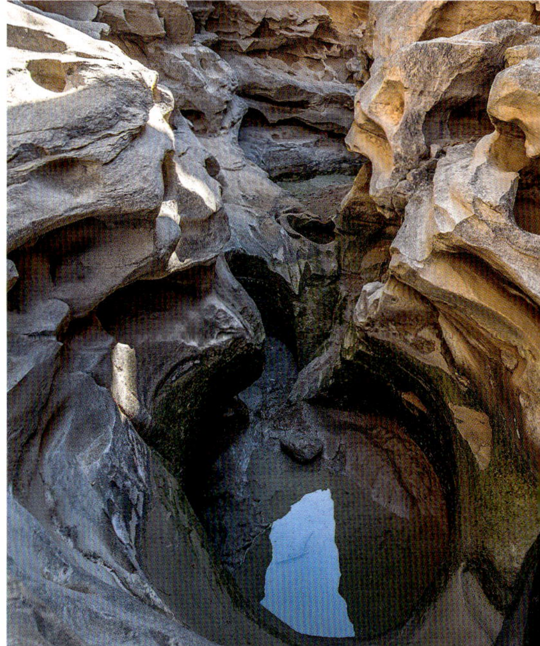

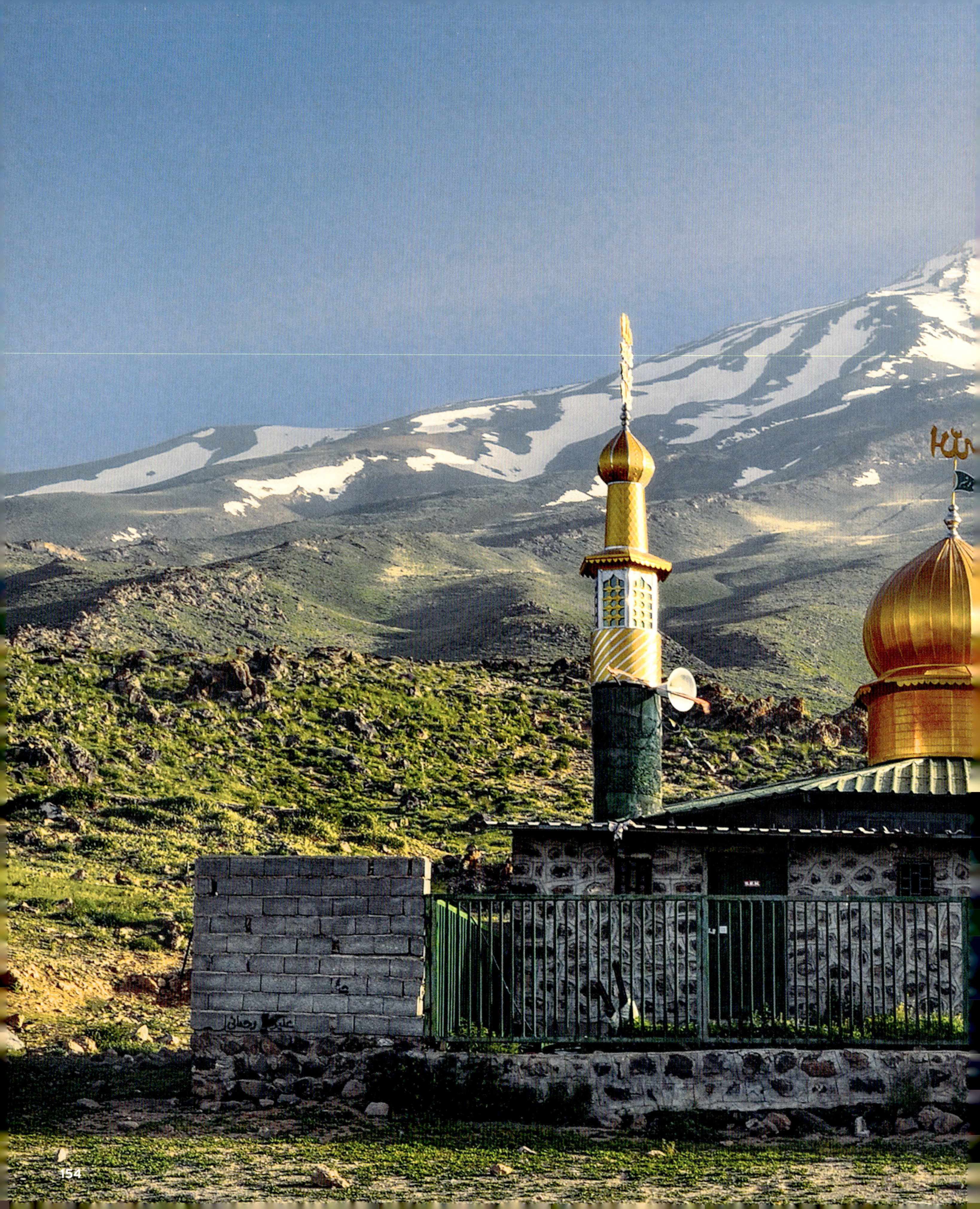

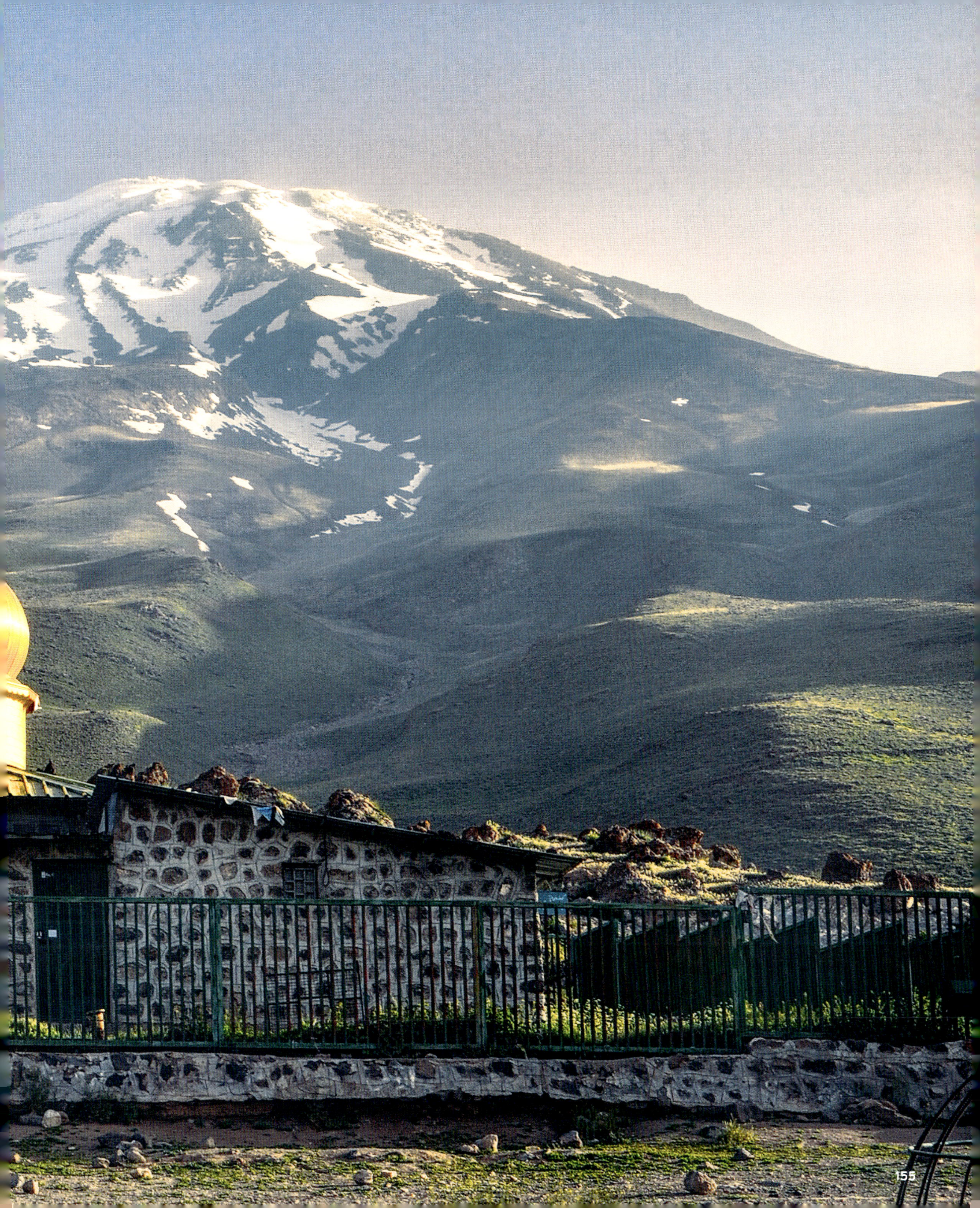

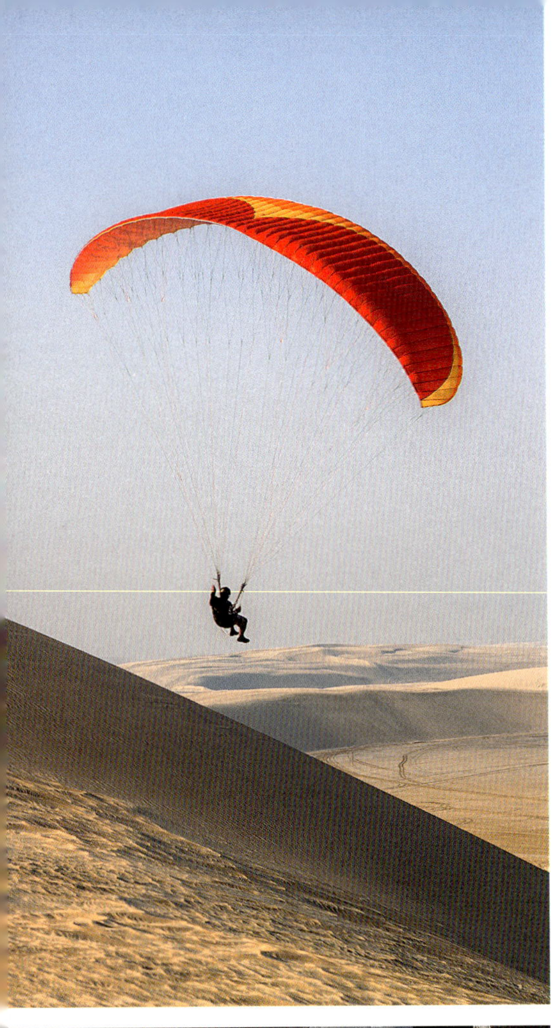

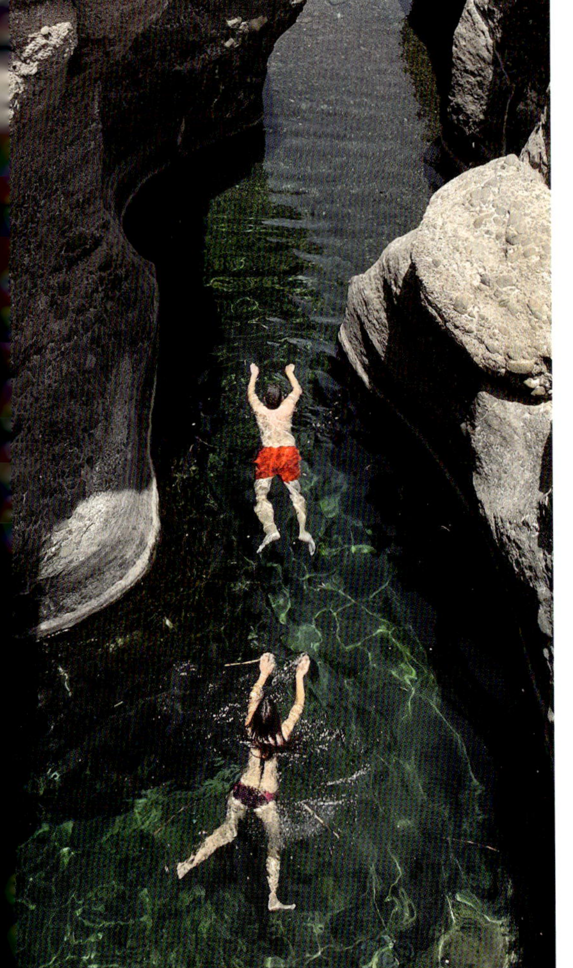

259. SWIMMING IN THE HATTA ROCK POOLS

Fata Morgana or a cool thing in the scorching heat? The Hatta Pools are in the southeastern Arabian Peninsula. This natural phenomenon of sharply broken rocks, with its crystal-clear, turquoise-colored water, appears completely unreal at first sight. Otherwise, there isn't a single drop of natural cool water far and wide, but here—what a treat, what a miracle from the Thousand and One Nights! You can reach the springs by a short hike from the parking lot. In addition to swimming in one of the beautiful pools, a short "canyoning tour" is a good idea: an inflatable sleeping mat served as a raft for my daughter, and we swam to the end of the big, steep canyon. The waters of this impressive natural wonder are surprisingly cold, making this an even greater outdoor adventure for kids.

Length / time it takes: 1.86 mi. (3 km) / 2 hours
Starting point: Hatta, UAE
Pools: are in the territory of Oman
Equipment: inflatable sleeping mats
Attention: take note of the current entry regulations for the UAE and Oman
Information: www.visithatta.com

260. HOT-AIR BALLOON FLIGHT OVER THE DUBAI DESERT

Just take off—in the middle of the endless expanse of the Arabian Desert and right at the gates of Dubai. There is a festive atmosphere early in the morning as the balloons are filled with hot air in the dark, and only the imposing gas flames illuminate their immediate surroundings. Right on time for the sunrise, it's off into the air—with an experienced pilot, of course. The entire beauty of the never-ending dunes of the desert opens up only from a dizzying height. When the view is clear, you can even see the skyline of the megacity on the horizon. But not only that: a falconer and his faithful bird provide a special kind of experience with their show at about the same altitude. This brings you a bit closer to the culture of this desert state many feet high up in the air above it.

Pickup time: 4 or 5 a.m.
Flight duration: 60 minutes
Starting point: Dubai
Season: October 1 to May 15
Information: visit-dubai.de

261. TANDEM PARAGLIDING OVER RUB AL-CHALI (THE EMPTY QUARTER)

Out and about as a high-flyer over the dunes! Desert extremes are turned head over heels on a tandem paragliding flight over Moreeb Dune in the Rub al-Chali Desert: this sand dune is 5,249 ft. (1,600 m) long and 394 ft. (120 m) high and has a slope of up to 50 degrees. The Rub al-Chali, also known as the "Empty Quarter," is simply the largest sand desert on Earth. Straight roads lead from the big city to the Moreeb Dune Area, where people often indulge in desert-typical motor sports such as dune racing. It's worth it to get up early: in the early morning the magical silence is interrupted only by the whispering of the palm trees, the rustling of the sand, and the flapping of your paraglider. The otherwise very strong thermals also keep themselves in check early in the morning. A fantastic flight experience.

Time it takes: 1 hour
Starting altitude: 394 ft. (120 m)
Starting point: Moreeb Dune, Liwa Oasis, Abu Dhabi, UAE
Elevation difference: depending on the thermals
Information: www.middleeastexperience.com

262. WATER FUN IN WADI ADVENTURE PARK

The "Wadi Adventure Park" in Al-Ain is probably the craziest place on Earth to indulge in water sports. An artificial whitewater world has been built in the middle of the Arabian Peninsula desert. Anyone visiting Dubai or one of the metropolises on the Arabian Peninsula should—especially in the consistent heat of the Arabian summer—consider taking an excursion into the world of wet-element experiences: these Middle East attractions include rafting, kayaking, and wave riding, along with the pool and wakeboard facility. Outdoor fun at the foot of Jebel Hafeet mountain is not exactly cheap, but it is as unusual as the entire artificial world of the United Arab Emirates. Utterly "strange," but that's why this is one visit you will remember.

Time it takes: day trip
Starting point: Al Ain, United Arab Emirates
Levels of difficulty: WW 2–4
Best time of year: all year round
Information: www.global-kayak.com

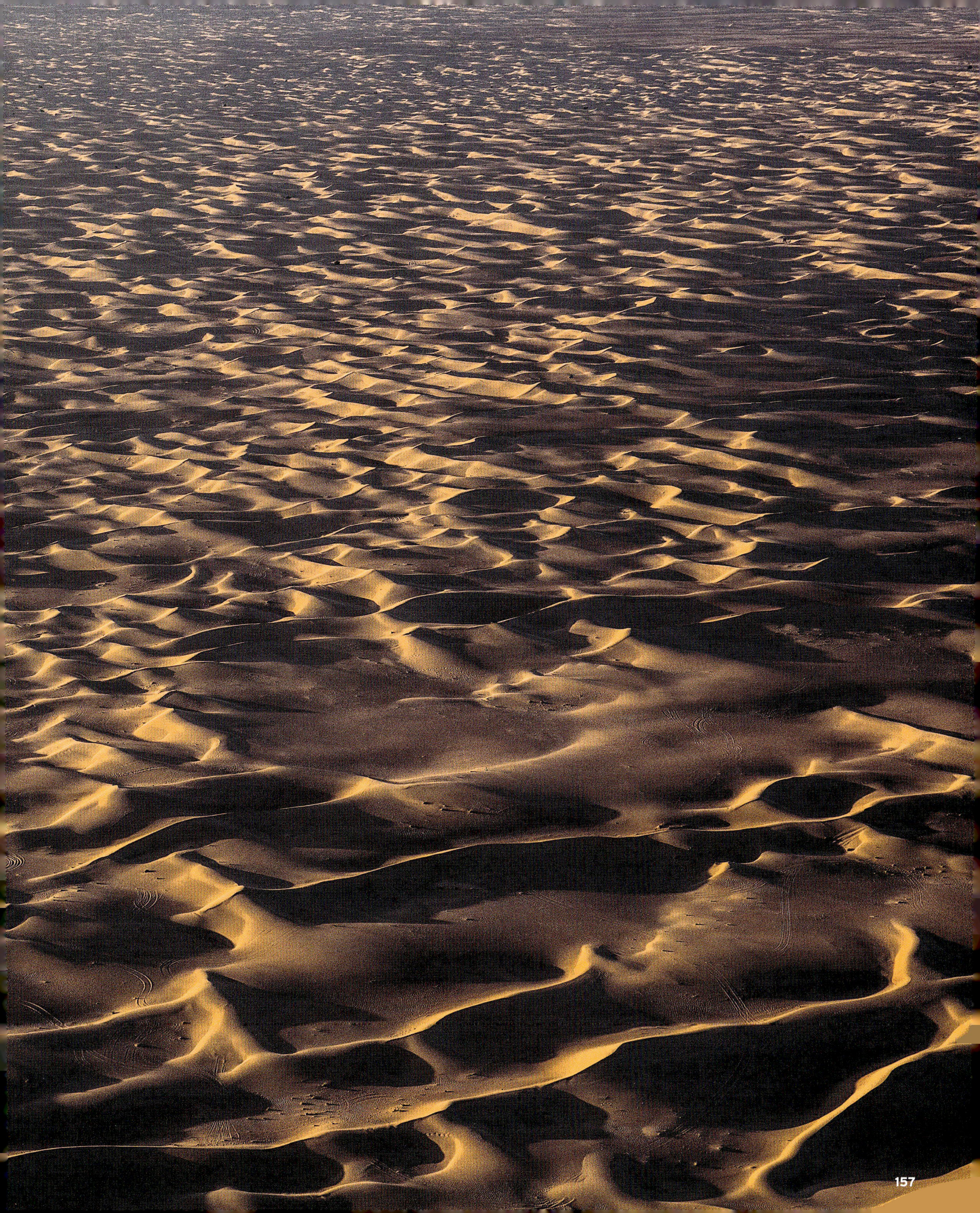

ASIA

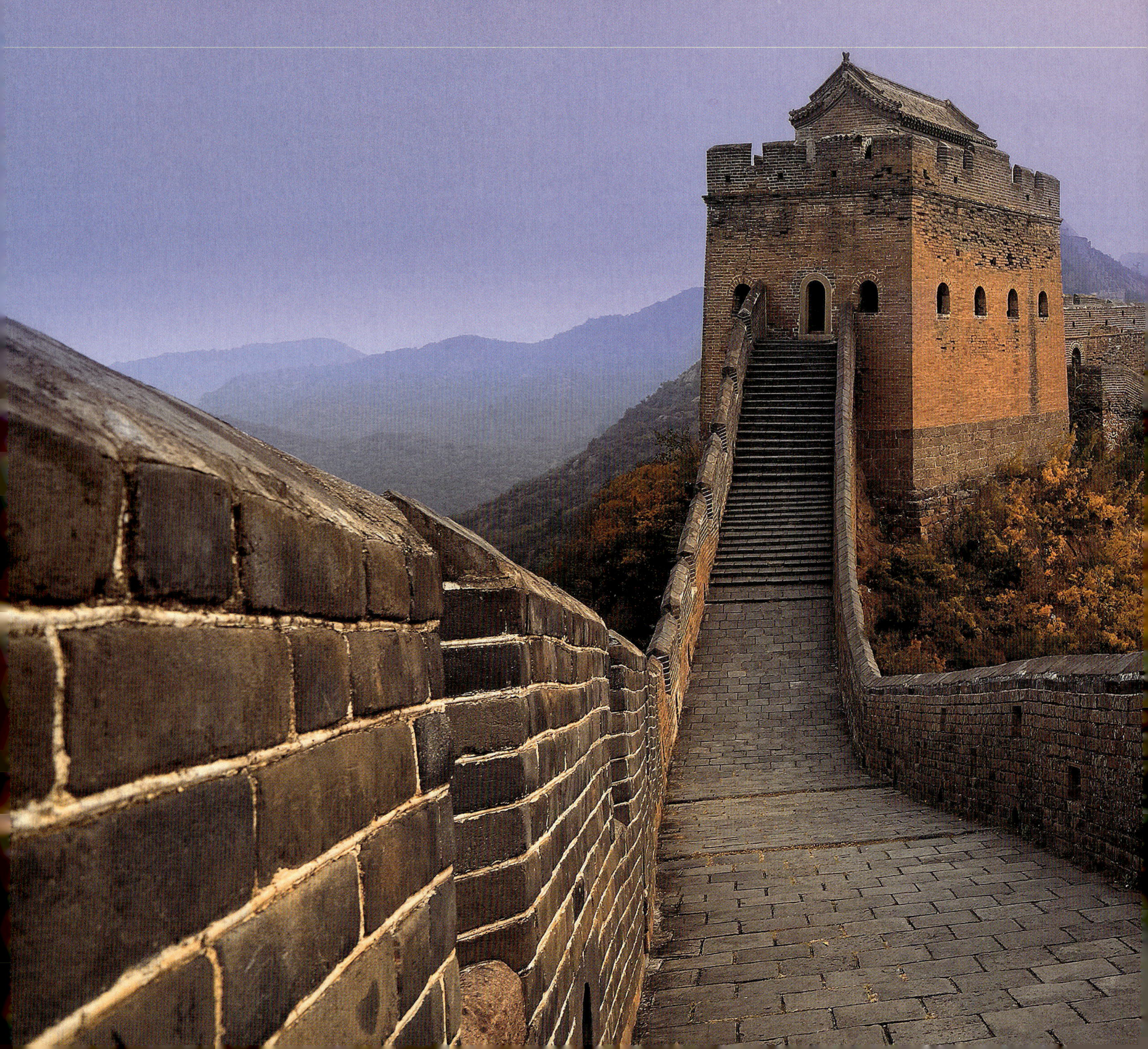

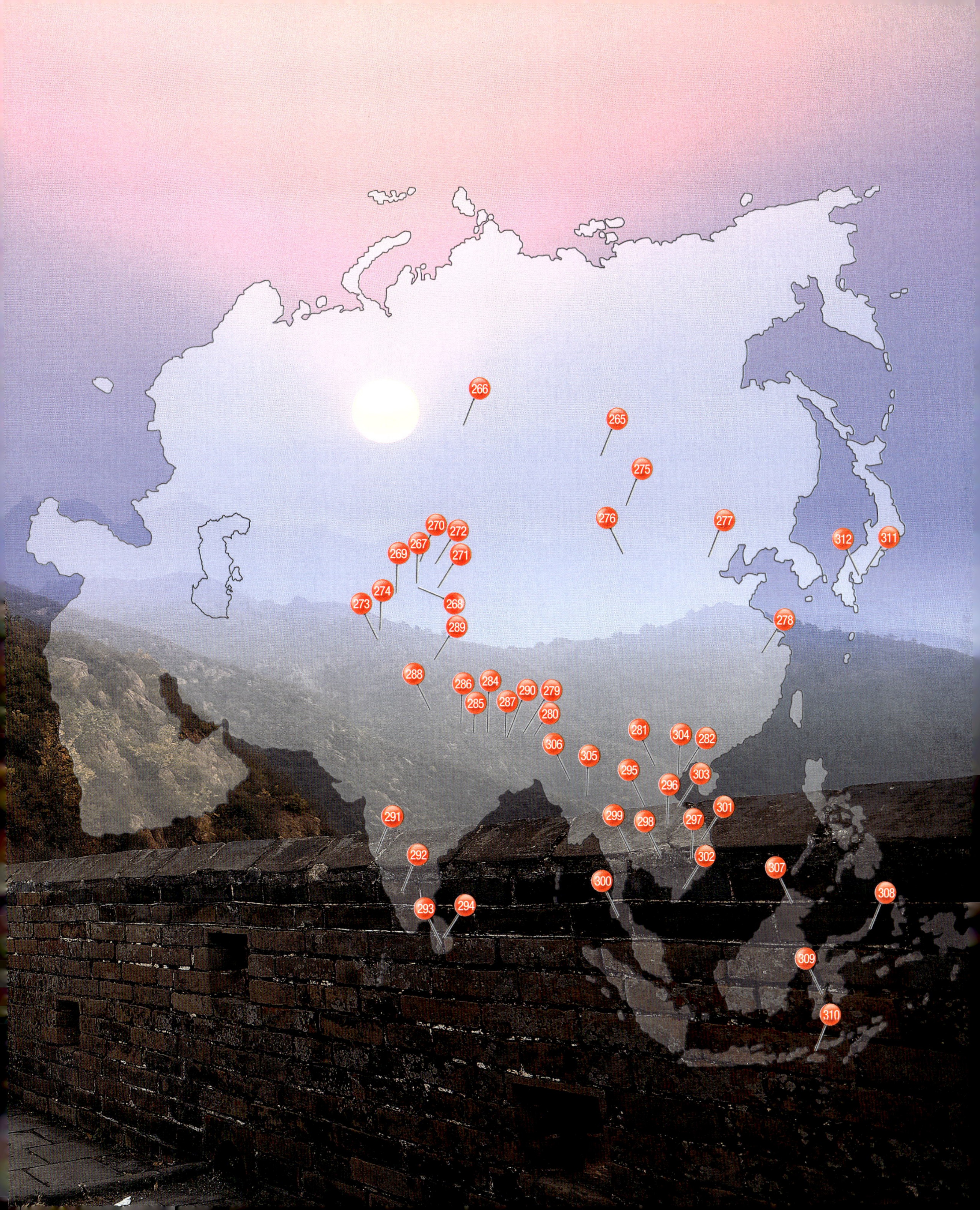

266
265
275
270
272
276
277
267
269
271
312
311
274
273
268
278
289
288
286
284
290
279
285
287
280
281
304
282
306
305
295
303
296
301
291
299
298
297
292
302
307
300
293
294
308
309
310

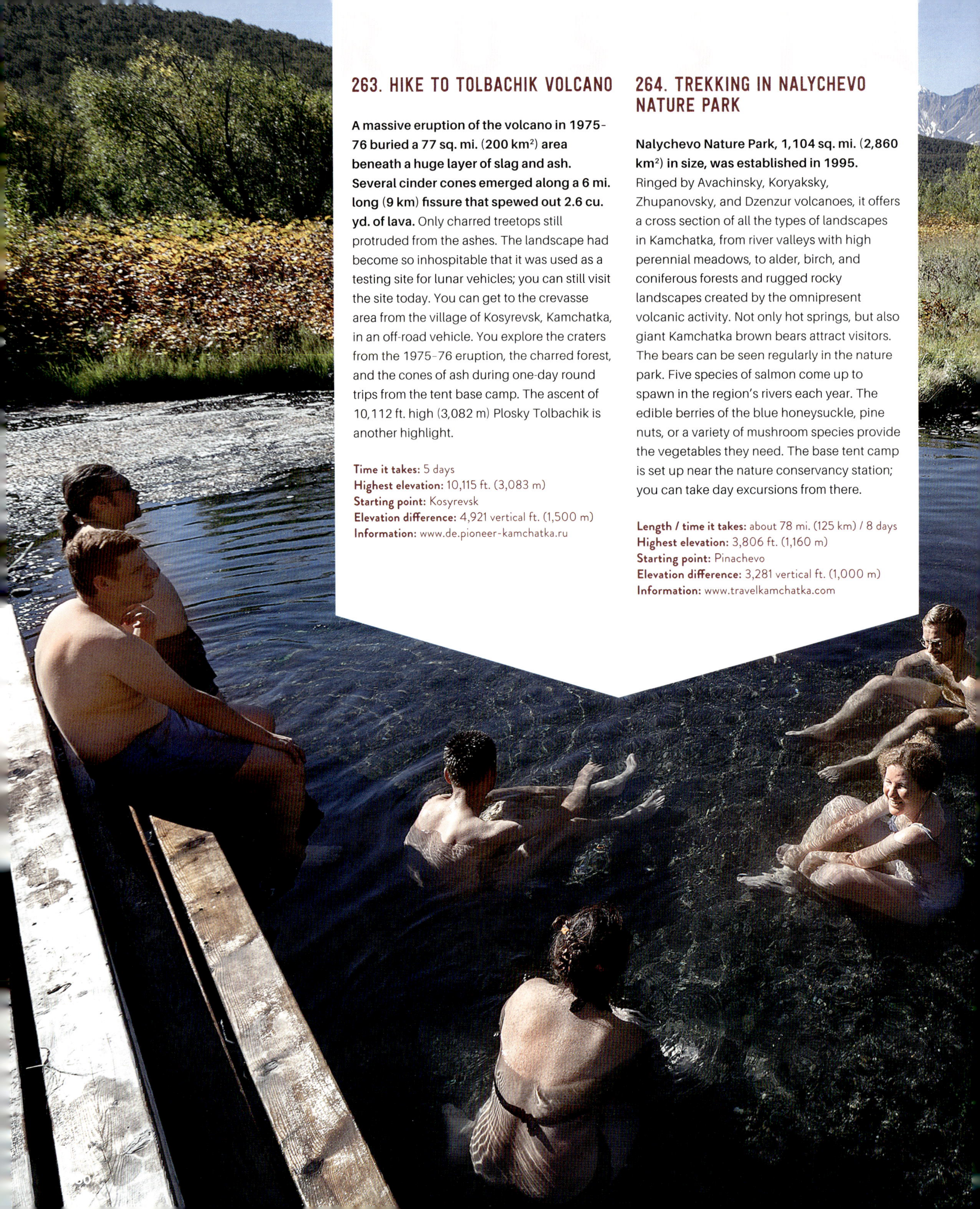

263. HIKE TO TOLBACHIK VOLCANO

A massive eruption of the volcano in 1975–76 buried a 77 sq. mi. (200 km^2) area beneath a huge layer of slag and ash. Several cinder cones emerged along a 6 mi. long (9 km) fissure that spewed out 2.6 cu. yd. of lava. Only charred treetops still protruded from the ashes. The landscape had become so inhospitable that it was used as a testing site for lunar vehicles; you can still visit the site today. You can get to the crevasse area from the village of Kosyrevsk, Kamchatka, in an off-road vehicle. You explore the craters from the 1975–76 eruption, the charred forest, and the cones of ash during one-day round trips from the tent base camp. The ascent of 10,112 ft. high (3,082 m) Plosky Tolbachik is another highlight.

Time it takes: 5 days
Highest elevation: 10,115 ft. (3,083 m)
Starting point: Kosyrevsk
Elevation difference: 4,921 vertical ft. (1,500 m)
Information: www.de.pioneer-kamchatka.ru

264. TREKKING IN NALYCHEVO NATURE PARK

Nalychevo Nature Park, 1,104 sq. mi. (2,860 km^2) in size, was established in 1995. Ringed by Avachinsky, Koryaksky, Zhupanovsky, and Dzenzur volcanoes, it offers a cross section of all the types of landscapes in Kamchatka, from river valleys with high perennial meadows, to alder, birch, and coniferous forests and rugged rocky landscapes created by the omnipresent volcanic activity. Not only hot springs, but also giant Kamchatka brown bears attract visitors. The bears can be seen regularly in the nature park. Five species of salmon come up to spawn in the region's rivers each year. The edible berries of the blue honeysuckle, pine nuts, or a variety of mushroom species provide the vegetables they need. The base tent camp is set up near the nature conservancy station; you can take day excursions from there.

Length / time it takes: about 78 mi. (125 km) / 8 days
Highest elevation: 3,806 ft. (1,160 m)
Starting point: Pinachevo
Elevation difference: 3,281 vertical ft. (1,000 m)
Information: www.travelkamchatka.com

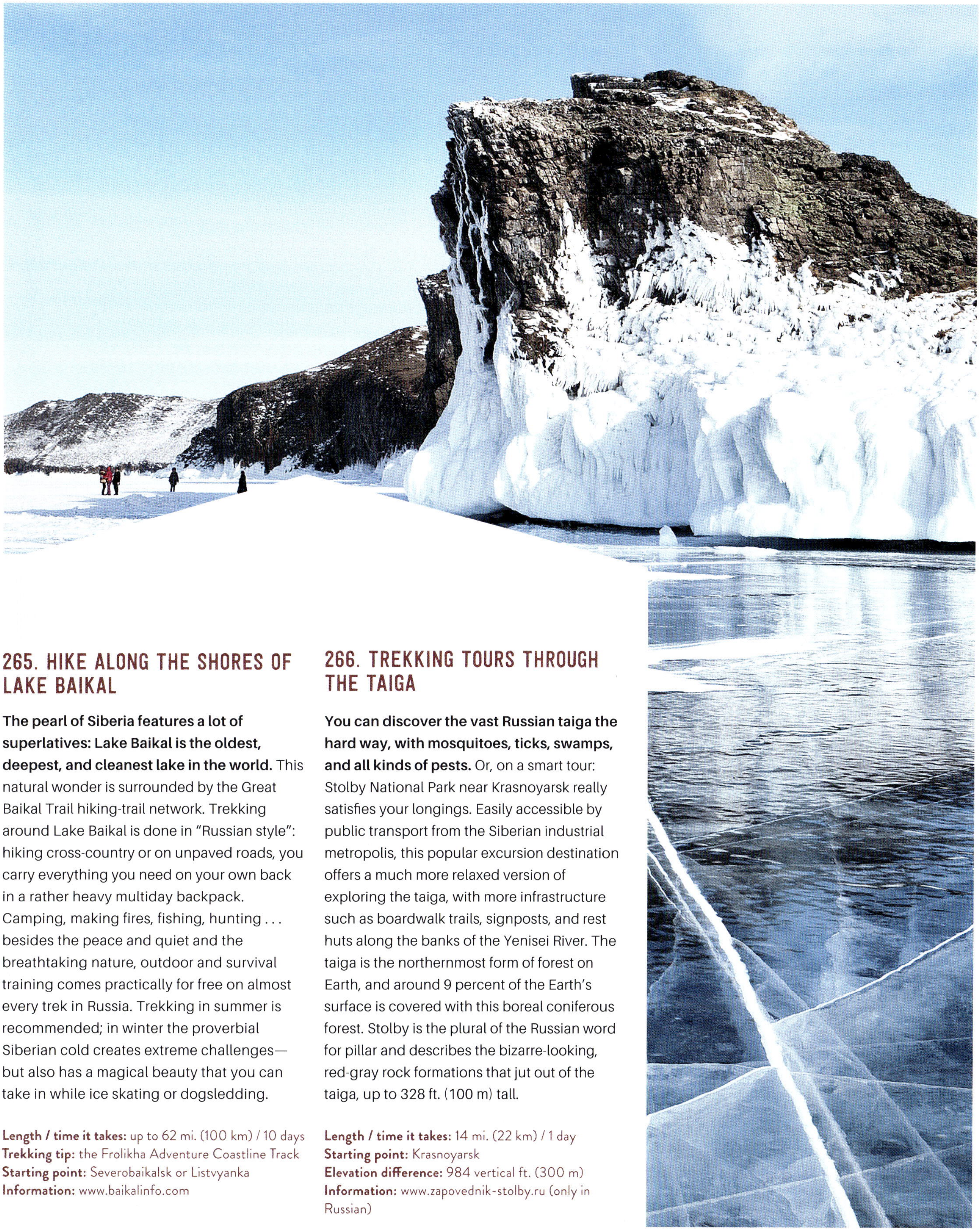

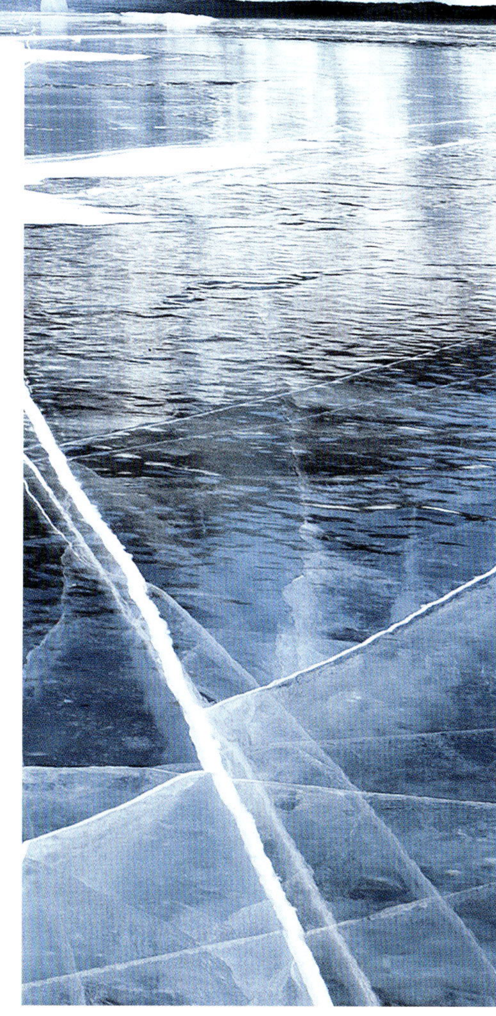

265. HIKE ALONG THE SHORES OF LAKE BAIKAL

The pearl of Siberia features a lot of superlatives: Lake Baikal is the oldest, deepest, and cleanest lake in the world. This natural wonder is surrounded by the Great Baikal Trail hiking-trail network. Trekking around Lake Baikal is done in "Russian style": hiking cross-country or on unpaved roads, you carry everything you need on your own back in a rather heavy multiday backpack. Camping, making fires, fishing, hunting . . . besides the peace and quiet and the breathtaking nature, outdoor and survival training comes practically for free on almost every trek in Russia. Trekking in summer is recommended; in winter the proverbial Siberian cold creates extreme challenges—but also has a magical beauty that you can take in while ice skating or dogsledding.

Length / time it takes: up to 62 mi. (100 km) / 10 days
Trekking tip: the Frolikha Adventure Coastline Track
Starting point: Severobaikalsk or Listvyanka
Information: www.baikalinfo.com

266. TREKKING TOURS THROUGH THE TAIGA

You can discover the vast Russian taiga the hard way, with mosquitoes, ticks, swamps, and all kinds of pests. Or, on a smart tour: Stolby National Park near Krasnoyarsk really satisfies your longings. Easily accessible by public transport from the Siberian industrial metropolis, this popular excursion destination offers a much more relaxed version of exploring the taiga, with more infrastructure such as boardwalk trails, signposts, and rest huts along the banks of the Yenisei River. The taiga is the northernmost form of forest on Earth, and around 9 percent of the Earth's surface is covered with this boreal coniferous forest. Stolby is the plural of the Russian word for pillar and describes the bizarre-looking, red-gray rock formations that jut out of the taiga, up to 328 ft. (100 m) tall.

Length / time it takes: 14 mi. (22 km) / 1 day
Starting point: Krasnoyarsk
Elevation difference: 984 vertical ft. (300 m)
Information: www.zapovednik-stolby.ru (only in Russian)

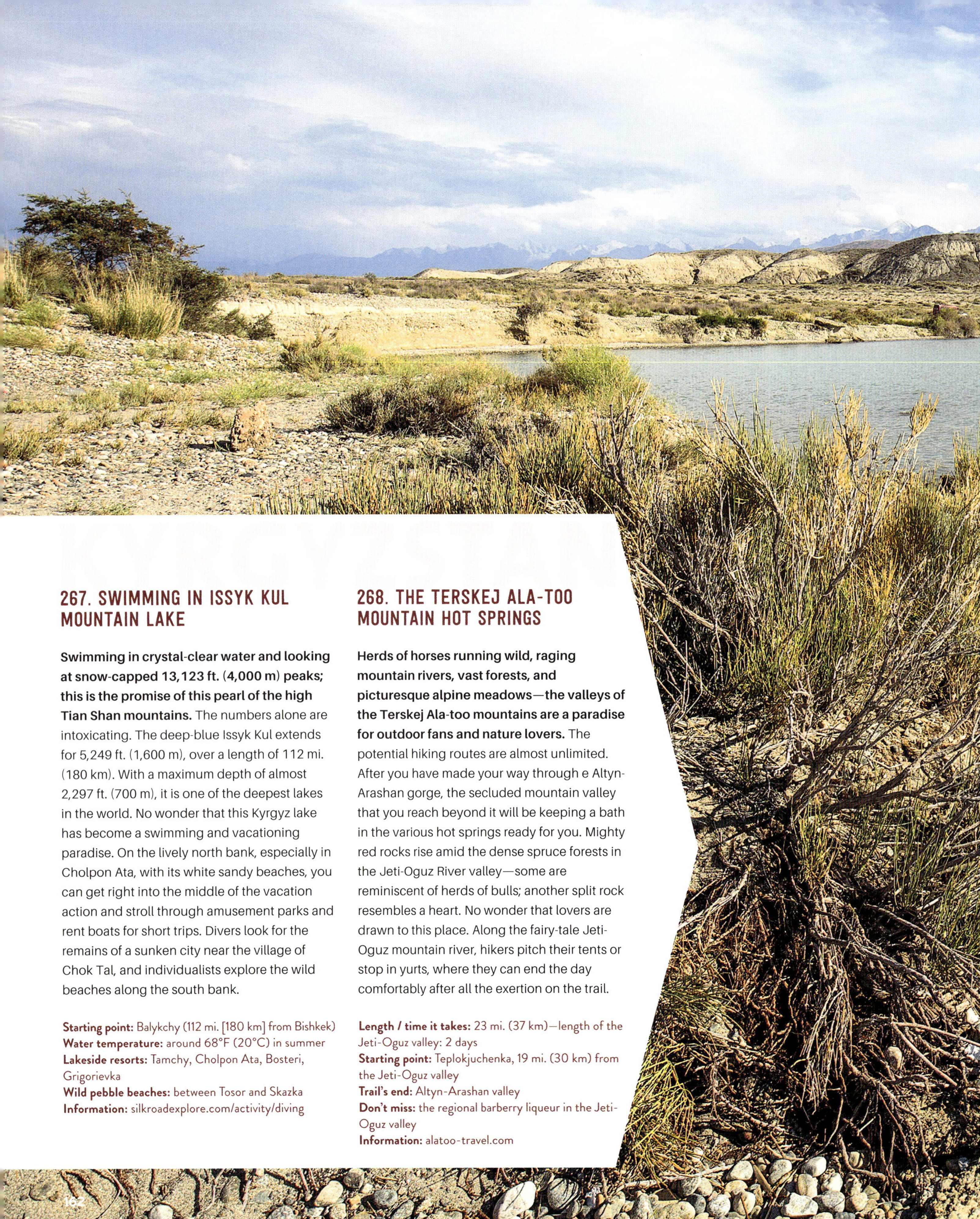

KYRGYZSTAN

267. SWIMMING IN ISSYK KUL MOUNTAIN LAKE

Swimming in crystal-clear water and looking at snow-capped 13,123 ft. (4,000 m) peaks; this is the promise of this pearl of the high Tian Shan mountains. The numbers alone are intoxicating. The deep-blue Issyk Kul extends for 5,249 ft. (1,600 m), over a length of 112 mi. (180 km). With a maximum depth of almost 2,297 ft. (700 m), it is one of the deepest lakes in the world. No wonder that this Kyrgyz lake has become a swimming and vacationing paradise. On the lively north bank, especially in Cholpon Ata, with its white sandy beaches, you can get right into the middle of the vacation action and stroll through amusement parks and rent boats for short trips. Divers look for the remains of a sunken city near the village of Chok Tal, and individualists explore the wild beaches along the south bank.

Starting point: Balykchy (112 mi. [180 km] from Bishkek)
Water temperature: around 68°F (20°C) in summer
Lakeside resorts: Tamchy, Cholpon Ata, Bosteri, Grigorievka
Wild pebble beaches: between Tosor and Skazka
Information: silkroadexplore.com/activity/diving

268. THE TERSKEJ ALA-TOO MOUNTAIN HOT SPRINGS

Herds of horses running wild, raging mountain rivers, vast forests, and picturesque alpine meadows—the valleys of the Terskej Ala-too mountains are a paradise for outdoor fans and nature lovers. The potential hiking routes are almost unlimited. After you have made your way through e Altyn-Arashan gorge, the secluded mountain valley that you reach beyond it will be keeping a bath in the various hot springs ready for you. Mighty red rocks rise amid the dense spruce forests in the Jeti-Oguz River valley—some are reminiscent of herds of bulls; another split rock resembles a heart. No wonder that lovers are drawn to this place. Along the fairy-tale Jeti-Oguz mountain river, hikers pitch their tents or stop in yurts, where they can end the day comfortably after all the exertion on the trail.

Length / time it takes: 23 mi. (37 km)—length of the Jeti-Oguz valley: 2 days
Starting point: Teplokjuchenka, 19 mi. (30 km) from the Jeti-Oguz valley
Trail's end: Altyn-Arashan valley
Don't miss: the regional barberry liqueur in the Jeti-Oguz valley
Information: alatoo-travel.com

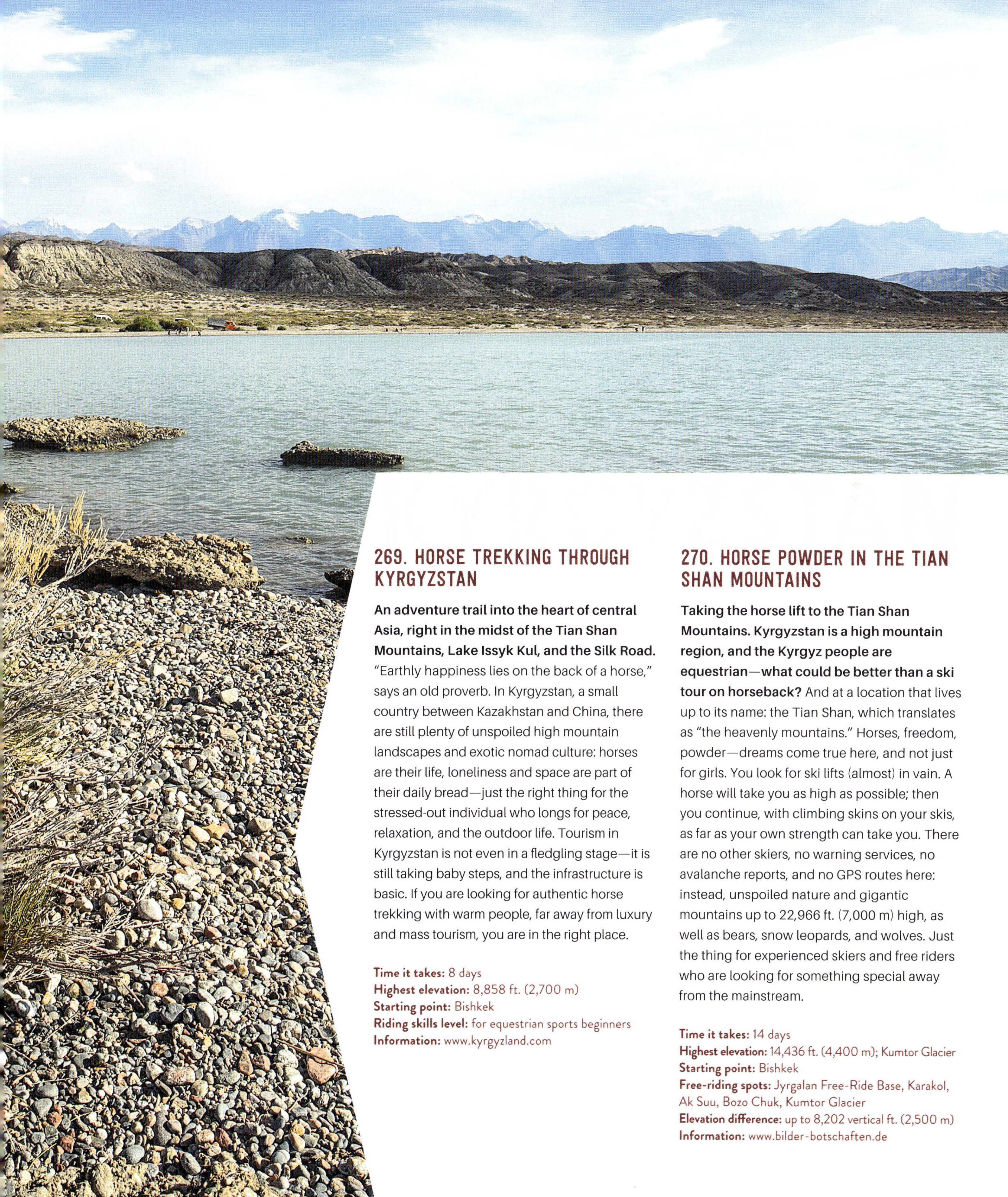

269. HORSE TREKKING THROUGH KYRGYZSTAN

An adventure trail into the heart of central Asia, right in the midst of the Tian Shan Mountains, Lake Issyk Kul, and the Silk Road. "Earthly happiness lies on the back of a horse," says an old proverb. In Kyrgyzstan, a small country between Kazakhstan and China, there are still plenty of unspoiled high mountain landscapes and exotic nomad culture: horses are their life, loneliness and space are part of their daily bread—just the right thing for the stressed-out individual who longs for peace, relaxation, and the outdoor life. Tourism in Kyrgyzstan is not even in a fledgling stage—it is still taking baby steps, and the infrastructure is basic. If you are looking for authentic horse trekking with warm people, far away from luxury and mass tourism, you are in the right place.

Time it takes: 8 days
Highest elevation: 8,858 ft. (2,700 m)
Starting point: Bishkek
Riding skills level: for equestrian sports beginners
Information: www.kyrgyzland.com

270. HORSE POWDER IN THE TIAN SHAN MOUNTAINS

Taking the horse lift to the Tian Shan Mountains. Kyrgyzstan is a high mountain region, and the Kyrgyz people are equestrian—what could be better than a ski tour on horseback? And at a location that lives up to its name: the Tian Shan, which translates as "the heavenly mountains." Horses, freedom, powder—dreams come true here, and not just for girls. You look for ski lifts (almost) in vain. A horse will take you as high as possible; then you continue, with climbing skins on your skis, as far as your own strength can take you. There are no other skiers, no warning services, no avalanche reports, and no GPS routes here: instead, unspoiled nature and gigantic mountains up to 22,966 ft. (7,000 m) high, as well as bears, snow leopards, and wolves. Just the thing for experienced skiers and free riders who are looking for something special away from the mainstream.

Time it takes: 14 days
Highest elevation: 14,436 ft. (4,400 m); Kumtor Glacier
Starting point: Bishkek
Free-riding spots: Jyrgalan Free-Ride Base, Karakol, Ak Suu, Bozo Chuk, Kumtor Glacier
Elevation difference: up to 8,202 vertical ft. (2,500 m)
Information: www.bilder-botschaften.de

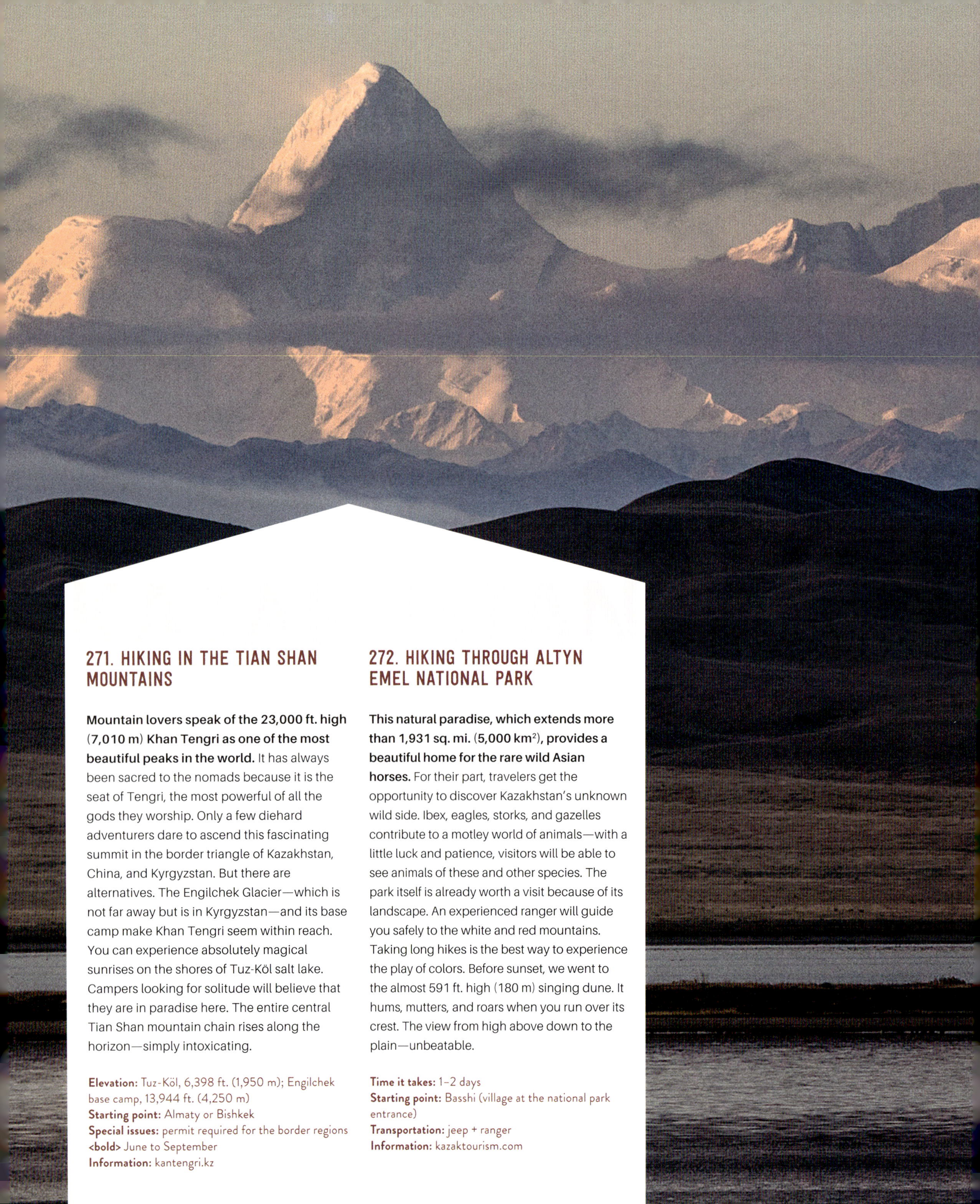

271. HIKING IN THE TIAN SHAN MOUNTAINS

Mountain lovers speak of the 23,000 ft. high (7,010 m) Khan Tengri as one of the most beautiful peaks in the world. It has always been sacred to the nomads because it is the seat of Tengri, the most powerful of all the gods they worship. Only a few diehard adventurers dare to ascend this fascinating summit in the border triangle of Kazakhstan, China, and Kyrgyzstan. But there are alternatives. The Engilchek Glacier—which is not far away but is in Kyrgyzstan—and its base camp make Khan Tengri seem within reach. You can experience absolutely magical sunrises on the shores of Tuz-Köl salt lake. Campers looking for solitude will believe that they are in paradise here. The entire central Tian Shan mountain chain rises along the horizon—simply intoxicating.

Elevation: Tuz-Köl, 6,398 ft. (1,950 m); Engilchek base camp, 13,944 ft. (4,250 m)
Starting point: Almaty or Bishkek
Special issues: permit required for the border regions
<bold> June to September
Information: kantengri.kz

272. HIKING THROUGH ALTYN EMEL NATIONAL PARK

This natural paradise, which extends more than 1,931 sq. mi. (5,000 km²), provides a beautiful home for the rare wild Asian horses. For their part, travelers get the opportunity to discover Kazakhstan's unknown wild side. Ibex, eagles, storks, and gazelles contribute to a motley world of animals—with a little luck and patience, visitors will be able to see animals of these and other species. The park itself is already worth a visit because of its landscape. An experienced ranger will guide you safely to the white and red mountains. Taking long hikes is the best way to experience the play of colors. Before sunset, we went to the almost 591 ft. high (180 m) singing dune. It hums, mutters, and roars when you run over its crest. The view from high above down to the plain—unbeatable.

Time it takes: 1–2 days
Starting point: Basshi (village at the national park entrance)
Transportation: jeep + ranger
Information: kazaktourism.com

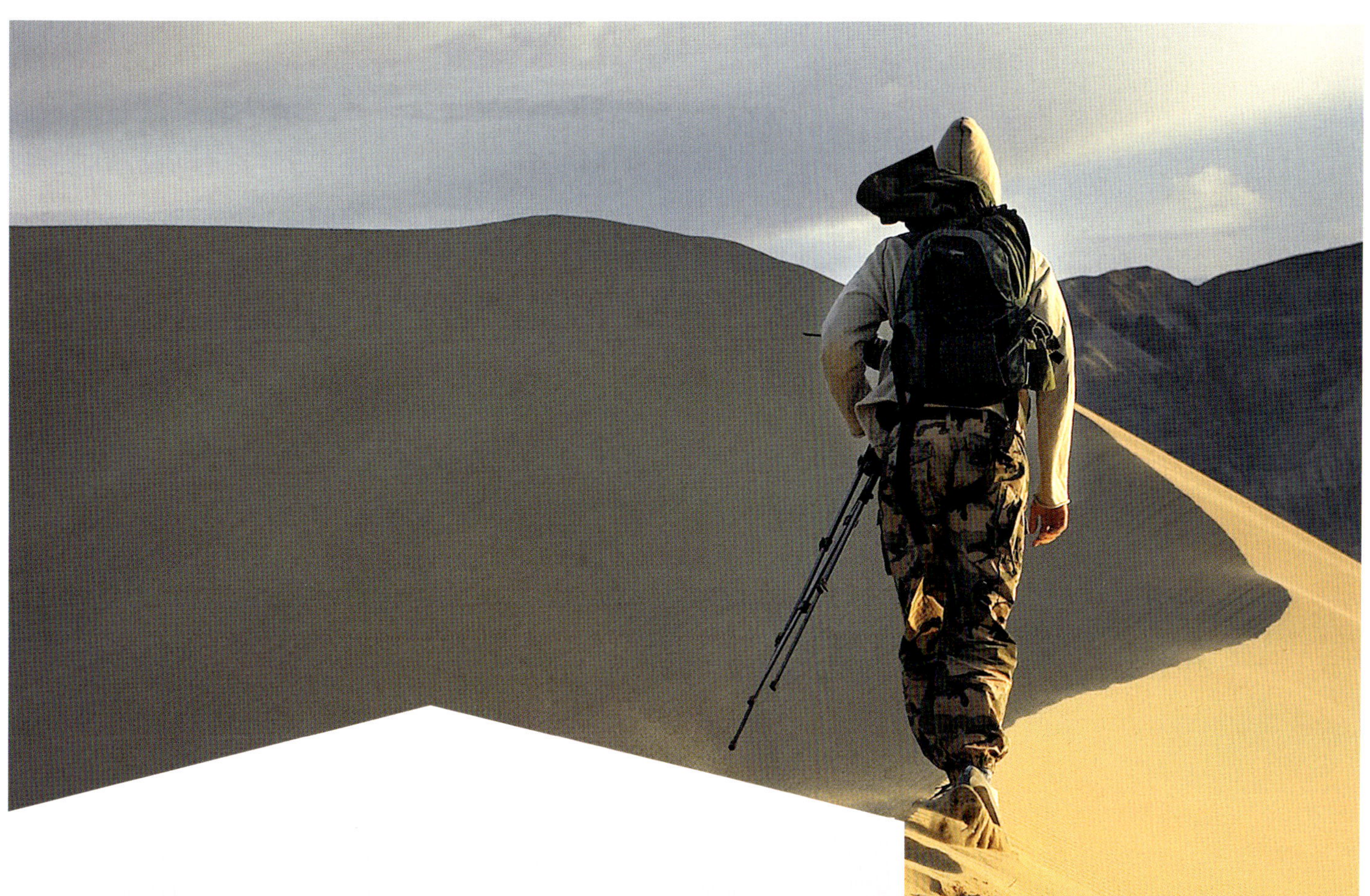

273. HIKING TO THE WAKHAN HOT SPRINGS

A picturesque valley, dizzyingly high mountains, and the torrential Panj border river—Wakhan already cast a spell over Marco Polo. It is not for nothing that the Marco Polo sheep that live here were named after him. The Pamir Mountains preside here, and the Hindu Kush greets you there. Travelers climbing in the ruins of the Tajik Yamchun and Vishim fortresses cast furtive glances at the Afghan side—so close and yet so far. The effort to climb to the high Yamchun fortress is worth it, not least because of the Bibi Fatima hot spring. Men and women take turns, but they habitually take off their clothes so that they can feel the water falling from the rocks on their bodies. Don't be shy about meeting the locals up close!

Time it takes: 1 day
Special issues: a GBAO permit in addition to a visa
Starting point: Vrang
Time to travel: May to October
Information: indy-guide.com

274. ALONG THE PAMIR HIGHWAY

A road that has long since become a myth. For many, driving along the Pamir Highway is their ultimate dream. Your route for fulfilling this dream begins in the capital, Dushanbe, and leads through clean-swept areas of the mountainous world to Murghab on its absolute periphery. Although you can find marshrutkas—shared minibus taxis that travel a set route—there, anyone who just wants to get from point A to point B will travel by jeep. The picturesque Bulunkul and Yashikul lakes should not be missed. An off-road track finally lures you through the wild side of the Pamirs to a geyser and finally to Alichur. Then the thing to do is breathe in the mountain air in Murghab, which lies at an incredible 11,870 ft. (3,618 m). Here you can experience an austere life in the cold Pamir desert.

Time it takes: 3–5 days
Starting point and trail's end: From Dushanbe to Murghab
Special issues: a GBAO permit in addition to a visa
Jeep costs: $15 for the driver; $0.70/mi.
Time to travel: June to September.
Information: worldroof-tours.com

275. HORSEBACK TOUR IN THE ORCHON VALLEY

Genghis Khan's humble riding arena: the 120,000-hectare meadow, made up of innumerable gently rolling green hills on the banks of the Orchon River, the longest river in Mongolia, is likely the oldest as well as the largest riding meadow in the world. The Orchon valley has been settled going back 100,000 years; the Mongolian horses living here are still half wild. The unique beauty of this ancient nomadic country is best explored on the only reasonable means of transport, exactly this four-legged Mongolian friend. You set off from the capital city, Ulaanbaatar. You ride for a long time, spend the night in picturesque yurts, and live the traditional way of the Mongolian nomads. An impressive and beautiful cultural landscape.

Length / time it takes: starting from 50 mi. (80 km) / 5 days
Starting point: Ulaanbaatar
Information: www.pferdesafari.de

276. TREKKING THROUGH THE GOBI DESERT

This moderately difficult hiking trip has a lot of geography to offer—from the Gobi Desert with its imposing sand dunes and broad steppes, to the bizarre rock formations of the Gobi Altai mountains, which look like the Alps, to the larch forests of the taiga—and carefully selected hiking treks, such as through the Khan Khentii Strictly Protected Area in Gorkhi-Terelj National Park (hiking time, four hours). There are more hiking trails in the Khogno-Khan Mountains (hiking time, three hours), near Bayanzag, where million-year-old remains of dinosaur skeletons lie (hiking time, two hours), and near Khongoryn Els, where hikers encounter Mongolia's highest sand dunes at up to 820 ft. (250 m) in elevation (hiking time, up to five hours) on their hikes through the majestic sandy landscape.

Length /time it takes: a variety of trips of different lengths / in total, 2.5 weeks

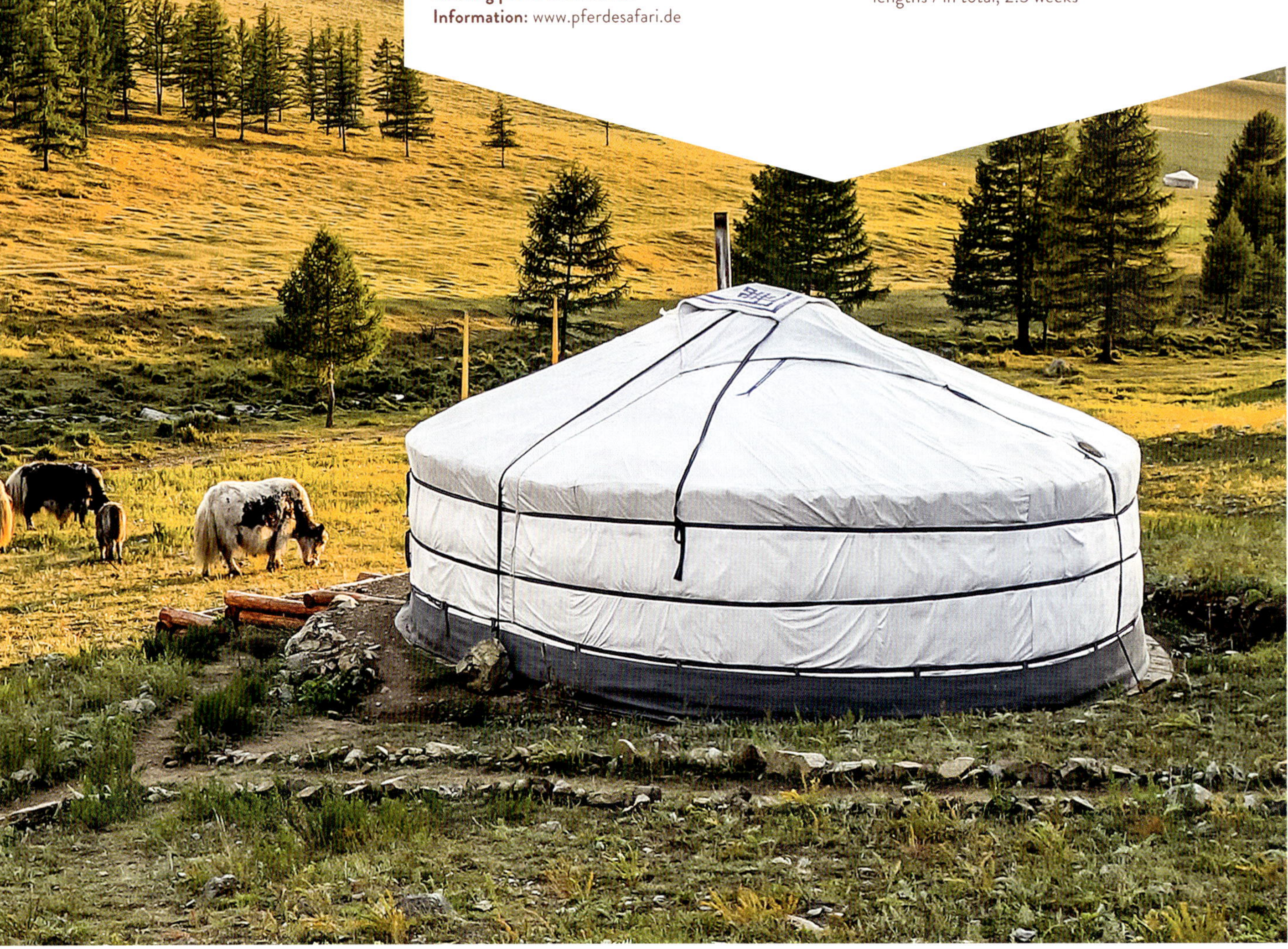

277. TREKKING ALONG THE GREAT WALL OF CHINA

This trip is also ideal for trekkers with orientation issues, because it is practically impossible to get lost, at least not on the well-restored part of the Great Wall of China around Beijing. In addition to the fascination of the wall in and of itself, there are sweeping views of wonderful nature from the longest human structure in the world; it measures a total of 13,171 ft. (21,196 km). The wall is continuously accessible in the area around Beijing. Here you can take day hikes from Gubeikou to Jinshanling, to Simatai (the most popular), to Jiankou, and to Mutianyu. While the wall is overcrowded along these sections during the day, things are quiet in the evening and the morning. In summer all you need is a sleeping bag and a sleeping mat, and you can spend the night in one of the watchtowers. Cool!

Length / time it takes: 42 mi. (67 km) / 4 days
Starting point: Beijing
Elevation difference: enormous; there are always stairs going up and down
Information: www.greatwalladventure.com

278. TREKS THROUGH THE HUANGSHAN MOUNTAINS

The Huangshan (Chinese for "Yellow Mountain") is the name of a mountain range in Anhui Province in the South of the People's Republic of China. Of the more than seventy peaks, the highest, Lianhuafen (Lotus Blossom peak), rises only to an elevation of 6,040 ft. (1,841 m). Nevertheless, hikers are enthusiastic about a natural region of steep, rocky mountains of bizarre shapes, where there are many routes for trekking through them. In 1990, the Huangshan region was declared a UNESCO World Heritage Site. Three different cable cars run to the summit, and you can hike up and down on superbly developed trails. The ascent, at five hours, lies in the shorter range, and thanks to the cable cars you can be very flexible and postpone your decision about how you will make the return trip until the last second.

Length: a variety of hikes of different lengths

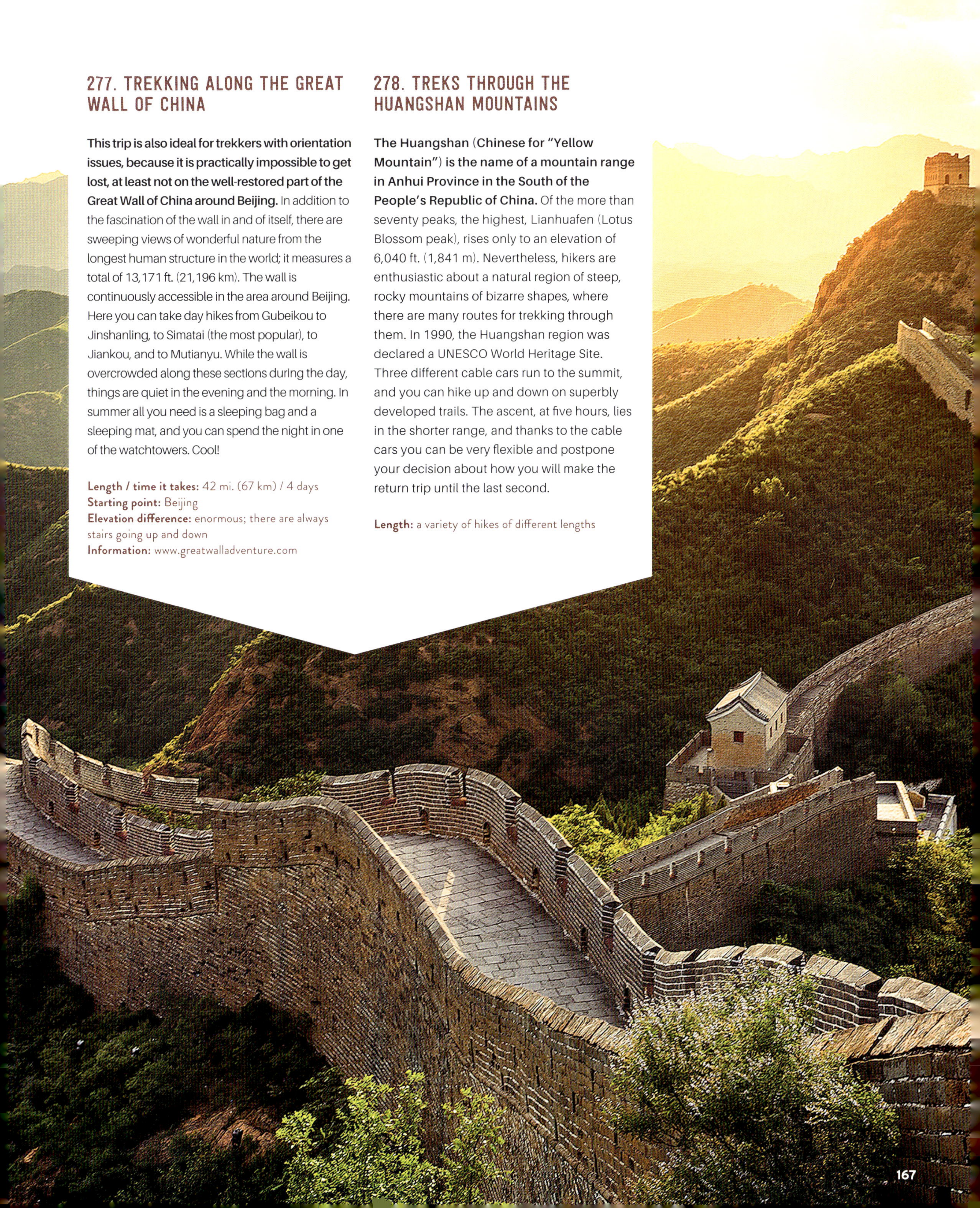

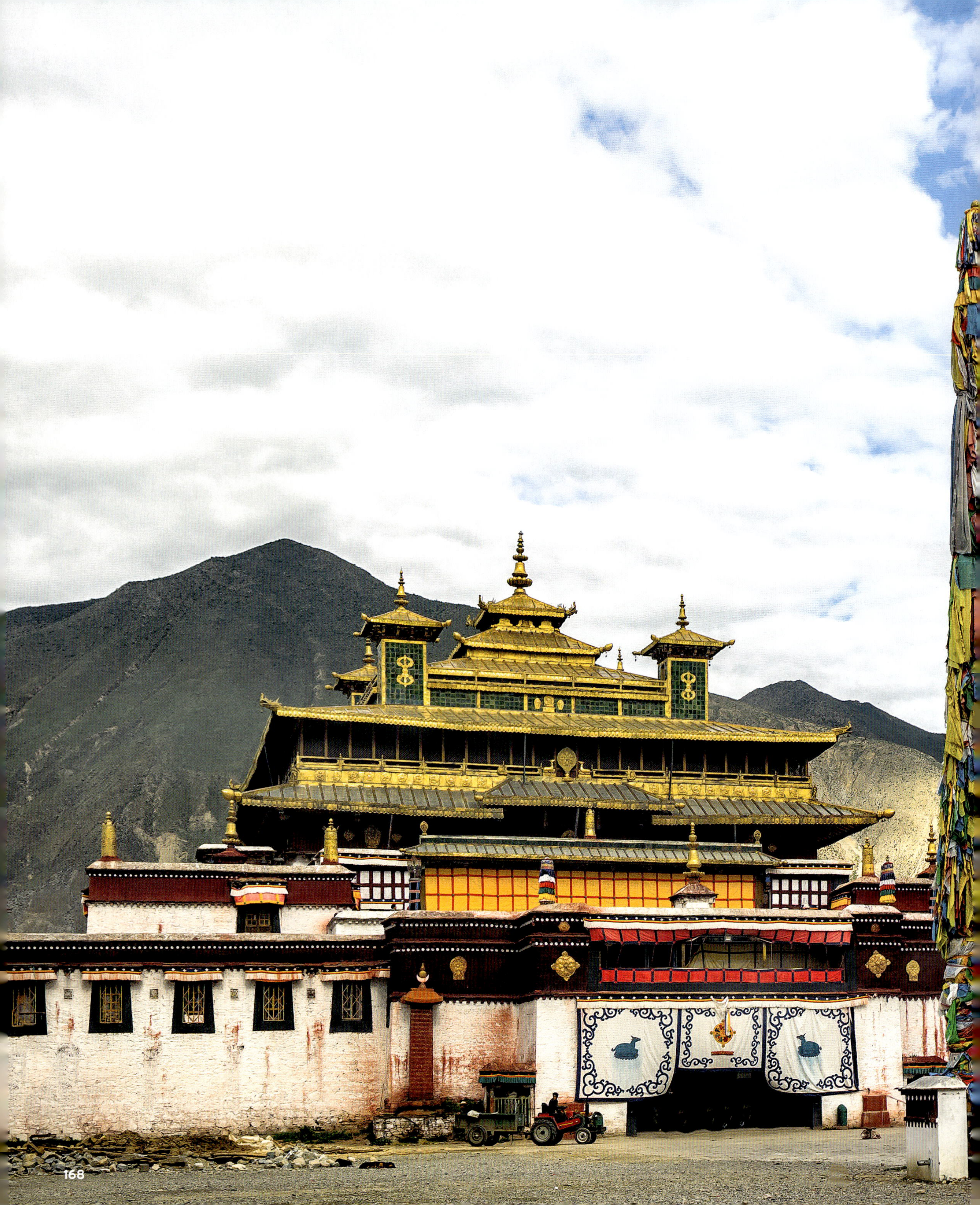

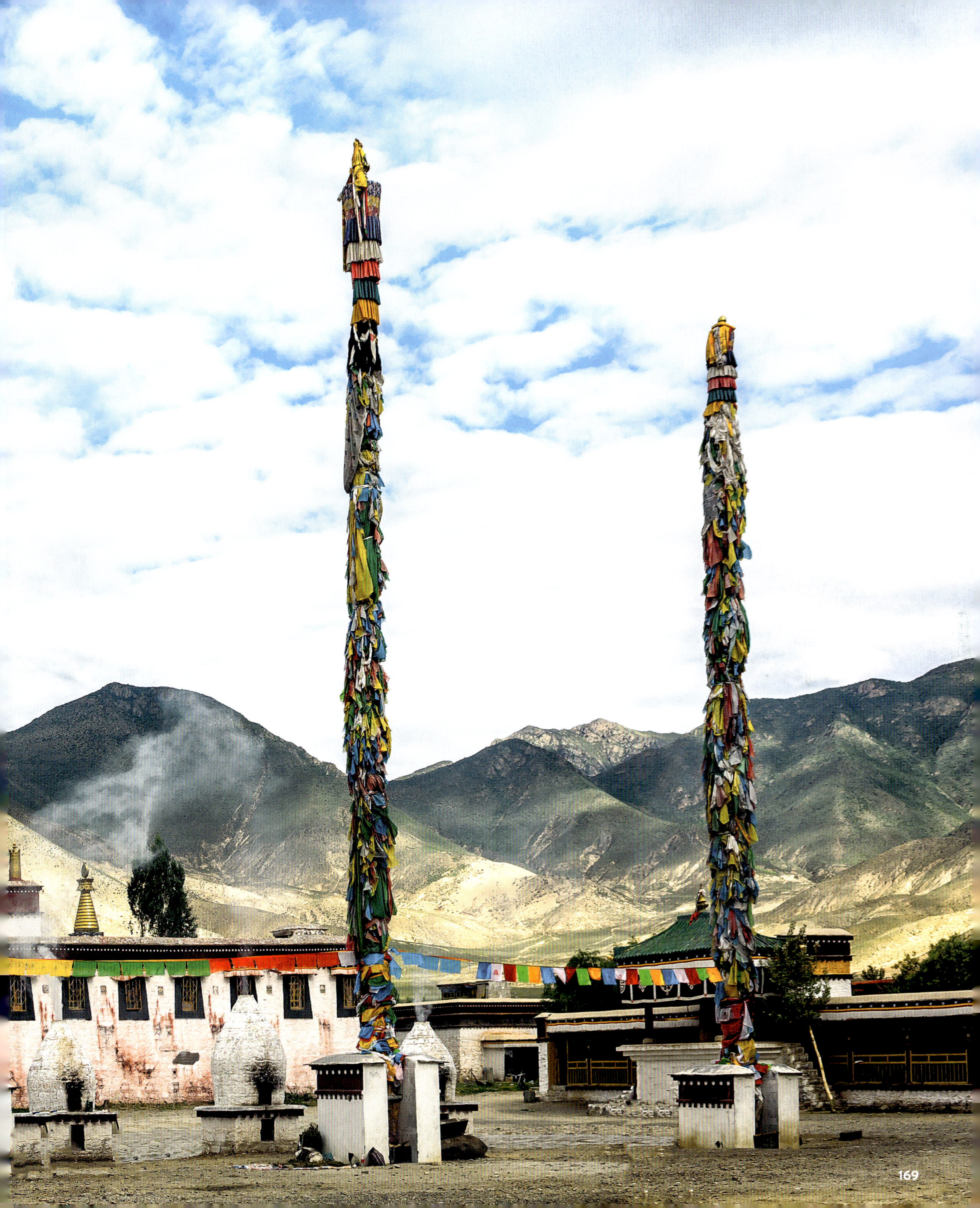

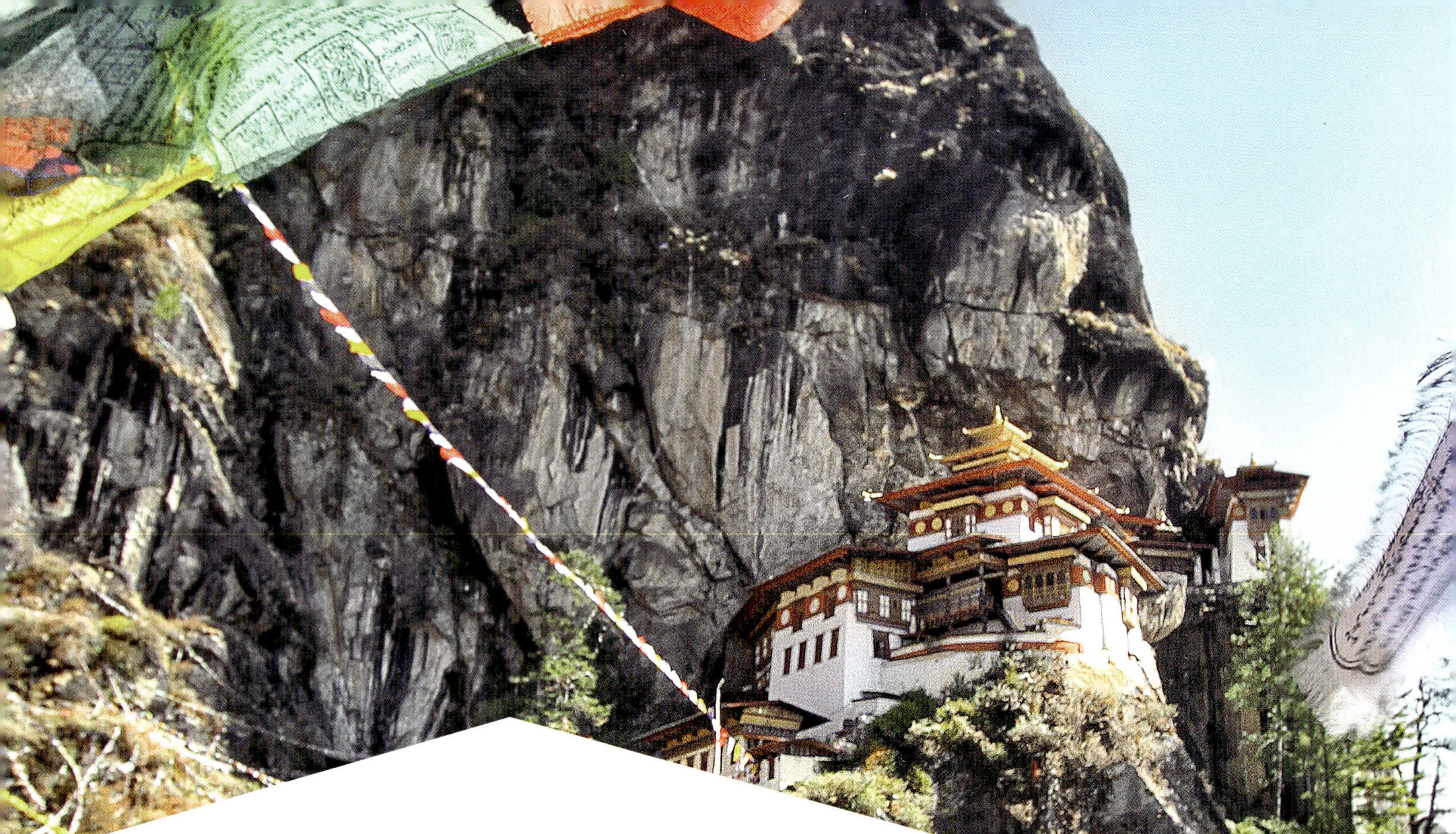

B H U T A N

279. ASCENT TO THE TIGER'S NEST MONASTERY

Colorful flags flutter in the wind everywhere along the way from Paro to the most famous monastery in the country, which was built entirely by hand on a rocky promontory in 1692, without any mechanical power. Legend has it that Guru Rinpoche flew up the mountain on the back of a tiger to convert the region to Buddhism. The strenuous ascent through dense forest begins on horseback, often with the guide on foot. After almost two hours of rocking back and forth, you reach the oasis: a cafe appears as if summoned from heaven. After that, it is only three-quarters of an hour on foot, passing by prayer flags, pine trees, and a waterfall, until you reach the monastery. You think: the way is the goal—in the end it is a path to your own blessedness.

Length / time it takes: 6.2 mi. (10 km) / 1 day
Highest elevation: 10,236 ft. (3,120 m)
Starting point: Paro
Elevation difference: 2,625 vertical ft. (800 m)
Information: tigersnestbhutan.com

280. CULTURE TREKKING IN THE BUMTHANG DISTRICT

Bumthang means "level country." In central Bhutan, this plain is framed by hills and mountains that are not very steep. More precisely, there are even four plains or valleys that lie at elevations between 8,530 and 13,123 ft. (2,600–4,000 m), where many villages and monasteries are located. The area thus offers its visitors the ideal situation to immerse themselves in the culture of Bhutan while taking relaxed hikes. So it is staying in the monasteries, which are still of great importance today, that is the highlight of every trip to Bhutan. A three-day hike leads through beautiful forests and valleys and follows the Chamkar Chu River, where it runs almost level on the first day. The second day leads over mountain meadows, 1,640 ft. (500 m) up to the Phephe La Pass at 11,030 ft. (3,362 m), which is the highest point of the trekking tour.

Time it takes: 3 days or day tours
Highest elevation: 11,030 ft. (3,362 m)
Starting point: Jakar
Best time to travel: March to April and October to November

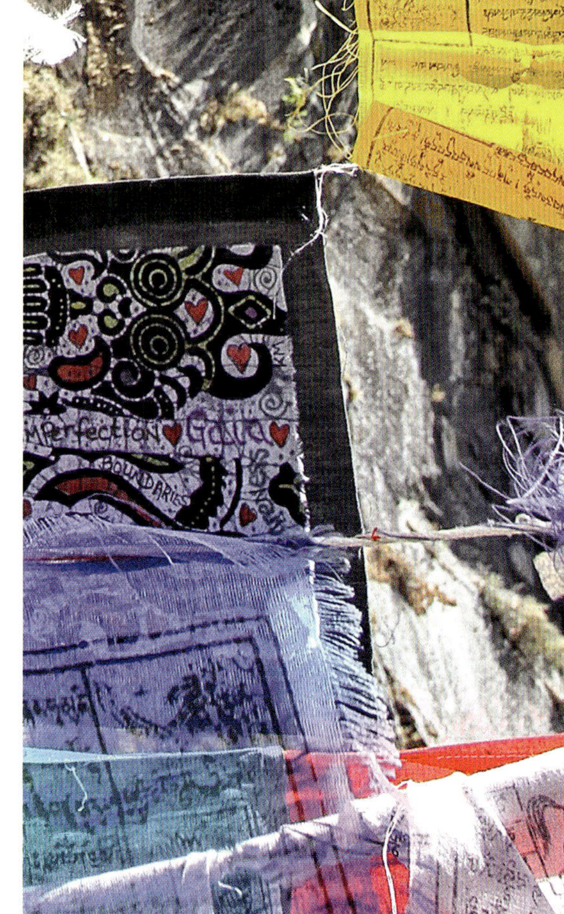

281. ON NOMAD'S TRAILS FROM GANDEN MONASTERY TO SAMYE MONASTERY

Dramatic mountain landscapes, fascinating monasteries, grazing yaks, and wandering nomads form the backdrop for this challenging trek. From picturesque Ganden Monastery, the old traders' and nomads' path leads right into the endless mountains of Tibet. Lonely high pastures, steep slopes, and sparse vegetation, but also a glaring sun and an occasional snow or rain shower will be your constant companions from now on. It will be necessary to cross two passes, the 17,192 ft. high (5,240 m) Shuge La and the 16,700 ft. high (5,090 m) Chitu La. Both are crowned by prayer flags fluttering in the everlasting wind, and heaps of stones left by the pilgrims to express their gratitude for reaching the passes. The destination is Samye Monastery, the oldest monastery in Tibet, on the lowlands of the Yarlung Tsangpo River.

Length / time it takes: 50 mi. (80 km) / 4–5 days
Highest elevation: 17,192 ft. (5,240 m)
Starting point: Ganden Monastery
Elevation difference: 3,084 vertical ft. (940 m)
Information: www.tibetctrip.com

282. ON THE PILGRIMAGE ROUTE AROUND MOUNT KAILASH

Mount Kailash is reminiscent of a temple of gigantic proportions. It is considered to be a natural temple by Hindus, Buddhists, and Bonpos alike. Tibetan Buddhists and Hindus try to make the Kora trek, the sacred circuit of the Kailash, at least once in their lives. The trail is not very demanding; rather, the thin mountain air is a challenge and creates problems for many pilgrims. Four prostration points mark those places where you look directly at Mount Kailash and where the pilgrims prostrate themselves reverently. Otherwise it hides behind high rock walls. Almost every stone, every rock, but also the course of the river is laden with symbols. Yet, the highlight is the Drölma La Pass, because whoever crosses it is freed from the sins of the past and can look forward to a new life.

Length / time it takes: 32 mi. (52 km) / 3 days
Highest elevation: 18,491 ft. (5,636 m)
Starting point: Darchen
Elevation difference: 2,362 vertical ft. (720 m)
Information: www.tibetctrip.com

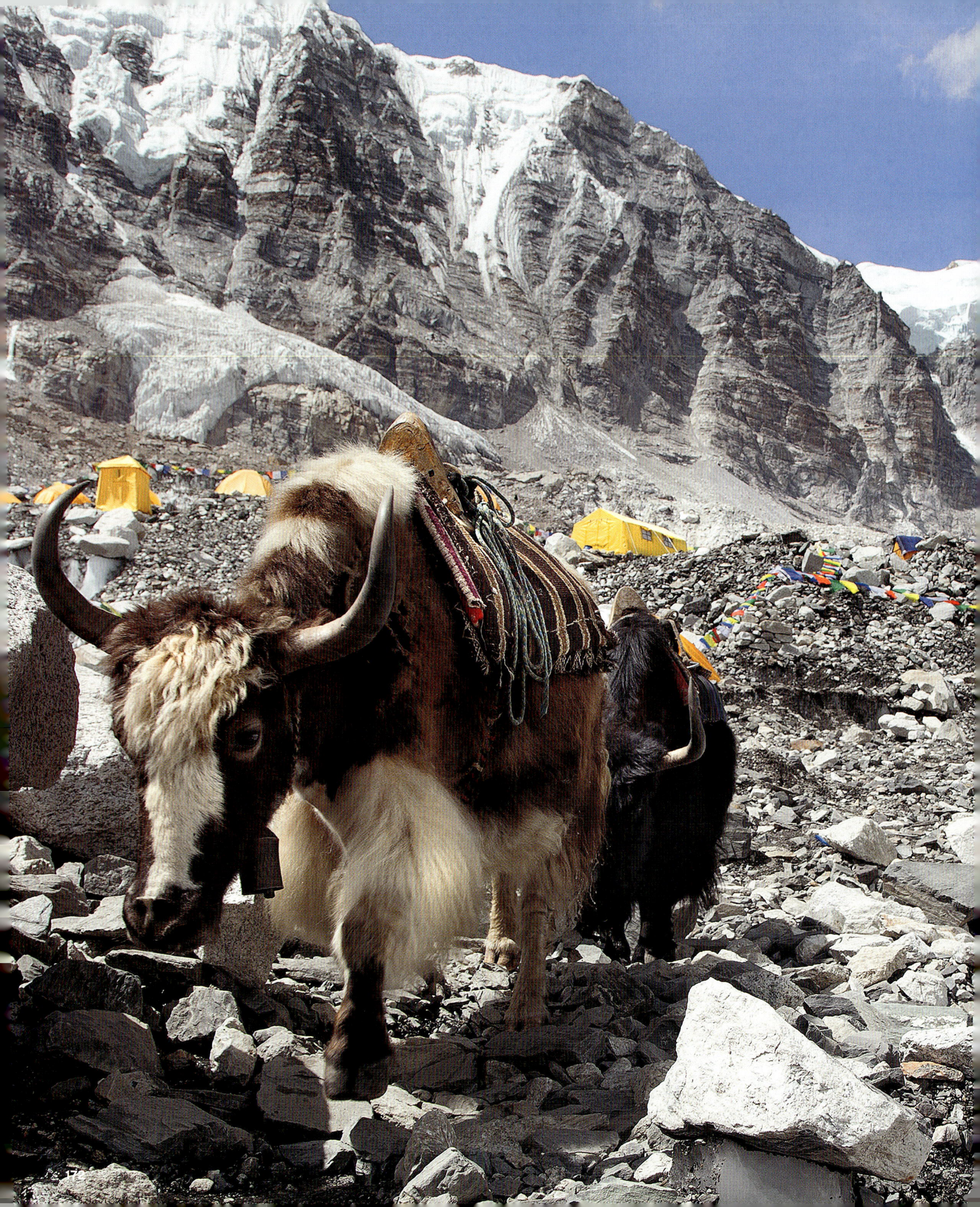

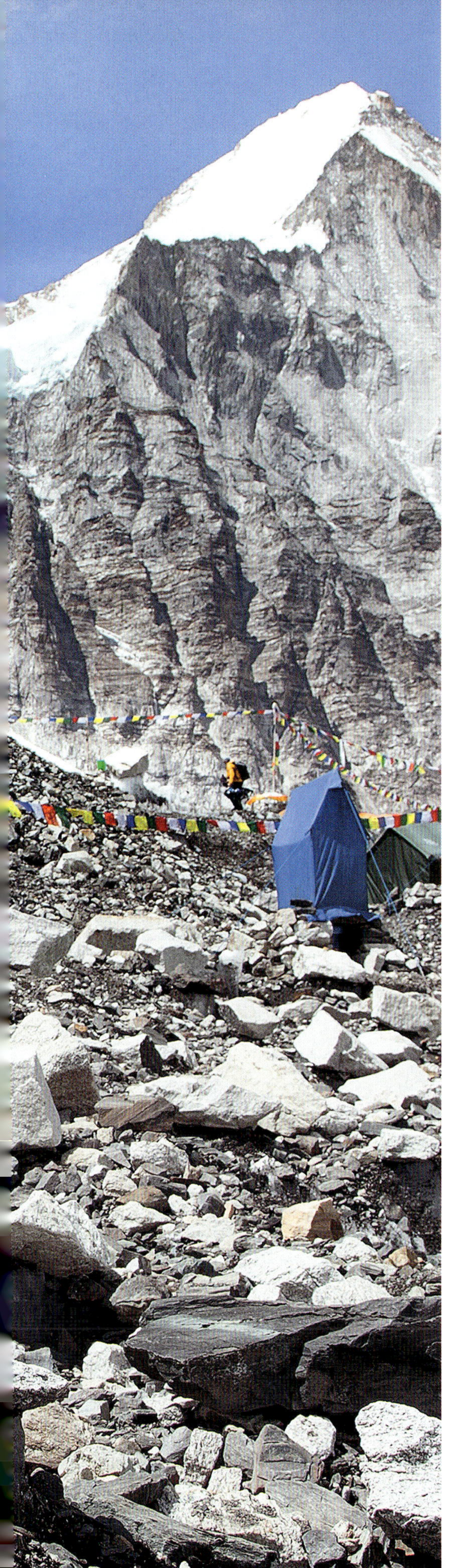

283. HIKE TO THE MOUNT EVEREST BASE CAMP

The Solu Khumbu district in the Nepalese Himalayas is what you can confidently call the "roof of the world"; there are a total of fourteen peaks more than 26,247 ft. (8,000 m) worldwide, and here you can see four of them from appropriate vantage points, including Everest, the highest mountain on Earth. And that is the goal of the hike—if you don't see the way itself as the goal. Because along the way, there are spectacular views of "small" mountains such as Ama Dablam (22,493 ft. [6,856 m]) and Nuptse (25,791 ft. [7,861 m]), which clearly surpass the rather hidden Everest in terms of attractiveness. From Gorak Shep, the last camp at 16,864 ft. (5,140 m) above sea level, you have the choice of walking to the base camp to see glaciers and tents or to the black hill of Kala Patthar (18,192 ft. [5,545 m]) for its far-better view of the mountains.

Time it takes: 12–15 days
Starting point: Lukhla
Elevation difference: 8,875 vertical ft. (2,705 m)
Best time to travel: March to April and October to November

284. THE HELAMBU TREK TO LANGTANG NATIONAL PARK

This trek, which starts right in the Kathmandu valley, lies in the shadow of the hiking trails to Nepal's icy giants. From Sunderjal, a long ascent through the forest of Shivapuri National Park leads to the first day's destination at Chipling. When the weather is good, Chipling offers a view of the main Himalayan ridge. The places along the way, inhabited by the Buddhist Tamang and Sherpa peoples, were hit hard by the earthquakes in April and May 2015—the residents are all the happier about the trekkers on the trail and give them a warm welcome. Because of the moss-covered rhododendron and oak forests in Langtang National Park, you are passing through a real fairy-tale forest, where, with a lot of luck, you might see a red panda. The highest point of the tour is the Tharepati, where there is a view toward the Everest massif.

Time it takes: 5–8 days
Starting point: Sunderjal
Elevation difference: 6,923 vertical ft. (2,110 m)
Best time to travel: March to April and October to November
Information: www.adventuresnepal.com/nepal/conservation-area/langtang-national-park.htm

285. HIKING AROUND THE ANNAPURNA GROUP MASSIF

The trek around the Annapurna Group is extremely varied and offers excellent views of majestic icy giants. Thanks to the continuous ascent, your altitude acclimatization is accomplished practically in passing. From the subtropical starting point at Jagat, the trail leads through a deep gorge on the back of the main Himalayan ridge, and thus from the Hindu lowlands to the Tibetan Buddhist highlands of Nepal. After crossing Thorong La Pass, with its highest point at 17,769 ft. (5,416 m), the trail leads through the deepest gorge in the world between Annapurna I and Dhaulagiri. From the Poon Hill station it runs, now in top condition, through rhododendron forest to Gorepani and up to the Annapurna Sanctuary, where the base camp for the Annapurna ascent lies in a glacier basin of immense dimensions.

Time it takes: 15–21 days
Starting point: Pokhara
Elevation difference: 15,276 vertical ft. (4,656 m)
Best time to travel: March to April and October to November

286. RAFTING AND KAYAKING ON THE KALI GANDAKI RIVER

Nepal is a fairy-tale country with mighty mountains, an exotic culture, colorful festivals, and medieval-looking villages. But it's not just the "roof of the world," the Himalayas, that attract outdoor explorers to Nepal. The unique multiday river expeditions in the paradisiacal landscape at the foot of these 26,247 ft. (8,000 m) peaks are largely unknown. A rafting or kayaking trip on the Kali Gandaki River takes outdoor enthusiasts off the beaten track on beautiful whitewater in areas where life has barely changed over recent decades. Camping in the evening on the sandbanks along the river, with a campfire under the starry sky, provides truly lasting experiences of nature far away from civilization and industry.

Length / time it takes: about 56 mi. (90 km) / 6 days
Level of difficulty: WW 2–3
Starting point: Ramdi, Darlamdanda
Best time of year: October and November
Information: www.global-kayak.comt

287. ON THE MOUNT KANGCHENJUNGA TREK

The thick moss growing on the trees shows that rain is not unusual in the region. Anyone who adjusts to this will experience wonderful, varied nature with magnificent forests and, in spring, can enjoy countless flowers, including the blooming rhododendrons. From Yuksom, the trail goes steeply up via Tshoka to Dzongri, a high-elevation pasture. Up there, a ridge lying above it allows an almost perfect view of the Kangchenjunga Group. From the camp in Thangsing, the ice walls of the southern foothills of Kangchenjunga appear close enough to touch. After another 3,281 ft. (1,000 m) of ascent, past Samity Lake and Onglakhing Glacier, you finally reach Gocha La. Your gaze passes over the Talung Glacier to Kangchenjunga, the third-highest mountain on Earth.

Time it takes: 1 day
Starting point: Yuksom
Elevation difference: 4,593 vertical ft. (1,400 m)
Best time to travel: June to October (spring)
Information: indiahikes.com

288. WALKING THROUGH THE VALLEY OF FLOWERS

The trail to the Valley of Flowers in northeastern Indi a is rather unusual for a trekking trip, because on the first day it runs along the same steps that countless Sikhs traverse on their way to the holy lake, Lake Hemkund. Old and young, women, children, and old men puff their way uphill. Those who are not fit enough let themselves be transported on a mule by mule drivers. Accommodation and restaurants are available in Ghangaria for pilgrims and trekkers. A detour leads through a gorge into the Valley of Flowers, high in the Himalayas. In summer the valley floor is covered with millions of flowering plants. The four-hour ascent to the holy lake, in whose water the Sikhs plunge for a cleansing bath, is then much more strenuous.

Time it takes: 5 days
Starting point: Govindghat
Elevation difference: 7,808 vertical ft. (2,380 m)
Best time to travel: June to October
Information: indiahikes.com/valley-of-flowers/, www.trekkingingarhwal.com

289. ON THE ZANSKAR TREK

The journey to Padum, over high passes along the main Himalayan ridge, is already driving up your adrenaline level. Far away in the Trans-Himalayas, the trek offers the hiker desertlike high mountains, where every village is an oasis. After two days along the Zanskar valley, the trail leads in a long, long ascent to the first high pass, the 15,499 ft. high (4,724 m) Hanuma La. From there you have a view of a valley basin and Lingshet. Behind Lingshet, two passes await: Singhe La and Sirsir La, which reach an altitude of just above and below 16,404 ft. (5,000 m), respectively, before the trail goes through the Hanupatta Gorge to Lamayuru. This route is almost impossible without a guide who knows the area and without pack animals. Incredible views and the fascinating Ladakhi culture make up for the effort in abundance.

Time it takes: 10 days
Starting point: Padum (11,680 ft. [3,560 m])
Elevation difference: 6,112 vertical ft. (1,863 m)
Best time to travel: July to August
Information: www.julehadventure.com/trekking_zanskar_padum_lamayuru.html

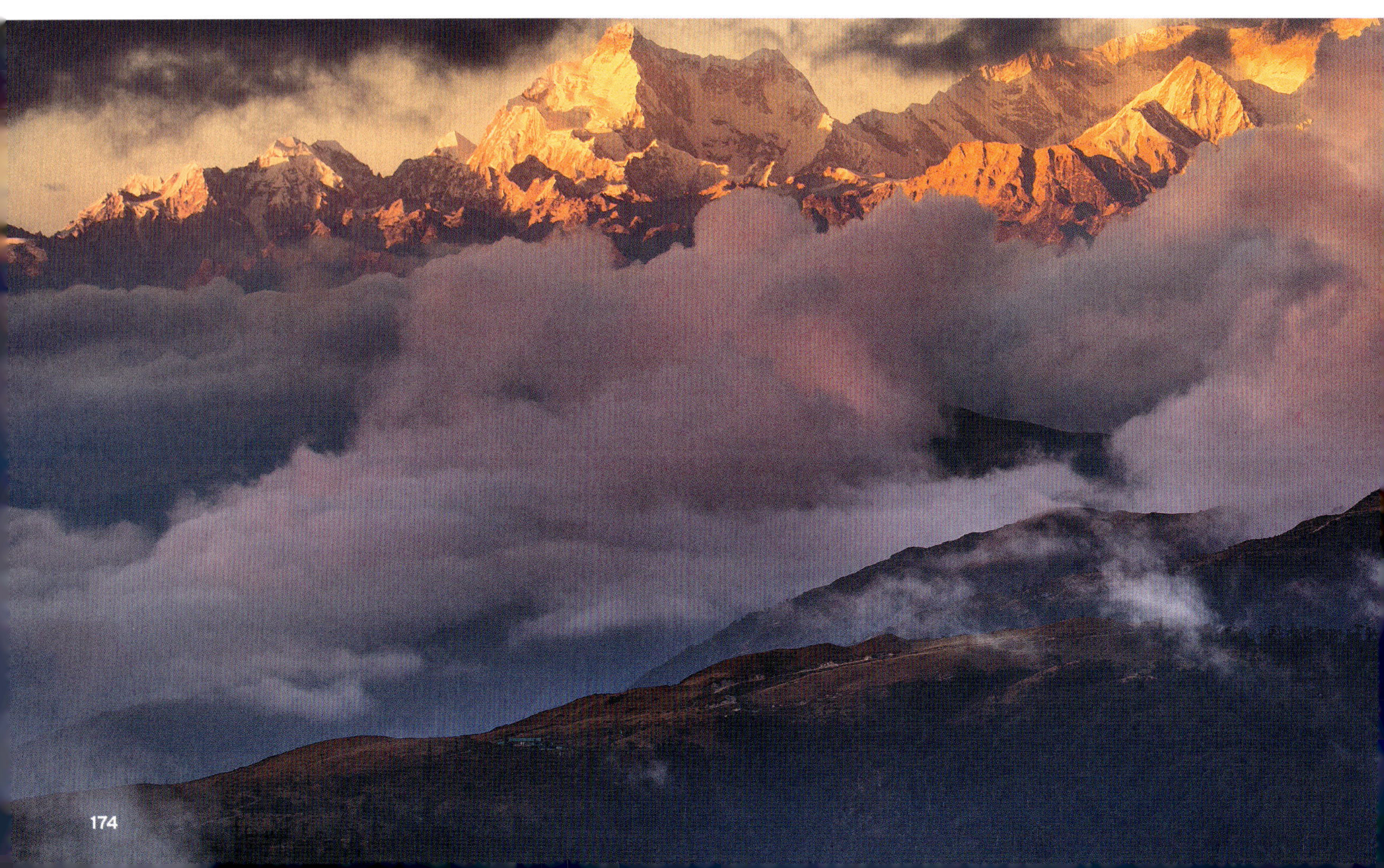

290. HIKING ALONG THE SINGALILA RANGE

From the rice terraces of India to where the snows are at home: "Hima Alaya" means "abode of snow" in Sanskrit, the ancient language of the Indian Brahmins. In complete seclusion, the trek along the Singalila range leads from the rice terraces of the tropical lowlands up into the Himalayas, to the foot of the 28,209 ft. high (8,598 m) Kangchenjunga. Thanks to the accompanying Sherpas, the trekker can walk with a light pack on the hidden trails, along high ridges, over wild passes, past holy lakes, and through deep valleys. At the same time, they always have the gigantic view of the highest mountains in the world before them; the majestic beauty and the wild splendor of the sublime landscape make both the effort and the sensation of cold during this secluded high-altitude trekking fade away into comparatively nothing.

Time it takes: 10–12 days
Highest elevation: 14,862 ft. (4,530 m)
Starting point: Uttarey, India
Information: www.dav-summit-club.de

291. KAYAKING IN THE BACKWATERS OF KERALA

The backwaters of Kerala are a network of canals and lakes parallel to the Malabar Coast from Kochi to Kollam. Large parts of the 734 sq. mi. (1,900 km^2) area can be reached only by boat. With twenty-nine larger lakes, forty-four rivers, and countless small canals lined with fields and palm trees, it has the impact of a hidden paradise. If you don't just want to chug through the larger canals, you can plunge into the world of the smaller and smallest canals in a kayak amid water hyacinths, coconut palms, and tropical vegetation. Like all of Kerala, the islands in the backwaters are densely populated, so that everywhere you can meet fishermen, farmers, or school children who walk on the paths along the canals or jump into the water to swim in the evening.

Time it takes: 4 hours–several days
Starting point: Kochi
Information: www.keralatourism.org/destination/backwater

292. HIKING IN THE GREEN NILGIRIS HILLS

Just getting to Ooty (Udagamandalam) by the narrow-gauge railroad, the Nilgiri Mountain Railway, is a special pleasure. The area is known for its tea cultivation, so you will find tea plantations and tropical forest covering this South Indian mountainous country when you arrive. The British military once set up many cooler retreats, so-called hill stations, to escape the searing heat before the monsoon. The highlands around Ooty were the most popular hill station in India. In the highlands, with the up to 8,652 ft. high (2,637 m) mountains, there are many opportunities for hiking to waterfalls, reservoirs, and lonely high pastures, past coffee plantations and tea gardens. Six-day treks lead from Ooty to Madumalai National Park, where, with a lot of luck, tigers can be seen—but encounters with deer, elephants, and monkeys such as the Nilgiri langurs are more likely.

Time it takes: 1–2 days
Starting point: Ooty
Elevation difference: 1,312 vertical ft. (400 m)
Information: indiahikes.com/nilgiri-hills-trek/

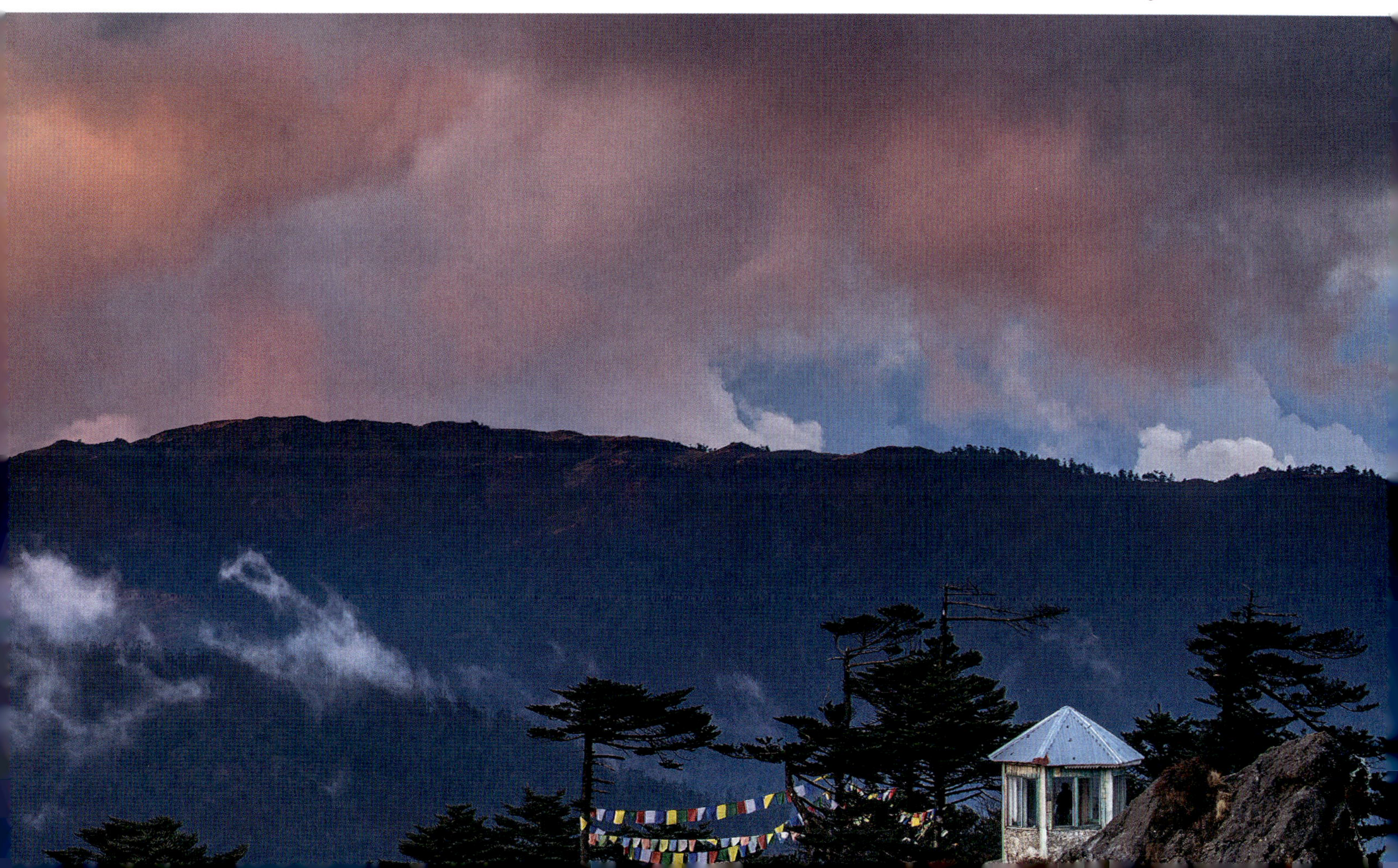

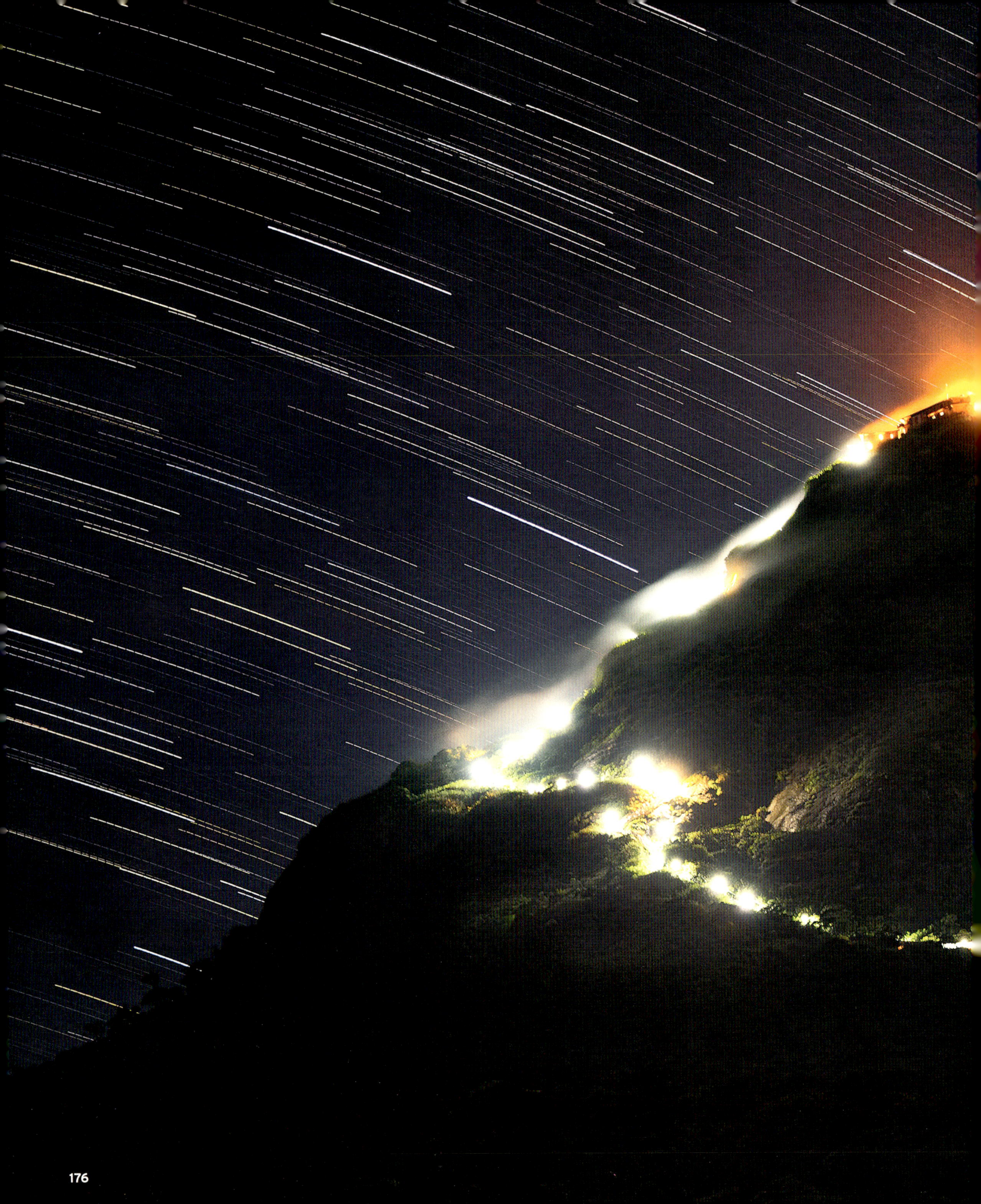

293. NIGHT HIKE TO ADAM'S PEAK

Adam's Peak, better known in Sri Lanka as Sri Pada, is sacred to all four of the island's religions. For Buddhists, the footprint of the buddha is to be seen on the summit; for Hindus, the giant imprint was left by Shiva; for Muslims, this was where Adam took his first step after being expelled from Paradise; and for Christians it was St. Thomas. In any case, if the weather is right, there will be plenty of people gathered on the mountain at sunrise: then the mountain casts a perfect triangular shadow on the highlands. It is best to start the ascent between 2:00 and 2:30 at night. You toil your way upward over illuminated steps, dripping with sweat, until you suddenly plunge into cold air currents. The view from above is simply stunning in the first light of day!

Time it takes: 5–8 hours
Starting point: Dalhousie
Best time to travel: December to April
Elevation difference: 3,428 vertical ft. (1,045 m)
Information: www.srilanka-reise.info/adams-peak/

294. HIKES IN THE KNUCKLES RANGE

Only about an hour away from Kandy, a centrally located city in the Sri Lankan highlands, you can immerse yourself in tea gardens and jungle at Elkaduwa. The steep mountain slopes are covered with giant trees, and rock walls tower above the valley. The landscape at Hunas Falls is particularly beautiful. You get to the foot of the waterfall through a spice garden. Above the waterfall, the hotel of the same name is on an artificial lake that also feeds the waterfall. From here, you can continue to ascend to the tea plantations. But there are also possibilities to make your way through the jungle on more-strenuous trails, such as climbing Dothalugala Mountain. From Corbetts Gap there is a beautiful but quite sweat-generating route to the Nitro Cave, which is home to thousands of bats.

Time it takes: 3–7 hours
Starting point: Kandy
Lowest elevation: 1,476 ft. (450 meters)
Highest elevation: 7,365 feet (2,245 m)
Best time to travel: February to September
Information: www.srilanka-reiseziel.com/hunas-falls.html, www.yohobed.com/blog/hiking-in-the-knuckles-mountain-range/

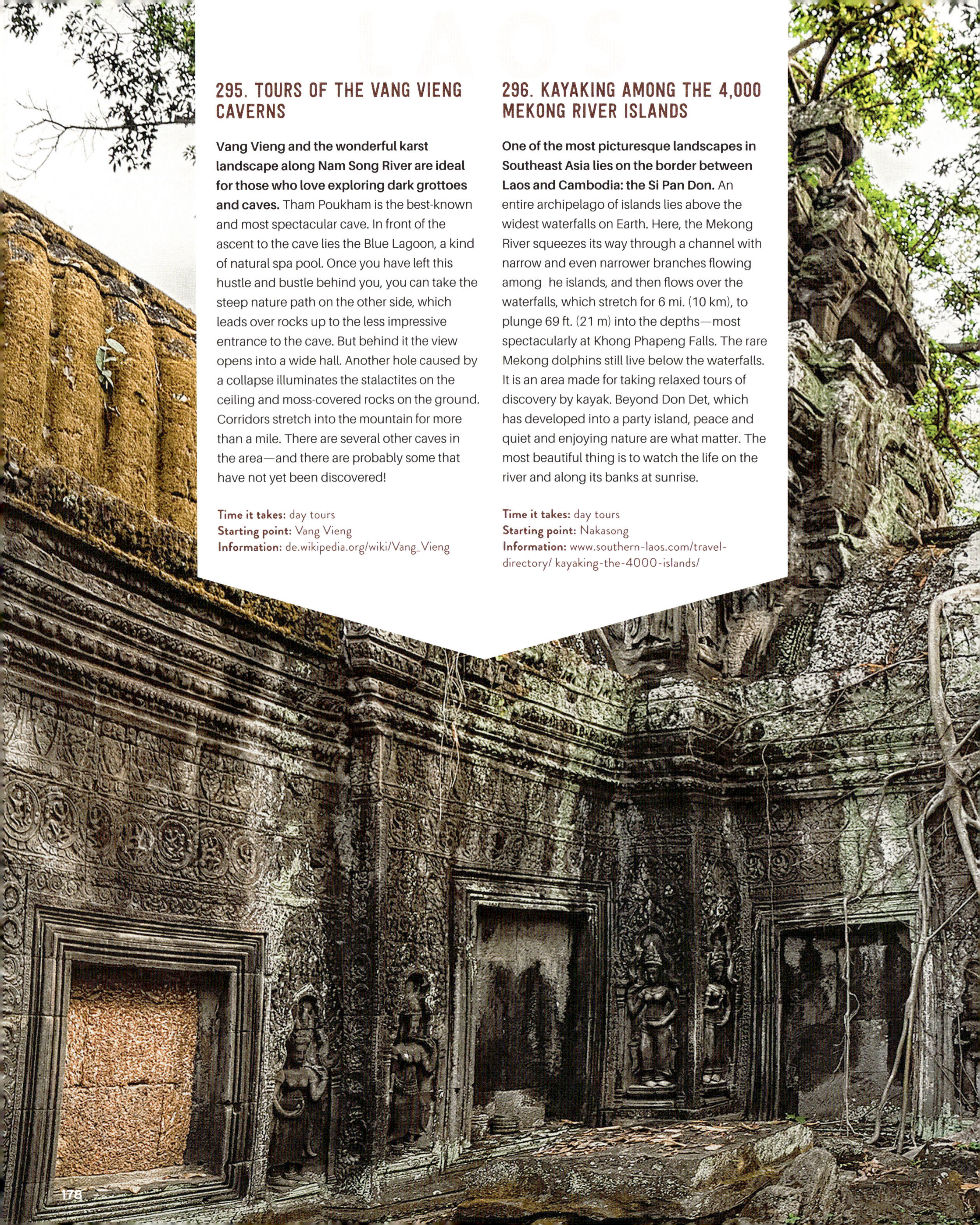

295. TOURS OF THE VANG VIENG CAVERNS

Vang Vieng and the wonderful karst landscape along Nam Song River are ideal for those who love exploring dark grottoes and caves. Tham Poukham is the best-known and most spectacular cave. In front of the ascent to the cave lies the Blue Lagoon, a kind of natural spa pool. Once you have left this hustle and bustle behind you, you can take the steep nature path on the other side, which leads over rocks up to the less impressive entrance to the cave. But behind it the view opens into a wide hall. Another hole caused by a collapse illuminates the stalactites on the ceiling and moss-covered rocks on the ground. Corridors stretch into the mountain for more than a mile. There are several other caves in the area—and there are probably some that have not yet been discovered!

Time it takes: day tours
Starting point: Vang Vieng
Information: de.wikipedia.org/wiki/Vang_Vieng

296. KAYAKING AMONG THE 4,000 MEKONG RIVER ISLANDS

One of the most picturesque landscapes in Southeast Asia lies on the border between Laos and Cambodia: the Si Pan Don. An entire archipelago of islands lies above the widest waterfalls on Earth. Here, the Mekong River squeezes its way through a channel with narrow and even narrower branches flowing among he islands, and then flows over the waterfalls, which stretch for 6 mi. (10 km), to plunge 69 ft. (21 m) into the depths—most spectacularly at Khong Phapeng Falls. The rare Mekong dolphins still live below the waterfalls. It is an area made for taking relaxed tours of discovery by kayak. Beyond Don Det, which has developed into a party island, peace and quiet and enjoying nature are what matter. The most beautiful thing is to watch the life on the river and along its banks at sunrise.

Time it takes: day tours
Starting point: Nakasong
Information: www.southern-laos.com/travel-directory/ kayaking-the-4000-islands/

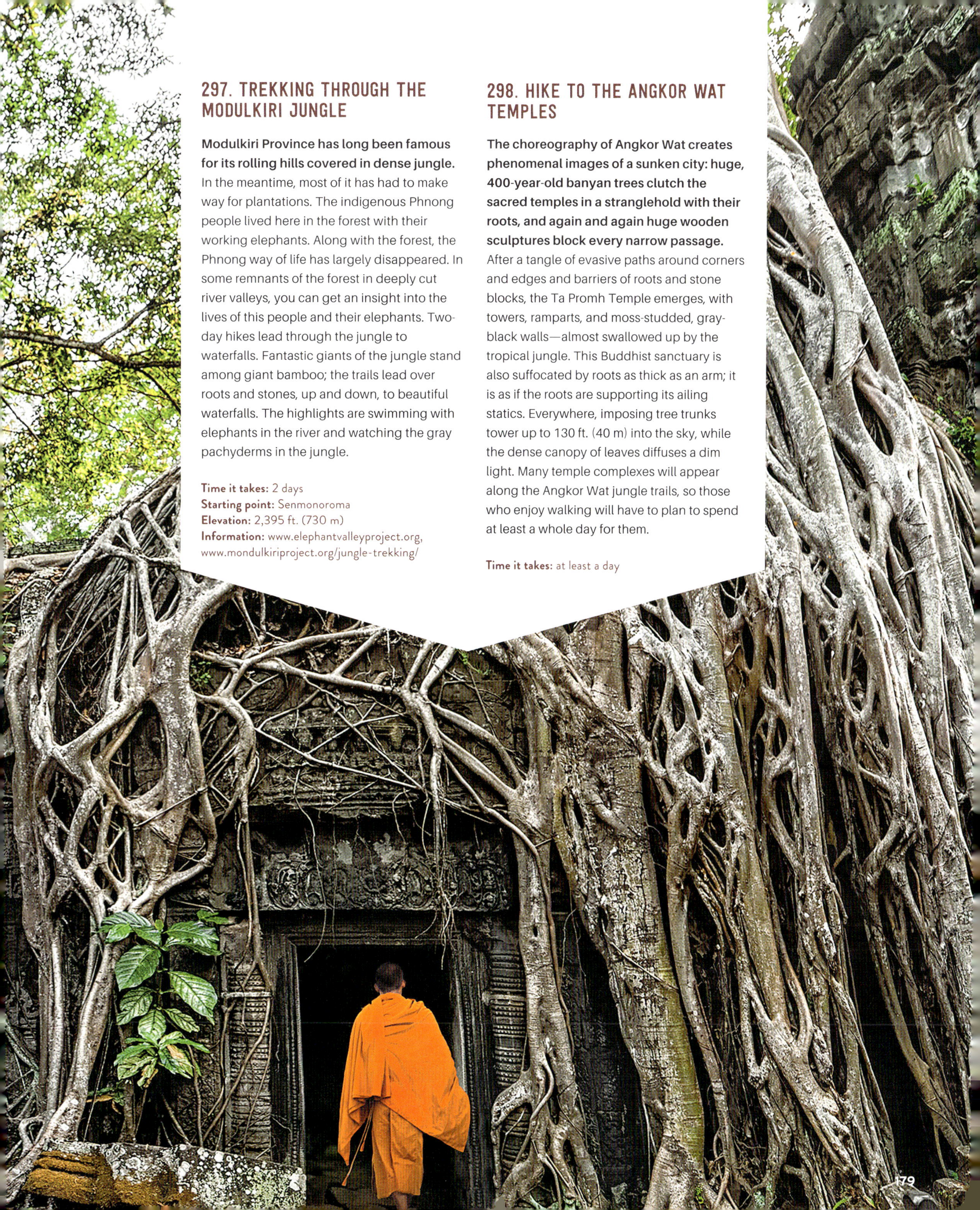

297. TREKKING THROUGH THE MODULKIRI JUNGLE

Modulkiri Province has long been famous for its rolling hills covered in dense jungle. In the meantime, most of it has had to make way for plantations. The indigenous Phnong people lived here in the forest with their working elephants. Along with the forest, the Phnong way of life has largely disappeared. In some remnants of the forest in deeply cut river valleys, you can get an insight into the lives of this people and their elephants. Two-day hikes lead through the jungle to waterfalls. Fantastic giants of the jungle stand among giant bamboo; the trails lead over roots and stones, up and down, to beautiful waterfalls. The highlights are swimming with elephants in the river and watching the gray pachyderms in the jungle.

Time it takes: 2 days
Starting point: Senmonoroma
Elevation: 2,395 ft. (730 m)
Information: www.elephantvalleyproject.org, www.mondulkiriproject.org/jungle-trekking/

298. HIKE TO THE ANGKOR WAT TEMPLES

The choreography of Angkor Wat creates phenomenal images of a sunken city: huge, 400-year-old banyan trees clutch the sacred temples in a stranglehold with their roots, and again and again huge wooden sculptures block every narrow passage. After a tangle of evasive paths around corners and edges and barriers of roots and stone blocks, the Ta Promh Temple emerges, with towers, ramparts, and moss-studded, gray-black walls—almost swallowed up by the tropical jungle. This Buddhist sanctuary is also suffocated by roots as thick as an arm; it is as if the roots are supporting its ailing statics. Everywhere, imposing tree trunks tower up to 130 ft. (40 m) into the sky, while the dense canopy of leaves diffuses a dim light. Many temple complexes will appear along the Angkor Wat jungle trails, so those who enjoy walking will have to plan to spend at least a whole day for them.

Time it takes: at least a day

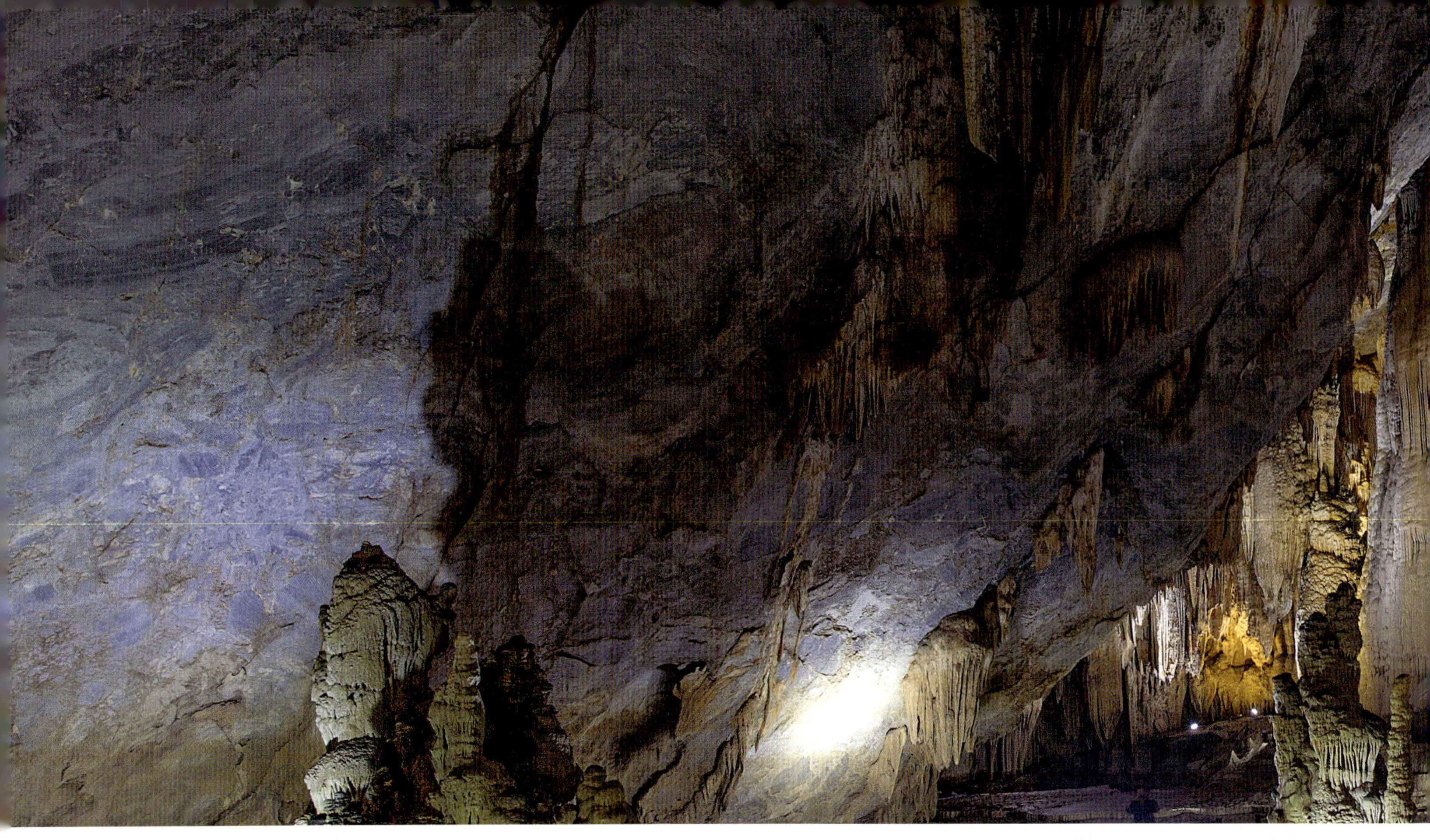

299. SEA KAYAKING IN ANG THONG NATIONAL MARINE PARK

The scenery is almost unbelievably beautiful. Author Alex Garland by no means exaggerates things in his book *The Beach*. Some 40 karst rocks rise from the sea and form an enchanted archipelago of uninhabited islands. Fishermen used to be the only visitors to this archipelago, which lies west of Ko Samui Island in the Gulf of Thailand. Today, vacationers who want to see this piece of paradise up close also come. You can enjoy paddling through the turquoise water, under the breathtaking overhanging karst rocks, and slip through narrow inlets and passages in your sea kayak. Caves, hongs, and bizarre rocks form a landscape so fascinating that you could hardly imagine it being any more so. You can get there by boat with a local tour operator such as Blue Stars, which will bring the sea kayaks along.

Time it takes: 8 hours
Starting point: Ko Samui
Best time to travel: April to October
Information: www.bluestars.info

300. BOAT TOURS OF AO PHANG NGA NATIONAL MARINE PARK BAY

The Hollywood movie *The Beach* is set in Ang Thong, but the film was shot by the Andaman Sea. And indeed there are even nicer beaches and lagoons here than on the gulf. It's a bizarre, crazy world that emerges as you paddle along in a sea kayak. Even from the outside, these rocks rising high out of the sea look impressive. Inside there are also hidden lagoons, which can be reached by kayak through long, dark caves when the water level is right. What looks like a solid rock from the outside is usually a tangle of folded ridges with lagoons and gorges in between them—accessible only to birds or through caves more than 328 ft. (100 m) long. Guides know the right entrances and the right time to go through them.

Time it takes: 8 hours
Starting point: Phuket
Best time to travel: November to April
Information: www.bluestars.info

301. VISITING THE BAHNAR MOUNTAIN PEOPLE

Not far from the national borders with Laos, in the central highlands of Vietnam, you will find the remote villages of the Bahnar, the people of the mountains. When you are not far from Kon Tum, your moped passes one of the many villages of the minority peoples who live here. But things get really secluded only after the spectacular suspension bridge and the village Kon Brap Dzu. When visiting a village of the Bahnar people, whose society is organized on matriarchal lines, you drink rice wine with the mayor. Over the riverbank and through the rice fields, the trail goes up into the mountains. There you accompany locals on a bird hunt and make your way through the jungle. At night, during a communal meal in one of the huts, the elders volunteer the stories of their youth and of the last tigers in the area.

Time it takes: 3 days
Transport: by moped from Kon Tum to Kon Brap Dzu, then on foot
Tour guide: Mr. Anh, owner of the Eva Cafe in Kon Tum
Information: instagram.com/evacafekontum

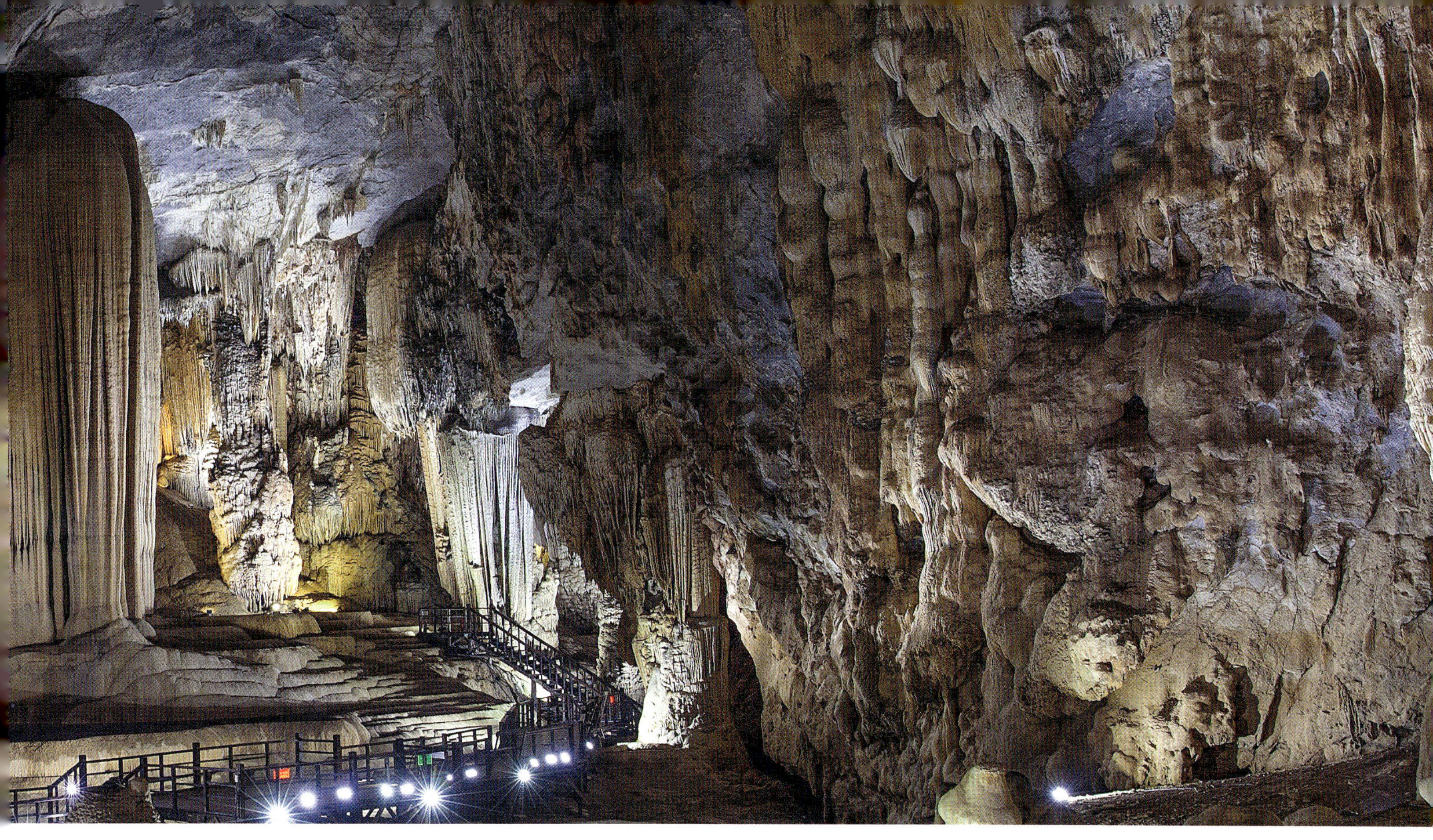

302. BOAT TRIP TO CAN THO FLOATING MARKET

The Mekong—the mighty lifeline of Southeast Asia. Not only does the river carry unbelievable amounts of water—traditional floating markets also enliven and are a special feature of the Mekong delta. Early in the morning especially, the many large and small boats of local traders cavort on the water in the light of the first rays of sunshine. The boat trip from Can Tho, the largest city in the delta and the fourth largest in Vietnam, to the Cai Rang Market takes just under an hour. Fruit and vegetables are transported from one floating platform to another. The trip continues past the lush green palm trees on the riverbank to Phong Dien Market. The boat women skillfully maneuver travelers through the colorful market bustle before leading them into the paradisiacal orchards on the side channels.

Time it takes: 4–6 hours
Starting point: Can Tho
Pickup time: 5 a.m.
Organization: local boat women along the city promenade
Information: vietnam-reiseprof.com

303. CAVE TOUR IN PHONG NHA KE BANG NATIONAL PARK

The oldest karst region in Asia attracts adventurers as well as cave fans—it contains many mysterious caves, including the largest in the world. The national park, with its lively rainforest, where animals can be heard loudly calling attention to themselves, would be beautiful in itself, but nothing that special if it weren't for hundreds of caves. A world of its own with mystical grottoes and underground rivers and waterfalls. Some 1,969 ft. (600 m) of the more than 3 mi. long (5 km) Phong Nha Cave can be navigated by dragon boats. Yet, the Paradise or Thien Duong Cave beats the Phong Nha, with its length of 19 mi. (31 km). Travelers stroll on foot for half a mile through the well-lit wonderland full of bizarre stalactites and stalagmites.

Time it takes: 1 day
Transportation: Rent a moped in Dong Hoi or book an organized tour
Starting point: Dong Hoi, 28 mi. (45 km) away from the national park
Best time to travel: April to October
Information: phongnhacavestour.com

304. BIKE AND BOAT TOUR OF NINH BINH

Rugged limestone cliffs cut through endless rice fields and create an unparalleled fairy-tale landscape near Ninh Binh, the "dry Ha Long Bay." Nothing is better than a bike for exploring and marveling at the narrow paths along the rice fields. Those who take on the exertion of climbing the countless stairs of the Hang Mua karst rock will be richly rewarded, not only by the pretty and graceful pagoda that towers up there, but also by an unforgettable view of the plains from above. But even those who like things to be more comfortable will feel at home here. Locals take travelers along with them in their traditional wooden boats and glide calmly past the bizarre rocks along the Ngo Dong River.

Time it takes: 1 day
Starting point: Ninh Binh boat pier: in Tam Coc
Information: ninhbinh-tour.com

305. HIKE TO INLE LAKE

The hiking trail from Kalaw to Inle Lake inspires less by any magnificent landscapes than by providing a look behind the scenes into the incredible tranquility of village life. The best thing is to organize the hike in Kalaw at a small trekking agency, preferably during November to February. It is advisable to hike with a guide. Without a guide, it will be difficult to find your way along the many trails that connect the villages. In addition to the guide, the team even includes a cook, who hurries ahead of the group and welcomes you with a wonderful meal at each respective stop. After the hike, you can observe life in the villages: cattle and bullock carts return along the dusty paths to the village in the evening light, children play by the monastery, and everything looks extremely peaceful.

Time it takes: 2–3 days
Starting point: Kalaw
Elevation difference: 4,265 ft. (1,300 m)
Information: a1trekking.blogspot.com/p/blog-page_7725.html

306. TREKKING ON MOUNT VICTORIA

A trip to Chin State still requires a little adventurous spirit. There has not been any road development in most of the Chin State, and there are only sparse bus connections. The trip to Kanpelet, on the edge of Mount Victoria National Park, is an adventure in itself. But there are plenty of accommodations available in wooden bungalows there. And here you can also find a guide for short or long hikes on the mountain and to the summit, or to the Chin villages in the area. A permit for ten dollars is sold at the entrance to the national park, but unfortunately, hunting and logging still go on in the park. While the forests still look tropical up to about 6,562 ft. (2,000 m) in elevation, mainly pine trees grow in the upper area of the mountain; parts of the mountain are covered with grassland.

Time it takes: 1–3 days
Starting point: Kanpelet; minimum elevation: 4,429 ft. (1,350 m)
Highest elevation: 10,070 ft. (3,070 m)
Information: trekking-myanmar.com/hiking-chin-state-mt-victoria/, www.go-myanmar.com/mount-victoria-nat-ma-taung-kanpetlet-mindat

307. ASCENDING MOUNT KINABALU

Mount Kinabalu is the highest mountain within a radius of 1,553 mi. (2,500 km). The lonely 13,435 ft. high (4,095 m) mountain rises from a beautiful rainforest. Even at the starting point at the park administration, there are various options for taking impressive hikes in the jungle. A wide footpath leads up the mountain over rocky steps, roots, and loam through the cloud forest. Over the course of the ascent, these giant trees shrink to the size of scrawny apple trees hung with long beards of moss. At an elevation of 10,827 ft. (3,300 m) there are accommodations with a view of the lower-lying clouds. From here, the ascent starts in the middle of the night on an easy but secured path over bare rock to the summit, which is about 2,624 ft. (800 m) higher.

Time it takes: 2 days
Starting point: Kota Kinabalu
Lowest elevation: 5,249 ft. (1,600 m)
Highest elevation: 13,435 ft. (4,095 m)
Permit: best to apply 6 months in advance!
Information: www.sabahtourism.com/destination/kinabalu-national-park/

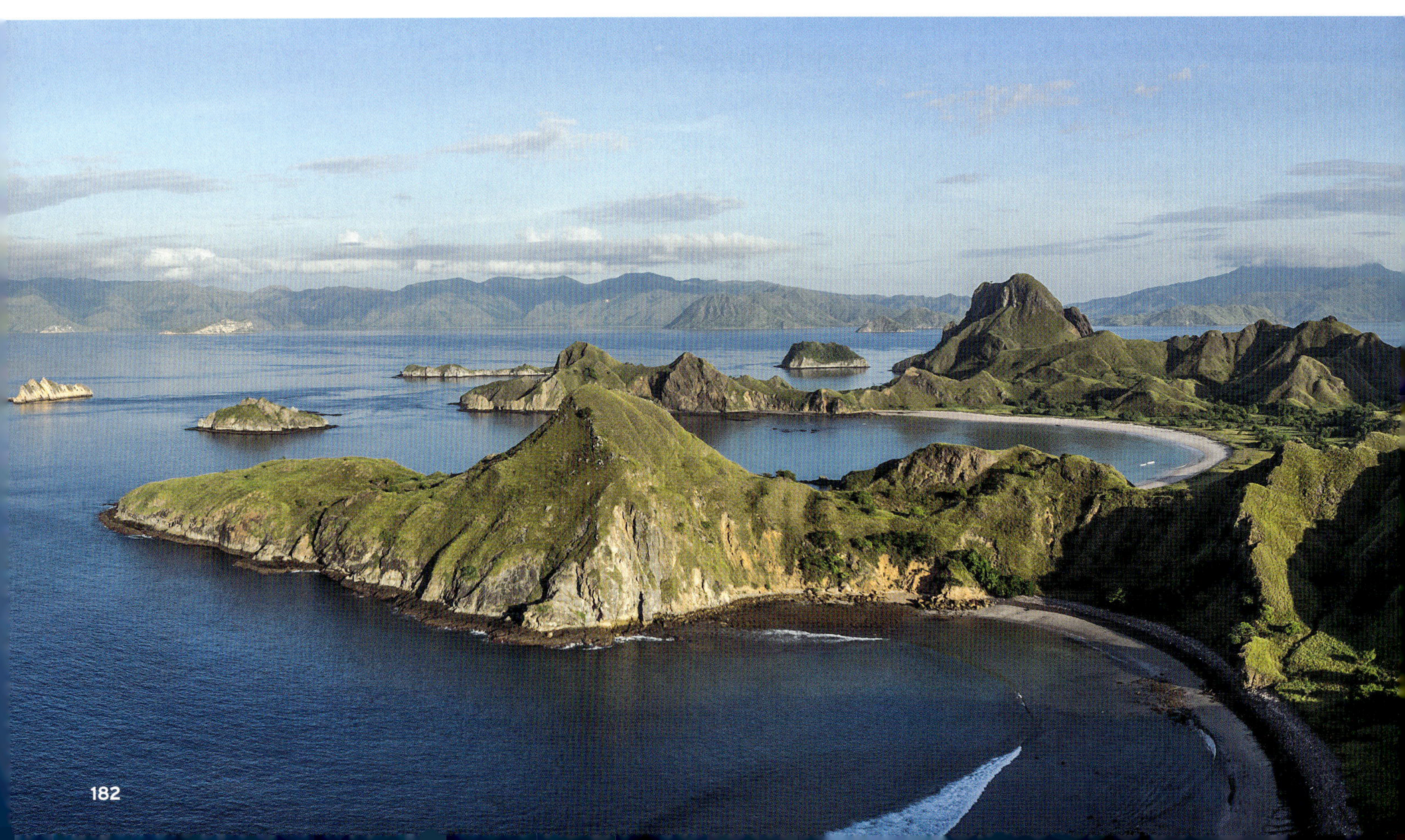

308. DIVING OFF BORNEO

The island of Sipadan is one of the most spectacular diving destinations in the world and is high up in the rankings of specialist magazines. Sipadan is to the east of Sabah, 22 mi. (36 km) off the coast of Semporna. This is a volcanic peak that barely protrudes from the sea, which is around 1,969 ft. (600 m) deep. The volcanic pipe took on the strange shape of a mushroom because of the coral that has grown on it. Since Jacques Cousteau raved about the island, every diver has been dreaming of diving here among the great fish, such as hammerhead sharks, whale sharks, barracudas, and many other species. In the "hanging coral gardens," you can always find turtles gliding majestically and relaxed through the water, and huge schools of fish feeding on small particles floating in the water before Barracuda Point.

Time it takes: 2 days
Starting point: Semporna
Permit: apply at least 2 months in advance!
Information: www.amazingborneo.com/sabah/diving/sites # diving-sites

309. HIKING THROUGH GREEN RICE TERRACES OF TANA TORAJA

The Toraja highlands, in the center of Sulawesi Island, offer one of the most enchanting landscapes in Indonesia: islands with karst rocks and giant bamboo protrude from a sea of green rice fields; in between are the villages of the Toraja people, with their uniquely constructed houses. The gables of the Tongkonan houses rise up like the prows of ships, while burial sites in caves and trees lie along the way. A hike on Mount Sesean promises the best view overlooking the valley of Rantepao. Your overnight stays are simple home stays with members of a family—which is part of the fun. While Mount Sesean can be easily explored on your own, a local guide is essential for longer hikes. This way, immersing yourself in the landscape and culture becomes a special experience.

Time it takes: 2–7 days
Starting point: Rantepao
Lowest elevation: 2,559 ft. (780 m)
Highest elevation: 6,890 ft. (2,100 m)
Information: de.wikivoyage.org/wiki/Tana_Toraja

310. SEA KAYAKING IN KOMODO NATIONAL PARK

Komodo Marine National Park, with its famous giant lizards, is between Flores and Sumbawa. Besides the Komodo dragons, the 702 sq. mi. (1,817 km^2) national park offers many small islets with a wonderful underwater world. With its grassy, wrinkled escarpment, Komodo looks like the perfect location for Jurassic Park. You would be only too happy to meet the last dinosaurs here—but the monitor lizards are just lizards, up to 10 ft. (3 m) long, that dive off the coast of Komodo and Rinca to hunt. You can experience this fantastic underwater world while snorkeling—ideally between May and September. It is incomparably beautiful to float on the turquoise-green water in a sea kayak and then spend the nights in a tent on the small, uninhabited islands.

Time it takes: 1–10 days
Starting point: Labuhan Bajo Island
Lowest elevation: sea level
Highest elevation: about 328 ft. (100 m)
Information: www.komodoseekajak.de, www.komodonationalpark.org

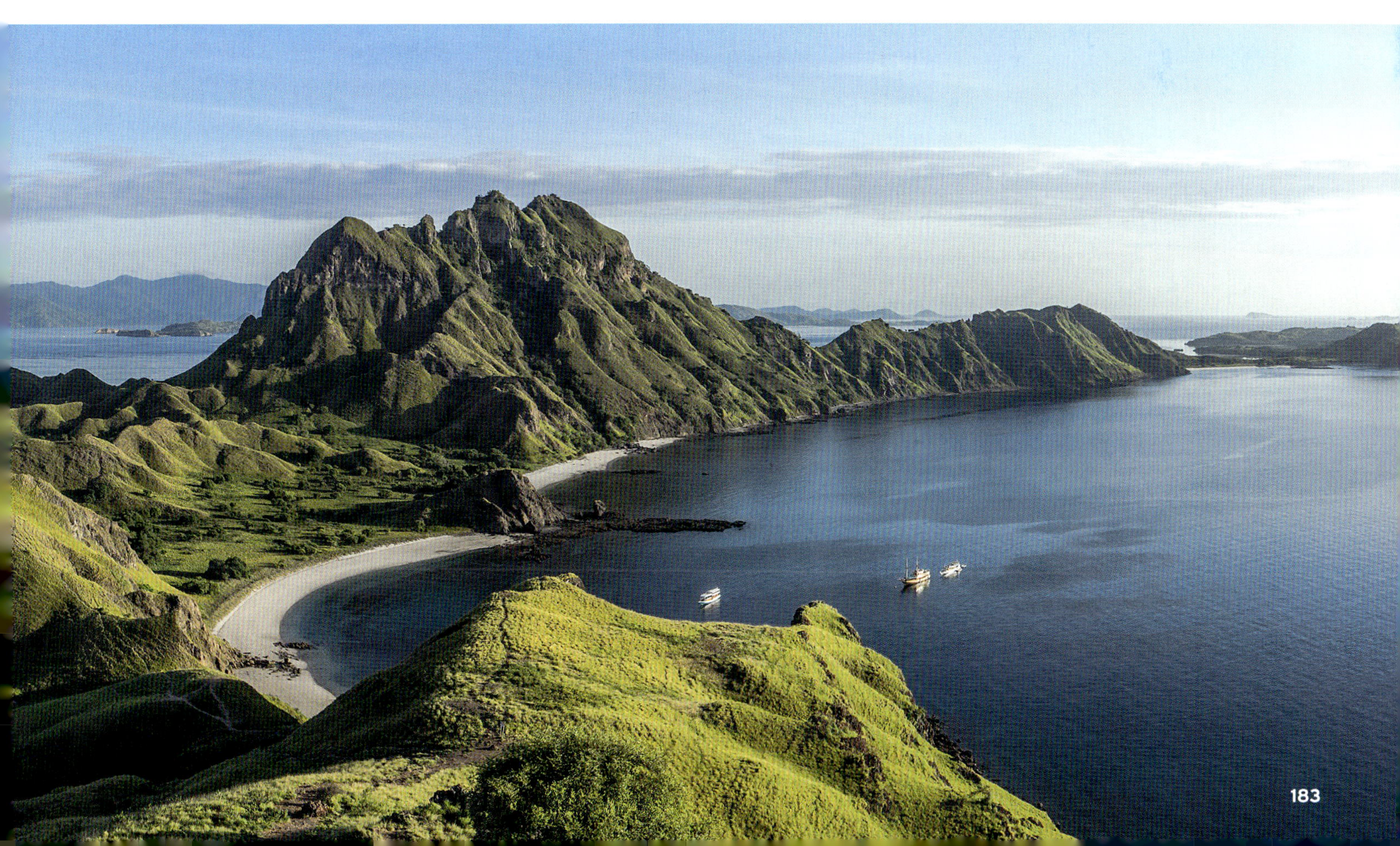

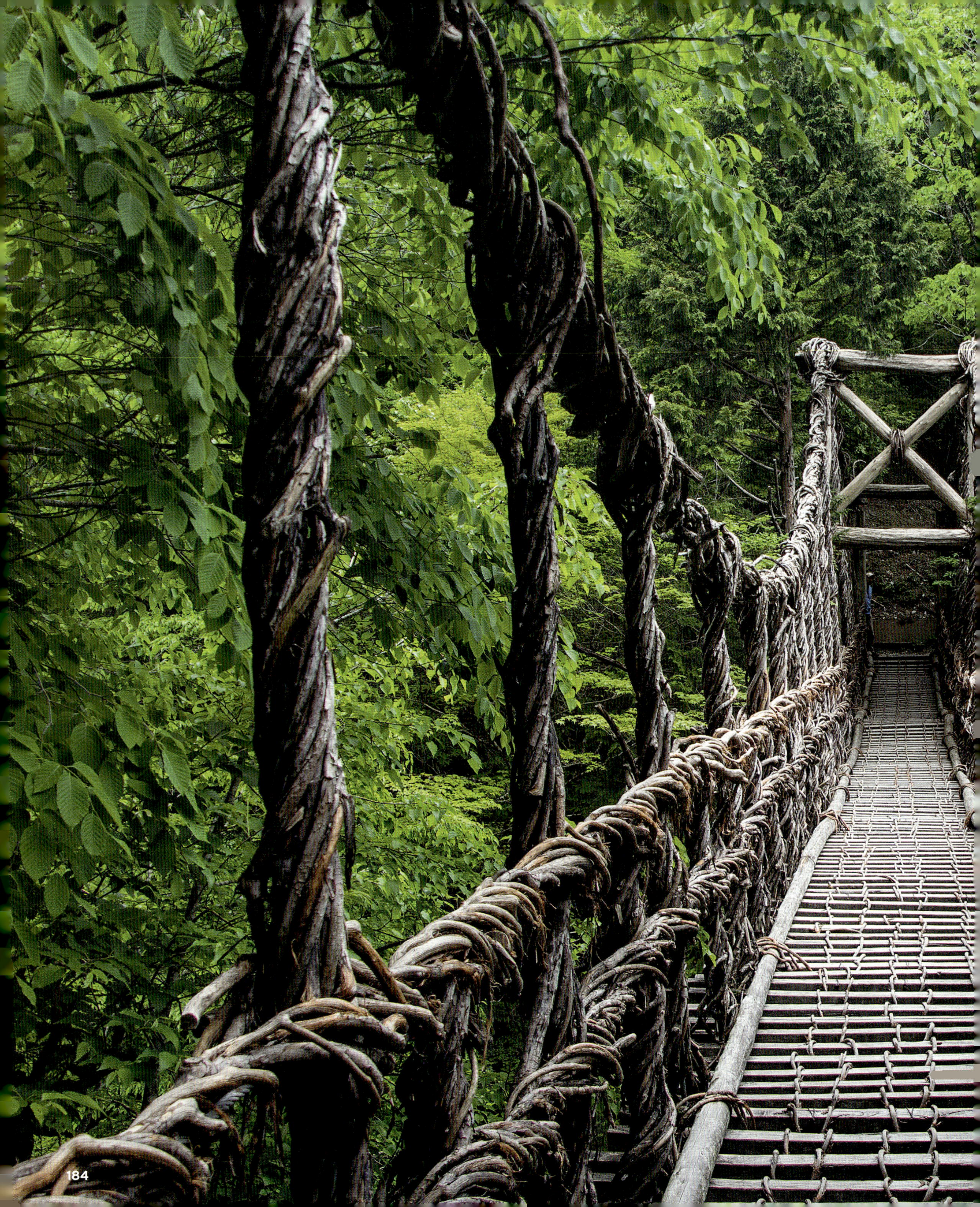

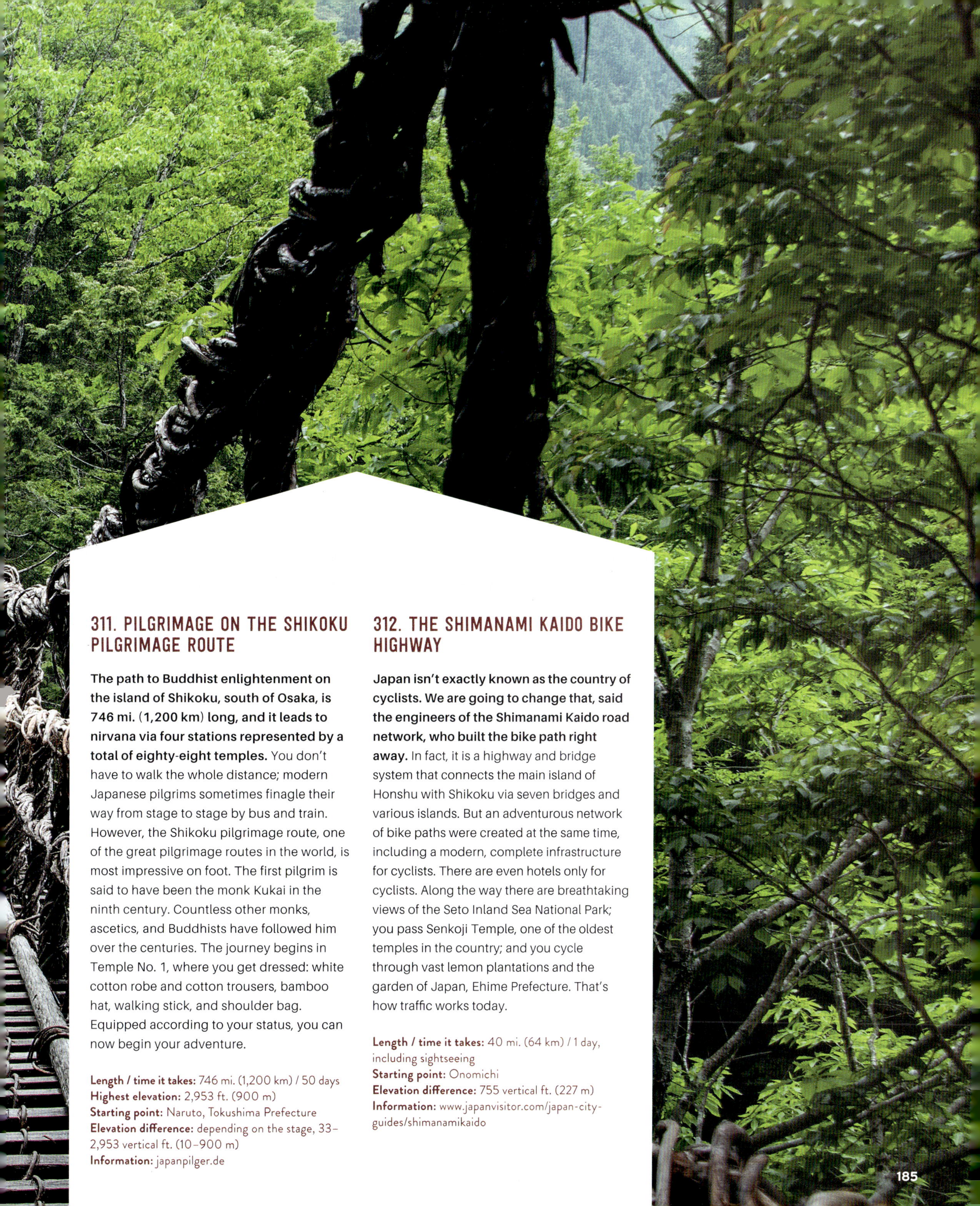

311. PILGRIMAGE ON THE SHIKOKU PILGRIMAGE ROUTE

The path to Buddhist enlightenment on the island of Shikoku, south of Osaka, is 746 mi. (1,200 km) long, and it leads to nirvana via four stations represented by a total of eighty-eight temples. You don't have to walk the whole distance; modern Japanese pilgrims sometimes finagle their way from stage to stage by bus and train. However, the Shikoku pilgrimage route, one of the great pilgrimage routes in the world, is most impressive on foot. The first pilgrim is said to have been the monk Kukai in the ninth century. Countless other monks, ascetics, and Buddhists have followed him over the centuries. The journey begins in Temple No. 1, where you get dressed: white cotton robe and cotton trousers, bamboo hat, walking stick, and shoulder bag. Equipped according to your status, you can now begin your adventure.

Length / time it takes: 746 mi. (1,200 km) / 50 days
Highest elevation: 2,953 ft. (900 m)
Starting point: Naruto, Tokushima Prefecture
Elevation difference: depending on the stage, 33–2,953 vertical ft. (10–900 m)
Information: japanpilger.de

312. THE SHIMANAMI KAIDO BIKE HIGHWAY

Japan isn't exactly known as the country of cyclists. We are going to change that, said the engineers of the Shimanami Kaido road network, who built the bike path right away. In fact, it is a highway and bridge system that connects the main island of Honshu with Shikoku via seven bridges and various islands. But an adventurous network of bike paths were created at the same time, including a modern, complete infrastructure for cyclists. There are even hotels only for cyclists. Along the way there are breathtaking views of the Seto Inland Sea National Park; you pass Senkoji Temple, one of the oldest temples in the country; and you cycle through vast lemon plantations and the garden of Japan, Ehime Prefecture. That's how traffic works today.

Length / time it takes: 40 mi. (64 km) / 1 day, including sightseeing
Starting point: Onomichi
Elevation difference: 755 vertical ft. (227 m)
Information: www.japanvisitor.com/japan-city-guides/shimanamikaido

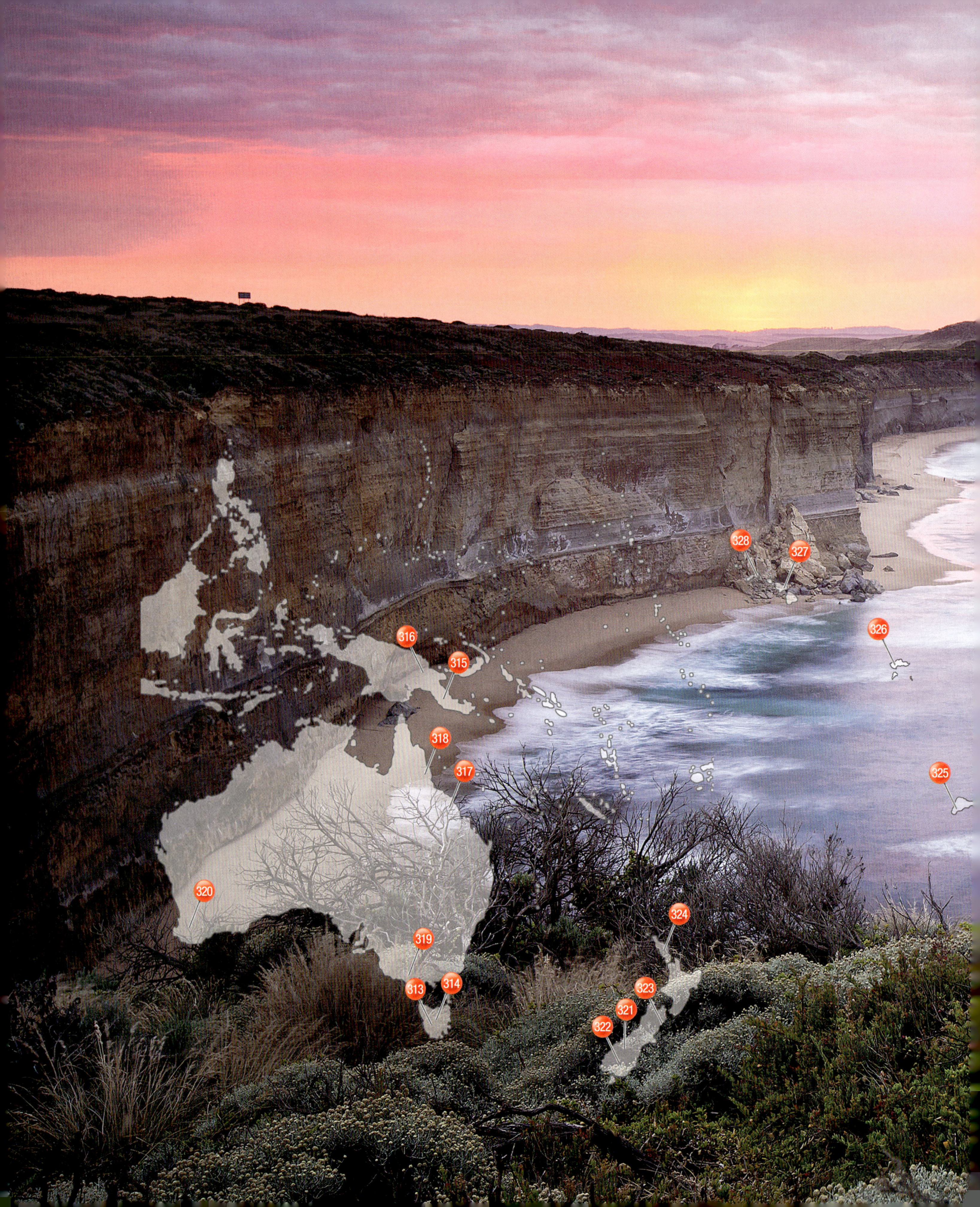
328
327
326
316
315
318
317
325
320
324
319
313
314
323
321
322

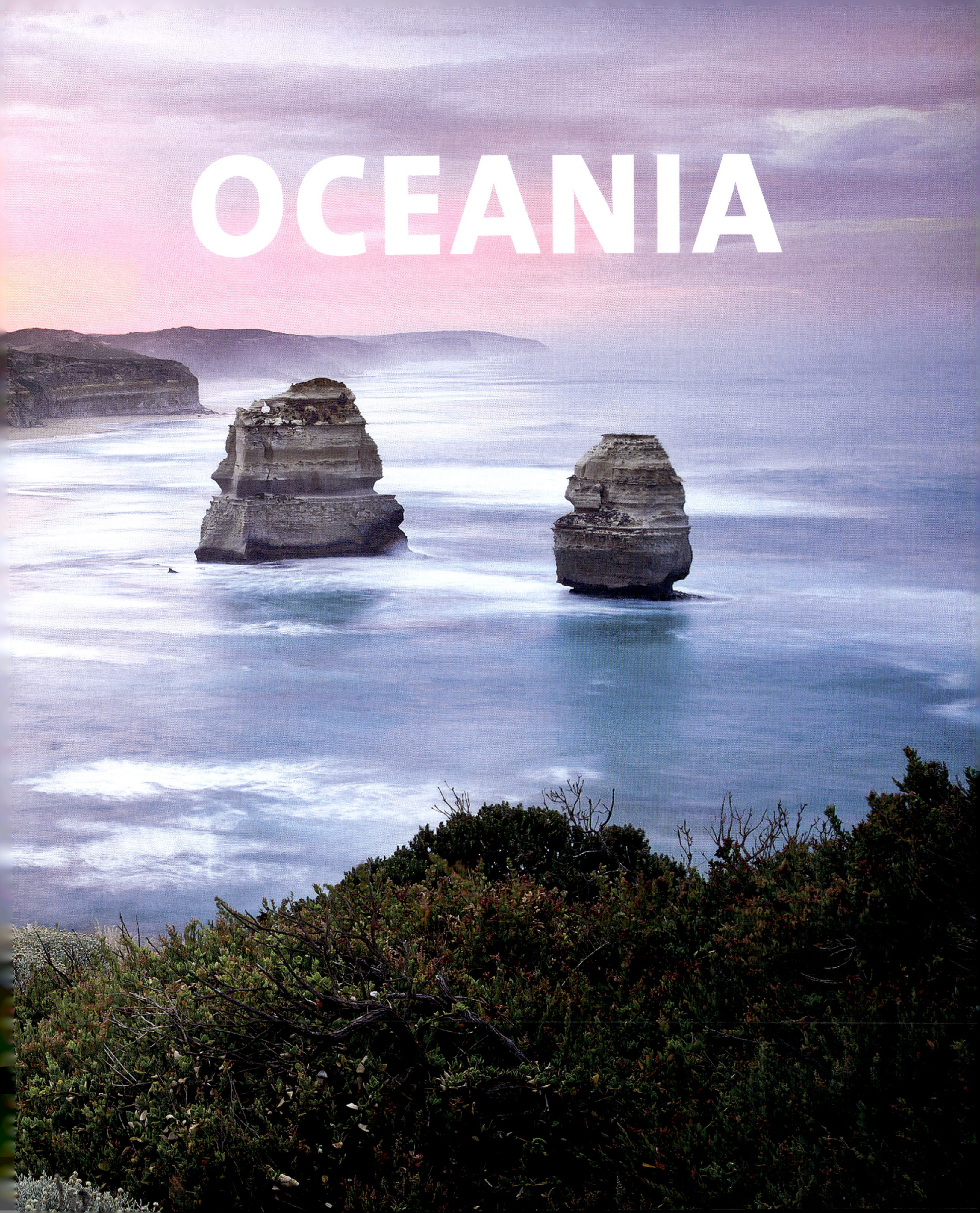
OCEANIA

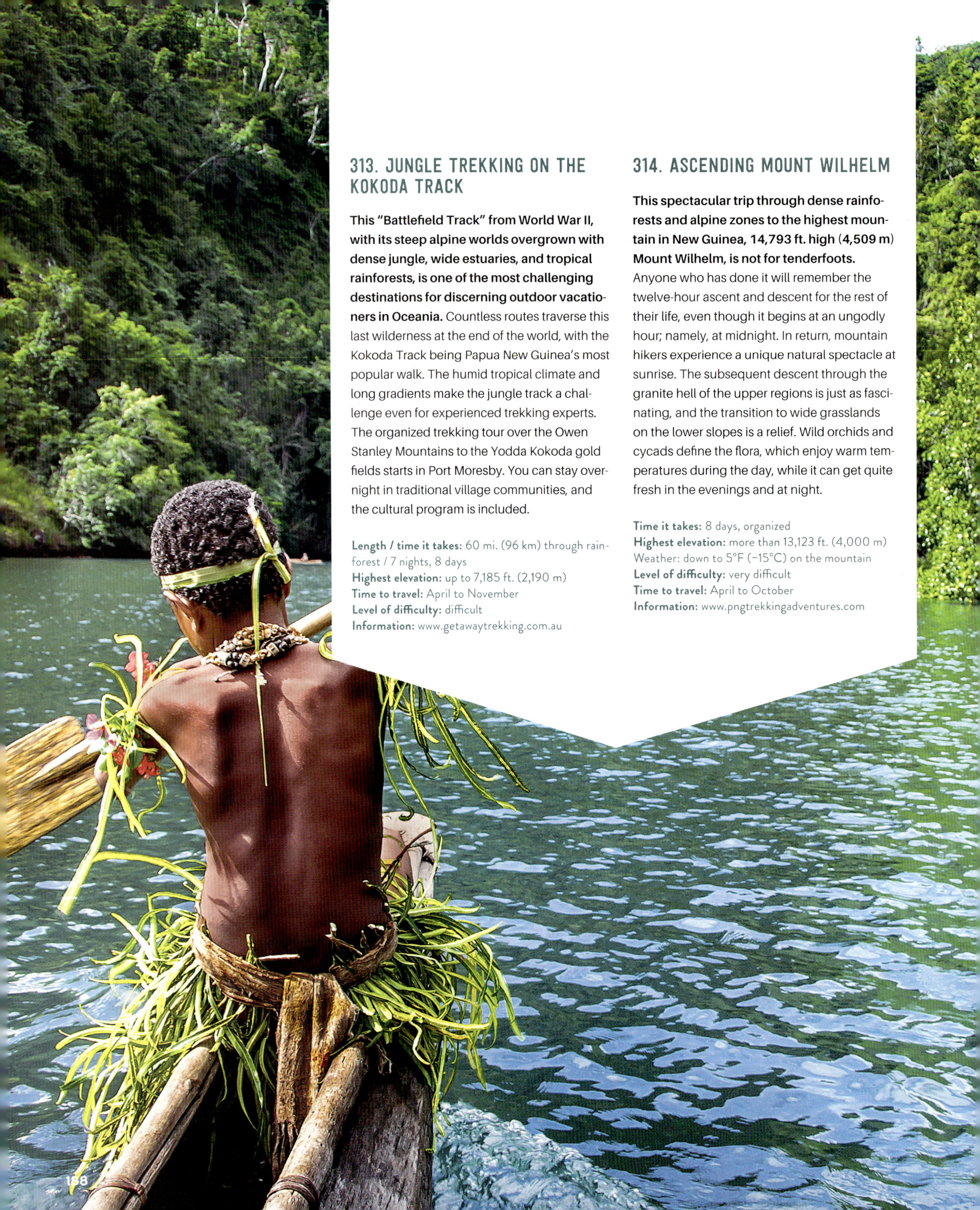

313. JUNGLE TREKKING ON THE KOKODA TRACK

This "Battlefield Track" from World War II, with its steep alpine worlds overgrown with dense jungle, wide estuaries, and tropical rainforests, is one of the most challenging destinations for discerning outdoor vacationers in Oceania. Countless routes traverse this last wilderness at the end of the world, with the Kokoda Track being Papua New Guinea's most popular walk. The humid tropical climate and long gradients make the jungle track a challenge even for experienced trekking experts. The organized trekking tour over the Owen Stanley Mountains to the Yodda Kokoda gold fields starts in Port Moresby. You can stay overnight in traditional village communities, and the cultural program is included.

Length / time it takes: 60 mi. (96 km) through rainforest / 7 nights, 8 days
Highest elevation: up to 7,185 ft. (2,190 m)
Time to travel: April to November
Level of difficulty: difficult
Information: www.getawaytrekking.com.au

314. ASCENDING MOUNT WILHELM

This spectacular trip through dense rainforests and alpine zones to the highest mountain in New Guinea, 14,793 ft. high (4,509 m) Mount Wilhelm, is not for tenderfoots. Anyone who has done it will remember the twelve-hour ascent and descent for the rest of their life, even though it begins at an ungodly hour; namely, at midnight. In return, mountain hikers experience a unique natural spectacle at sunrise. The subsequent descent through the granite hell of the upper regions is just as fascinating, and the transition to wide grasslands on the lower slopes is a relief. Wild orchids and cycads define the flora, which enjoy warm temperatures during the day, while it can get quite fresh in the evenings and at night.

Time it takes: 8 days, organized
Highest elevation: more than 13,123 ft. (4,000 m)
Weather: down to 5°F (−15°C) on the mountain
Level of difficulty: very difficult
Time to travel: April to October
Information: www.pngtrekkingadventures.com

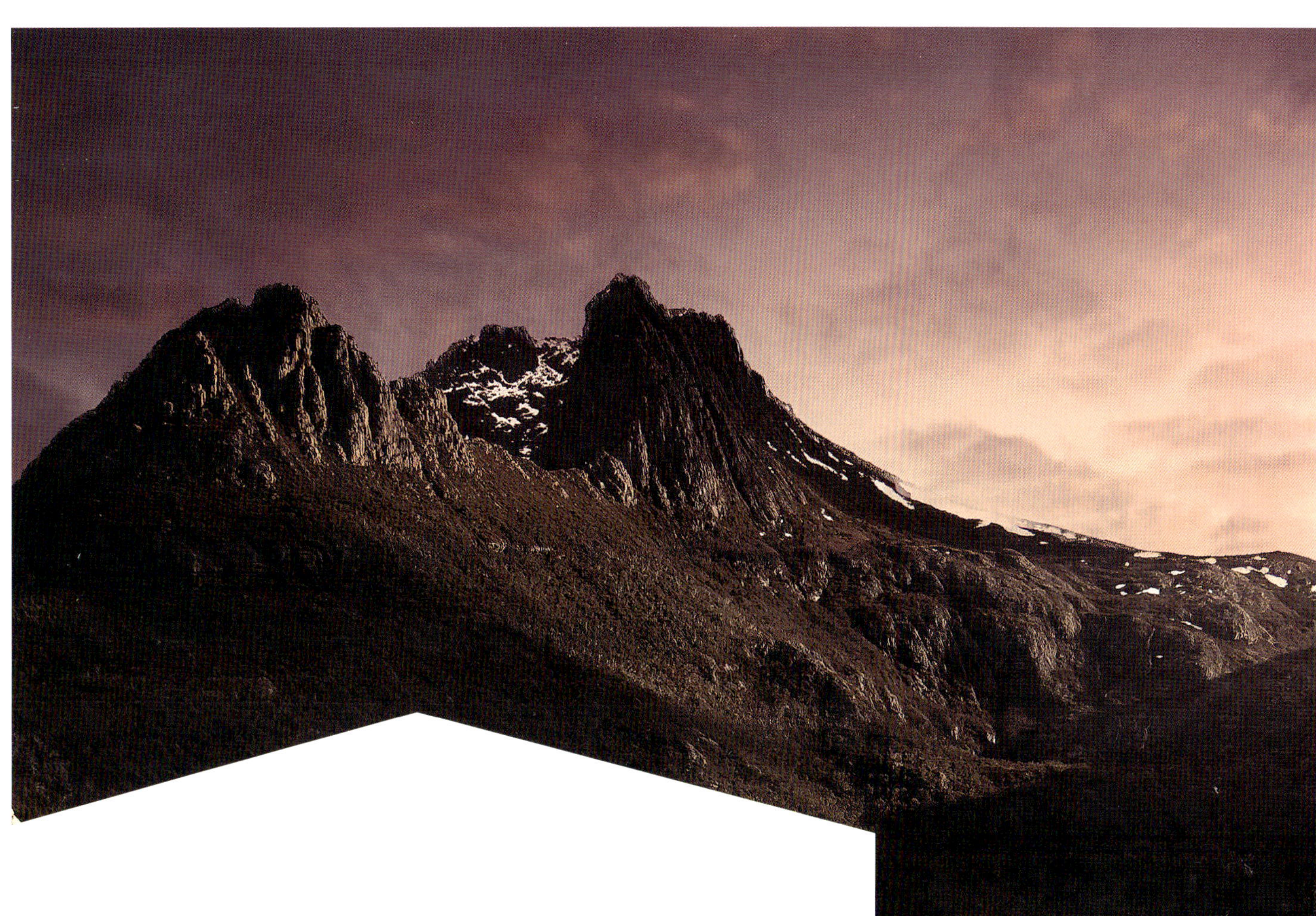

315. HIKING THE FRANKLIN RIVER NATURE TRAIL

The Franklin River is one of the most famous rivers in Wild Rivers National Park. As soon as they arrive, visitors are introduced to the green waters of the Franklin between Queenstown and Derwent Bridge. The river rolls through pristine rainforest and offers a picture of long-forgotten natural perfection. Where else would a hiking track run if not along these wild banks, which offer a more exciting wide-angle panorama after every bend in the river? It is hard to believe that this pearl was once the focus of harsh disputes, when conservationists fought against the construction of a dam and led a courageous battle against the energy industry—with a heavenly outcome. The highlight of the Franklin River is whitewater rafting.

Time it takes: 10- or 12-day group expeditions
Requirements: no rafting experience necessary
Equipment: will be provided, full board
Time to travel: October to April (spring/autumn)
Information: www.franklinriverrafting.com

316. ASCENDING CRADLE MOUNTAIN

National parks unite this Australian state that is "out to sea." They transform it into a single, large nature conservation paradise. Hikers will find ideal terrain everywhere, with overnight huts as well as campsites along the most-beautiful routes. One of the most exciting and popular tracks is the ascent to Cradle Mountain in Cradle Mountain–Lake St Clair National Park. Here, the "Tassies" offer their visitors (and themselves) an outstanding route—one through the middle of the wilderness of a particularly fascinating ecosystem. Those who love paradisiacal landscapes, with spectacular mountain peaks, picture-perfect waterfalls, and extraordinary flora—that is, everything that goes with it—will find their mecca here.

Time it takes: 8 hours
Requirements: Bad-weather equipment
Elevation difference: 1,969 vertical ft. (600 m)
Hiking elevation: 2,953–5,069 ft. (900–1,545 m)
Information: www.parks.tas.gov.au/?base=1362

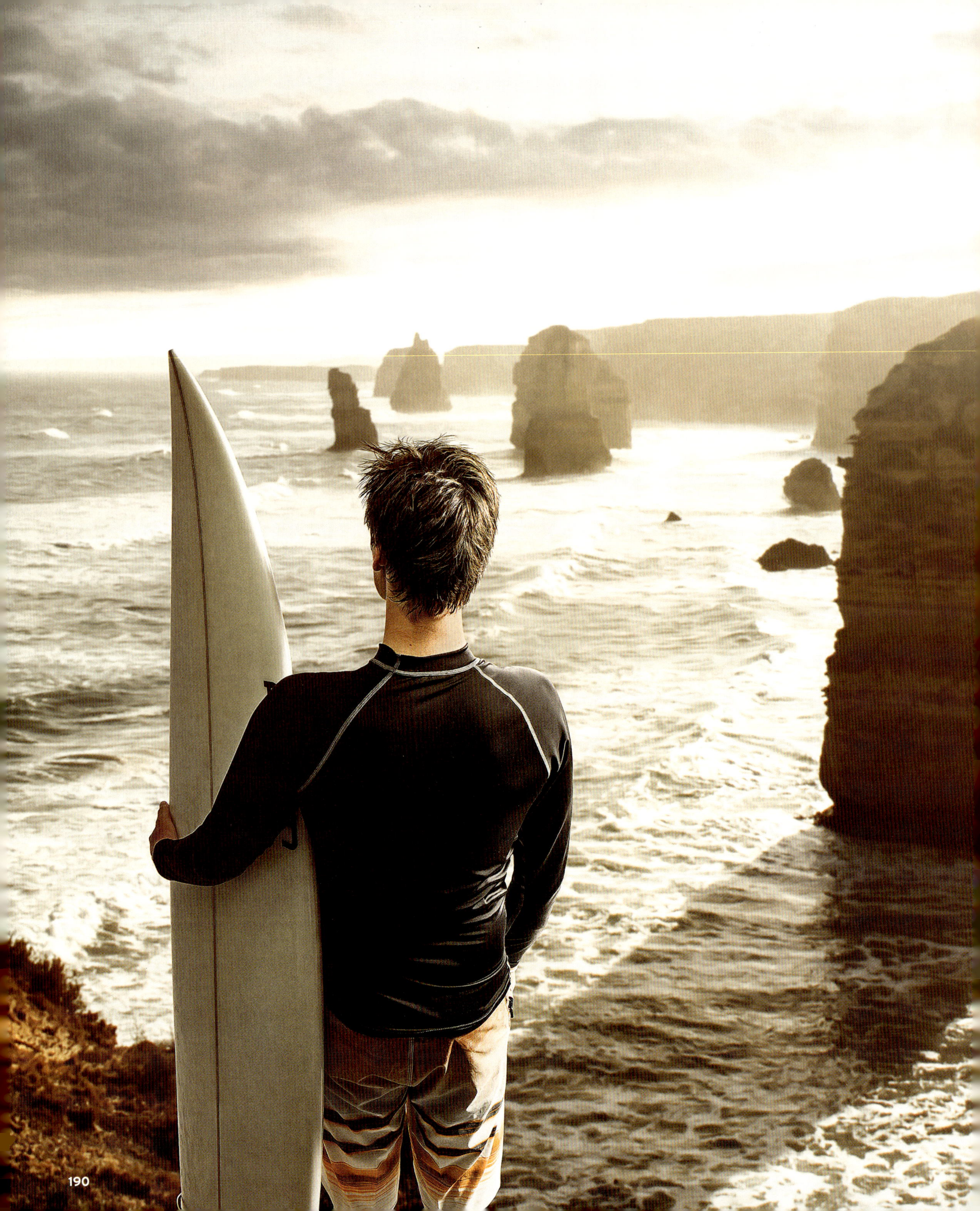

317. SNORKELING OFF LIZARD ISLAND

Lizard Island, which can be reached from Cairns by a Cessna flight, is one of the most exclusive Great Barrier Reef islands in the state of Queensland. Because it is barely 12 mi. (20 km) from the spectacular diving grounds of the "Outer Reef" and because the local "Cod Hole" is considered one of the five best diving spots in the world, it is a true paradise for snorkelers and diving fans. If you wade just 33 ft. (10 m) into the water, you will find the life-filled reef before your snorkeling goggles. There is good reason for the fact that the island has a marine biology research station that lists well-known German scientists in its guest book, as well as Prince Charles and Tom Cruise. On August 12, 1770, Capt. James Cook climbed Lizard's highest point (1,207 ft. [368 m]), which has since then been called "Cook's Look."

Getting there: flight from Cairns
Diving: Cod Hole: 45 minutes by boat
Snorkeling: right from the beach
Water sports: glass boat, sea kayaking, SUP
Information: www.lizardisland.com.au

318. OVER SUSPENSION BRIDGES THROUGH WOOROONOORAN NATIONAL PARK

Walking on suspension bridges high above the rainforest is not for everyone, and a head for heights is a great advantage. The impressive infrastructure that stretches through the treetops of the jungle in Wooroonooran National Park, an hour-and-a-half drive south of Cairns, is an educational component of the Wet Tropics World Heritage Area. The Mamu Canopy Walkway offers experienced, sure-footed hikers a great all-around overview of everything that the Australian rainforest has to offer in terms of flora and fauna, from the ground up to the treetops. The Forest Walk, which even wheelchair users can manage, is more down to earth; then comes the Elevated Walkway with its suspension bridges that stretch through the jungle, dizzyingly high up in the treetops.

Length: 1,148 ft. (350 m)
Walkway elevation: 49 ft. (15 m)
Highest elevation: 115 ft. (35 m), observation deck
Hiking: 7 mi. (11 km) of forest walks
Information: www.mamutropicalskywalk.com.au

319. ON THE GREAT OCEAN WALK TO THE TWELVE APOSTLES

The most in-demand walk in South Australia begins at Apollo Bay, a three-hour drive southwest of Melbourne. The hiking programs include paradisiacal natural enclaves such as Milanesia Beach or Moonlight Head, with the highest cliffs in Australia. One natural spectacle follows another over a distance of about 62 mi. (100 km), and of course the bizarre rock formations of the Twelve Apostles are the goal of every step. An uplifting moment when you really see it before you. Along the way, your camera lenses will find lush rainforests, thundering waterfalls, steep cliffs made of sandstone, and beautiful heathland. Due to the good infrastructure, you can "cherry-pick" the most-beautiful sections, and almost everywhere there are bed-and-breakfast accommodations available nearby for hikers.

Length / time it takes: 102 mi. (164 km) / 1 week
Starting point: Apollo Bay
Level of difficulty: easy to moderate
Information: www.greatoceanwalk.com.au

320. ON THE BIBBULMUN TRACK THROUGH WESTERN AUSTRALIA

The Bibbulmun Track is one of the greatest long-distance hiking trails in the world. It is named after the indigenous people who live in the southwestern state of Western Australia. It runs from Perth for almost 621 mi. (1,000 km) to Albany, on the southwest coast. Along the way, the track goes through small, secluded villages and picturesque coastal towns. Dwellingup, Collie, Balingup, Donnelly River, Pemberton, Walpole, and Denmark are the names that stick with you. Hiking from place to place is the privilege of the pilgrim, who can enjoy their overnight stay in a simple tent in the middle of nature and should take time to spare along with them. However, bed-and-breakfast accommodations are also available everywhere. During the summer, shorter sections should be sufficient; the Australian autumn and spring are ideal hiking times.

Length: 621 mi. (1,000 km) long-distance hiking trail
Starting point: Kalamunda, east of Perth
Time to travel: all year round, climatic spring/winter
Requirements: be adventurous
Information: www.bibbulmuntrack.org.au

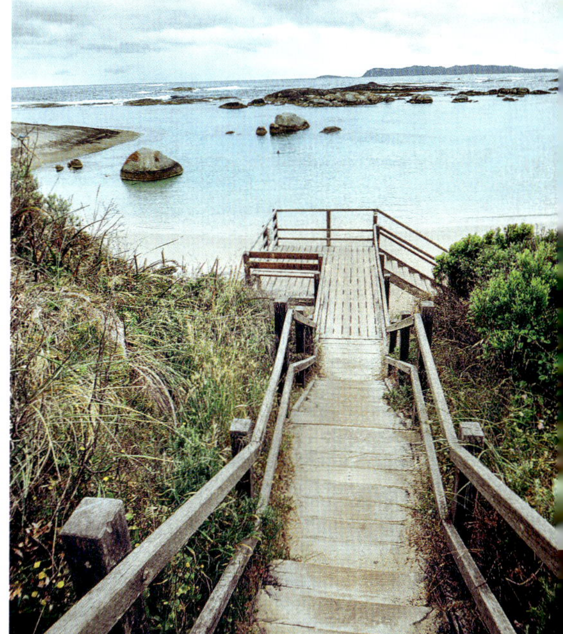

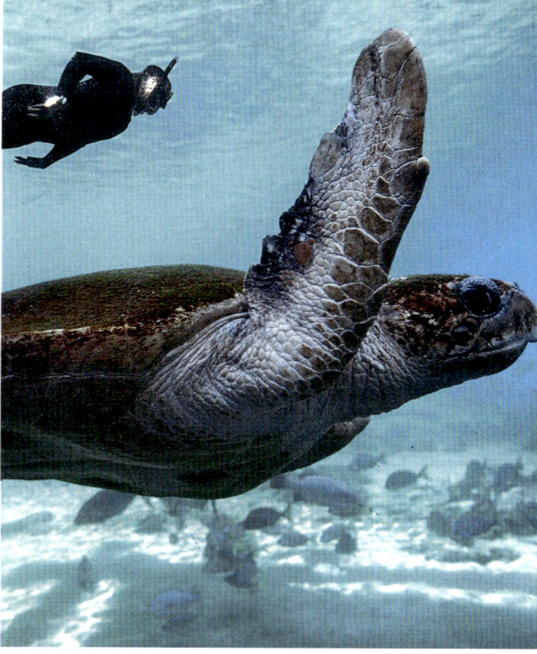

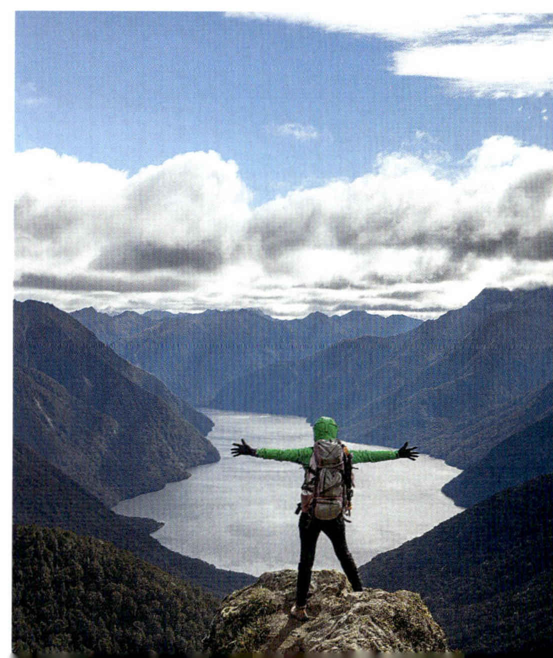

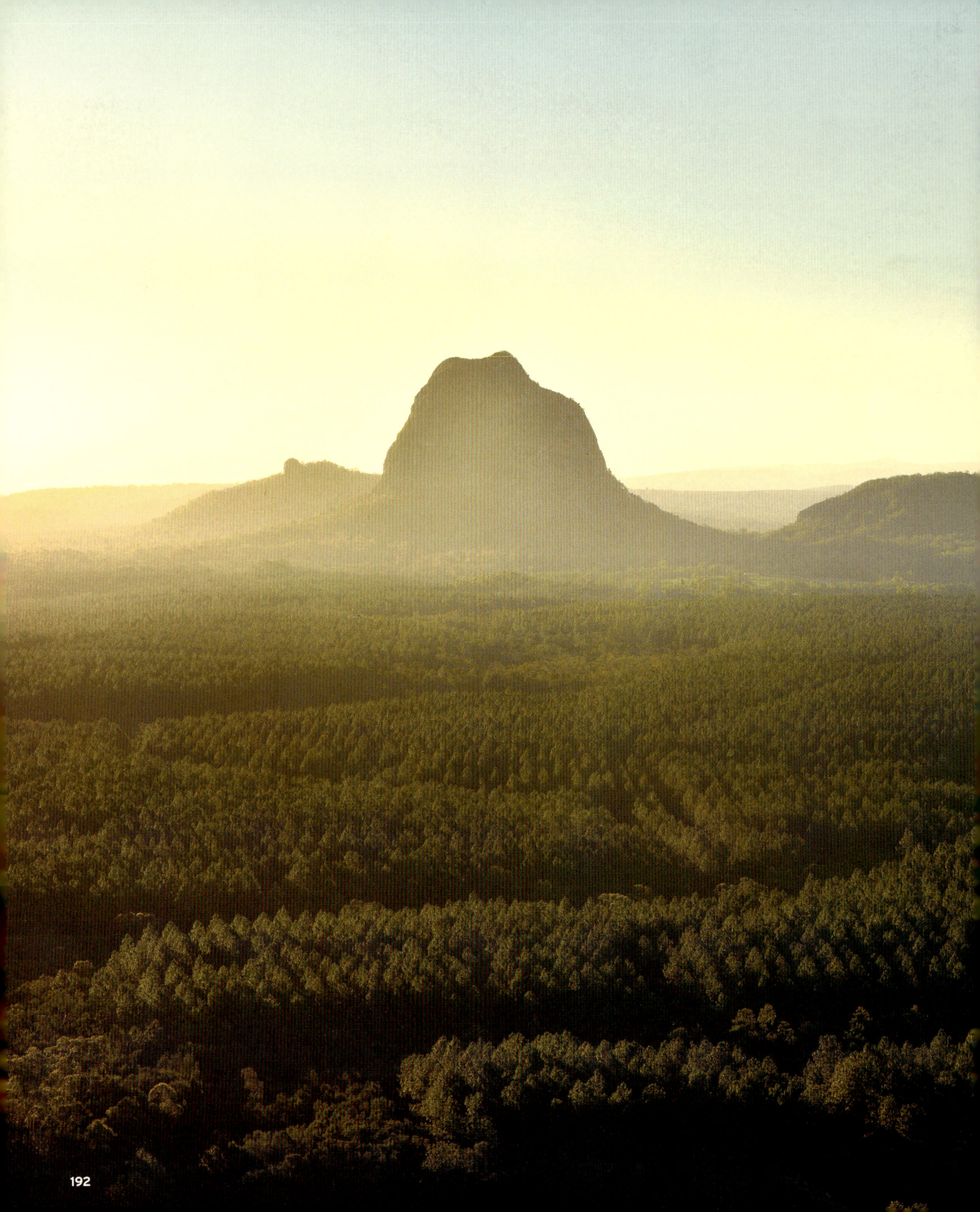

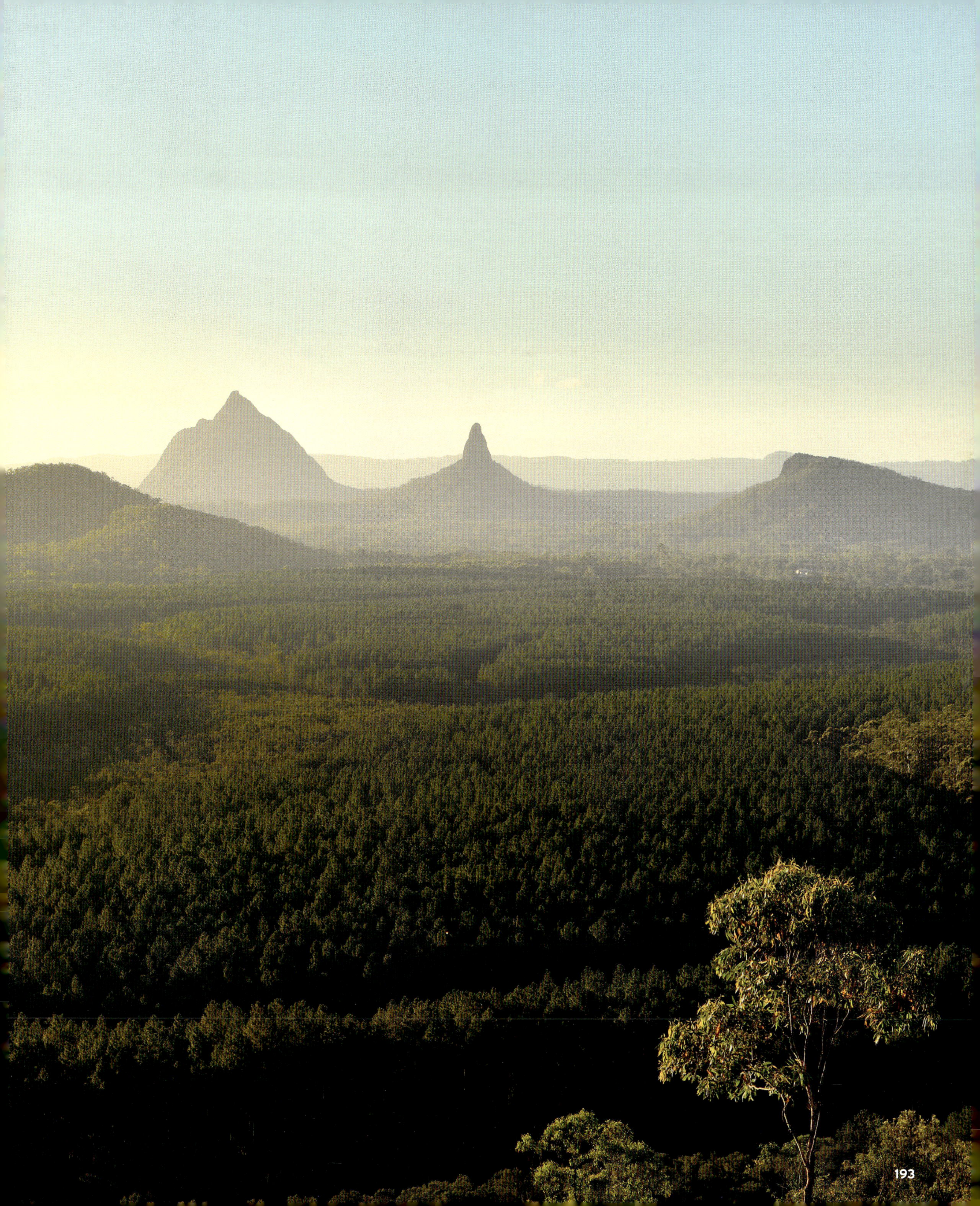

321. WALKING THE MILFORD TRACK

Milford Sound, a 10 mi. long (16 km) coastal fjord in the southwestern part of New Zealand's South Island, is bordered by mountains rising steeply from the sea to up to 6,562 ft. (2,000 m). Here a cruise comes highly recommended, as a multiday trek on one of the most beautiful hiking trails in Kiwi Land, the 33 mi. long (53 km) Milford Track. The highlight of the scenario is the 5,814 ft. high (1,772 m) summit of Miter Peak, which drops almost vertically into the steel-blue sea. Countless waterfalls and high levels of precipitation make the Milford Track a unique natural spectacle that will certainly involve penguins, seals, and playful dolphins. Anyone who manages to avoid the main travel season will find a paradise of the first order.

Length / time it takes: 27 mi. (43 km) / 4 days
Time to travel: end of October to end of April (summer)
Vertical meters: 2,297 vertical ft. (700 m)
Information: www.milfordtrack.net

322. THROUGH FIORDLAND NATIONAL PARK

If you want to get to know the intoxicating highlights of New Zealand's fjordland in one fell swoop, the best way to hike the area is on the Kepler Track, opened in 1988. This track includes the Lake Te Anau and Lake Manapouri lake district in the "Te Wāhipounamu-South West New Zealand World Heritage Area." The track runs through beech forests and lush grasslands, over jagged mountain ranges, and past rushing waterfalls and wide mudflats and provides for trouble-free hiking: there are bridges for crossing rivers, wooden boardwalks cover any boggy surfaces, and there are stairs to manage gradients. Every year, the Kepler Challenge takes place here at the beginning of December: the goal is to cover the 37 mi. (60 km) in less than five hours!

Length / time it takes: 37 mi. (60 km) loop trail / 3–5 days
Time to travel: end of October to end of April (summer)
Requirements: no dogs; no children under 10 years old
Information: www.doc.govt.nz/keplertrack

323. WHALE WATCHING AT KAIKOURA

Enterprising Maori people have converted the train station for the former whaling station—which is fortunately no longer in use—to modernized style and founded the Whale Watch company here. Initially ridiculed, Whale Watch, a purely Maori-run company, has now grown into a solid commercial force and serves as a successful example of integration for the indigenous people. Ships take visitors close to the giant marine mammals; they can also watch by plane and helicopter. In view of this perfectly organized marketing machine, the beautiful "Peninsular Walk," a coastal hiking trail around the Kaikoura Peninsula, appears to come from another time, but you are moving a bit more and will see the same marine life as those on the many whale-watch boats leaving the port.

Time it takes: 2.5 hours by ship
Departures: daily at 7:15 a.m., 10:00 a.m., 12:45 p.m.
Starting point: Whaleway Station, Kaikoura
Peninsula walk: easy coastal loop
Information: www.whalewatch.co.nz

324. SKYWALKING AND SKY JUMPING IN AUCKLAND

The platform that runs around the Auckland Sky Tower observation and telecommunications tower is only 47 in. (120 cm) wide. The Sky Tower rises 1,076 ft. (328 m) into the sky in the middle of the business district and is the highest television tower in the entire Southern Hemisphere. As narrow as the loop is, there is all the more space for going down: there is a steep, 630 ft. (192 m) drop for bungee freaks that really gets the adrenaline junkies jumping. Skywalkers can still feel halfway relaxed as they circle the Sky Tower, secured with belts on the skywalks. Anyone who finds they aren't on the right track if there isn't a railing can watch the brave ones with a head for heights through the windows of the restaurant on the fifty-fourth floor.

Location: Sky Tower, Auckland
Total elevation: 1,076 ft. (328 m)
Sky jump: bungee jump 53 floors down
Skywalk: loop walk on a 360° platform
Information: www.bungy.co.nz/auckland/sky-tower

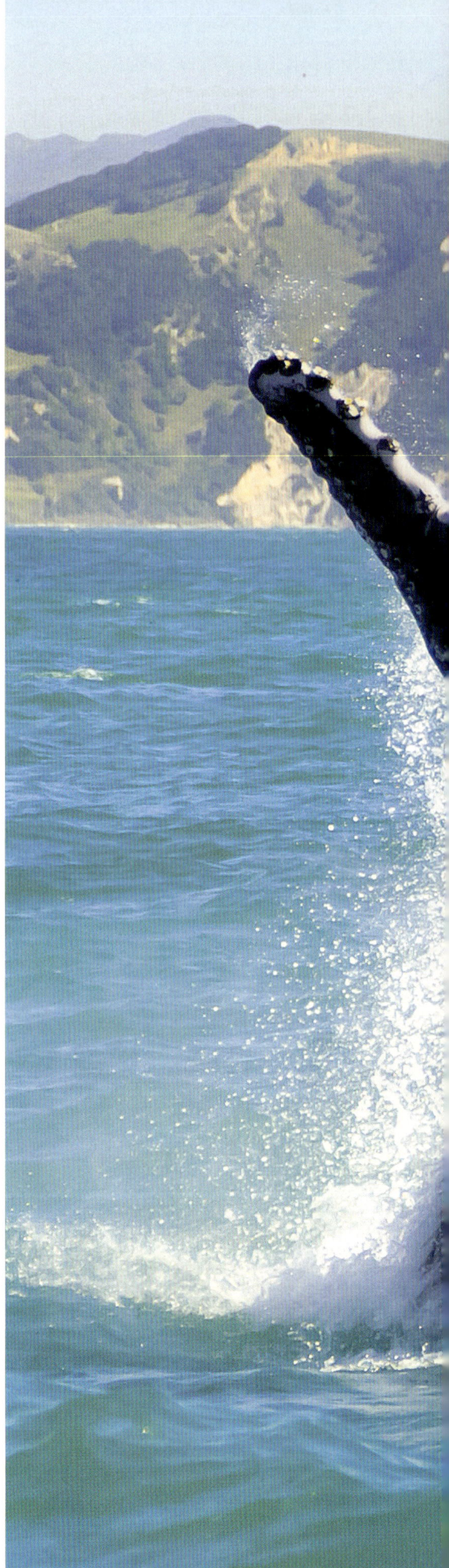

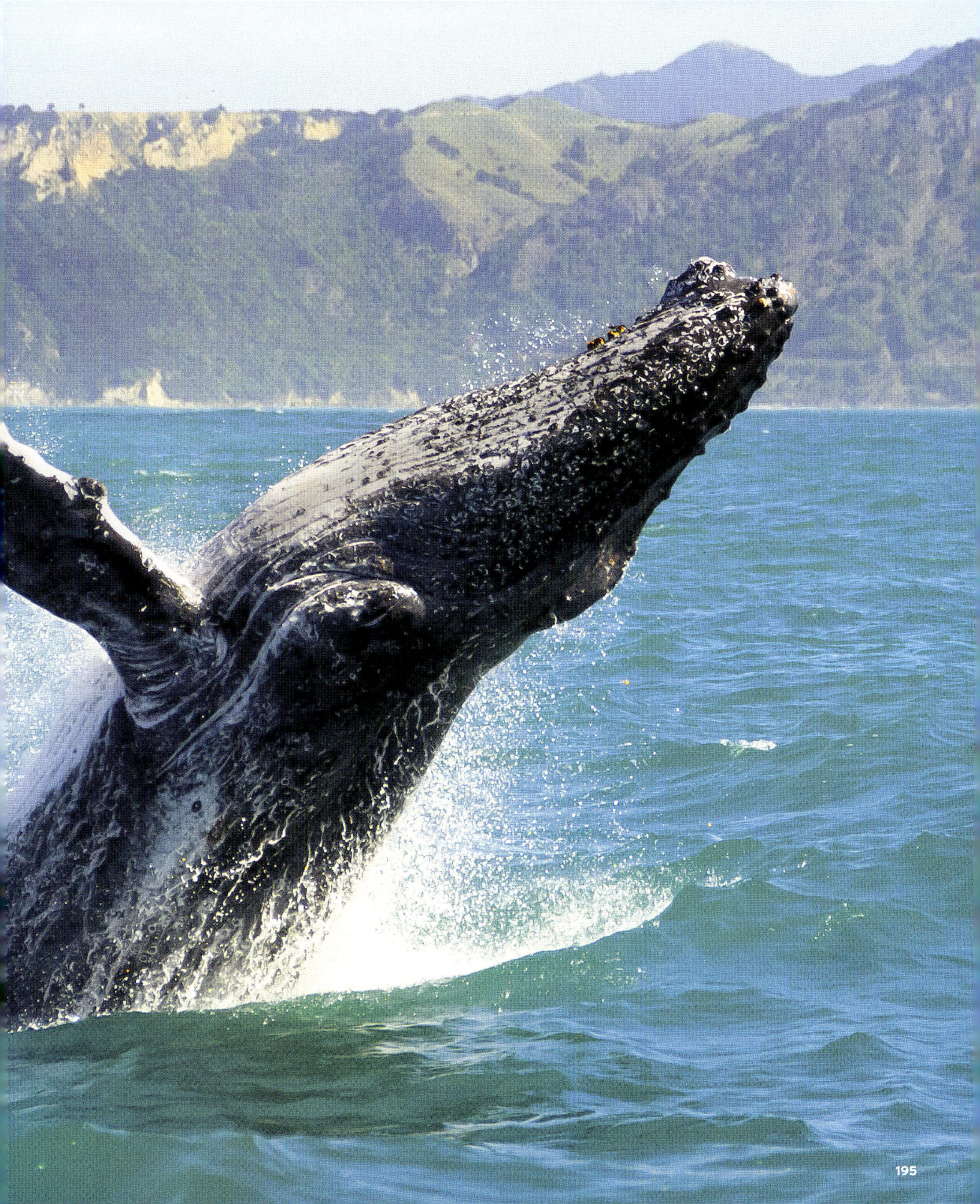

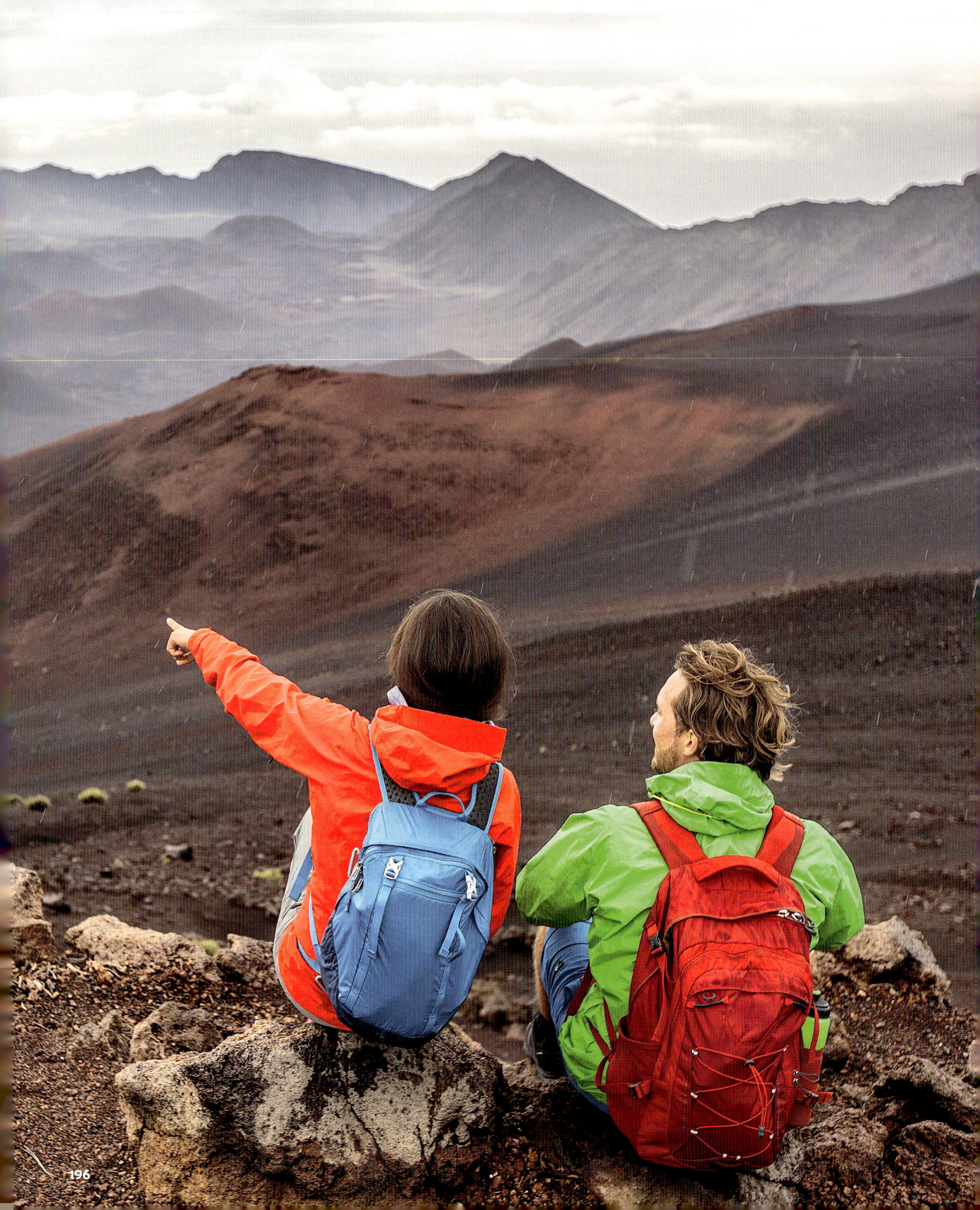

325. ASCENDING RANO KAU VOLCANO ON EASTER ISLAND

The walk to the extinct Rano Kau volcano takes about five hours; initially it runs quite harmlessly and level over fields and pastureland. At the same time, you always have a view of the surf-swept coast of Rapa Nui, as the locals call Easter Island. Those who have made it up to the summit on the moderately difficult track can enjoy the breathtaking view over the entire island. That can also be topped when you reach the rim of the Rano Kau crater, with its view of the crater lake overgrown with totora reeds. The way back includes a visit to the ceremonial village of Orongo, where petroglyphs and other archeological treasures can be seen, as well as Ana Kai Tangata cave with its interesting cave paintings.

Length / time it takes: 19 mi. (30 km) / 5.5 hours, there and back
Access: via the capital Hanga Roa
Level of difficulty: easy to moderate
Elevation difference: 1,047 vertical ft. (319 m)
Information: www.chile.travel/de

326. BY PASSENGER FREIGHTER THROUGH THE MARQUESAS ISLANDS

Huge basalt columns jut out of the sea like the rocks of the North Cape. This is the island of Ua Pou, part of French Polynesia, the jagged one that thrusts impressive igneous rock formations out of the green jungle landscape. On Hiva Oa, mountain ranges stretch up to more than 3,281 ft. (1,000 m). It was here in this remote, exotic paradise where French painter Paul Gauguin ended up in the nineteenth century. He has lain buried in Atuona town cemetery since 1903. There are many loading stops on various islands where passengers can go ashore. The program includes sports activities such as mountain trekking, swimming, and snorkeling, as well as tours and cultural events before one of the most spectacular boat trips in the world comes to an end.

Length / time it takes: 2,485 mi. (4,000 km)—both archipelagos) / 14 days from Tahiti, including the Tuamotu atolls
Short trip: 7 days from the Marquesas, excluding Tuamotu
Time to travel: March to October
Information: www.aranui.com

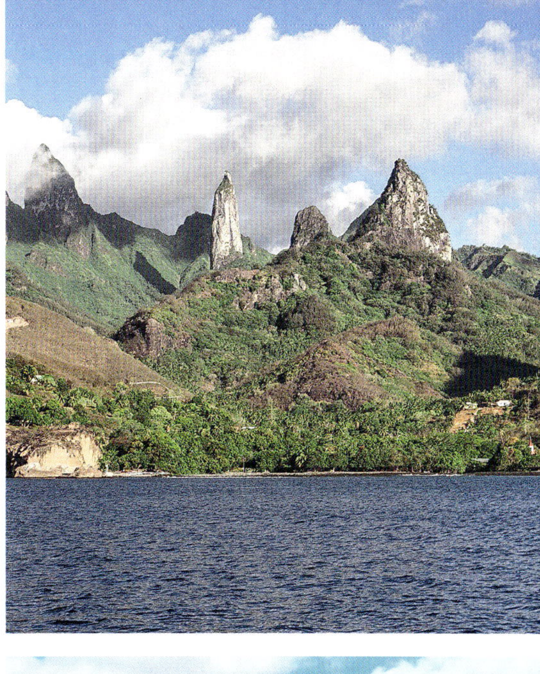

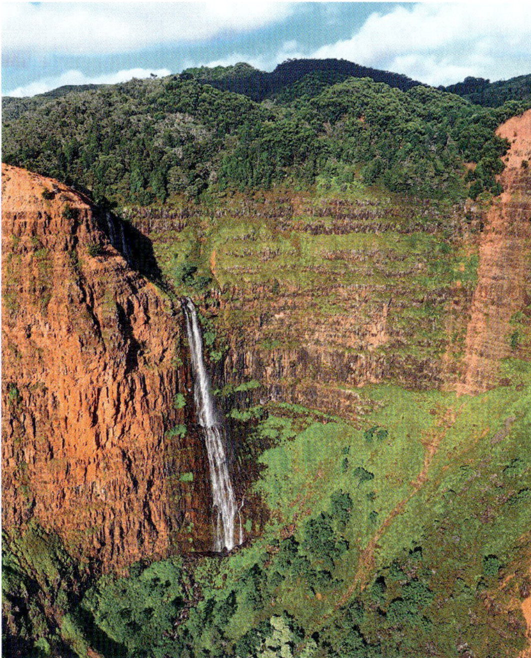

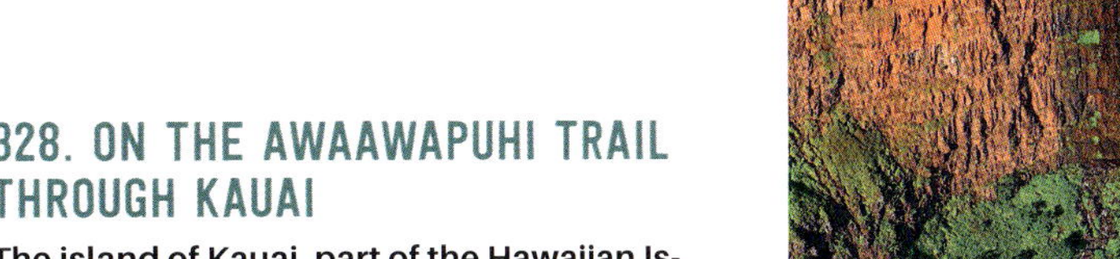

327. COASTAL TREK ON MAUI

The almost 3 mi. long (5 km) trip on the coastal track of the Ke Ala Loa O Maui / Piilani Trail keeps hikers on their toes. The one-and-a-half-hour hike between rocky bays and rugged cliffs runs initially rather moderately up and down; only your sure-footedness is put at risk now and then by the spectacular views of the coast near Hana and the rolling hills of Haleakala! Access to the beautiful coastal climbing tour is complex: to make the 53 mi. (85 km) trip there on Highway 360, you need to travel a three-hour "Coastal Ride" from Kahului Airport, which is worth the trip by itself. The starting point on Waianapanapa Road is a perfect place for swimming, camping, and picnicking.

Length / time it takes: 3–4 mi. (5–6 km) / 3–4 hours
Route: coastal area with lots of ups and downs
Level of difficulty: easy to moderate
Elevation difference: 197 vertical ft. (60 m)
Information: www.dlnr.hawaii.gov/dsp

328. ON THE AWAAWAPUHI TRAIL THROUGH KAUAI

The island of Kauai, part of the Hawaiian Islands and also known as the "garden island," lists thirty-four trails from which hikers can choose the most-suitable routes, depending on requirements and the landscape. For example, the Awaawapuhi Trail, 3 mi. (5 km) long in Na Pali Kona Forest Reserve in Kōke'e State Park, puts high demands on the hiker and leads to an elevation of 2,500 ft. (762 m). All the travail is rewarded with spectacular panoramic views of the Awaawapuhi and Nualolo valleys, as well as the vast steel blue of the Pacific Ocean. At the track's destination point, a steep wall descends hundreds of yards, which could cause you to make the return trip much too fast if you take a wrong step.

Length: 6.2 mi. (10 km)
Location: Kōke›e State Park
Elevation difference: 1,181 vertical ft. (360 m)
Level of difficulty: difficult
Information: www.dlnr.hawaii.gov/dsp

NORTH AMERICA

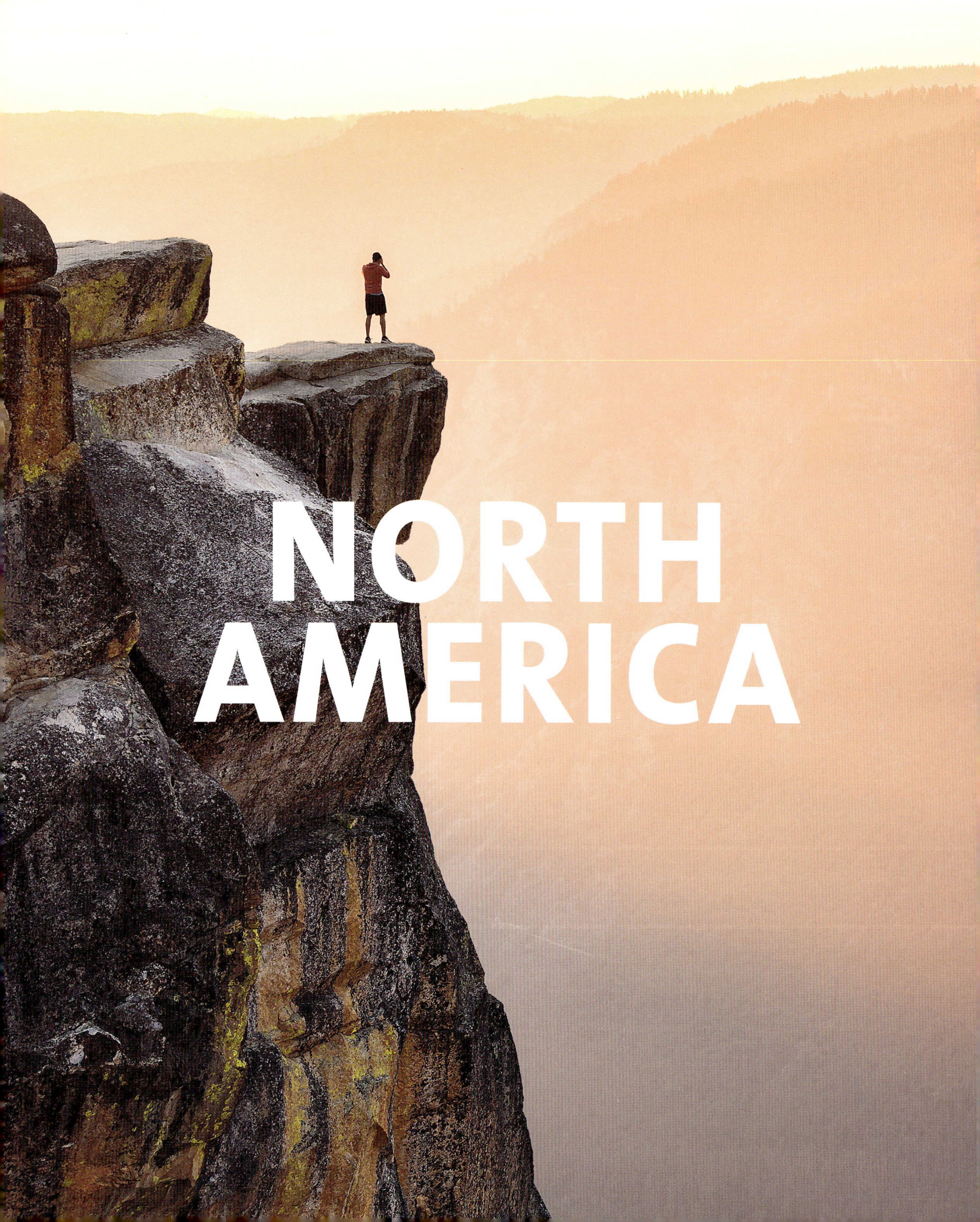

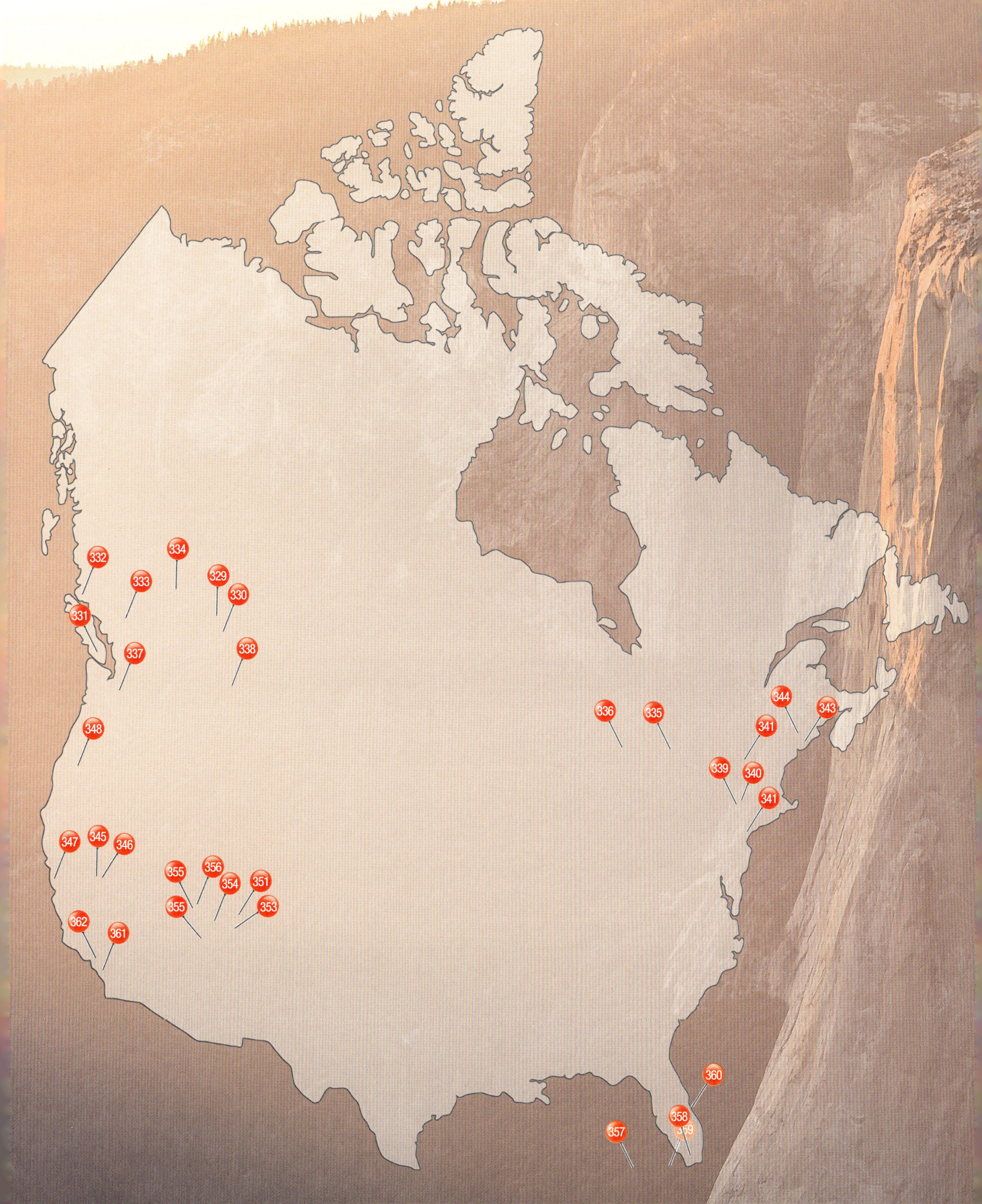

332
334
333
329
330
331
337
338
336
335
344
343
341
348
339
340
341
347
345
346
356
355
354
351
355
353
362
361
360
358
357

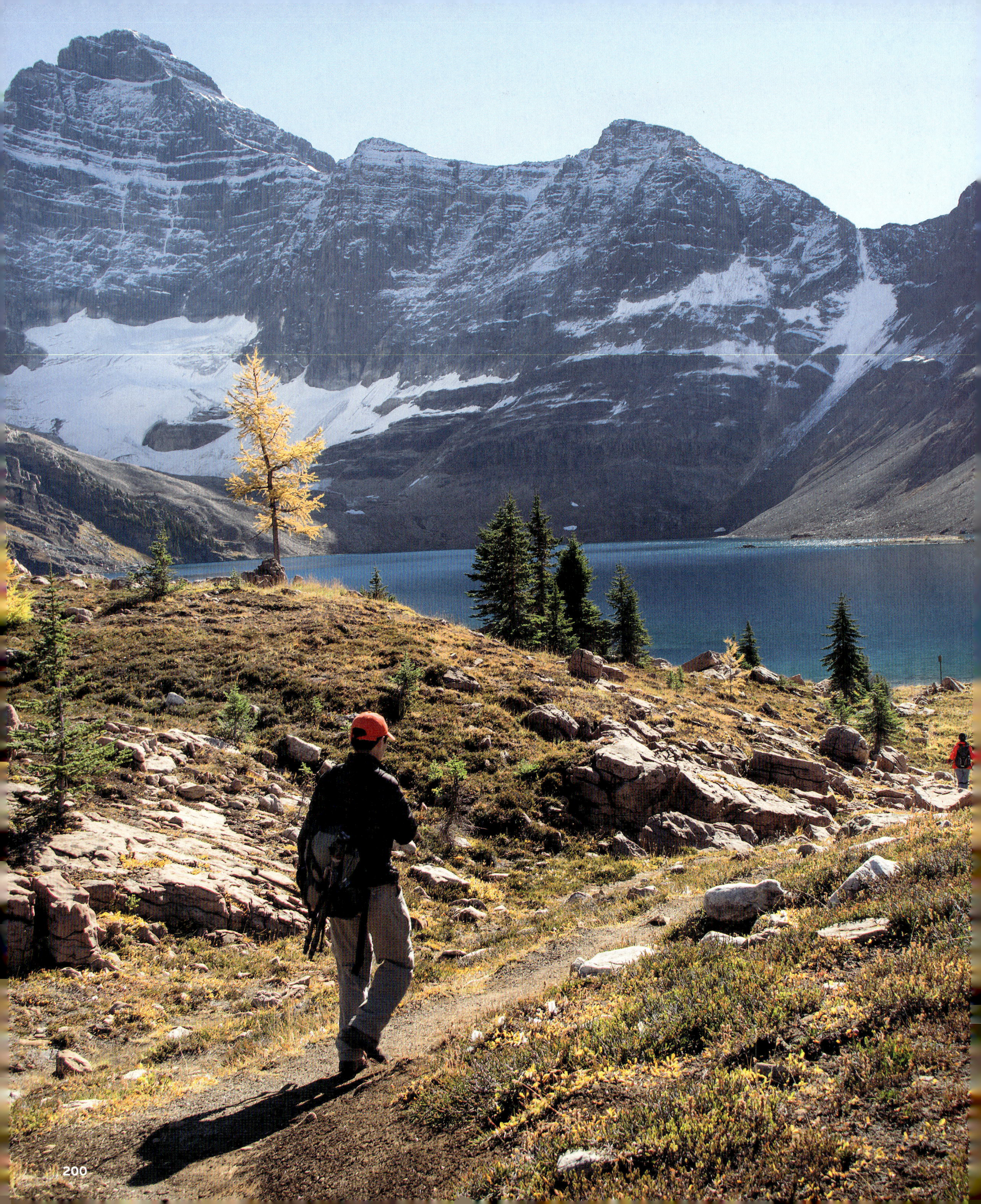

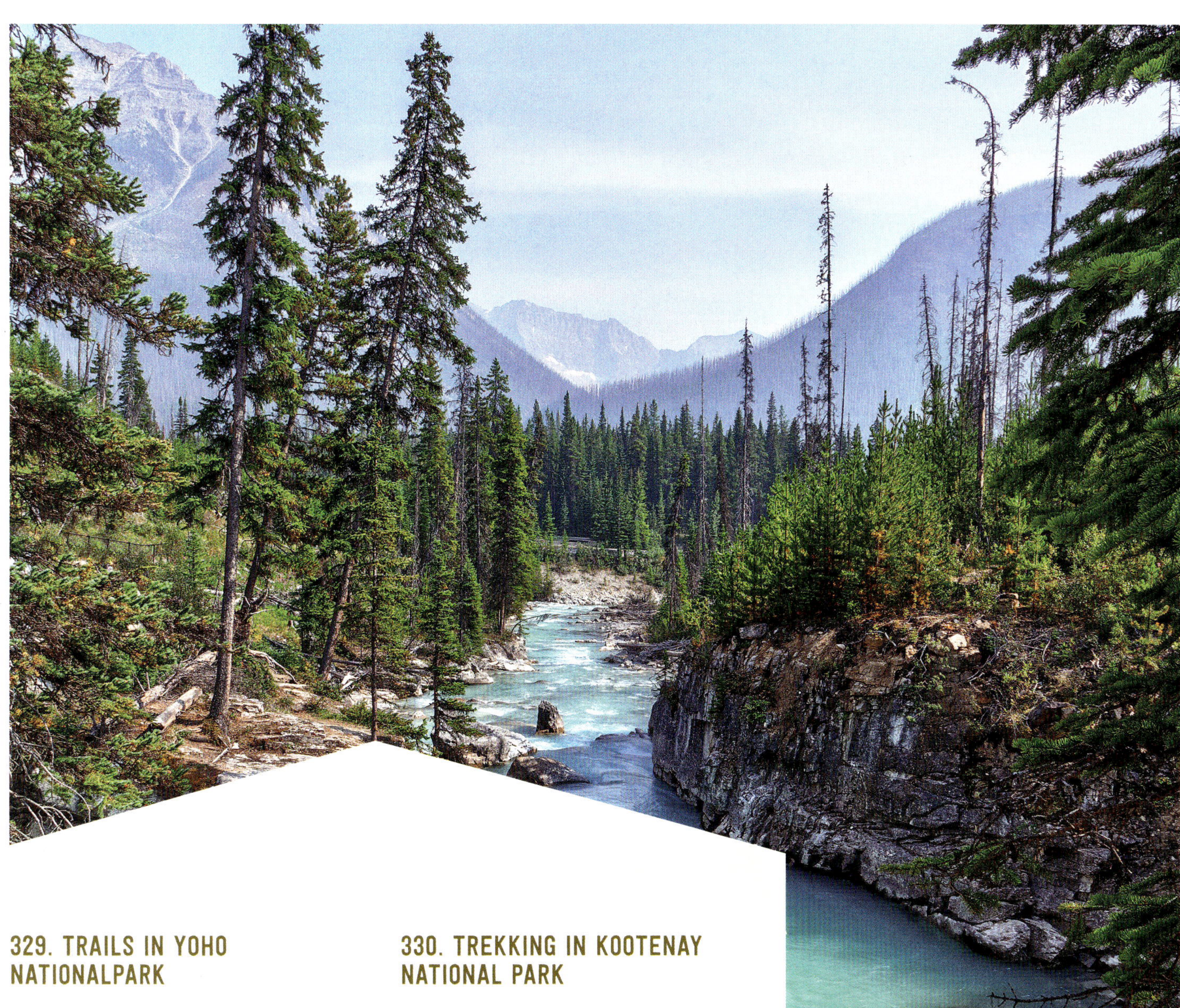

329. TRAILS IN YOHO NATIONALPARK

The name "Yoho" comes from the Cree word for awe, and no wonder: twenty-eight peaks rise here more than 9,842 ft. (3,000 m) and sit enthroned over turquoise-blue glacial lakes; there are some of the highest waterfalls in Canada, as well as dense forests of red cedar and fir trees. The park offers a network of trails with hikes of all levels of difficulty, ranging from the easy Wapta Falls Trail (1 hour), to the falls of the Kicking Horse River, to the moderate Yoho Valley Trail (5 mi. [7.9 km], 3 hours), which leads to the spectacular Twin Falls. Yoho borders on Banff and Kootenay National Parks on the Trans-Canada Highway.

Length / time it takes: trail network totaling 249 mi. (400 km) / starting from 1 hour
Highest elevation: 9,842 ft (3,000 m)

330. TREKKING IN KOOTENAY NATIONAL PARK

Kootenay National Park, on the border between British Columbia and Alberta, protects an area that exemplifies the landscape and ecology of the southwestern Rocky Mountains: mountain glaciers, wooded valleys, semiarid grasslands, canyons, rivers, and streams and waterfalls create a spectacular impact. You should definitely visit the Paint Pots, where the iron-rich spring water produces a strong orange color. The Floe Lake Trail (difficult, 6.6 mi. [10.7 km]; easy, 5 hours) leads to the beautiful lake of the same name, and the Rockwall Trail, with its peaks and rocks, is the park classic for a multiday hike!

Length / time it takes: from 6.2 mi. (10 km) / from 5 hours

331. HIKING THE WEST COAST TRAIL

On the Pacific's Vancouver Island, which lies to the west off the coast opposite the city of Vancouver, hikers are attracted by what is not only the most strenuous trek in Canada, but also one of the most beautiful in the world. The trail runs through Pacific Rim National Park along the southwest coast of Vancouver Island. Along the way in the wilderness, everything is on offer, from a wild, steep coast, to a temperate rainforest, to rocky and swampy areas that can be conquered only with the help of ladders. It runs through rivers and over gorges with "cable cars"—it is impossible to find any more adrenaline and variety! The trail originally functioned as a rescue route for people who had been shipwrecked, because shipwrecks were the order of the day around 1900. Thus, this hike always has a touch of adventure to it, in addition to the always fascinating nature.

Length / time it takes: 47 mi. (75 km) / 6–8 days
Highest elevation: 755 ft. (230 m)
Starting point: northern trailhead: Pachena Bay; southern trailhead: Gordon River
Elevation difference: 1,821 vertical ft. (555 m)
Information: hikewct.com

332. PADDLING WITH ORCAS

The kayak doesn't offer much protection if a killer whale appears right next to you, but it is not as bloodthirsty as its name suggests, and luckily he likes a salmon better than plastic. In the Johnstone Strait off the north coast of Vancouver Island, you can paddle in a kayak with orcas. In the summer, these mighty mammals of the ocean cavort in this narrow strait to do just that: enjoy the plentiful salmon. The best paddling trips go near Robson Bight; here the orca populations are particularly high along the rocky coast and in the waters around the bay. North Island Kayak Trips offers kayaking tours in Johnstone Strait. With a guide, you are going into an adventure amid the killer whales!

Time it takes: at least 1 day
Starting point: Telegraph Cove
Information: kayakbc.ca/johnstone-strait

333. MTB TOURS THROUGH WHISTLER PARK

This is the mecca of downhill mountain bikers from all over the world: legendary Whistler Bike Park. The park has many fantastic routes at all levels of difficulty for its visitors. With a day ticket, you can run down the slopes as often as you want and be transported back up by the lift or gondola. Bikes with good extra equipment are available for rent from local suppliers in Whistler Village, where they also offer cycling courses. So the park is not only attractive to real professionals, but also to newbies—the guides adapt the courses to the participant's skill level. The athletics are flanked by a sensationally beautiful landscape that hosted the 2010 Winter Olympics!

Time it takes: at least 1 day
Starting point: Whistler Mountain
Information: www.whistlerblackcomb.com

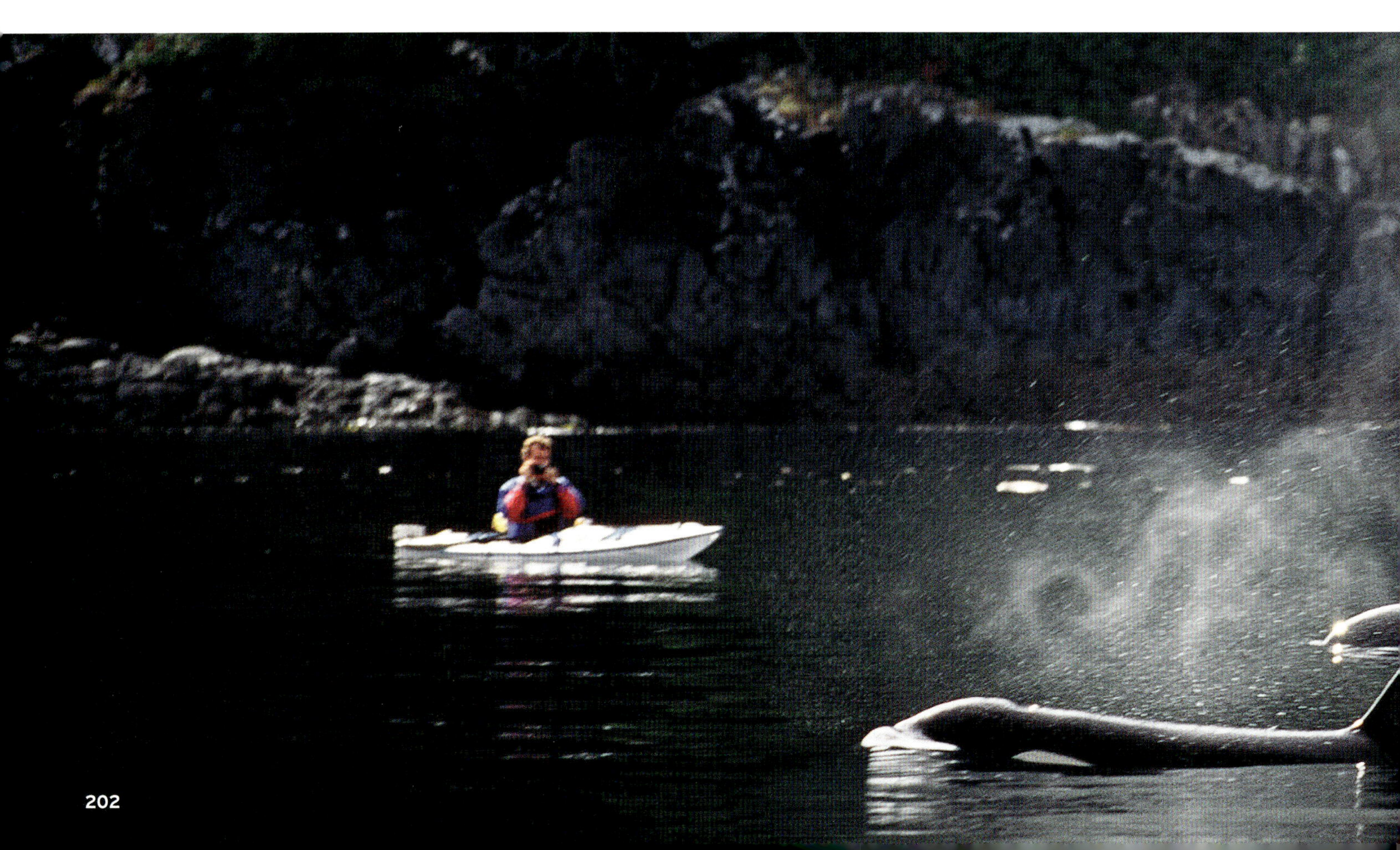

334. HIKING IN WELLS GRAY PARK

Sky-high mountains, thundering waterfalls where salmon leap, and rustling forests: the almost unspoiled Wells Gray Provincial Park is a huge outdoor playground for hikers and canoeists. It is home to the really big four-legged creatures, such as moose, caribou, and bears, but also smaller ones, such as wolves, pumas, lynxes, and wolverines—in any case, there are plenty of the dangerous ones. The Wells Gray Corridor is the starting point for many hikes, including the Helmcken Falls Rim Trail (easy, 5 mi. [8 km]) through the forested valley of the Murtle River, or the Horseshoe Trail (easy, 0.6 mi. [1 km]) to the Clearwater River, where you can watch the salmon spawning (be careful; the bears like to hang out here too!), or Chain Meadows (moderate, 9 mi. [14 km]) along Clearwater Lake to Osprey Falls.

Length: a variety of hikes of different lengths

335. CANOEING IN ALGONQUIN PROVINCIAL PARK

This park, which is imposing for its size alone, has an even more impressive number of lakes—perfect for a canoe adventure. It unfolds on a journey of discovery of the countless lakes and waterways in Algonquin Provincial Park. Athletic paddling joins with unspoiled nature, and with a little luck you can spot moose, herons, and beavers. You can travel on the waters on your own or explore the area as part of a guided tour. There are two tour operators, as well as canoe rental companies within the park. They provide you with not only the canoe and equipment, but also any other equipment you might need for a several-day trip with or without a guide. A night at an isolated campsite in this lake wilderness rounds off the ultimate adventure.

Time it takes: at least half a day
Starting point: Lake Opeongo access point
Information: www.algonquinpark.on.ca

336. TREKKING ON THE SILHOUETTE TRAIL LOOP

In Ontario, on the north shore of Lake Huron, Killarney Provincial Park measures 249 sq. mi. (645 km^2) and offers an embarrassment of riches. It provides hikers and paddlers with everything that Canada's wilderness has to offer: deep-blue lakes in the dense forest green, where small, wild islands create a scenery of peculiar beauty. Georgian Bay lies on Lake Huron to the northeast, and the Silhouette Trail Loop leads through the rugged LaCloche Mountains. The trail is recommended only for wilderness hikers who have experience in such rough terrain. From George Lake Campground, the trail runs to Baie Fine, and from there along picturesque lakes and over exposed ridges to Silver Peak, and across Heaven Lake back to the starting point.

Length / time it takes: 62 mi. (100 km) / 7–10 days
Starting point / trail's end: George Lake

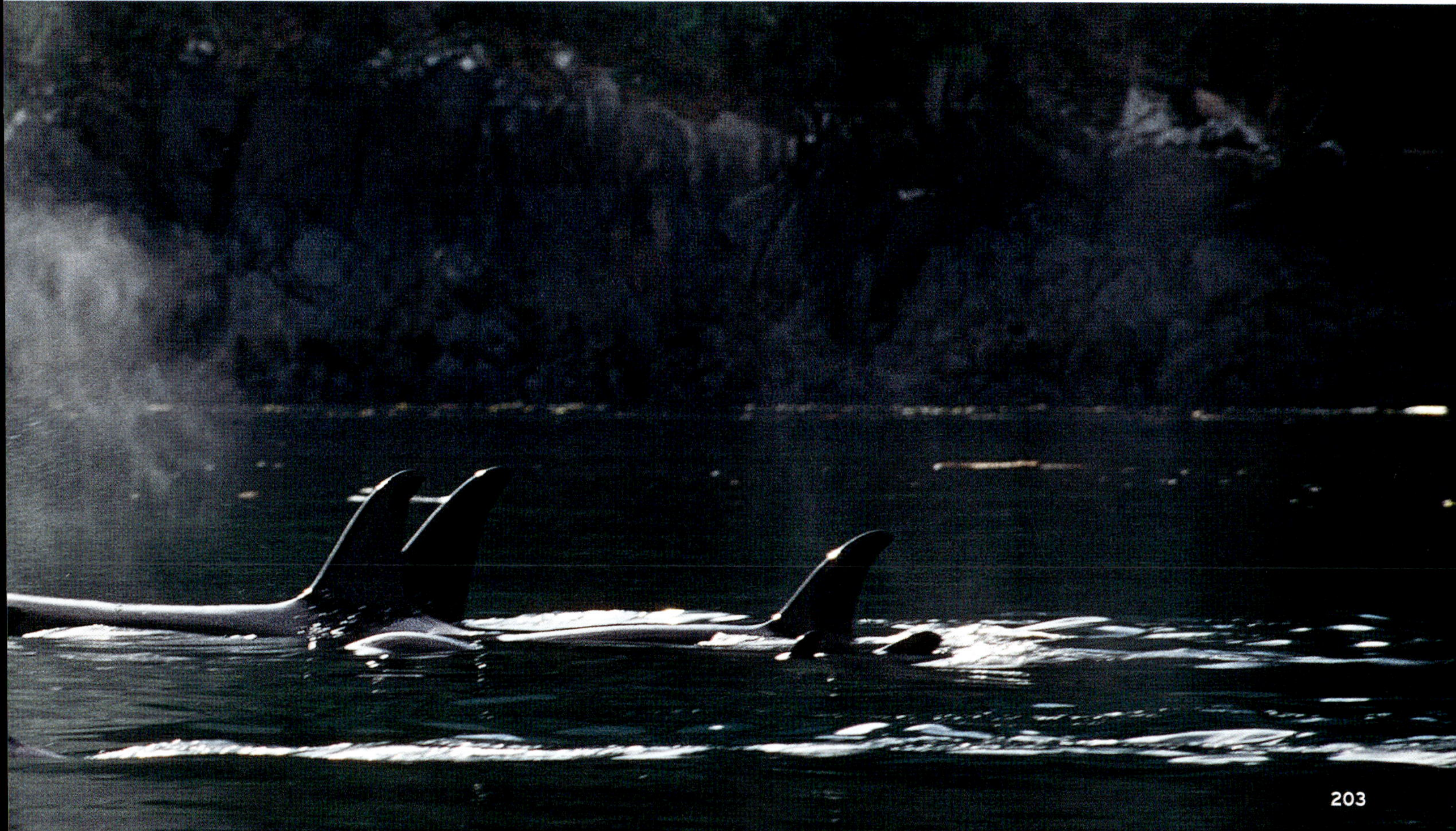

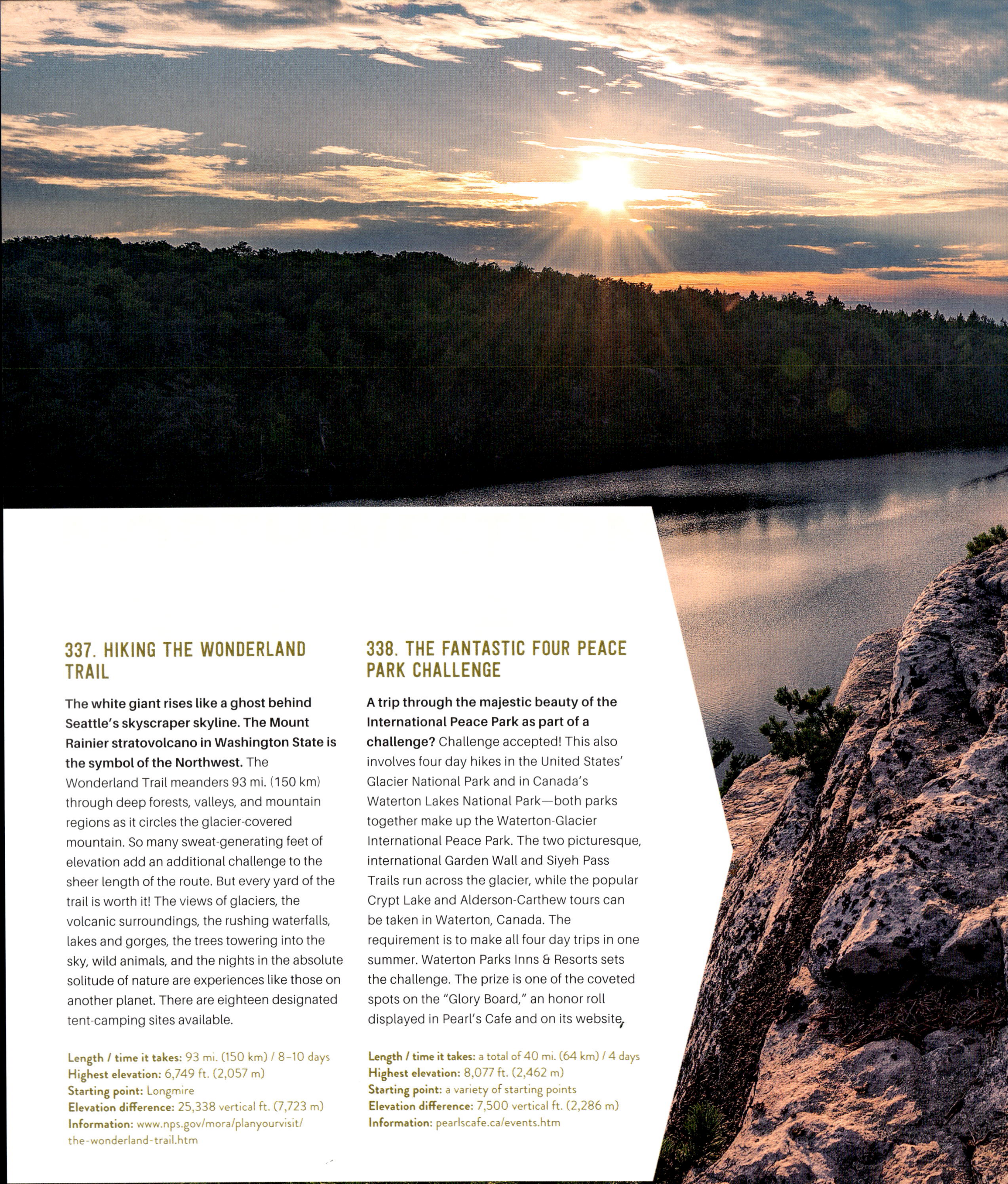

337. HIKING THE WONDERLAND TRAIL

The white giant rises like a ghost behind Seattle's skyscraper skyline. The Mount Rainier stratovolcano in Washington State is the symbol of the Northwest. The Wonderland Trail meanders 93 mi. (150 km) through deep forests, valleys, and mountain regions as it circles the glacier-covered mountain. So many sweat-generating feet of elevation add an additional challenge to the sheer length of the route. But every yard of the trail is worth it! The views of glaciers, the volcanic surroundings, the rushing waterfalls, lakes and gorges, the trees towering into the sky, wild animals, and the nights in the absolute solitude of nature are experiences like those on another planet. There are eighteen designated tent-camping sites available.

Length / time it takes: 93 mi. (150 km) / 8–10 days
Highest elevation: 6,749 ft. (2,057 m)
Starting point: Longmire
Elevation difference: 25,338 vertical ft. (7,723 m)
Information: www.nps.gov/mora/planyourvisit/the-wonderland-trail.htm

338. THE FANTASTIC FOUR PEACE PARK CHALLENGE

A trip through the majestic beauty of the International Peace Park as part of a challenge? Challenge accepted! This also involves four day hikes in the United States' Glacier National Park and in Canada's Waterton Lakes National Park—both parks together make up the Waterton-Glacier International Peace Park. The two picturesque, international Garden Wall and Siyeh Pass Trails run across the glacier, while the popular Crypt Lake and Alderson-Carthew tours can be taken in Waterton, Canada. The requirement is to make all four day trips in one summer. Waterton Parks Inns & Resorts sets the challenge. The prize is one of the coveted spots on the "Glory Board," an honor roll displayed in Pearl's Cafe and on its website.

Length / time it takes: a total of 40 mi. (64 km) / 4 days
Highest elevation: 8,077 ft. (2,462 m)
Starting point: a variety of starting points
Elevation difference: 7,500 vertical ft. (2,286 m)
Information: pearlscafe.ca/events.htm

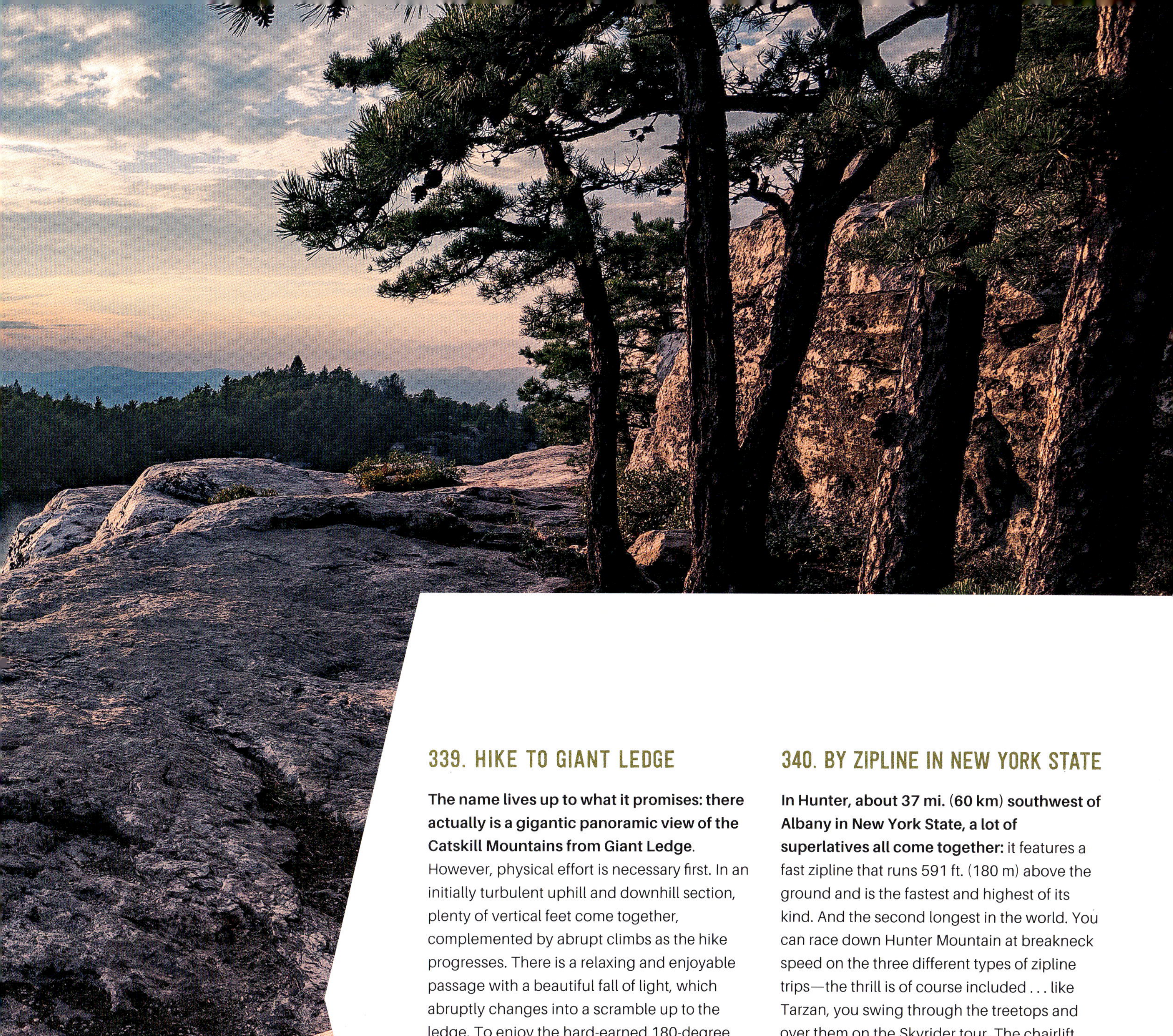

339. HIKE TO GIANT LEDGE

The name lives up to what it promises: there actually is a gigantic panoramic view of the Catskill Mountains from Giant Ledge. However, physical effort is necessary first. In an initially turbulent uphill and downhill section, plenty of vertical feet come together, complemented by abrupt climbs as the hike progresses. There is a relaxing and enjoyable passage with a beautiful fall of light, which abruptly changes into a scramble up to the ledge. To enjoy the hard-earned 180-degree view, everyone looks for a secluded picnic spot on one of the rocky plateaus—where you are presiding, almost floating free, above the abyss. If you haven't had enough after all the legwork, you can hike on for 3 mi. (5 km) and up 755 vertical ft. (230 m) to Panther Mountain.

Length / time it takes: 7.5 mi. (12 km) / 4–5 hours
Highest elevation: 3,182 ft. (970 m)
Starting point: Day parking lot at Woodland Valley Campground
Elevation difference: 1,969 vertical ft. (600 m)
Information: http://catskillmountaineer.com/SMW-giant.html

340. BY ZIPLINE IN NEW YORK STATE

In Hunter, about 37 mi. (60 km) southwest of Albany in New York State, a lot of superlatives all come together: it features a fast zipline that runs 591 ft. (180 m) above the ground and is the fastest and highest of its kind. And the second longest in the world. You can race down Hunter Mountain at breakneck speed on the three different types of zipline trips—the thrill is of course included . . . like Tarzan, you swing through the treetops and over them on the Skyrider tour. The chairlift takes you up the mountain; you then zip downhill on one of five ziplines (the first is the longest; afterward, your heart gradually starts beating in its normal rhythm again). All are different from each other. The biggest challenge: the step taking you off the platform!

Time it takes: 3 hours
Location: Hunter Mountain Ski Bowl
Requirements: no fear of heights!
Times: daily all year round
Information: www.zipenewyork.com

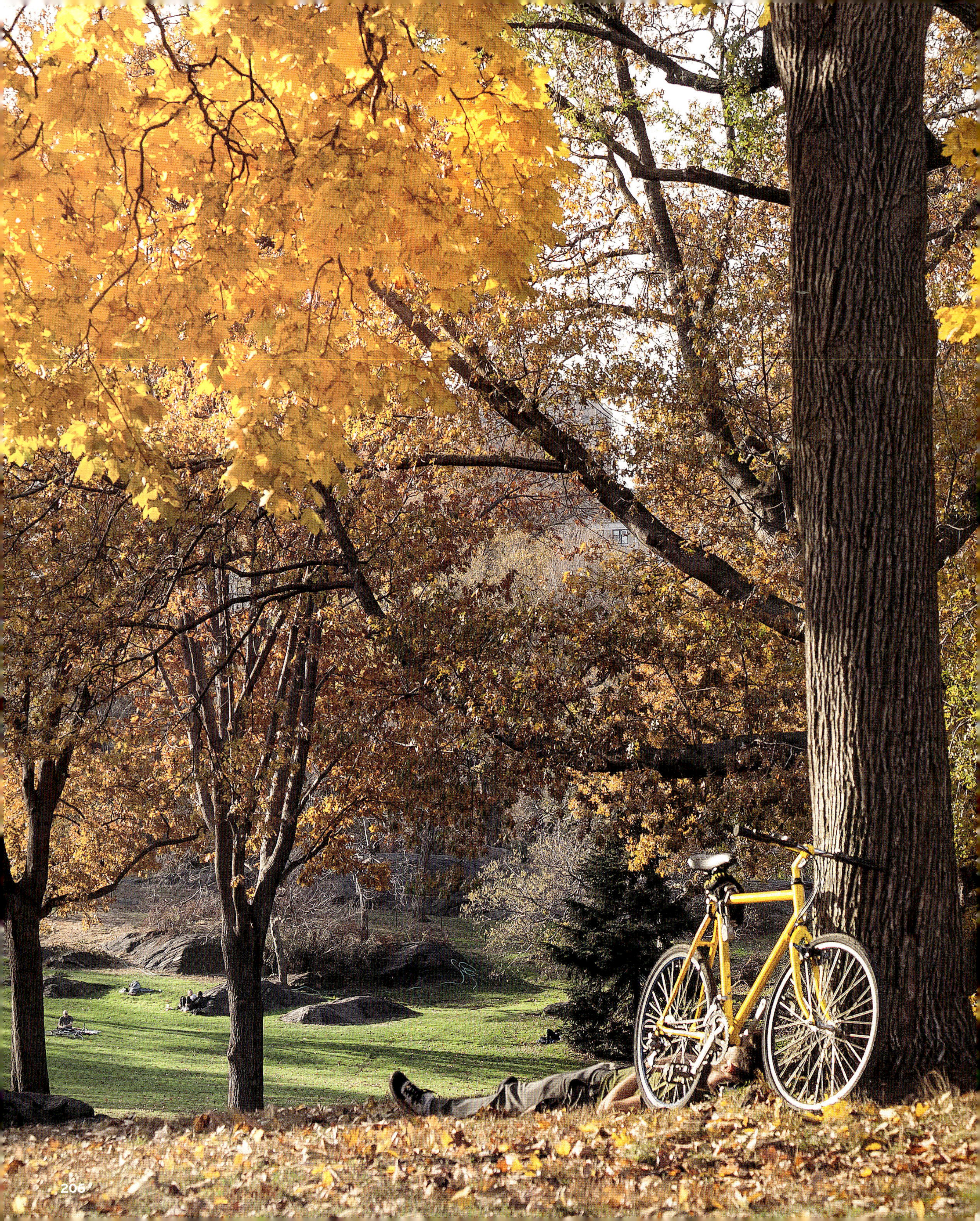

341. BIKING THROUGH CENTRAL PARK

This New York City park in the heart of Manhattan is not only an often-seen setting for many films and series, but it has long since become famous for its own sake. And there is much more than just scenes from your favorite movies to discover; this 340-hectare area is simply a magnetic attraction right in the middle of cosmopolitan New York City. Here everyone can find an island of tranquility amid the hustle and bustle of the big city—and so that you can see as much of this park as possible, a bike tour is ideal. Cyclists can let themselves drift through this stage set of green meadows, sculptures and statues, fountains, playgrounds and fields, lakes, romantic bridges, and fairgrounds in a relaxed way on specially created bike paths. There are plenty of rental bike suppliers all around the park.

Time it takes: half a day to one day
Starting point: as desired, at any park entrance
Information: www.centralparknyc.org

342. DOGSLEDDING ON THE ICE

When Mirror Lake, in the small ski resort of Lake Placid, New York, is frozen over, the most awesome activities are on the program. By far the coolest thing you can do in this weather is a dogsled ride, and it is something you will long remember. With up to eight Alaskan malamutes or Siberian huskies already in harness, you can take a rapid ride either over the icy surface of the lake or through the Adirondack Mountains wilderness. Considering driving the energetic dog team as a musher yourself? It looks easy but is a real athletic challenge—and has nothing in common with a romantic ride in a horse-drawn carriage. But the nice thing is, you don't have to be a real Inuit to be able to drive the team!

Time it takes: half a day to one day
Highest elevation: 5,344 ft. (1,629 m)
Starting point: at High Peaks Resort
Information: visitadirondacks.com/recreation/thunder-mountain-dog-sled-tours

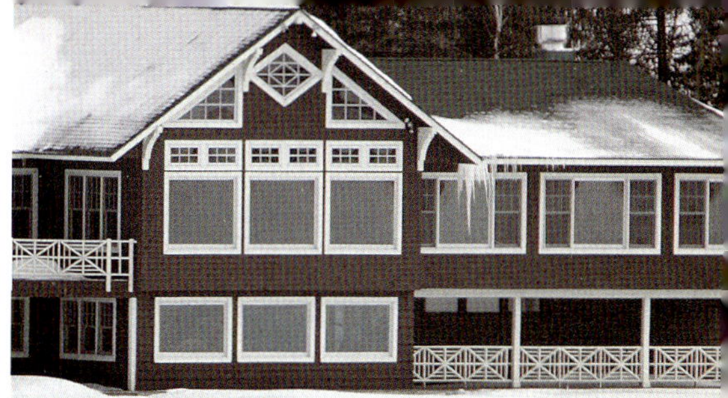

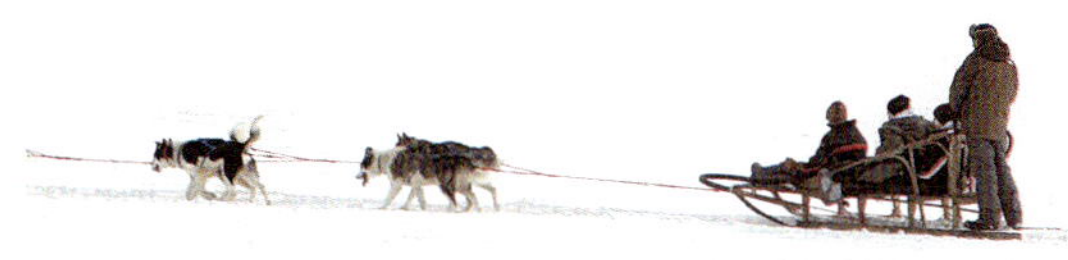

343. HIKE ON ACADIA'S PRECIPICE TRAIL

Acadia National Park is on an island off the state of Maine and offers a whole range of tours, including the Precipice Trail hike. It's considered the most demanding, but it's also great fun! After enterprising climbing and clambering passages, the final climb is straight upward. To help you manage it, there are iron rungs set into the rock for your hands and feet, then a field of scree and finally iron stairs and ladders (i.e., via ferrata). The trail keeps running steeply upward, and with every drop of sweat the view becomes more magnificent. Once at the top of the cliffs, you are allowed to take a short break. Then a moderate hike leads to the summit of Champlain Mountain, with a spectacular view of the Atlantic Ocean directly below it.

Length / time it takes: 3 mi. (5 km) / 3–5 hours
Highest elevation: 1,070 ft. (326 m)
Starting point: 2 mi. (3 km) from the Sieur de Monts park entrance on Park Loop Road
Elevation difference: 1,148 vertical ft. (350 m)
Information: www.nps.gov/acad/planyourvisit/hike-summit.htm

344. CYCLING IN ACADIA NATIONAL PARK

The old horse-drawn carriage roads have been converted and now allow unrestricted cycling fun in Acadia National Park, which is on an island. Only hikers, cyclists, and horseback riders are allowed to use the roads, which are closed to car traffic—that means sustainable and relaxing vacations. All kinds of enjoyable bike tours are possible amid the picturesque landscapes on roads running through the park in all directions! The network of trails covers 45 mi. (72 km). Every time the bike paths meet the Park Loop Road, pretty old cobblestone bridges offer wonderful photo opportunities and places to rest. Acadia Bike Rentals in Bar Harbor supplies all kinds of bicycles for exploring the national park, which at 74 sq. mi. (190 km^2) is rather small by American standards.

Time it takes: at least 1 day
Highest elevation: 1,529 ft. (466 m)
Starting point: a variety of starting points on Mount Desert Island
Information: www.nps.gov/acad/planyourvisit/bicycling.htm

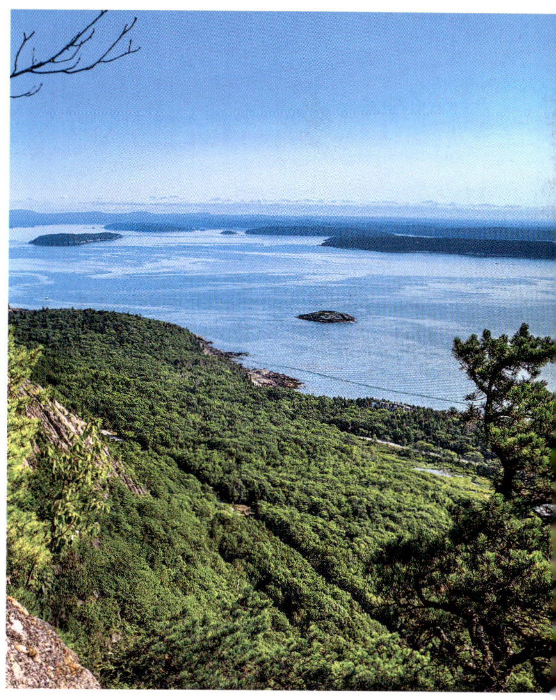

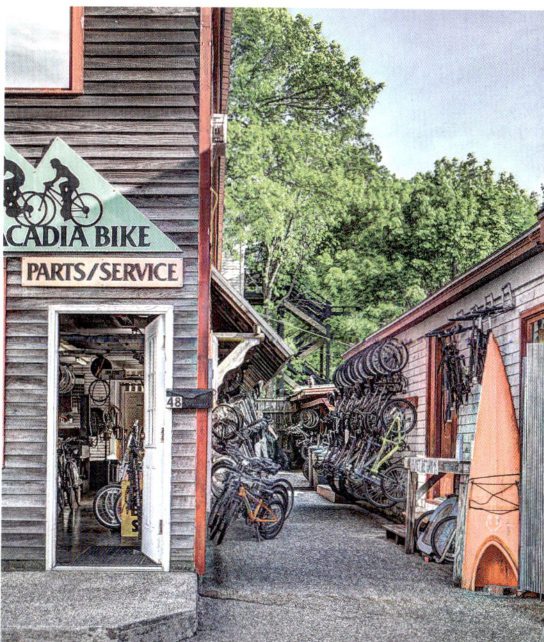

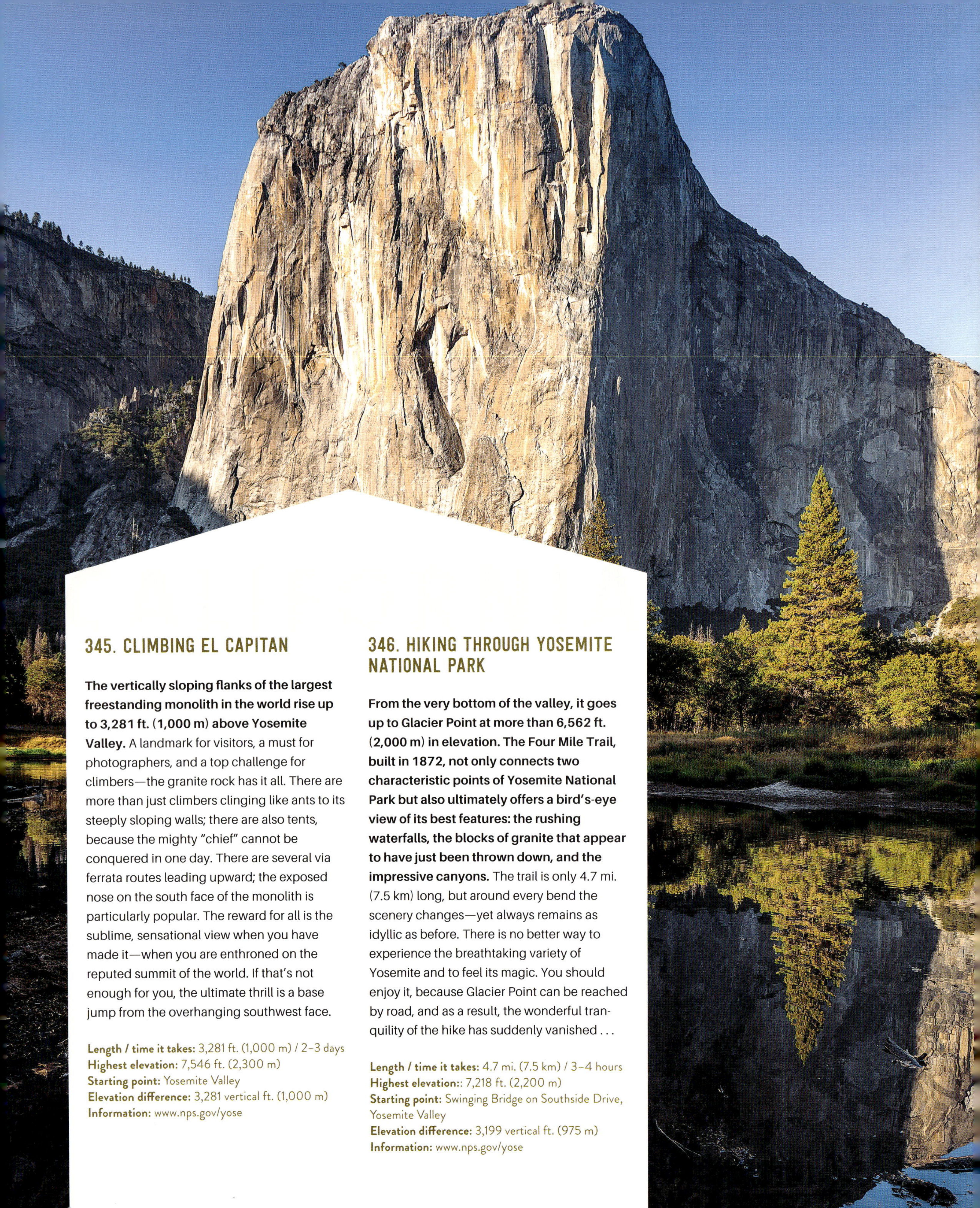

345. CLIMBING EL CAPITAN

The vertically sloping flanks of the largest freestanding monolith in the world rise up to 3,281 ft. (1,000 m) above Yosemite Valley. A landmark for visitors, a must for photographers, and a top challenge for climbers—the granite rock has it all. There are more than just climbers clinging like ants to its steeply sloping walls; there are also tents, because the mighty "chief" cannot be conquered in one day. There are several via ferrata routes leading upward; the exposed nose on the south face of the monolith is particularly popular. The reward for all is the sublime, sensational view when you have made it—when you are enthroned on the reputed summit of the world. If that's not enough for you, the ultimate thrill is a base jump from the overhanging southwest face.

Length / time it takes: 3,281 ft. (1,000 m) / 2–3 days
Highest elevation: 7,546 ft. (2,300 m)
Starting point: Yosemite Valley
Elevation difference: 3,281 vertical ft. (1,000 m)
Information: www.nps.gov/yose

346. HIKING THROUGH YOSEMITE NATIONAL PARK

From the very bottom of the valley, it goes up to Glacier Point at more than 6,562 ft. (2,000 m) in elevation. The Four Mile Trail, built in 1872, not only connects two characteristic points of Yosemite National Park but also ultimately offers a bird's-eye view of its best features: the rushing waterfalls, the blocks of granite that appear to have just been thrown down, and the impressive canyons. The trail is only 4.7 mi. (7.5 km) long, but around every bend the scenery changes—yet always remains as idyllic as before. There is no better way to experience the breathtaking variety of Yosemite and to feel its magic. You should enjoy it, because Glacier Point can be reached by road, and as a result, the wonderful tranquility of the hike has suddenly vanished . . .

Length / time it takes: 4.7 mi. (7.5 km) / 3–4 hours
Highest elevation:: 7,218 ft. (2,200 m)
Starting point: Swinging Bridge on Southside Drive, Yosemite Valley
Elevation difference: 3,199 vertical ft. (975 m)
Information: www.nps.gov/yose

347. CYCLING ACROSS GOLDEN GATE BRIDGE

Just the sight of the most striking suspension bridge in the world alone already gives you goose bumps. Peddling a bike across is fantastic. With the mouth of San Francisco Bay deep below, and the orange-red bridge piers at eye level and separated from the traffic, this bike tour is a unique experience. At Fisherman's Wharf, suppliers such as Blazing Saddles make this dream come true with the aid of a rental bike. The trip runs along the water toward the bridge; the chilled tour is garnished with views of Alcatraz Prison island. Riding across the bridge becomes an unforgettable experience when you see dolphins, whales, or seals far below in the bay. The charming town of Sausalito, on the other side of the bay, and other idyllic towns in Marin County are the icing on the cake for this bike tour.

Length / time it takes: 8 mi. (13 km) / 1.5 hours to Sausalito
Highest elevation: 745 ft. (227 m)
Starting point: Fisherman's Wharf, San Francisco
Information: www.blazingsaddles.com

348. BIKE TOUR ON CRATER LAKE RIM DRIVE

This deep-blue pearl owes its existence to the eruption of Mount Mazama volcano. The lake—the deepest in the United States—is 1,946 ft. (593 m) deep, and the rim of the crater runs 984 ft. (300 m) above it. Crater Lake is surrounded by picturesque hills and snow-capped volcanic peaks, as well as evergreen forests. This natural spectacle becomes a special experience when circuiting the crater by bike on the Ride the Rim tour. In a challenging mix of steep climbs and a long route, cyclists experience the breathtaking beauty of the magnificent mountain lake: the contrast between the bright-blue water and the dark-green conifer trees is almost unreal. The East Rim is car free, and that is almost 25 mi. (40 km) of the total lap. Twice a year the route is restricted just to cyclists and hikers.

Length / time it takes: 33 mi. (53 km) / 1 day
Highest elevation: 7,631 ft. (2,326 m)
Starting point: North Junction
Elevation difference: 3,500 vertical ft. (1,067 m)
Information: ridetherimoregon.com

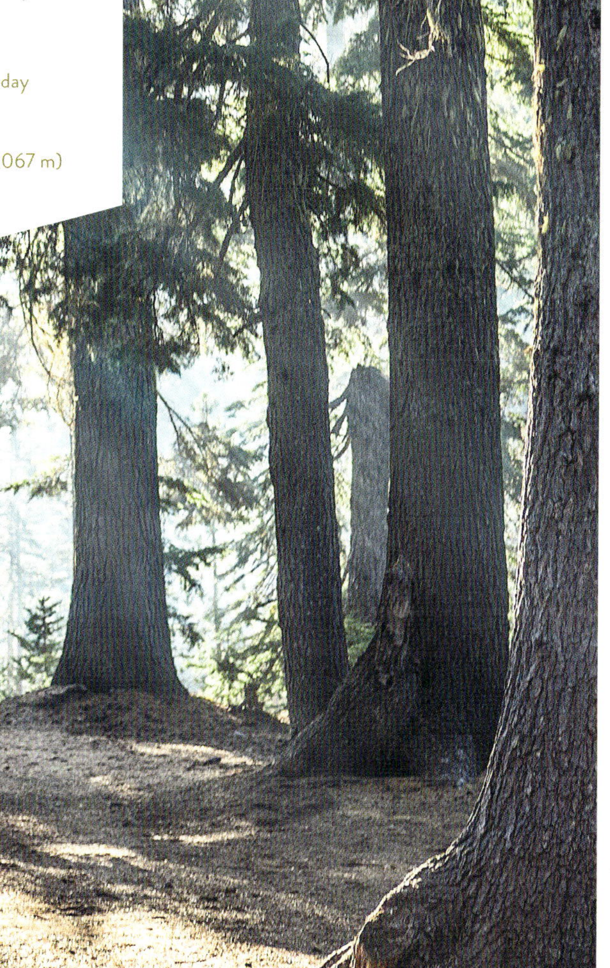

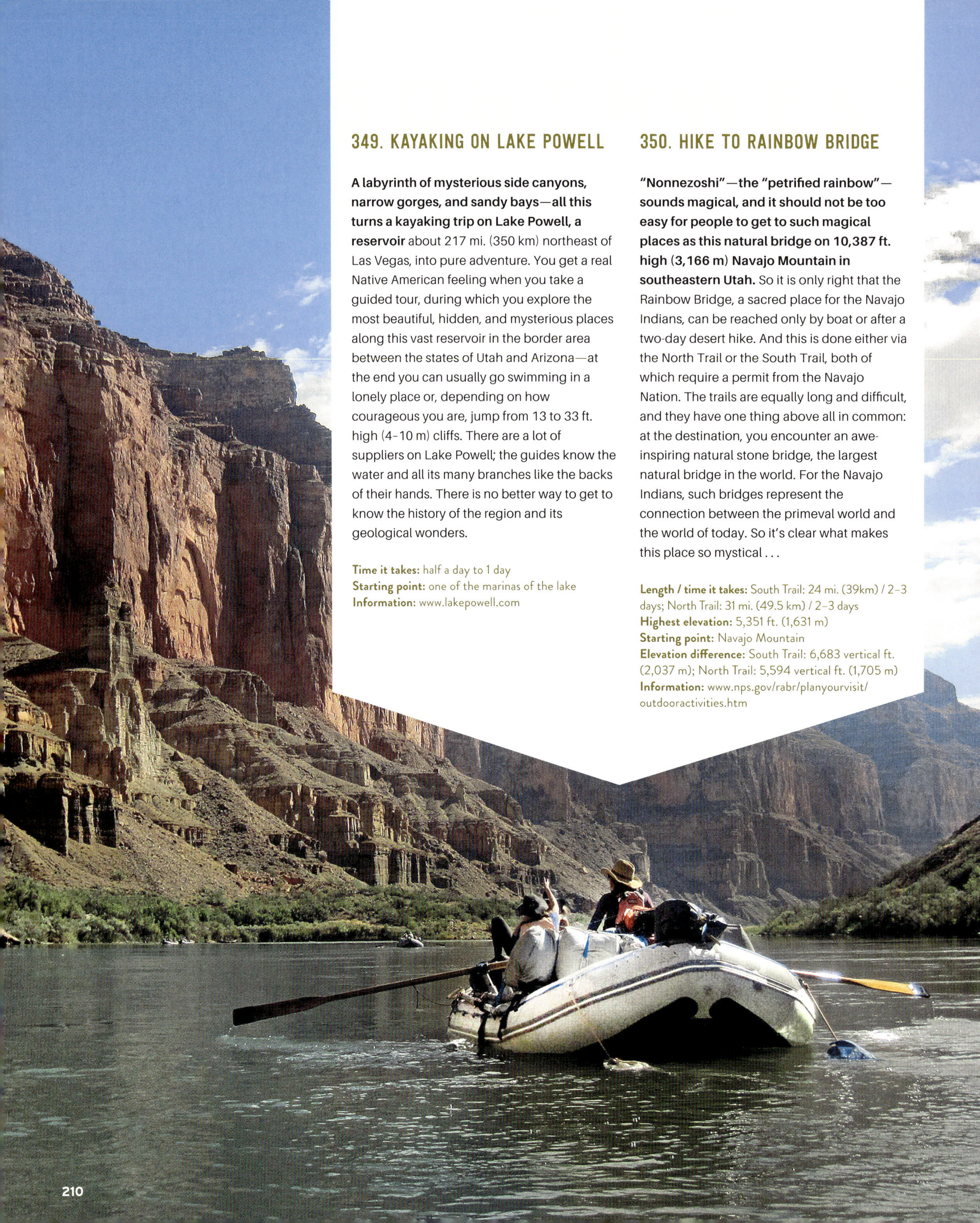

349. KAYAKING ON LAKE POWELL

A labyrinth of mysterious side canyons, narrow gorges, and sandy bays—all this turns a kayaking trip on Lake Powell, a reservoir about 217 mi. (350 km) northeast of Las Vegas, into pure adventure. You get a real Native American feeling when you take a guided tour, during which you explore the most beautiful, hidden, and mysterious places along this vast reservoir in the border area between the states of Utah and Arizona—at the end you can usually go swimming in a lonely place or, depending on how courageous you are, jump from 13 to 33 ft. high (4–10 m) cliffs. There are a lot of suppliers on Lake Powell; the guides know the water and all its many branches like the backs of their hands. There is no better way to get to know the history of the region and its geological wonders.

Time it takes: half a day to 1 day
Starting point: one of the marinas of the lake
Information: www.lakepowell.com

350. HIKE TO RAINBOW BRIDGE

"Nonnezoshi"—the "petrified rainbow"—sounds magical, and it should not be too easy for people to get to such magical places as this natural bridge on 10,387 ft. high (3,166 m) Navajo Mountain in southeastern Utah. So it is only right that the Rainbow Bridge, a sacred place for the Navajo Indians, can be reached only by boat or after a two-day desert hike. And this is done either via the North Trail or the South Trail, both of which require a permit from the Navajo Nation. The trails are equally long and difficult, and they have one thing above all in common: at the destination, you encounter an awe-inspiring natural stone bridge, the largest natural bridge in the world. For the Navajo Indians, such bridges represent the connection between the primeval world and the world of today. So it's clear what makes this place so mystical . . .

Length / time it takes: South Trail: 24 mi. (39km) / 2–3 days; North Trail: 31 mi. (49.5 km) / 2–3 days
Highest elevation: 5,351 ft. (1,631 m)
Starting point: Navajo Mountain
Elevation difference: South Trail: 6,683 vertical ft. (2,037 m); North Trail: 5,594 vertical ft. (1,705 m)
Information: www.nps.gov/rabr/planyourvisit/outdooractivities.htm

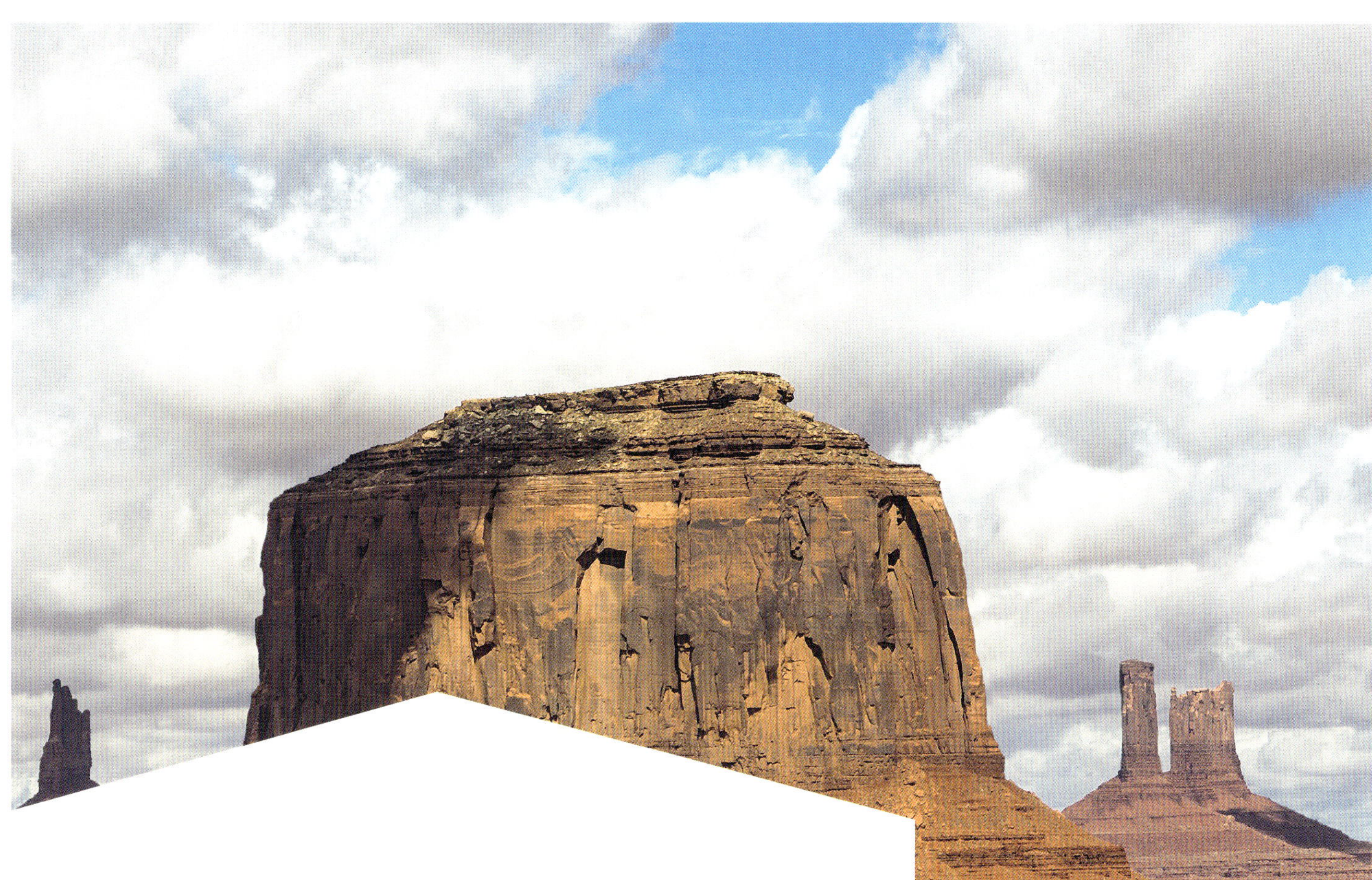

351. RAFTING AND KAYAKING THROUGH THE GRAND CANYON

Anyone can look into the Grand Canyon from above. But experiencing the grandeur of the canyon and violence of the Colorado River up close on the water is a once-in-a-lifetime experience. This trip is never boring; the 135 mighty rapids in the Colorado River alone ensure that: thundering rapids with yard-high waves that will never cease to amaze even diehard whitewater kayakers and toughened outdoor travelers. The 39 ft. long (12 m) inflatable rafts are floating hotels and guarantee five-star catering and luxury camping under the starry sky. Cool: the many hikes in the beautiful side canyons. This natural wonder that is the Grand Canyon, consisting of rock, sand, and water, is the ideal place to really leave everyday life behind.

Length / time it takes: 224 mi. (360 km) / 12 days
Level of difficulty: WW 4, turbulent water
Starting point: Lee's Ferry, Arizona
Best time of year: End of August to early September
Information: www.global-kayak.com

352. HORSEBACK RIDING THROUGH MONUMENT VALLEY

There is only one way in the world to get the ultimate "Wild West" feeling: a ride through Monument Valley on the Colorado Plateau in the very south of Utah. This is a setting just right for filming, which has already been proven in countless western movies. To suddenly sit in a western saddle yourself, trotting together with Native American guides through the famous scenery of glowing red-sandstone formations, flanked by dogs to chase away rattlesnakes in this unspoiled wilderness—this is more than you can take in at once, and the fulfillment of a dream. At the same time, no equestrian skills are required; the horses are good and sure-footed. Any ambitious riders are also allowed to gallop. At the least, this will take you to complete rapture . . .

Time it takes: at least 2 hours
Starting point: visitor center at the View Hotel
Information: monumentvalleyview.com/navajo-guided-tours

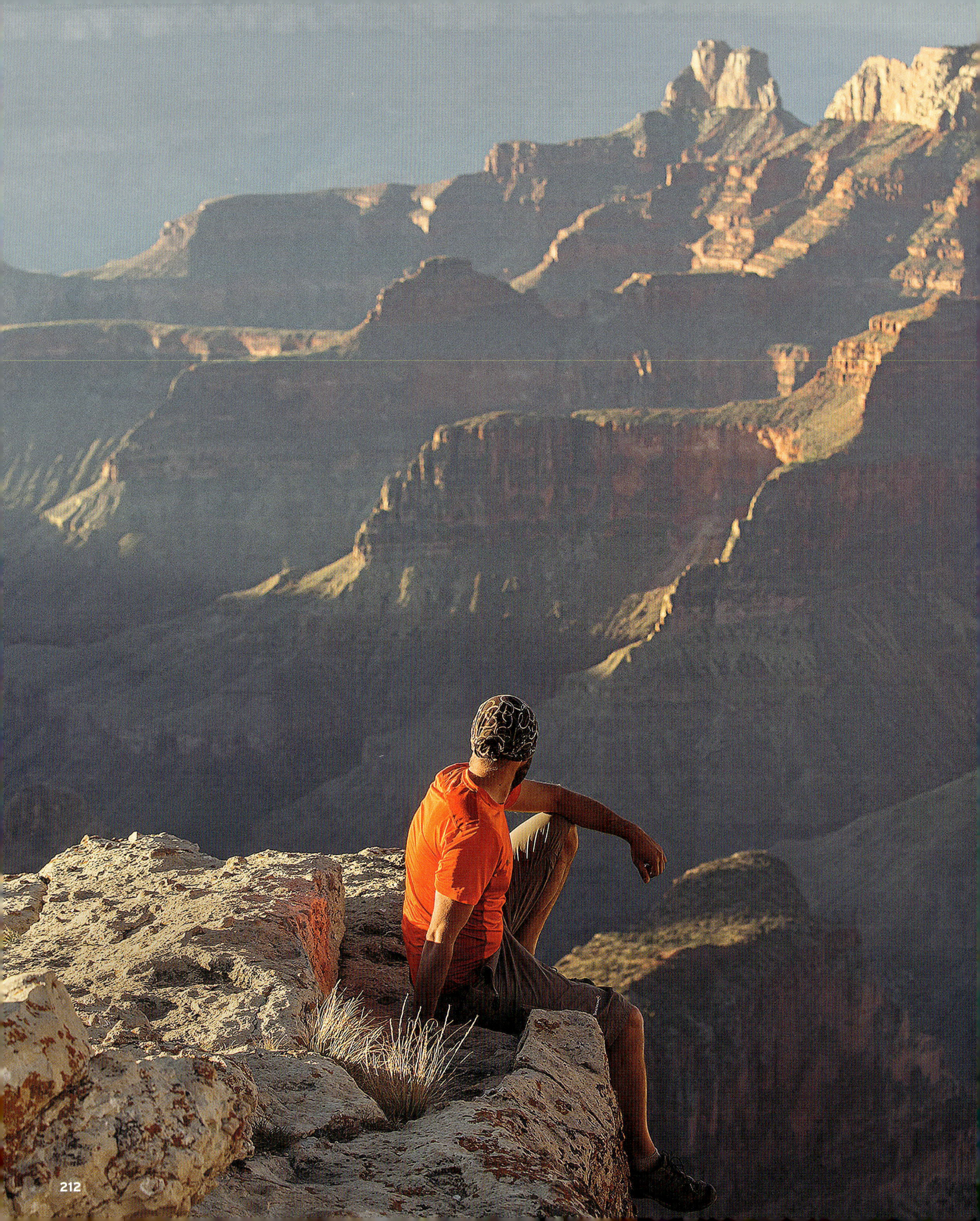

353. RIM-TO-RIM HIKE

Descending once into the depths of the legendary canyon—every hiker's dream! This becomes a much-cooler adventure if you don't have to hike back up the same trail again after a night on the Colorado River. From the North Rim, hovering high above the Grand Canyon, the route runs into the canyon along the North Kaibab Trail; its counterpart on the South Rim is the South Kaibab Trail. The northern version is 14 mi. (22 km) long and extends 5,906 vertical ft. (1,800 m); from the South Rim it is 9 mi. (15 km) and 4,593 vertical ft. (1,400 m) down into the canyon. Regardless of the direction from which you experience one of the greatest and most unforgettable adventures there is—you don't have to march back over it again but can be comfortably driven back to the starting point in a rim-to-rim bus.

Length / time it takes: 23 mi. (37 km) / at least 2 days
Highest elevation: 8,241 ft. (2,512 m)
Starting point: North Kaibab Trailhead or South Kaibab Trailhead
Elevation difference: 10,499 vertical ft. (3,200 m)
Information: www.trans-canyonshuttle.com

354. MULE RIDE THROUGH THE GRAND CANYON

The floor of the majestic Grand Canyon, down there along the Colorado River, seems almost inaccessible. In fact, there are a lot of vertical feet between the South Rim and the Grand Canyon floor. But they can't harm a mule. Sure-footed and laden with riders and packs, entire caravans of these well-behaved beasts of burden trot down the Bright Angel Trail and up the South Kaibab Trail the next day. That sounds more comfortable than it is, because the many hours in the saddle over impassable terrain, either heading steep downhill or uphill, take their toll. But it has a totally convincing advantage: you can experience the fascination of this unbelievably large and impressive canyon with all your senses, unencumbered by heavy backpacks or blisters on your feet.

Length / time it takes: 15.5 mi. (25 km) / 2 days
Highest elevation: 6,998 ft. (2,133 m)
Starting point: The corral at Bright Angel Lodge
Elevation difference: 10,364 vertical ft. (3,159 m)
Information: www.grandcanyonlodges.com/plan/mule rides

355. TRAIL TO OBSERVATION POINT, ZION NATIONAL PARK

A bird's-eye view of the vast, wild Zion Canyon starts from a very special place: Observation Point. A sweaty but incredibly beautiful hike leads up to the roof of Zion National Park in southwestern Utah, from where you can also see the popular but therefore hopelessly overrun Angels Landing lookout point. The steep trip to Observation Point zigzags up the rocks and impresses hikers with an enchanted landscape of rugged and brightly colored rocks, magically beautiful slot canyons, and sensational views. The trail is worth every drop of sweat. It ends at a scenic outlook almost 6,562 ft. (2,000 m) high; eastern Zion Canyon lies below, at the observer's feet.

Length / time it takes: 8 mi. (13 km) / 4–6 hours
Highest elevation: 6,519 ft. (1,987 m)
Starting point: Weeping Rock shuttle stop
Elevation difference: 2,297 vertical ft. (700 m)
Information: www.nps.gov/zion/planyourvisit/zion-canyon-trail-descriptions.htm

356. ON THE NARROWS RIVERSIDE WALK

A first-class river hike through the Virgin River, a tributary of the Colorado River, surrounded by steep cliffs: hiking the Narrows is a highlight of the Southwest. Narrow passages and lofty, almost 2,624 ft. high (800 m) canyon walls alternate with sunny sections—but one thing always stays the same: when the paved, almost 1 mi. long (1.5 km) Riverside Walk ends, you wade ankle to knee deep through the Virgin River, which flows through the states of Utah, Arizona, and Nevada. You can enjoy this adventure for as long as you like, passing by waterfalls and side branches that want to be explored. But you should at least make it to the narrowest bottleneck, called "Wall Street," which is 1.6 mi. (2.5 km) from the entry point; then you have already had a very intense experience.

Length / time it takes: as desired; at least 5.3 mi. (8.5 km) / 1 day
Starting point: Temple of Sinawava shuttle stop
Elevation difference: 197 vertical ft. (60 m)
Information: www.nps.gov/zion/planyourvisit/thenarrows.htm

357. SNORKELING IN THE GULF OF MEXICO

Anyone who thinks that this might be the end of the world and that there is nothing going on is very wrong. Dry Tortugas National Park does indeed lie more than 62 mi. (100 km) west of Key West and in the middle of the Gulf of Mexico, but this location also makes all water lovers' hearts beat faster. Because this is where many inhabitants of the sea have their habitats, and wrecked ships lie sunken on the seabed all around the national park. Snorkel gear on, and in you go. Colorful corals shine brightly, and tropical fish and turtles do their laps—the strong colors stand in spectacular contrast to the snow-white sandy beach! Whether you arrive by the excursion ferry or by seaplane, you will be provided with equipment and tips on where to find the best spots.

Time it takes: half a day to 1 day
Starting point: Key West ferry terminal or airport
Information: www.nps.gov/drto

358. TREKKING ON THE FLORIDA TRAIL

The long-distance hiking trail through Florida—just like the North Country Trail, and as if it were its counterpart—is a national scenic trail that runs from the national park between Miami and Naples to the Gulf Islands National Seashore on Pensacola Beach. You start in the "Florida Panhandle" on the Gulf of Mexico and hike on through pine forests and past sandy beaches, through the hills, lakes, and rivers of northern Florida to the extensive marshland in the south, to the Everglades with its rich flora and fauna.

Length: 1,429 mi. (2,300 km)
Starting point: Big Cypress National Preserve
Final goal: Fort Pickens

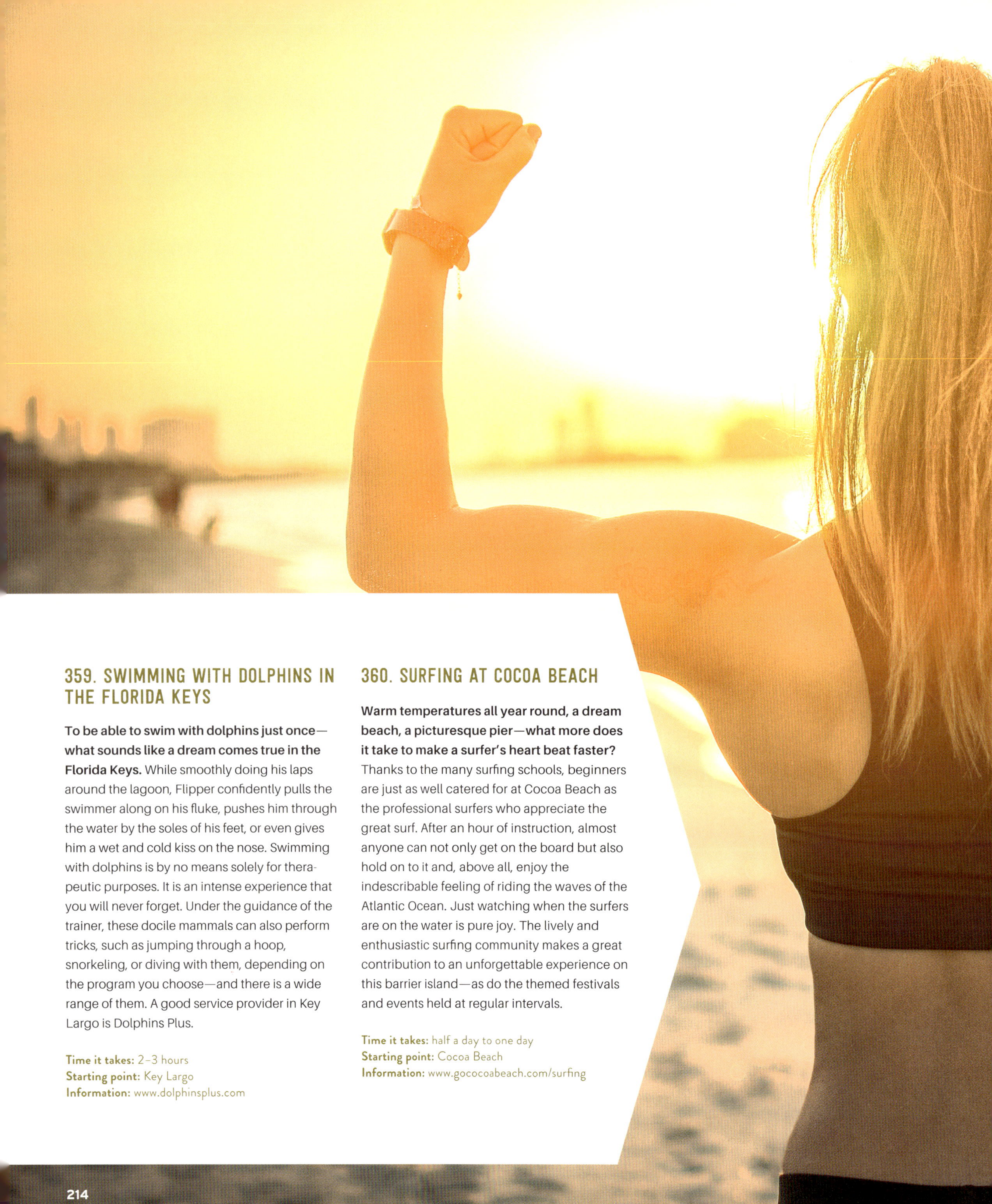

359. SWIMMING WITH DOLPHINS IN THE FLORIDA KEYS

To be able to swim with dolphins just once—what sounds like a dream comes true in the Florida Keys. While smoothly doing his laps around the lagoon, Flipper confidently pulls the swimmer along on his fluke, pushes him through the water by the soles of his feet, or even gives him a wet and cold kiss on the nose. Swimming with dolphins is by no means solely for therapeutic purposes. It is an intense experience that you will never forget. Under the guidance of the trainer, these docile mammals can also perform tricks, such as jumping through a hoop, snorkeling, or diving with them, depending on the program you choose—and there is a wide range of them. A good service provider in Key Largo is Dolphins Plus.

Time it takes: 2–3 hours
Starting point: Key Largo
Information: www.dolphinsplus.com

360. SURFING AT COCOA BEACH

Warm temperatures all year round, a dream beach, a picturesque pier—what more does it take to make a surfer's heart beat faster? Thanks to the many surfing schools, beginners are just as well catered for at Cocoa Beach as the professional surfers who appreciate the great surf. After an hour of instruction, almost anyone can not only get on the board but also hold on to it and, above all, enjoy the indescribable feeling of riding the waves of the Atlantic Ocean. Just watching when the surfers are on the water is pure joy. The lively and enthusiastic surfing community makes a great contribution to an unforgettable experience on this barrier island—as do the themed festivals and events held at regular intervals.

Time it takes: half a day to one day
Starting point: Cocoa Beach
Information: www.gococoabeach.com/surfing

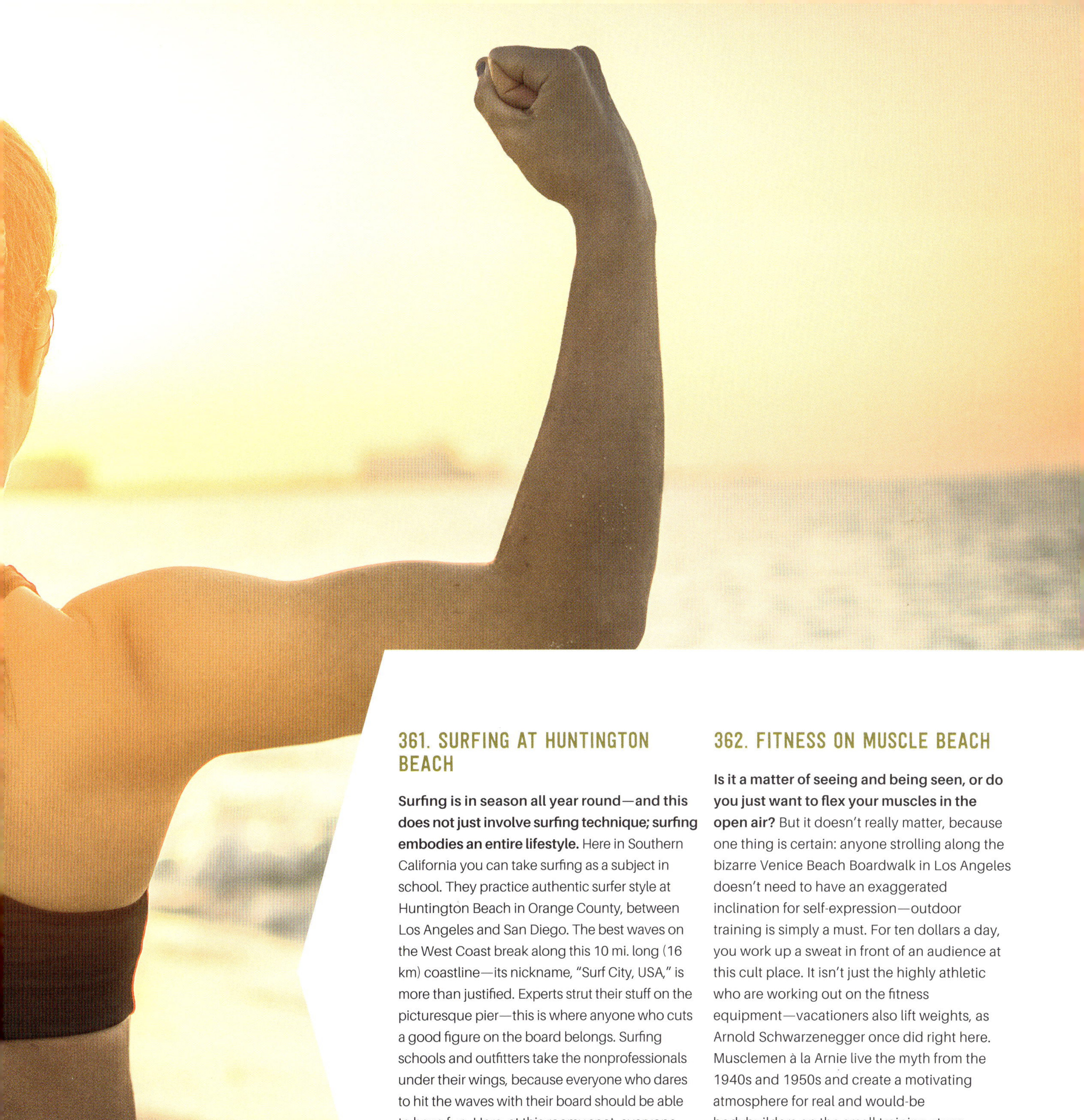

361. SURFING AT HUNTINGTON BEACH

Surfing is in season all year round—and this does not just involve surfing technique; surfing embodies an entire lifestyle. Here in Southern California you can take surfing as a subject in school. They practice authentic surfer style at Huntington Beach in Orange County, between Los Angeles and San Diego. The best waves on the West Coast break along this 10 mi. long (16 km) coastline—its nickname, "Surf City, USA," is more than justified. Experts strut their stuff on the picturesque pier—this is where anyone who cuts a good figure on the board belongs. Surfing schools and outfitters take the nonprofessionals under their wings, because everyone who dares to hit the waves with their board should be able to have fun. Here, at this roomy spot, everyone will find the right class of wave.

Length / time it takes: as desired
Location: Huntington Beach Pier
Information: www.surfcityusa.com

362. FITNESS ON MUSCLE BEACH

Is it a matter of seeing and being seen, or do you just want to flex your muscles in the open air? But it doesn't really matter, because one thing is certain: anyone strolling along the bizarre Venice Beach Boardwalk in Los Angeles doesn't need to have an exaggerated inclination for self-expression—outdoor training is simply a must. For ten dollars a day, you work up a sweat in front of an audience at this cult place. It isn't just the highly athletic who are working out on the fitness equipment—vacationers also lift weights, as Arnold Schwarzenegger once did right here. Musclemen à la Arnie live the myth from the 1940s and 1950s and create a motivating atmosphere for real and would-be bodybuilders on the small training stage.

Time it takes: half a day
Location: Venice Beach Boardwalk
Information: www.venicebeach.com/muscle-beach-gym

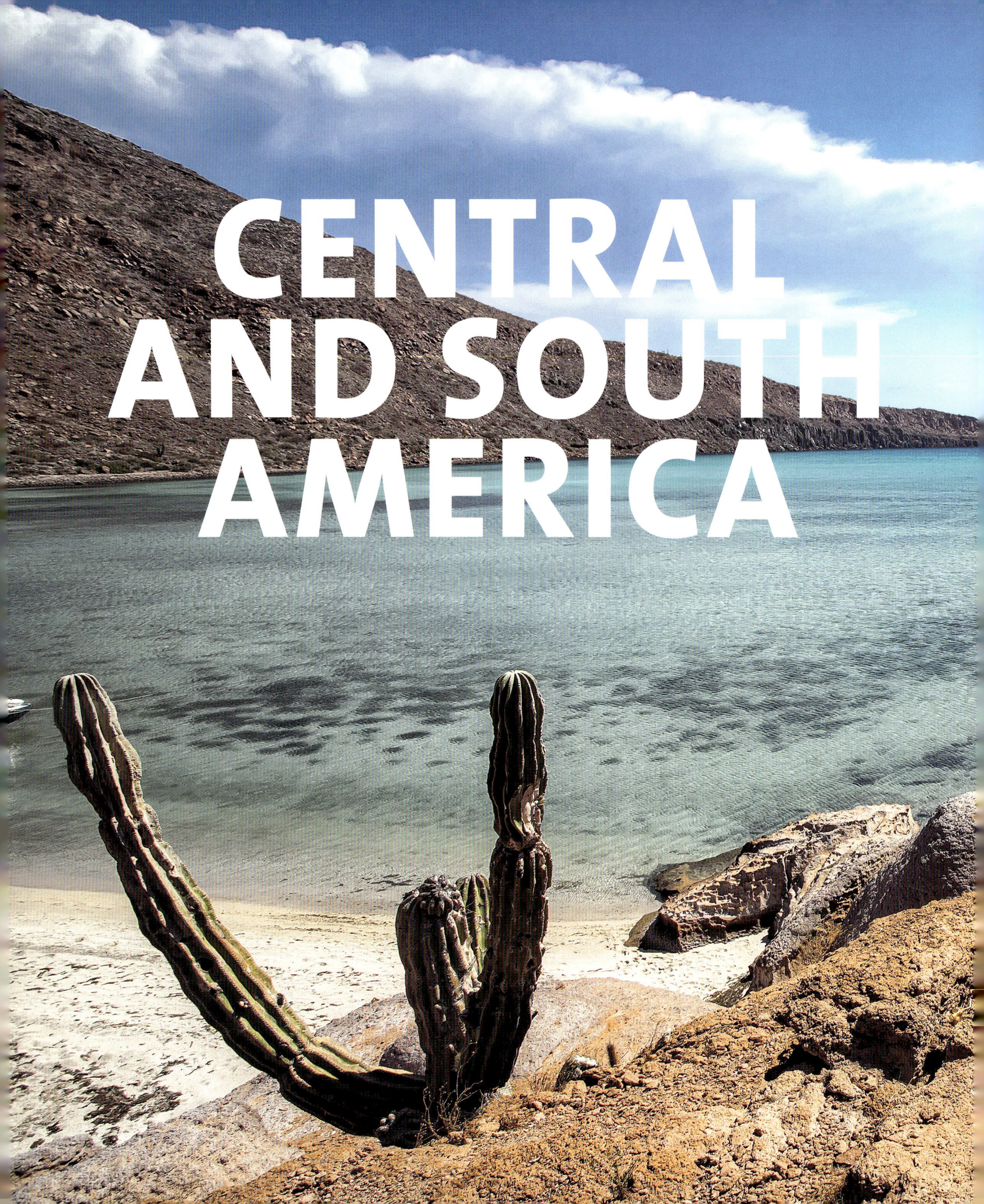

CENTRAL AND SOUTH AMERICA

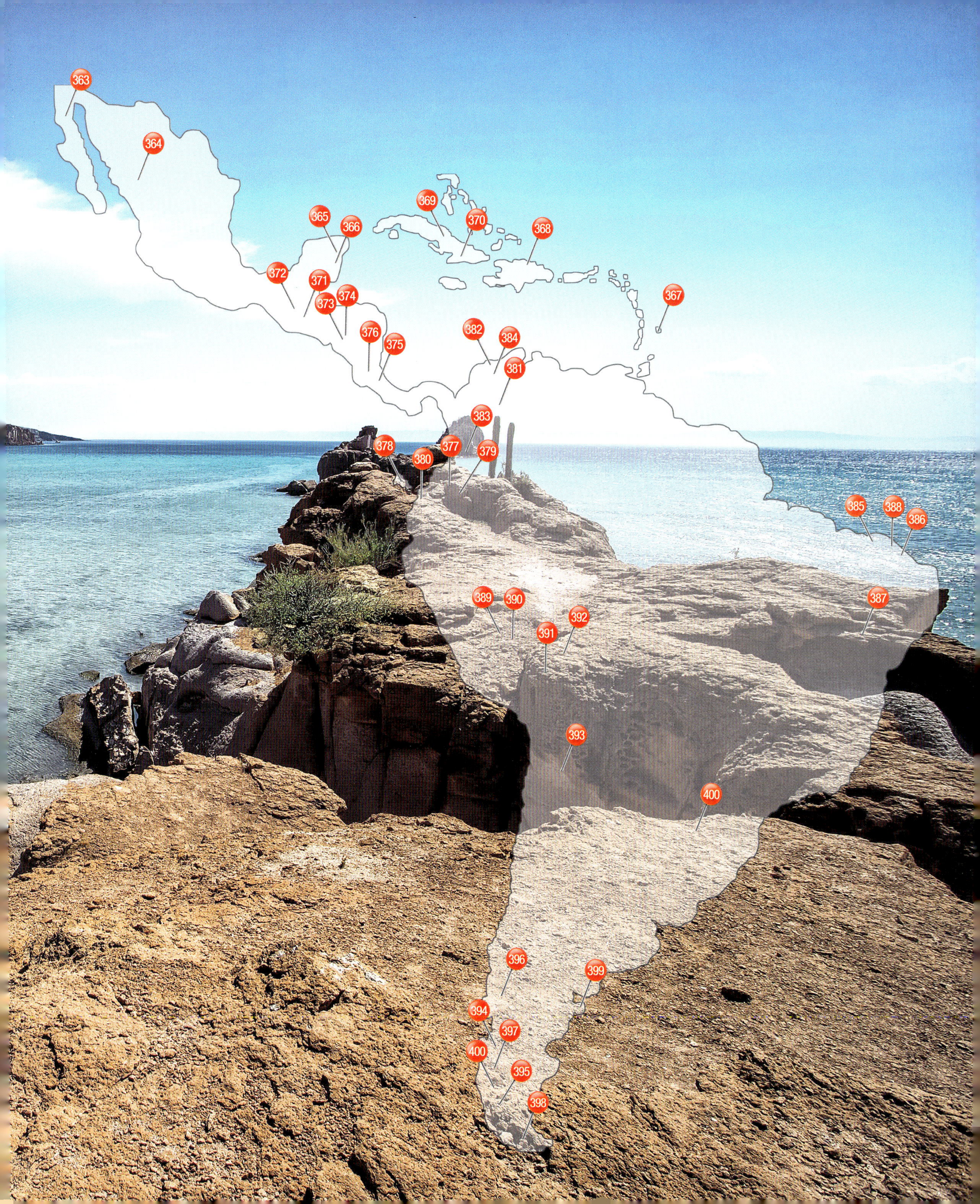
363
364
369
365
366
370
368
372
371
374
373
367
382
384
376
375
381
383
378
377
380
379
385
388
386
389
390
392
391
387
393
400
396
399
394
397
400
395
398

363. SEA KAYAKING BY BAJA CALIFORNIA

Fissured earth, brown on brown. Thornbushes, cacti, heat shimmering over the sunburned ground—and the dark blue of the Sea of Cortez. These are typical compositions on Baja California south of Tijuana. As a geographical appendage of Mexico, it thrusts more than 746 mi. (1,200 km) deep into the sea and ends in the exclusive holiday region of Los Cabos. A sea kayak is a great option for exploring small sections of the Sea of Cortez; you can circumnavigate headlands, bays, and rockfalls in your kayak. And in between, your guide Lalo calls out to the group: "Now everyone out here to snorkel. I'll hold the kayaks together with ropes!" It can also happen here that sea lions swim up and circle around the masked bipeds until their curiosity fades away.

Information: see, for example, www.bajawild.com

364. HIKING THROUGH THE BARRANCAS DEL COBRE CANYONS

The Barrancas del Cobre (Copper Canyons) canyons are more extensive than the Grand Canyon to the north and are one of the greatest natural wonders on the continent. It is a breathtaking system of canyons in the Sierra Madre Occidental, home of the Tarahumara Indians. Up to today, there is only one railroad line that crosses the region of the six canyons. You should undertake the hike only with a guide; without them, you would not be able to find the few springs along the way. This demanding trek leads through the narrow canyons, and the many ascents and descents of more than 3,281 vertical ft. (1,000 m) will drain your strength. While a wintry wind is blowing above, bananas and citrus fruits grow on the canyon floors! The hike is a great nature experience, with no tourism at all.

Time it takes: 5–10 days
Starting point: Creel
Hire a guide: in Creel or Batopilas
Information: www.visitmexico.com/es/destinosprincipales/ chihuahua/barrancas-del-cobre

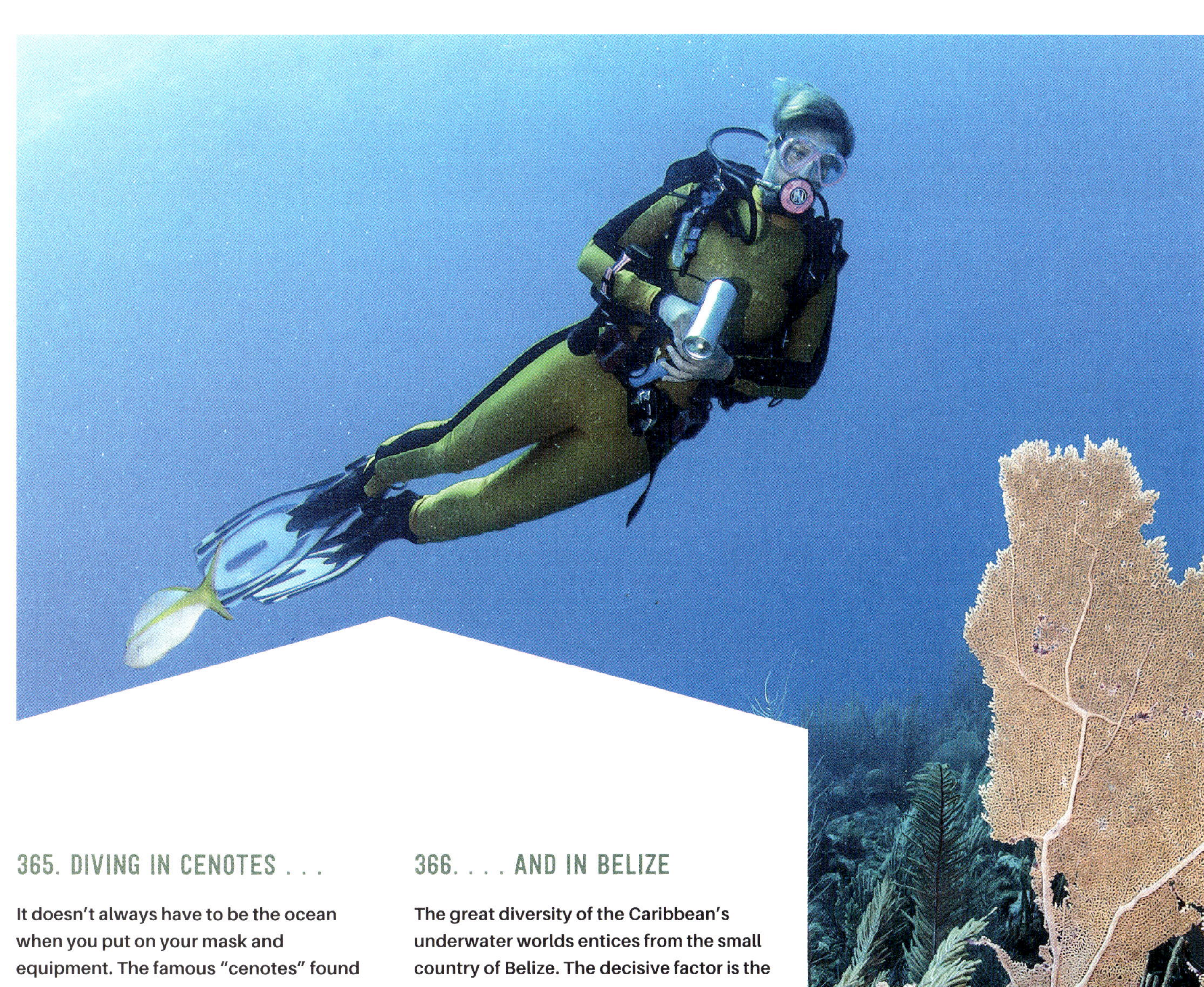

365. DIVING IN CENOTES . . .

It doesn't always have to be the ocean when you put on your mask and equipment. The famous "cenotes" found on the Yucatán Peninsula are some very special diving grounds. These cenotes are huge, doline (sinkhole-like), freshwater reservoirs formed millions of years ago in limestone areas. Some of them are connected to underground cave and river systems. This offers a completely different thrill than snorkeling in the sea, where you should always expect animal companions. During Mayan times, some of the cenotes were used as sites for ritual sacrifices where people were drowned, but fortunately that was a long time ago. No one cenote is like any other; tour operators take visitors into Chichén Itzá, Chac Mool, Dos Ojos, Calavera, and many others. Visibility in the crystal-clear water can be more than 492 ft. (150 m).

Information: for example, www.playadivingcenter.com, www.mexicobluedream.com

366. . . . AND IN BELIZE

The great diversity of the Caribbean's underwater worlds entices from the small country of Belize. The decisive factor is the Belize Barrier Reef, the second-largest continuous reef system in the world after the Australian Great Barrier Reef. It is lined with atolls, and the coral gardens with dozens of different species are especially fascinating. Divers praise the good to very good visibility and rave about the varied fauna in this submarine paradise. The quarry of your underwater watching ranges from sea turtles to moray eels and nurse sharks to the real giants: whale sharks. Popular spots include Caye Caulker Marine Reserve and Hol Chan Marine Reserve. The Great Blue Hole, with its worlds of caves with stalagmites and stalactites, is a real stunner and the dream of every diver.

Information: for example, belizedivingservices.com, www.frenchiesdivingbelize.com

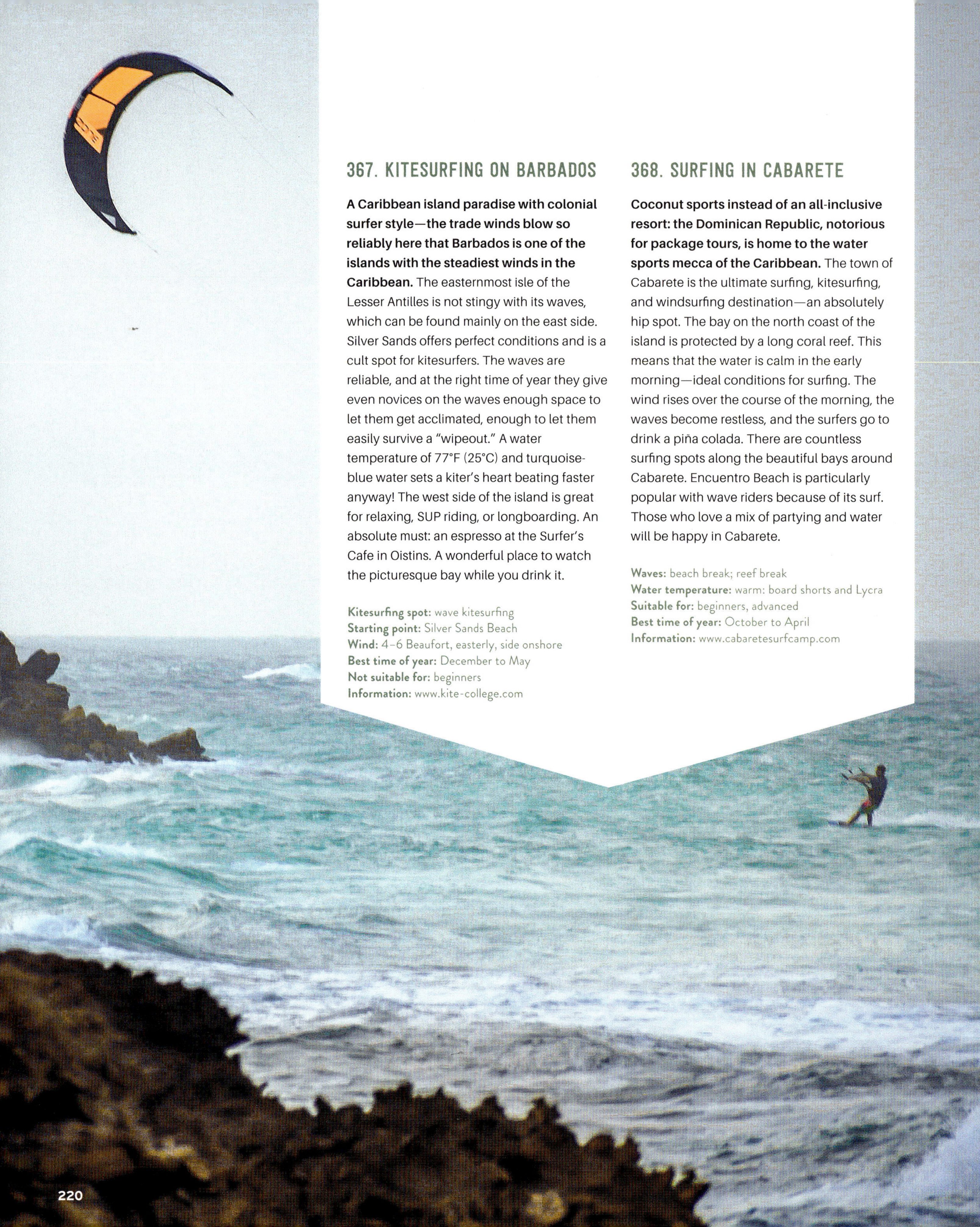

367. KITESURFING ON BARBADOS

A Caribbean island paradise with colonial surfer style—the trade winds blow so reliably here that Barbados is one of the islands with the steadiest winds in the Caribbean. The easternmost isle of the Lesser Antilles is not stingy with its waves, which can be found mainly on the east side. Silver Sands offers perfect conditions and is a cult spot for kitesurfers. The waves are reliable, and at the right time of year they give even novices on the waves enough space to let them get acclimated, enough to let them easily survive a "wipeout." A water temperature of 77°F (25°C) and turquoise-blue water sets a kiter's heart beating faster anyway! The west side of the island is great for relaxing, SUP riding, or longboarding. An absolute must: an espresso at the Surfer's Cafe in Oistins. A wonderful place to watch the picturesque bay while you drink it.

Kitesurfing spot: wave kitesurfing
Starting point: Silver Sands Beach
Wind: 4–6 Beaufort, easterly, side onshore
Best time of year: December to May
Not suitable for: beginners
Information: www.kite-college.com

368. SURFING IN CABARETE

Coconut sports instead of an all-inclusive resort: the Dominican Republic, notorious for package tours, is home to the water sports mecca of the Caribbean. The town of Cabarete is the ultimate surfing, kitesurfing, and windsurfing destination—an absolutely hip spot. The bay on the north coast of the island is protected by a long coral reef. This means that the water is calm in the early morning—ideal conditions for surfing. The wind rises over the course of the morning, the waves become restless, and the surfers go to drink a piña colada. There are countless surfing spots along the beautiful bays around Cabarete. Encuentro Beach is particularly popular with wave riders because of its surf. Those who love a mix of partying and water will be happy in Cabarete.

Waves: beach break; reef break
Water temperature: warm: board shorts and Lycra
Suitable for: beginners, advanced
Best time of year: October to April
Information: www.cabaretesurfcamp.com

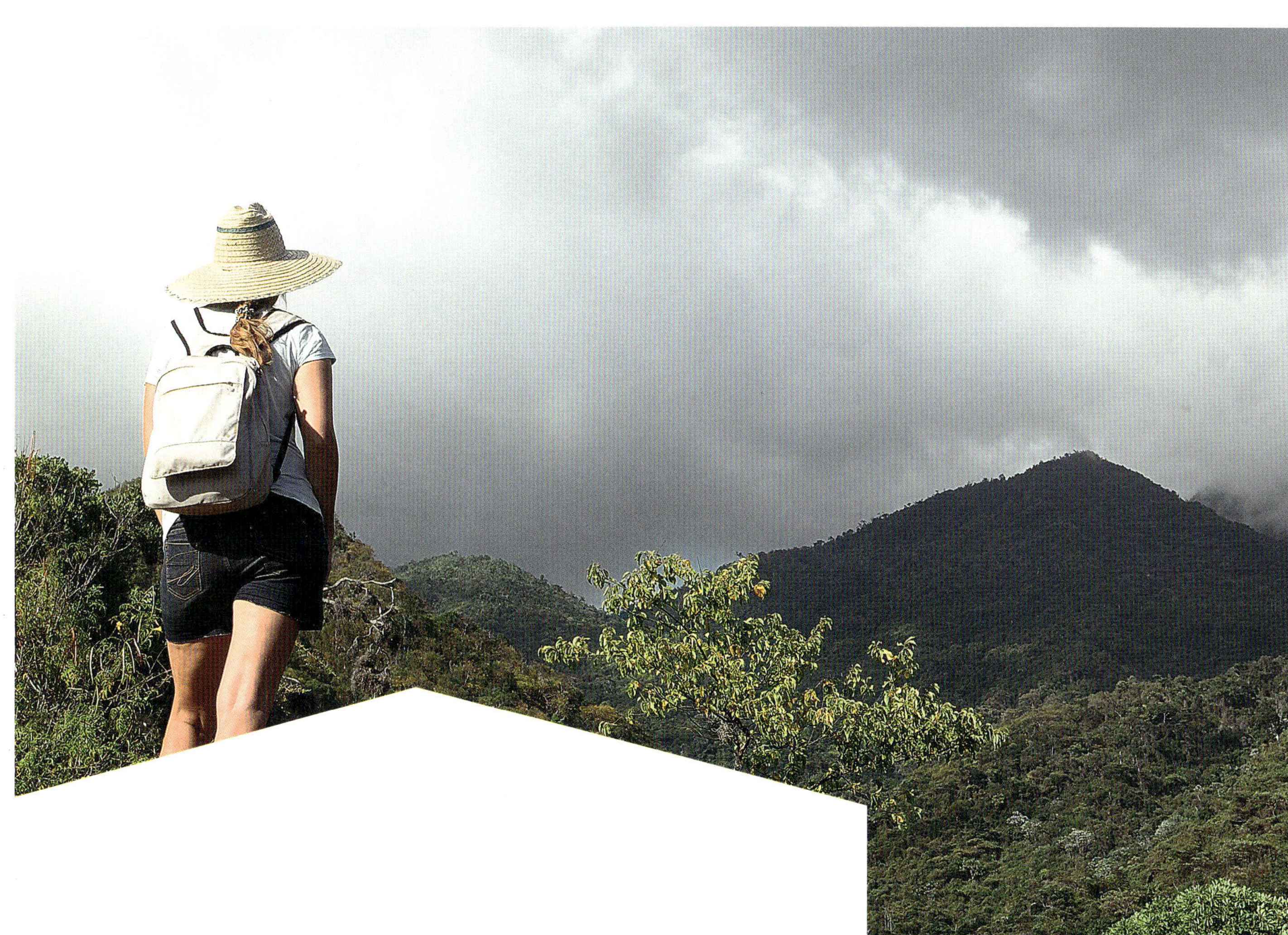

369. MTB TOUR OF CUBA

In the footsteps of Fidel and Che: through Cuba by mountain bike. The Cubans snippily call their "Made by socialism" bicycles "asthmáticas," because these sluggish workhorses gasp like an asthma sufferer even at the lowest speed. When mountain biking, you can traverse one or the other route together with the sociable (and panting) Cubans, chatting and laughing. Then you step on the gas again and experience Cuba's optimal mixture of athletic activity, nature, and culture; the Caribbean and joie de vivre; pulsating cities and quaint villages. Cuba has it all, and by bike is a perfect way to discover it on your own, either as a bike trip or MTB tour. Both are great.

Time it takes: day trip to up to 14 days
Tip: Ride an MTB in the wild east of Cuba, the Sierra Maestra
Starting point: Santiago de Cuba
Information: www.rad-reise-service.de, www.bike-adventure-tours.ch

370. TREK TO PICO REAL TURQUINO

Many people have already gone into rapture while on a mountain hike from Las Cuevas to Santo Domingo, not least among them the "discoverer" of America, Christopher Columbus. He waxed euphoric when he landed in Cuba in 1492: "This island is probably the most beautiful that human eyes have ever seen," he enthused. Mojito cocktails, Havana cigars, fine vintage cars, and white beaches—these are all familiar to us from Cuba, but the sugar cane island has much more to offer. It is the isle with the most diverse landscape in the Caribbean! During a trek through the beautiful, wooded Sierra Maestra, you can climb Pico Turquino (6,476 ft. [1,974 m]), the highest mountain on the island, on easy trails and admire an abundance of endemic plants and rare animal species during the hike.

Time it takes: 3 days
Highest elevation: 6,476 ft. (1,974 m)
Elevation difference: about 8,202 vertical ft. (2,500 m)

371. HIKING WITH A VIEW OF PACAYA ACTIVE VOLCANO

Pacaya, which lies south of Guatemala City, is quite simply one of the most active volcanoes in the world. In hundreds of daily eruptions, it hurls glowing lava bombs, slag, and ash several hundred yards into the air. These eruptions are called Strombolian activity, after the southern Italian volcano for which this type of eruption is typical. The magical spectacle is particularly effective at night, when the red-hot rays of fire stand out against the dark night sky—a sight that truly takes your breath away. Of course, all this has an onomatopoeic accompaniment: you can't ignore the rumbling thunder of the outbursts, sometimes every minute. A hike leads to the neighboring summit, which is the best place to observe and photograph the spectacle. Be sure to take a tripod with you!

Length / time it takes: 3.73 mi. (6 km) / 5 hours
Highest elevation: 8,399 ft. (2,560 m)
Starting point: San Vicente Pacaya
Elevation difference: about vertical 2,625 ft. (800 m)
Information: www.guatemalanadventure.com

372. TREKKING BY LAGO DE ATITLÁN

There are moments and places in life that will forever be etched in your memory. The first view of Lago de Atitlán in Guatemala is one of these. The deep, azure-blue lake has a water surface of 50 sq. mi. (130 km^2) and is framed by three mighty volcanoes. In the early morning, the surrounding green and fertile slopes and the ancient Mayan villages slowly and mystically emerge from the mist. Lago de Atitlán lies in a caldera, so it is a crater lake and is considered one of the wonders of our world. Even Alexander von Humboldt once called the lake "the most beautiful lake in the world." A trekking tour around the lake provides fantastic insights into traditional Mayan culture and the rural lakeside life. A place for arriving and never wanting to leave again.

Length / time it takes: 42 mi. (67 km) / 3 days
Highest elevation: 5,118 ft. (1,560 m)
Starting point: Panajachel
Information: www.guatemalanadventure.com

373. SANDBOARDING ON CERRO NEGRO VOLCANO

If you are setting out for Cerro Negro (the "Black Mountain"), you have something special in mind: sandboarding on volcanic ash! The 2,388 ft. high (728 m) volcano rises not far from the city of León. Once there, you have to drag the board to the summit yourself. It takes an hour, over dust and lava rubble. "When you get to the edge it will seem steep, extremely steep," says tour guide Marjiory about your imminent arrival above the western flank, where you are to whoosh down in a "schuss" downhill run. There, on the edge of the abyss, your heart slips into the protective suit. Scarf over your mouth, protective goggles on, and in sitting position on the back of the board. It picks up speed, and you can barely restrain it during the ride. Lava particles blast away all around, and there is a constant clacking on your glasses. Finally it comes to an end. Your face is black and your hair feels like straw—an unforgettable experience of an unusual kind.

Information: for example, www.nicaraguantrails.com

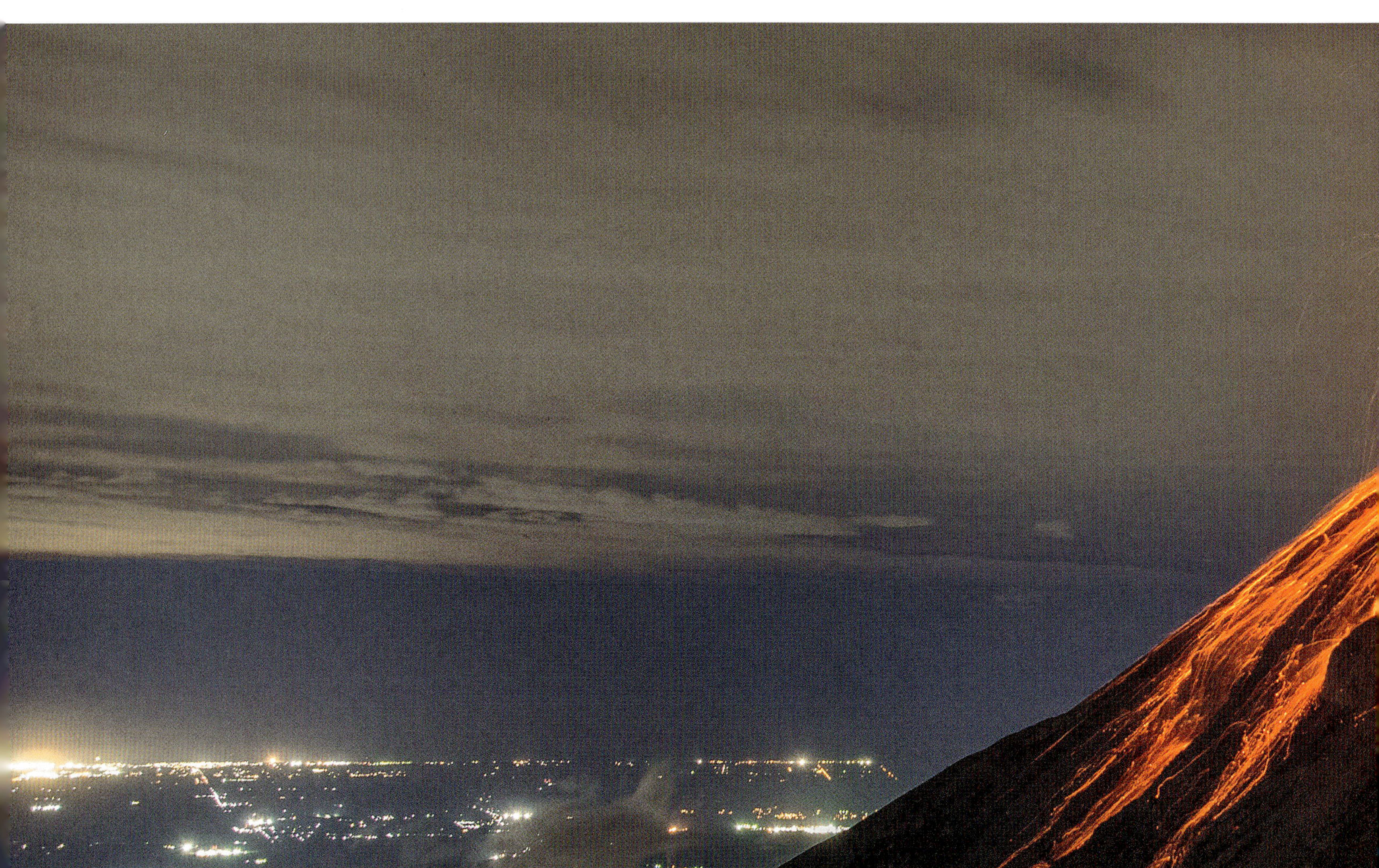

374. ON A TRAIL THROUGH MOMBACHO VOLCANO NATURE RESERVE

Nicaragua, the land between the Pacific and the Caribbean, is also the land of volcanoes. The short but intensive loop around the main crater of extinct Mombacho volcano provides a good impression of this. The volcano humps up south of the colonial city of Granada to well more than 4,265 ft. (1,300 m) high. The area is a nature conservancy, and the high elevations are covered with cloud forest. And sections of this easily accessible trail lead right through this forest. Trees are covered with epiphyte plants; the air smells of sulfur. There are bromeliads in abundance, yard-high ferns. Then it opens up a bit to reveal fantastic views of Lake Nicaragua and the Isletas de Granada archipelago. Incidentally, two other trails through the green of the Volcán Mombacho Nature Reserve, the El Puma and El Tigrillo trails, are longer and more demanding.

Time it takes: maximum of 1 hour with photo stops
Information: www.mombacho.org

375. RAFTING ON THE RÍO PACUARE

Outdoor magazines have already rated Río Pacuare, in the eastern part of Costa Rica, as one of the top rafting destinations in Latin America. It is a relatively short river that has it all. It rises on an elevation of about 9,842 ft. (3,000 m) in the Cordillera de Talamanca mountain range and makes its way toward the Caribbean. Rafting on the Pacuare is an ideal opportunity to experience the natural beauties of Costa Rica from a different angle: in between the rapids and smoother, gentle passages. Guaranteed are that it will cool you down, have a high thrill factor, and help you shed calories due to the vigorous paddling required to help the boat make it through foaming passages among rocks and walls of water. Tour operators offer one-day trips, but also two-day packages for those who can't get enough.

Information: www.ticoriver.com

376. ZIPLINING ON ARENAL VOLCANO

This zipline adventure within sight of the giant Arenal volcano in Costa Rica is not for those who are faint of heart or on a tight budget. The schuss runs on the static ropes are up to 2,461 ft. (750 m) long. First, a jungle cable car takes you up so you can really gain altitude. Guide Brandon instructs inexperienced participants: keep your body consistently in a lying (supine) position along the way, first cross your legs, then toward the end spread your legs apart and move the handlebar back and forth to slow down. The course is made up of seven routes, high above the rainforest canopy and over two deep-cut valleys, up to 656 ft. (200 m) above sea level. The most-spectacular sections are called "Big Daddy" and "Big Mama." There are mechanical hisses and buzzes. This is where you reach the maximum speed rush of 43.5 mph (70 km/h).

Information: skyadventures.travel

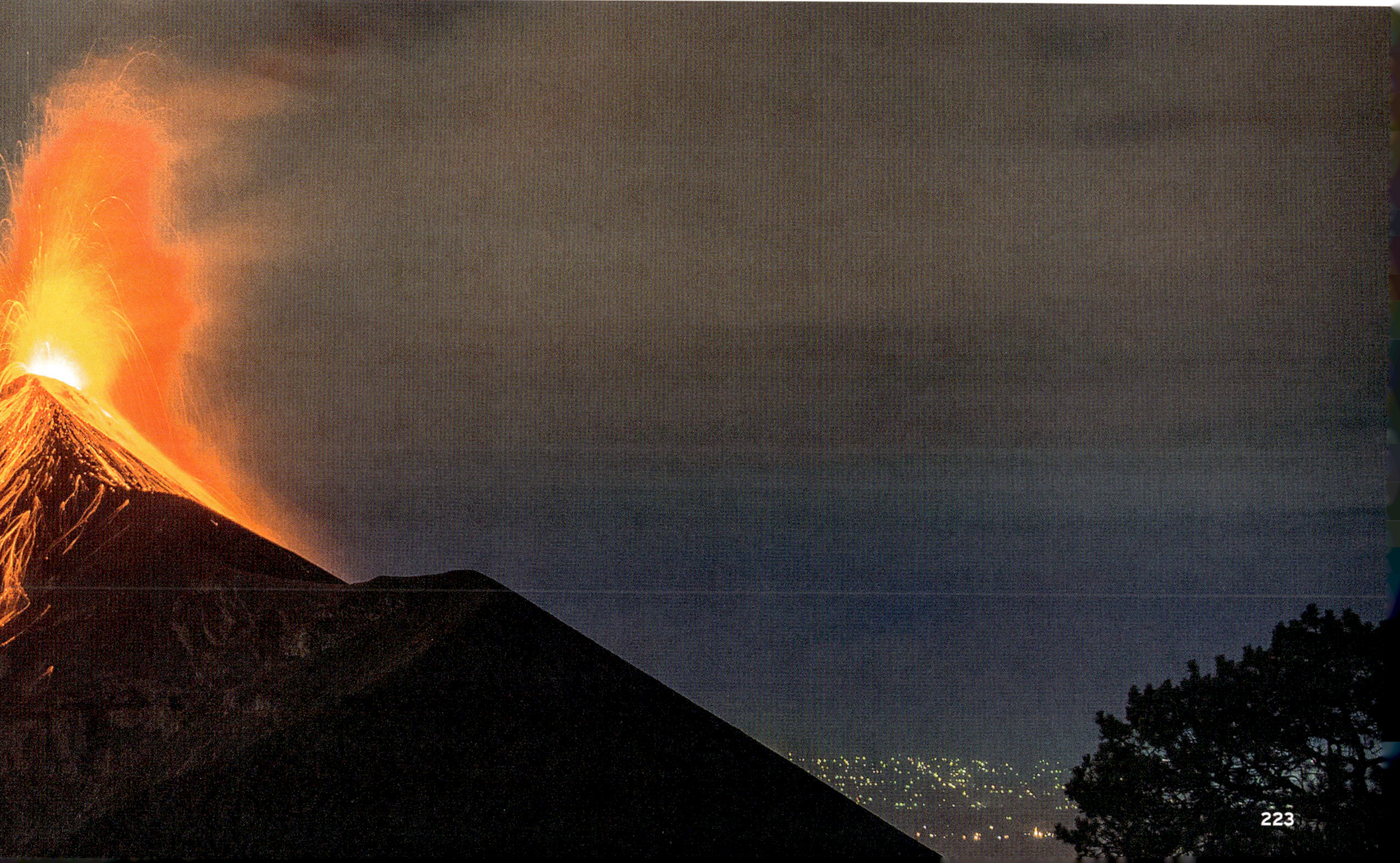

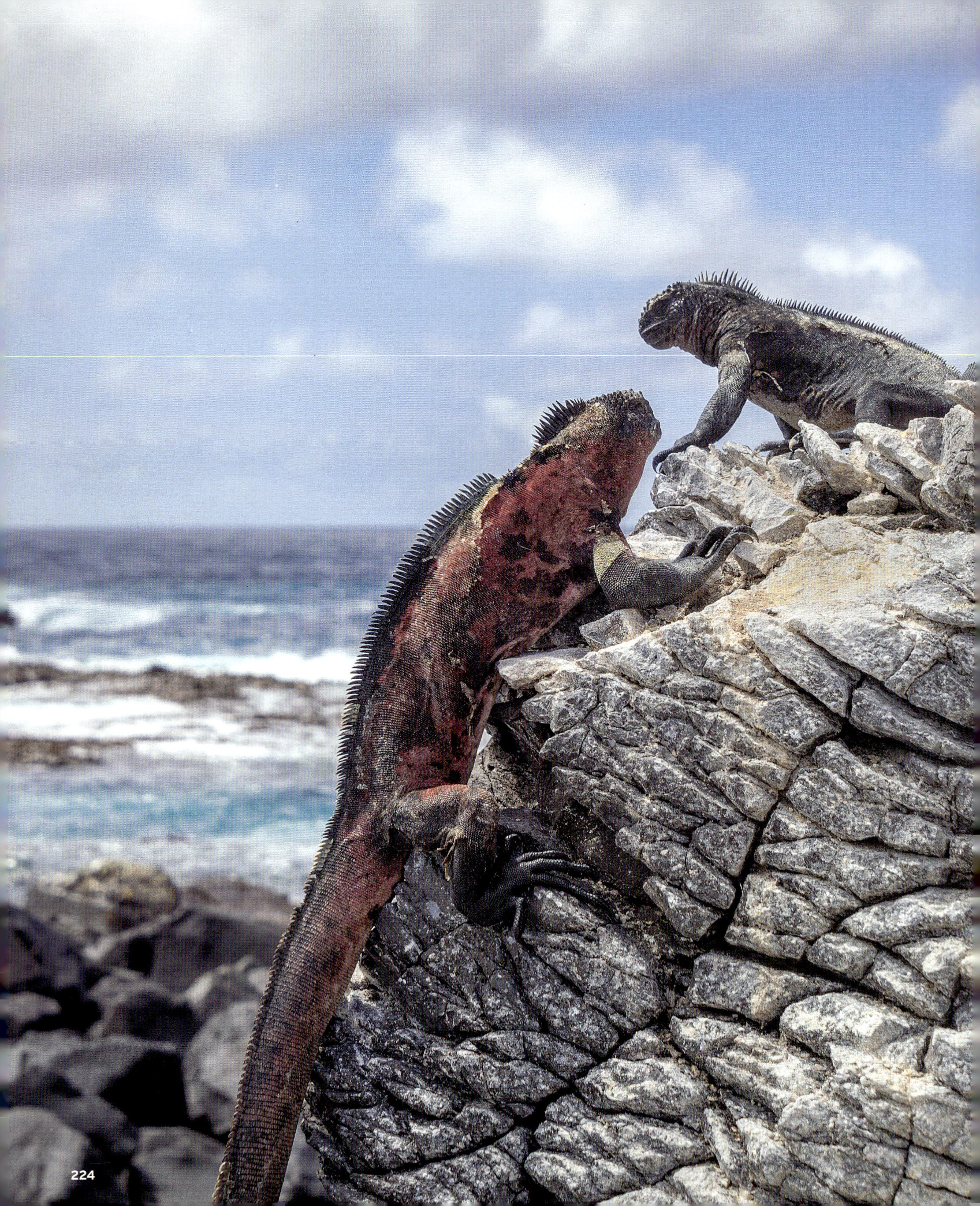

377. RAFTING AND KAYAKING IN TENA IN THE AMAZON RIVER BASIN

This tropical water playground is in the middle of the rainforest: the enormous density of rivers of all levels of difficulty around the town of Tena attracts kayakers and rafters from all over the world. In the evergreen rainforest, where the Andes merge into the Amazon, it is a perfect place to experience jungle and whitewater together perfectly—and in pleasantly warm water. Two rafting and kayak trips are particularly recommended: on the Río Jatunayacu, one of the headwaters of the Amazon, you can take a relaxed, easy trip in the wide-open valley, which is ideal for getting to know each other and paddling along. Then there is the combination of the Río Jondachi and the Río Hollín: an authentic and exotic whitewater adventure in a class of its own, where man and woman really have to actively paddle together.

Length / time it takes: each about 16 mi. (25 km) / 4–6 hours
Level of difficulty: WW 1–2/3–4
Starting point: Tena, Napo Province
Best time of year: November to January
Information: www.ecuador-kajak.com

378. WILDLIFE WATCHING ON THE ISLA DE LA PLATA

The "Galápagos for the poor": those who do not want or cannot afford the more expensive trip to Ecuador's Galápagos Islands have an alternative. On the Isla de la Plata, translated as Silver Island, which is part of the Machalilla National Park, you can get an idea of what it feels like to be on the real Galápagos: the birds and animals are close enough to touch and not afraid of people. Even with a wide-angle lens, you can get a full-frame shot of a blue-footed booby or a turtle. The boat trip from the mainland takes about an hour, and on one of the loop hikes around the island you can discover many of the birds' and animals' breeding grounds, nests, and caves. Nevertheless—with all the natural beauty of the Isla de la Plata—the fascination and unique biodiversity of the real Galápagos Islands remains above any of the competition.

Length / time it takes: hike: various loop trails of a total of 12.5 mi. (20 km) / 3 hours
Length / time it takes: boat trip: 14 mi. (23 km) one way / 1 hour
Elevation difference: 104 vertical ft. (167 m)
Starting point: Puerto Lopez
Information: www.visitpuertolopez.com

379. YASUNÍ WILDERNESS WATER TRAIL ON THE RÍO CURARAY

Yasuní National Park, in the Amazon basin of Ecuador, is one of the most biodiverse parks in the world; both jungle fauna and flora are represented here in abundance. The Río Curaray river marks the southern border of the protected region, declared a biosphere reserve by UNESCO. During an expedition in a dugout canoe from the village of Curaray (airstrip for small Cessnas) down to the Peruvian border, you dive into a wilderness—with no safety net. The river is home to caimans, anacondas, and piranhas, and the rainforest is home to everything you could imagine in your wildest jungle dreams. The Huaorani indigenous people also live here; they refuse any contact with the outside world and also make this clear by using spears or curare-poisoned arrows from their blowguns.

Length / time it takes: 149 mi. (240 km) / 10 days
Elevation difference: 295 vertical ft. (90 m)
Starting point: Shell/Puyo airport, flight to Curaray
Information: www.bilder-botschaften.de

380. SUP RIDING AND SURFING IN THE PACIFIC

Surf spot after surf spot line up like pearls along the Ruta del Sol, Ecuador's "road of the sun." The tropical west coast of this South American country on the Pacific attracts surfers and stand-up paddlers from all over the world. The great waves, the gently swaying palm trees, the pleasantly temperate water, the beach parties, and the cool drinks—these are the ingredients for an all-around successful vacation. Spots such as Montanita, Ayampe, La Canoa, and around Manta (San Mateo, San Lorenzo, Santa Marianita) are well known in surfing circles. The insider tip is Mompiche, a sleepy fishing village, with waves for professionals and for beginners and a potpourri of beautiful beaches. After the fun in the water, fortify yourself with a typical Ecuadorian dish: "pescado encocado," fresh fish in coconut sauce.

Waves: beach break, reef break
Water temperature: warm: board shorts and Lycra will do
Suitable for: beginners, advanced
Best time of year: July to December
Information: www.ecuador-kajak.com

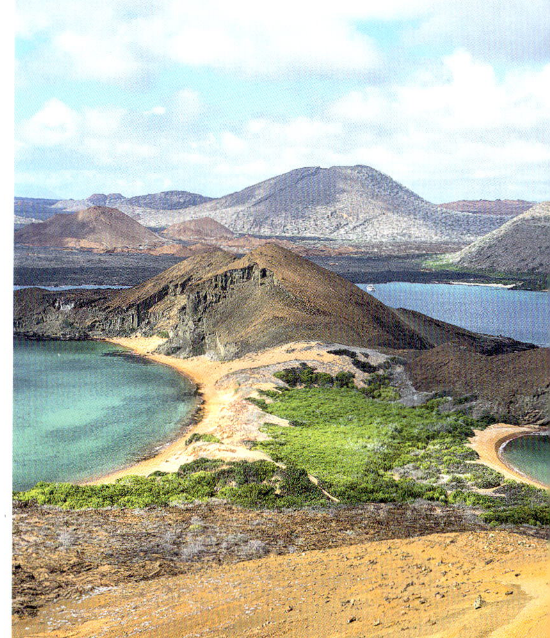

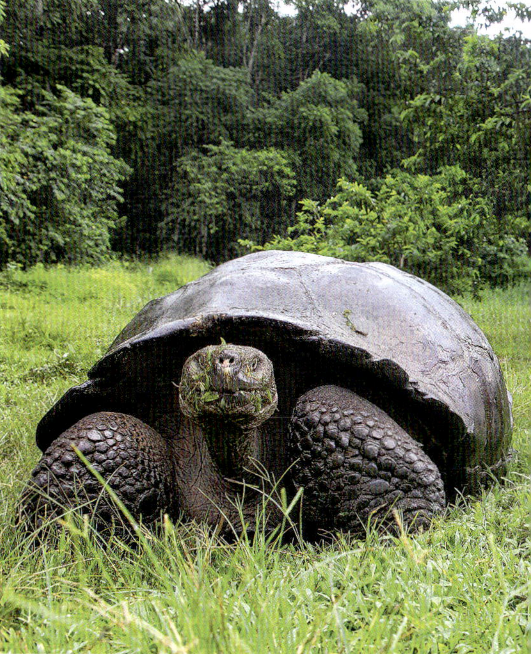

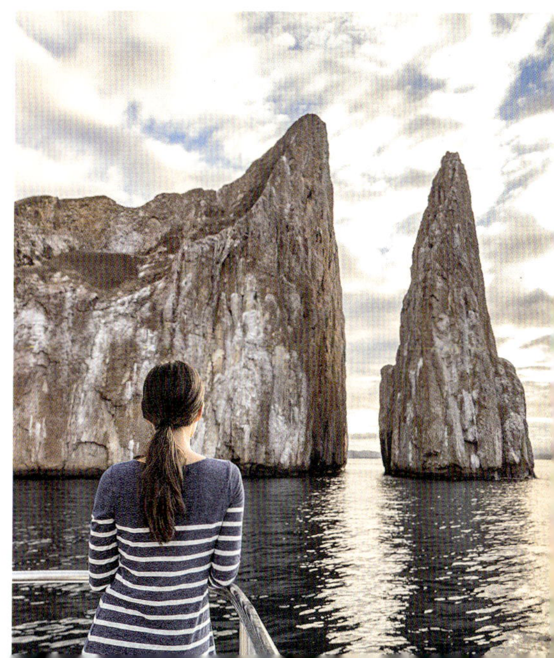

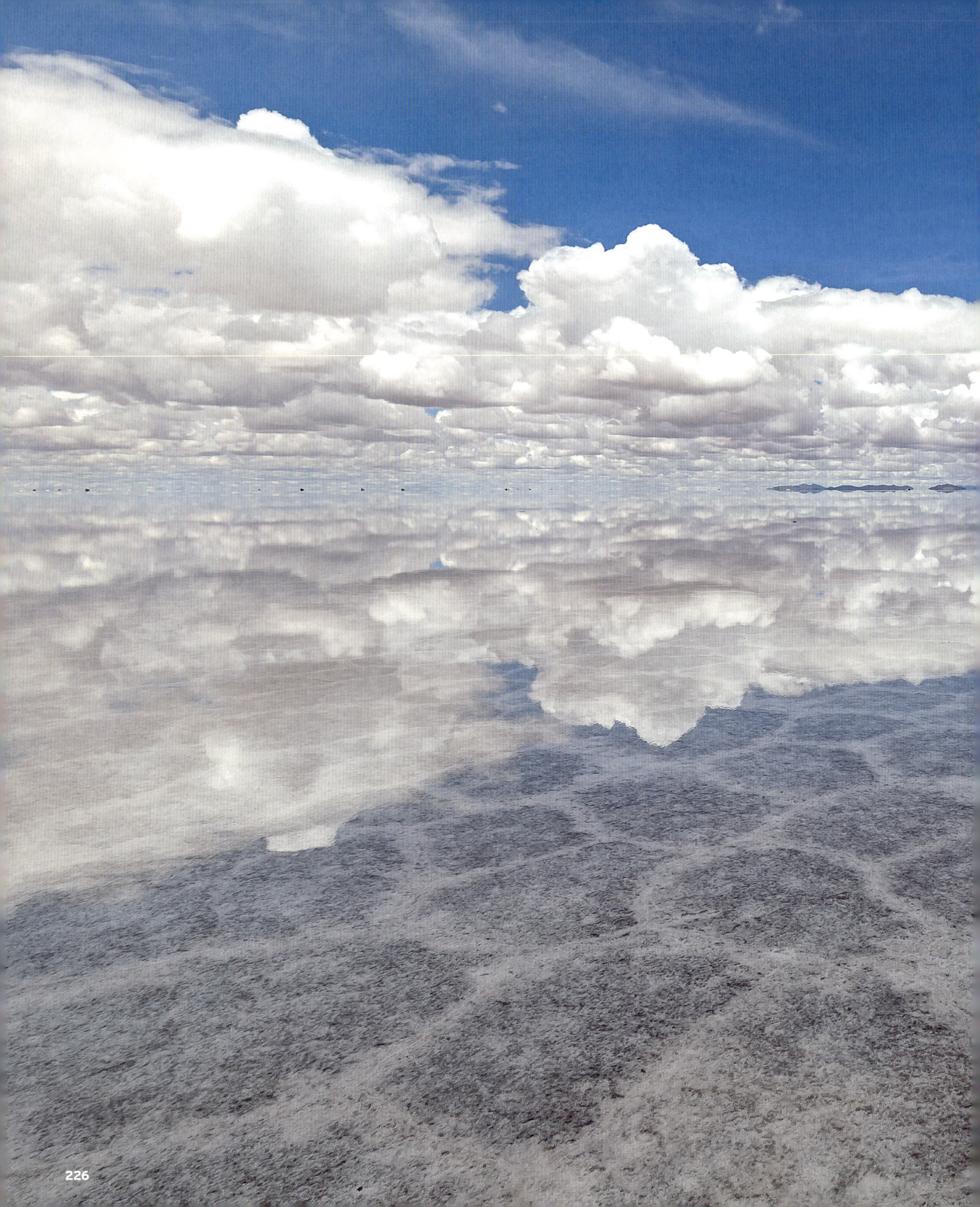

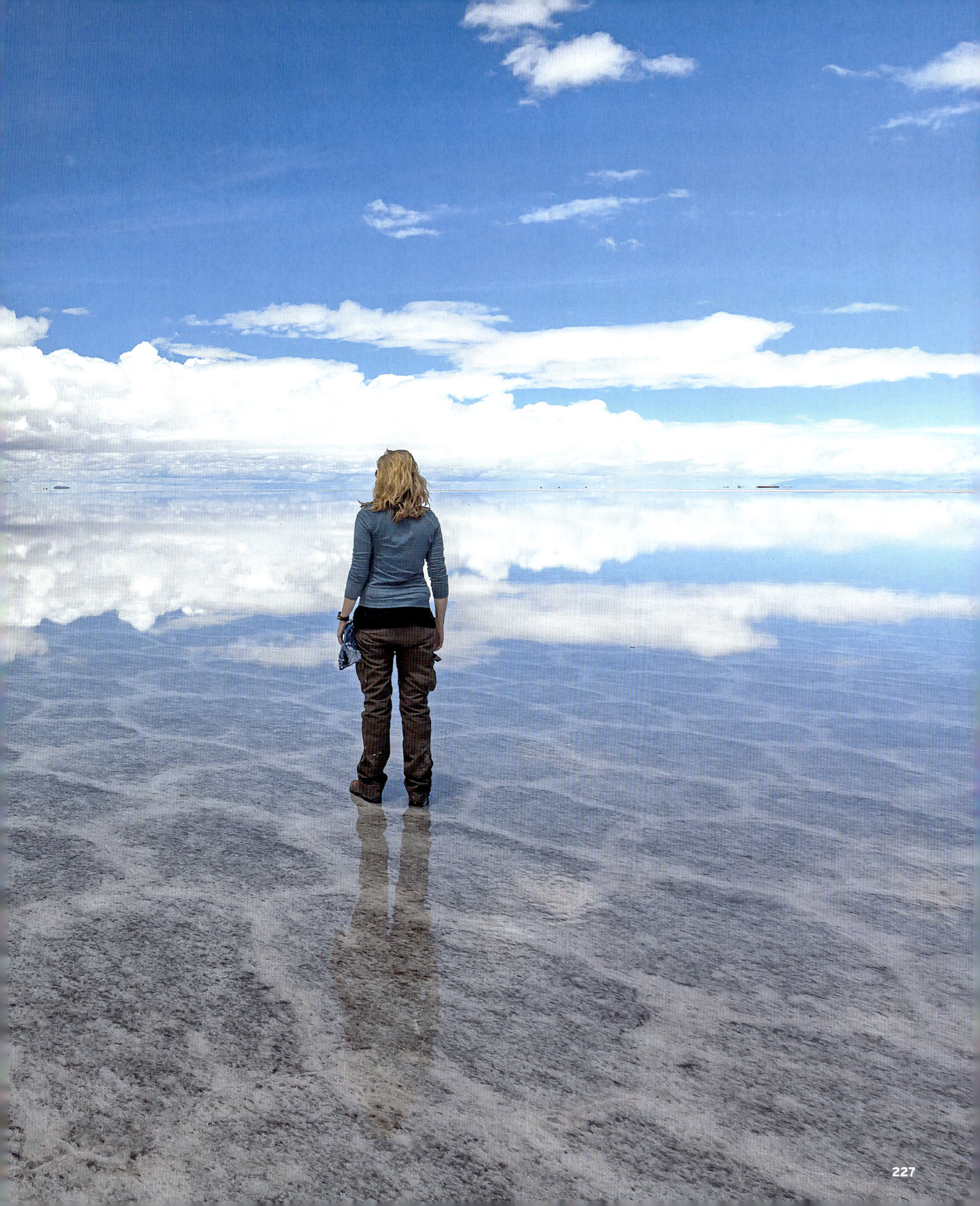

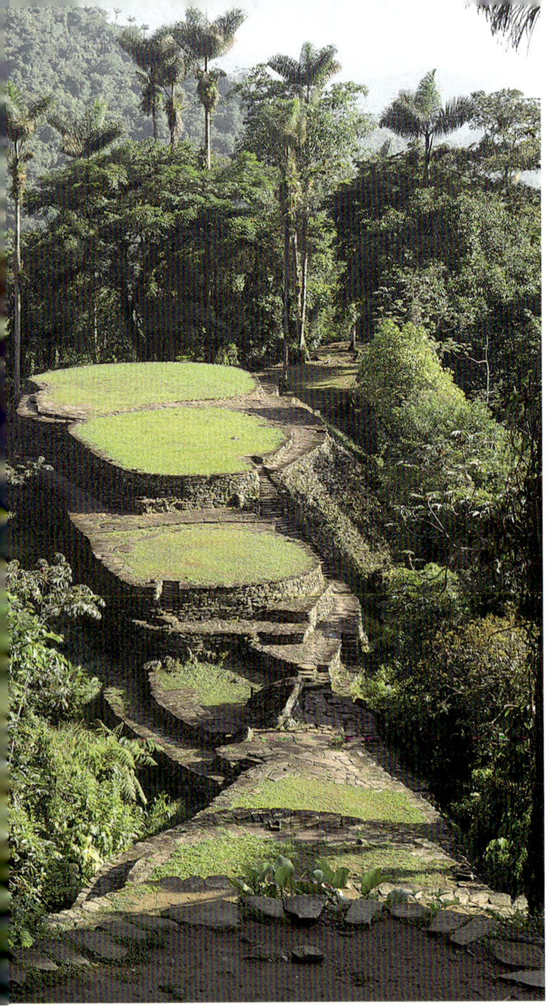

381. RAFTING ON THE RÍO FONCE

In the east-central part of Colombia lies San Gil, a lively, friendly town where the Río Fonce river flows past—and it has potential as rafting terrain. Guide Rodney prepares his group on the bank—including for the "washboiler," which appears soon after casting off. Then you suddenly think, "How on earth do you get through here?," and are already engulfed by the masses of water. You paddle like you're trying to cheat death. Other rapids await you later, then the drama subsides. Then there is time to jump into the brown river and drift down with your life jacket still on while Rodney follows with the boat. Shortly before arriving at San Gil, you are spinning around really comfortably. Tours are also available on the Río Suarez and Río Chicamocha.

Information: www.colombiarafting.com

382. HIKING IN TAYRONA NATIONAL PARK

Beyond Calabazo, an entry point for Tayrona National Park, streams and leafcutter ants will soon cross your path. Butterflies are dancing around and birds are chirping. Natural windows among the plants provide an open view into the hinterlands to the mountains. The trail climbs through dense vegetation; giant trees tower up an estimated 131 ft. (40 m). Pueblito, a historical settlement of the indigenous Kogi people, is a stopover. Later, the steep trail winds downward over slabs of rock and tangled roots while the sun casts a grid of shadows through the trees. A bright-blue morpho butterfly flutters past, and lizards disappear into crevices in the rock. Now the surf can be heard and the Caribbean shore appears. At Cape San Juan del Guía, hikers can then happily plunge into the water. Pelicans hover in the steel-blue sky.

Length / time it takes: 6.2 mi. (10 km) / half a day

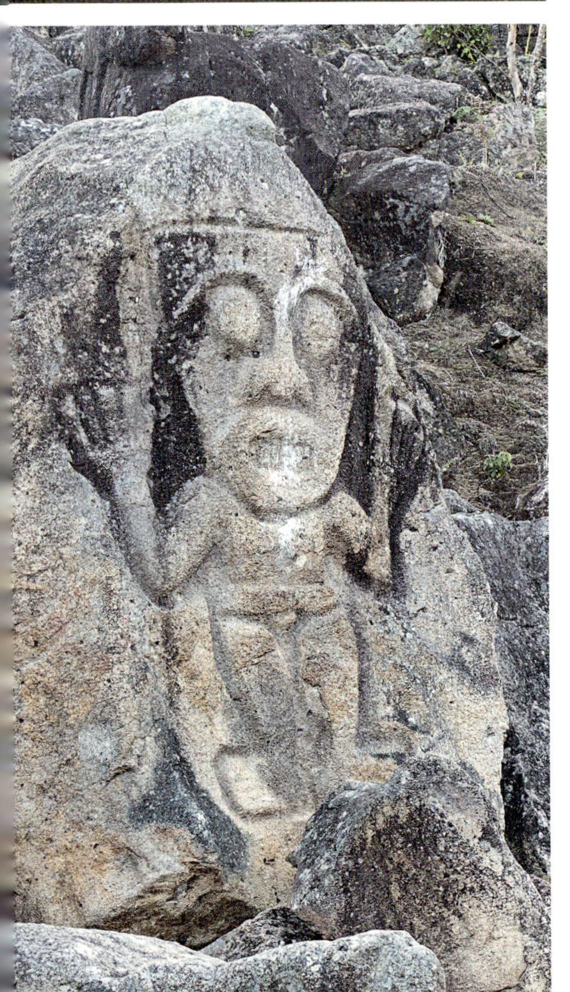

383. MOUNTAIN HIKE TO THE TIERRADENTRO TOMBS

For the conquistadors, Tierradentro meant "far inland" and deeply hidden. It has stayed that way. That makes a visit to this UNESCO World Heritage Site all the more attractive. These painted shaft chamber tombs of the long-vanished Tierradentro culture in southern Colombia are unparalleled in Latin America. Most of the caves are easily accessible near the town of San Andrés de Pisimbalá, but those on the El Aguacate ridge are the most difficult to access in the Tierradentro Archaeological Park. What starts on trail passages going by huts and slopes covered with coffee bushes ends in solitude in the mountains with a fantastic 360-degree panorama. Some of the entrances to the long-plundered tomb caves have fallen into disrepair, but the "salamander grave," where salamander designs creep over the walls, is a highlight.

Time it takes: at least 2.5 hours for there and back

384. TREKKING TO CIUDAD PERDIDA

The name alone arouses curiosity and exerts an irresistible attraction: Ciudad Perdida, Colombia's "lost city" in the northern part of the country. The area where these pre-Columbian ruins lie amid the green of the Sierra Nevada de Santa Marta range has no access to the road network. Getting there requires a grueling, four-day trek that involves sacrificing comfort. Nature and of course the site itself compensate for this. It was built by the Tayrona, representatives of an important civilization and ancestors of the current Kogi. According to sources, the origins of the "lost city" go back to around 700 CE. It was built during the sixteenth century, shortly after the invasion by the conquistadors. The extensive area was subsequently buried by plant life and was rediscovered only in the 1970s.

Information: expotur-eco.com

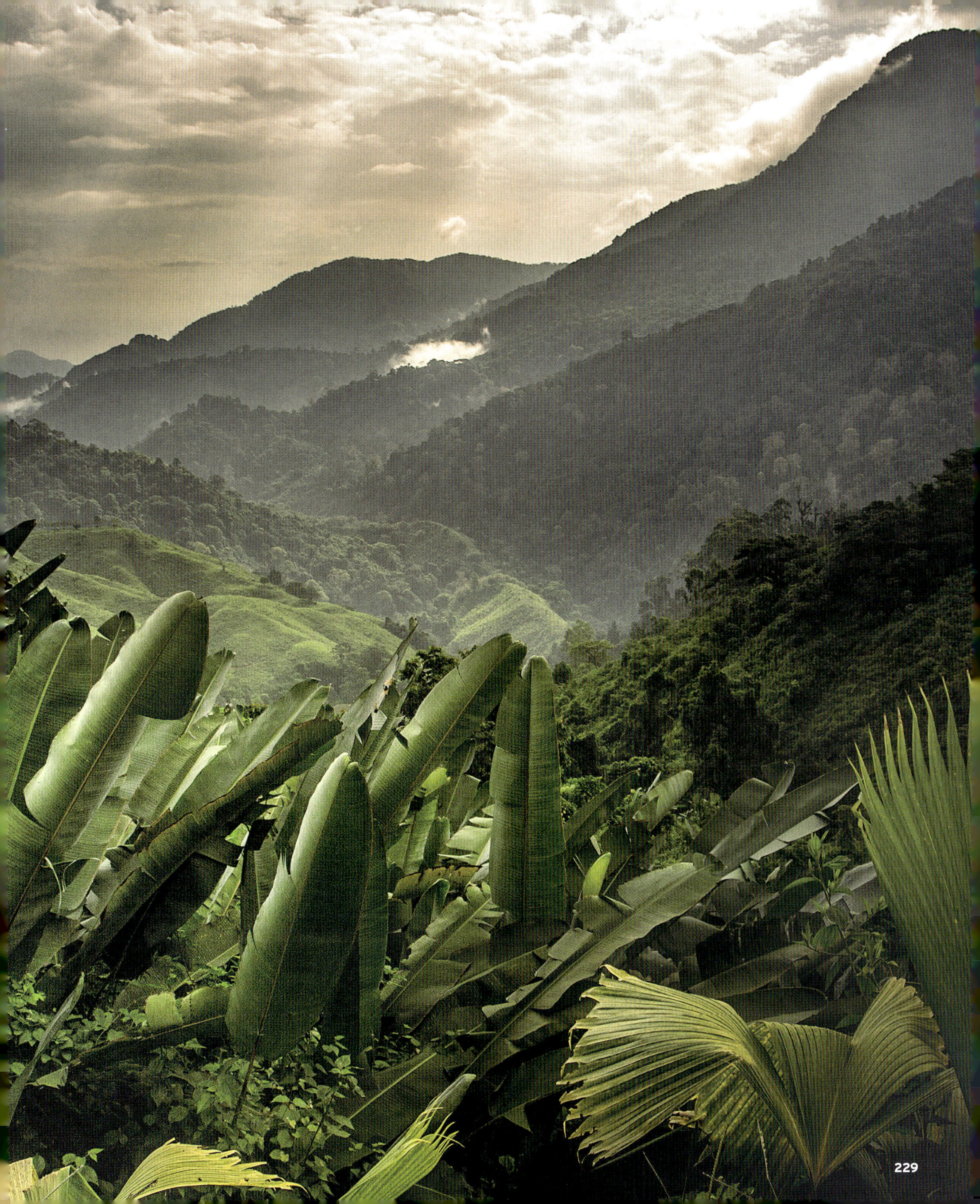

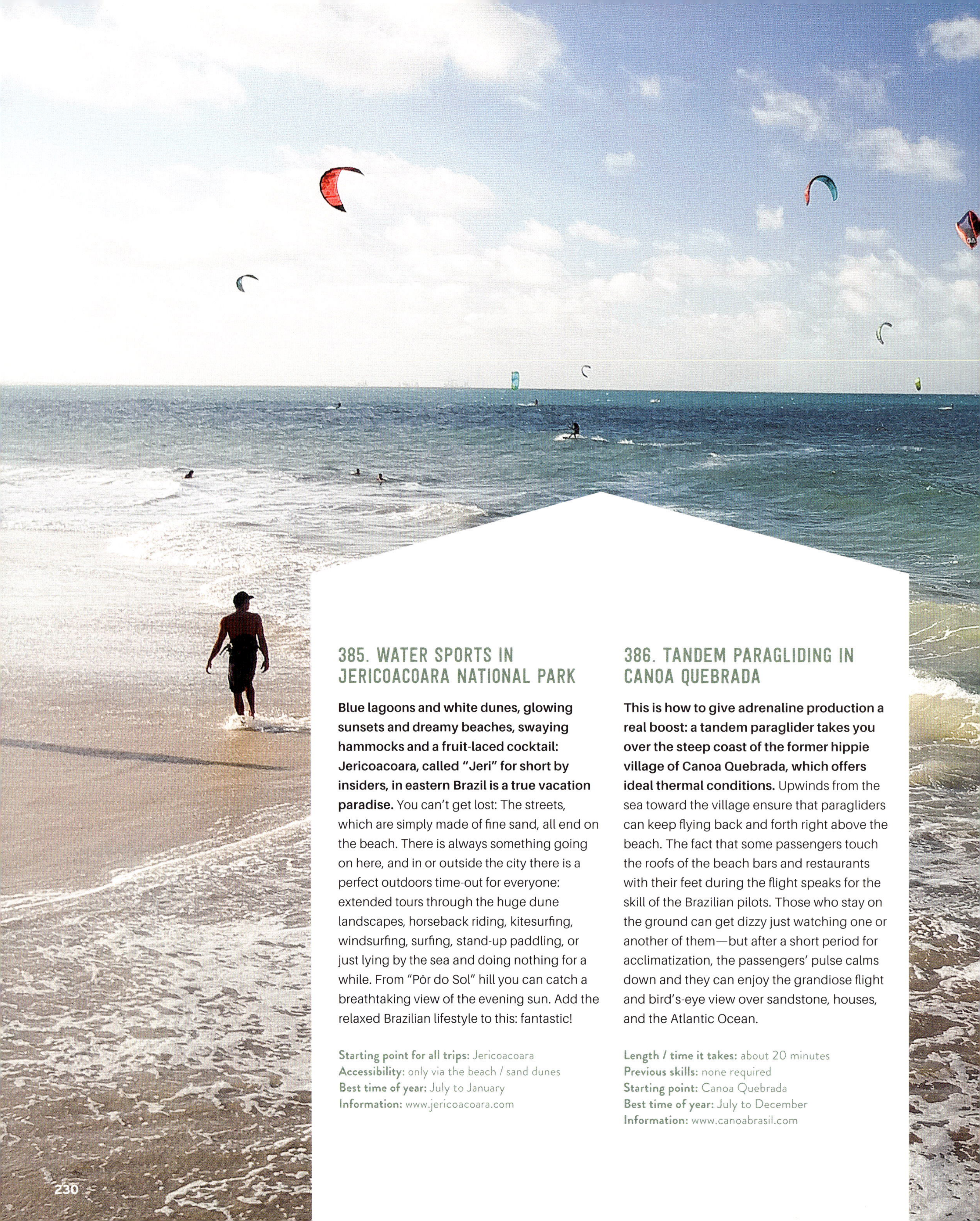

385. WATER SPORTS IN JERICOACOARA NATIONAL PARK

Blue lagoons and white dunes, glowing sunsets and dreamy beaches, swaying hammocks and a fruit-laced cocktail: Jericoacoara, called "Jeri" for short by insiders, in eastern Brazil is a true vacation paradise. You can't get lost: The streets, which are simply made of fine sand, all end on the beach. There is always something going on here, and in or outside the city there is a perfect outdoors time-out for everyone: extended tours through the huge dune landscapes, horseback riding, kitesurfing, windsurfing, surfing, stand-up paddling, or just lying by the sea and doing nothing for a while. From "Pôr do Sol" hill you can catch a breathtaking view of the evening sun. Add the relaxed Brazilian lifestyle to this: fantastic!

Starting point for all trips: Jericoacoara
Accessibility: only via the beach / sand dunes
Best time of year: July to January
Information: www.jericoacoara.com

386. TANDEM PARAGLIDING IN CANOA QUEBRADA

This is how to give adrenaline production a real boost: a tandem paraglider takes you over the steep coast of the former hippie village of Canoa Quebrada, which offers ideal thermal conditions. Upwinds from the sea toward the village ensure that paragliders can keep flying back and forth right above the beach. The fact that some passengers touch the roofs of the beach bars and restaurants with their feet during the flight speaks for the skill of the Brazilian pilots. Those who stay on the ground can get dizzy just watching one or another of them—but after a short period for acclimatization, the passengers' pulse calms down and they can enjoy the grandiose flight and bird's-eye view over sandstone, houses, and the Atlantic Ocean.

Length / time it takes: about 20 minutes
Previous skills: none required
Starting point: Canoa Quebrada
Best time of year: July to December
Information: www.canoabrasil.com

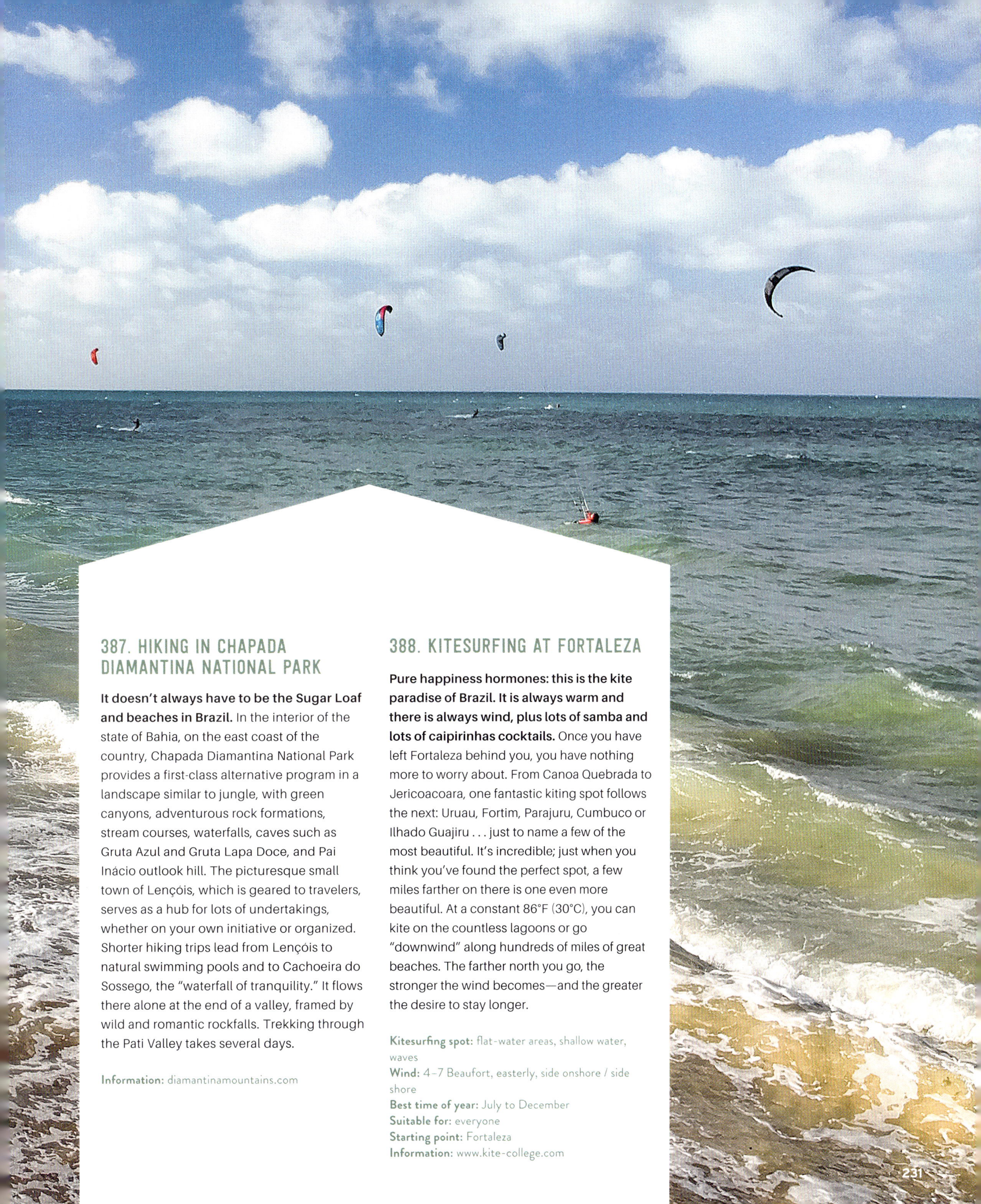

387. HIKING IN CHAPADA DIAMANTINA NATIONAL PARK

It doesn't always have to be the Sugar Loaf and beaches in Brazil. In the interior of the state of Bahia, on the east coast of the country, Chapada Diamantina National Park provides a first-class alternative program in a landscape similar to jungle, with green canyons, adventurous rock formations, stream courses, waterfalls, caves such as Gruta Azul and Gruta Lapa Doce, and Pai Inácio outlook hill. The picturesque small town of Lençóis, which is geared to travelers, serves as a hub for lots of undertakings, whether on your own initiative or organized. Shorter hiking trips lead from Lençóis to natural swimming pools and to Cachoeira do Sossego, the "waterfall of tranquility." It flows there alone at the end of a valley, framed by wild and romantic rockfalls. Trekking through the Pati Valley takes several days.

Information: diamantinamountains.com

388. KITESURFING AT FORTALEZA

Pure happiness hormones: this is the kite paradise of Brazil. It is always warm and there is always wind, plus lots of samba and lots of caipirinhas cocktails. Once you have left Fortaleza behind you, you have nothing more to worry about. From Canoa Quebrada to Jericoacoara, one fantastic kiting spot follows the next: Uruau, Fortim, Parajuru, Cumbuco or Ilhado Guajiru . . . just to name a few of the most beautiful. It's incredible; just when you think you've found the perfect spot, a few miles farther on there is one even more beautiful. At a constant 86°F (30°C), you can kite on the countless lagoons or go "downwind" along hundreds of miles of great beaches. The farther north you go, the stronger the wind becomes—and the greater the desire to stay longer.

Kitesurfing spot: flat-water areas, shallow water, waves
Wind: 4–7 Beaufort, easterly, side onshore / side shore
Best time of year: July to December
Suitable for: everyone
Starting point: Fortaleza
Information: www.kite-college.com

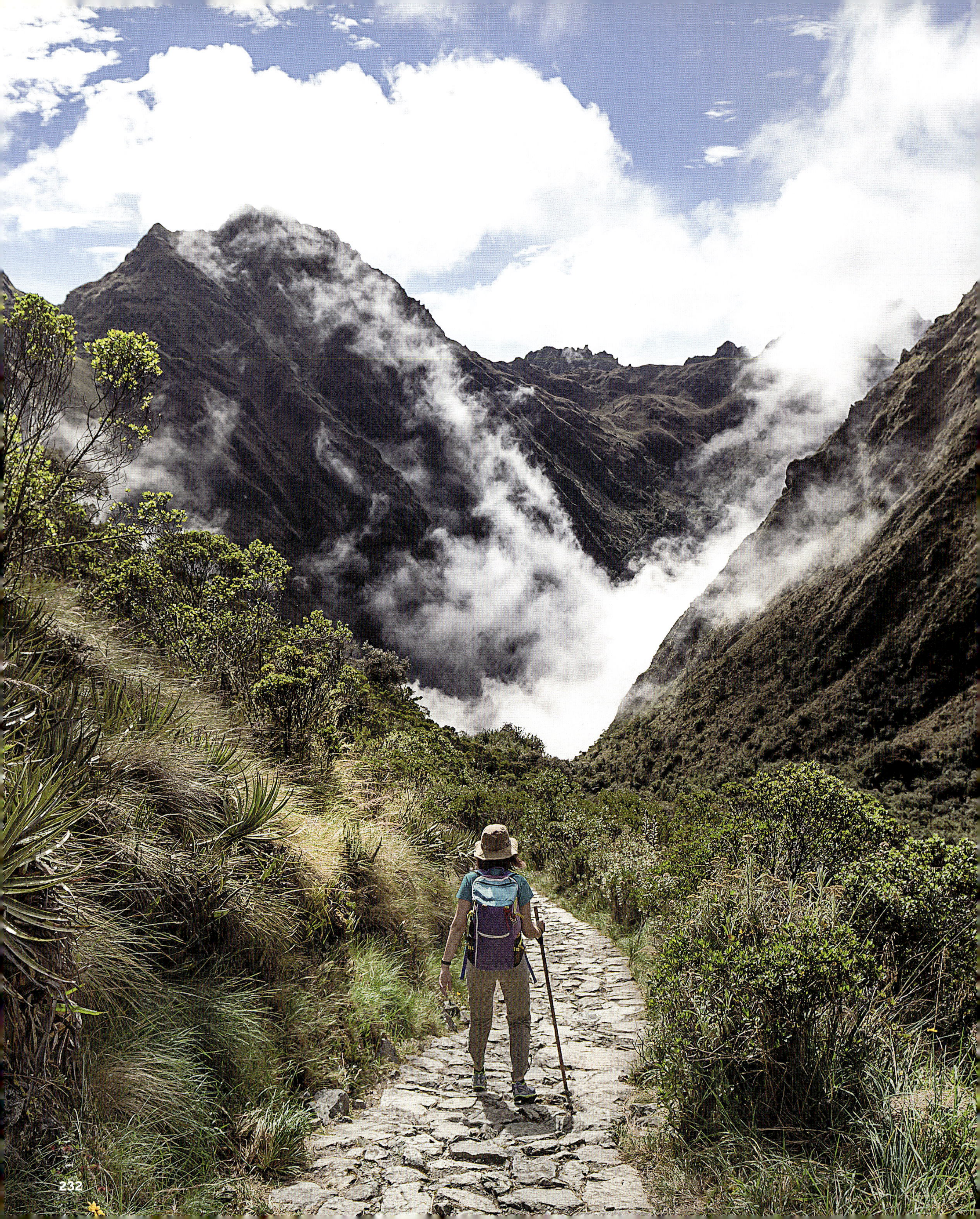

389. TREKKING ON THE INCA TRAIL

It remains to be seen whether the way is the goal here also, but this goal is a magical top spot on the continent: Machu Picchu, the "lost city of the Incas," which fortunately remained hidden from the Spaniards and was rediscovered only in 1911 by US archeologist Hiram Bingham with the help of the indigenous people. The Inca Trail there is only a tiny fraction of the vast network of trails that the Incas built to consolidate their power via the capability to transmit news by relay messengers. The trail to Machu Picchu runs through grandiose mountain scenery and passes the 13,780 ft. (4,200 m) mark. Here you have to be able to deal with high elevations well. When you finally stand at the Sun Gate, high above the ruined city, and look down at it, that's a wow effect. Access to the Inca Trail is limited. Tours are organized with guides, porters, and cooking staff.

Length / time it takes: 26 mi. (42 km) / 4 days
Information: for example, www.intrepidtravel.de

390. BIKING THROUGH THE SACRED VALLEY OF THE INCAS

The air is fresh and thin in this river valley of the Río Urubamba, which sometimes becomes constricted and sometimes broadens out considerably: the Valle Sagrado de los Incas, or "Sacred Valley of the Incas." Centuries ago, the Incas cultivated corn here, built farming terraces high up on the slopes of the Andes, and built fortresses. It is exciting to approach the Sacred Valley by mountain bike. A 38 mi. long (61 km) cycle route (Ciclovía) connects Pisaq with Ollantaytambo. Imposing fortifications preside above both places. The course runs along cornfields and adobe buildings, cacti, and agaves taller than a man. Everywhere you get friendly greetings. Again and again the trail runs around bends of the river, the lifeline of the valley and the turf of the Andean gulls, which are easy to watch.

Time it takes: at least 3 hours
Information: for example, www.peruculturetravel.com

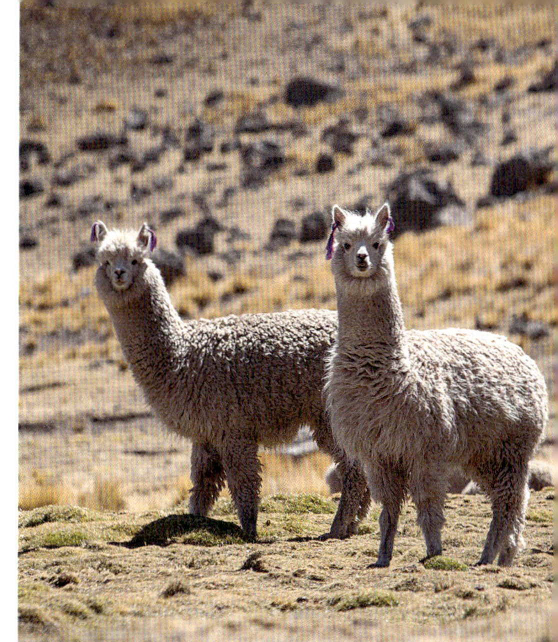

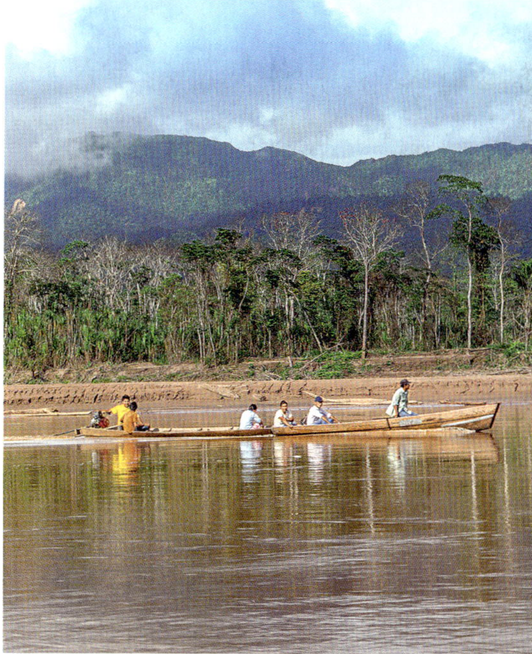

391. ASCENT TO CERRO CALVARIO ABOVE LAKE TITICACA

Copacabana sounds like washboard abs and thong-wearing beauties, but here you actually need a bear skin. This Copacabana is on the shores of ice-cold Lake Titicaca at an elevation of 12,467 ft. (3,800 m). The trail goes even higher up to Cerro Calvario (Calvary Hill). Anyone who has problems with altitude moves in slow motion toward the low-oxygen 13,123 ft. (4,000 m) mark, past Stations of the Cross and eucalyptus giants, over misshapen stone pavement and in some places extremely high steps. The reward for all the exertion is fantastic summit views over the place, with its shrine to the Virgin Mary and Copacabana Bay, dotted with dozens of boats. Arriving at sunset is ideal. Atop the mountain, ghostly candles flicker in rock niches, then the orange sphere sinks above the vastness of the lake.

Time it takes: at least 1 hour
Starting point / trail's end: Copacabana

392. JUNGLE HIKES

Bolivia's Amazonian lowlands are calling—and with them, the town of Rurrenabaque, which is connected to the domestic flight network, lies on the Río Beni and serves as an entry point for organized jungle tours, including overnight stays in eco-lodges. In the green wilderness of the Parque Nacional Madidi, indigenous communities live free from digital influences and are happy anyway—or just for that reason. The national park represents great biodiversity and is home to more than 180 mammal, eighty-two reptile, ninety-two amphibian, and more than 900 bird species. The chances are therefore excellent that you will see plenty of exotic wildlife on various hikes through largely unspoiled nature, be it howler monkeys, caimans, parrots, or capybaras. Encounters with spectacled bears and jaguars are rather unlikely.

Information: for example, www.viacha-tours.com, www.balatours.com

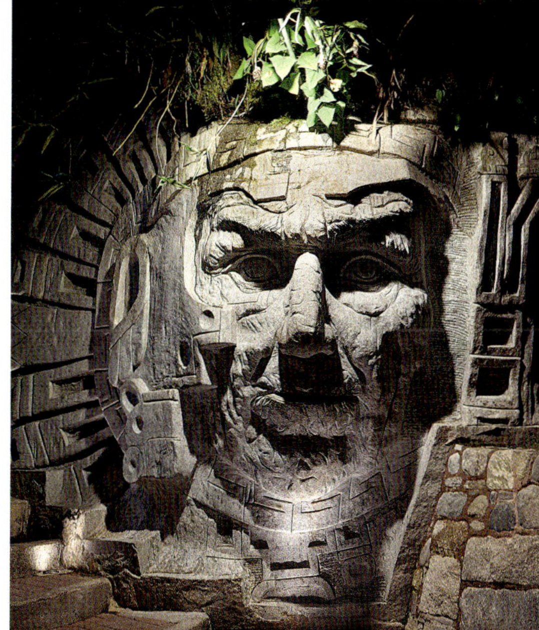

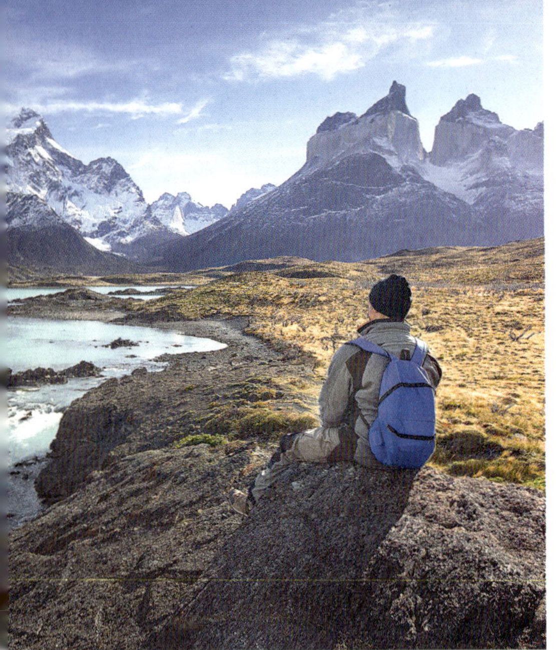

393. CYCLING IN THE ATACAMA DESERT

No, you don't have to tackle the entire Atacama Desert as a cyclist. That would be an "XL extreme" project, because there are scarcely any more-arid regions on Earth. But right in the middle is the large oasis of San Pedro de Atacama, an ideal starting point for tackling various routes on the mountain bike that you can rent. Short-term goals are the historic fortress ruins of Pukará de Quitor and various village communities (local term: *ayllu*, "family clan") such as Sequitor and Cucuter. About 7.5 mi. (12 km) away, the Valle de la Luna ("Valley of the Moon") opens up with a scene of bizarre rock formations and salt formations left by erosion. Minerals and salts glitter everywhere in the sunshine, and it is most beautiful in the morning or evening light. Huge sand dunes that are reminiscent of the Sahara complete the panoramas.

Length: 7.5 mi. (12 km)
Starting point: San Pedro de Atacama
Trail's end: Valle de la Luna
Information: www.sanpedrochile.com

394. ICE TREKKING ON CALLUQUEO GLACIER

In Chile's deep south, Jimmy Valdés conducts an extraordinary tour from his home town of Cochrane to Calluqueo Glacier, which rolls in a broad, bright band down from Monte Lorenzo, one of Patagonia's highest peaks. After a passage by boat across Lake Calluqueo, the trail runs on foot toward the glacier, which Valdés has seen retreat "by 200 meters [656 ft.]" in recent years due to global warming. The feasibility of an "ice trekking" always depends on current trail conditions. The trail cracks and crunches already at the access to the outermost extensions of the glacier tongue, then the white wilderness embraces you. After successfully completing the tour, Jimmy serves liqueur made from the fruits of the calafate bush—with glacier ice he cut himself.

Time it takes: at least 4 hours, there and back
Information: for example, lordpatagonia.wixsite.com

395. TREKKING IN TORRES DEL PAINE NATIONAL PARK

Like giant fingers, they jut up to 9,350 ft. (2,850 m) high into the sky over Patagonia: the granite Torres del Paine (Paine Towers), which gave their name to one of the most spectacular national parks in South America. In addition, glaciers, waterfalls, lakes, and icy rivers create images that will leave no one indifferent. Every trekker's dream is the loop hike around the Paine massif, where, with a bit of luck, you can watch guanacos, highland geese, foxes, or rheas. Overnight huts (*refugios*) and camping sites (*campamentos*) are available for the night; there is a compulsory reservation system. A demanding passage leads over the John Gardner Pass, where strong winds can come in and there can be snow flurries even in the Southern Hemisphere summer. Just the view of Gray Glacier makes up for all the exertion.

Length / time it takes: 58 mi. (93 km) / 8 days
Information: www.parquetorresdelpaine.cl

396. BIKING ON THE CARRETERA AUSTRAL HIGHWAY

Anyone who wants to tackle the Carretera Austral, Chile's legendary "Southern Highway," has a lot before them. Some 775 mi. (1,247 km) separate Puerto Montt from the shores of Lago O'Higgins. There are in fact some paved passages, but the majority of the route has gravel surfaces, meaning a good mountain bike is essential. The farther south it goes, the more unspoiled and wild nature becomes. Deep-blue lakes such as Lago General Carrera and milky turquoise rivers such as Rio Baker merge with the white of glaciers. Bands of clouds surround mountain silhouettes in the distance. In the foreground, rosehip bushes with fat, bright-red fruit stand out against the green of the southern beech forests. And the gunnera (giant rhubarb), with a diameter of 7 ft. (2 m), live up to their name. This is the essential Patagonia at its purest.

Length / time it takes: 775 mi. (1,247 km) / at least 4 weeks
Information: recorreaysen.cl (Aysén Region)

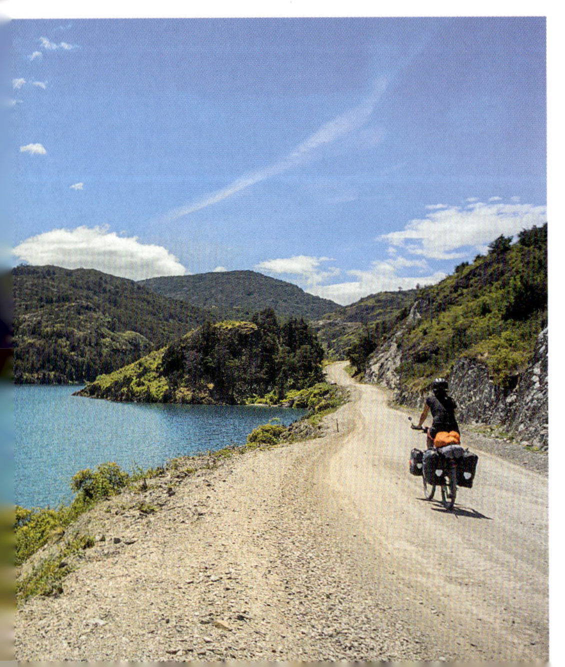

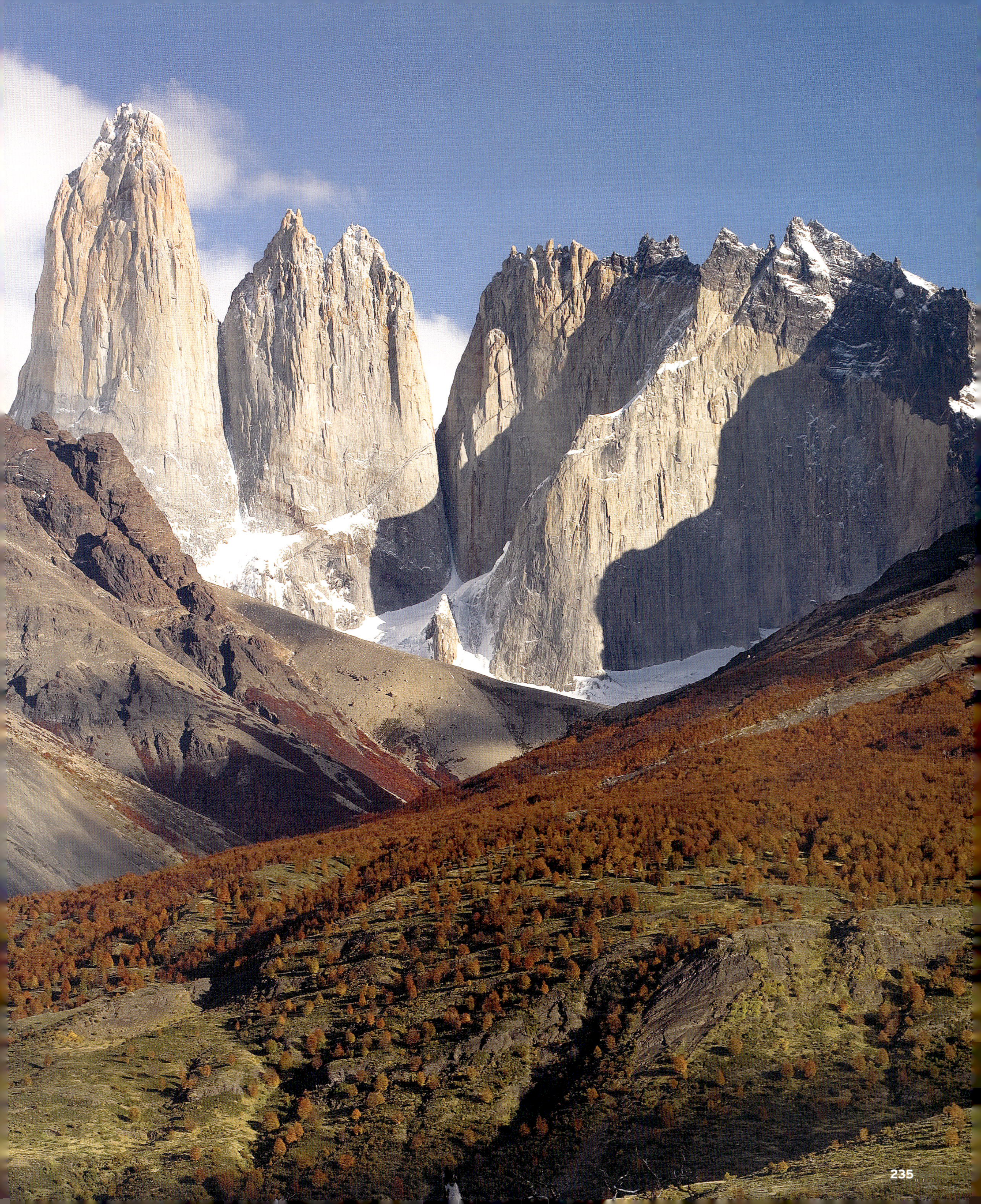

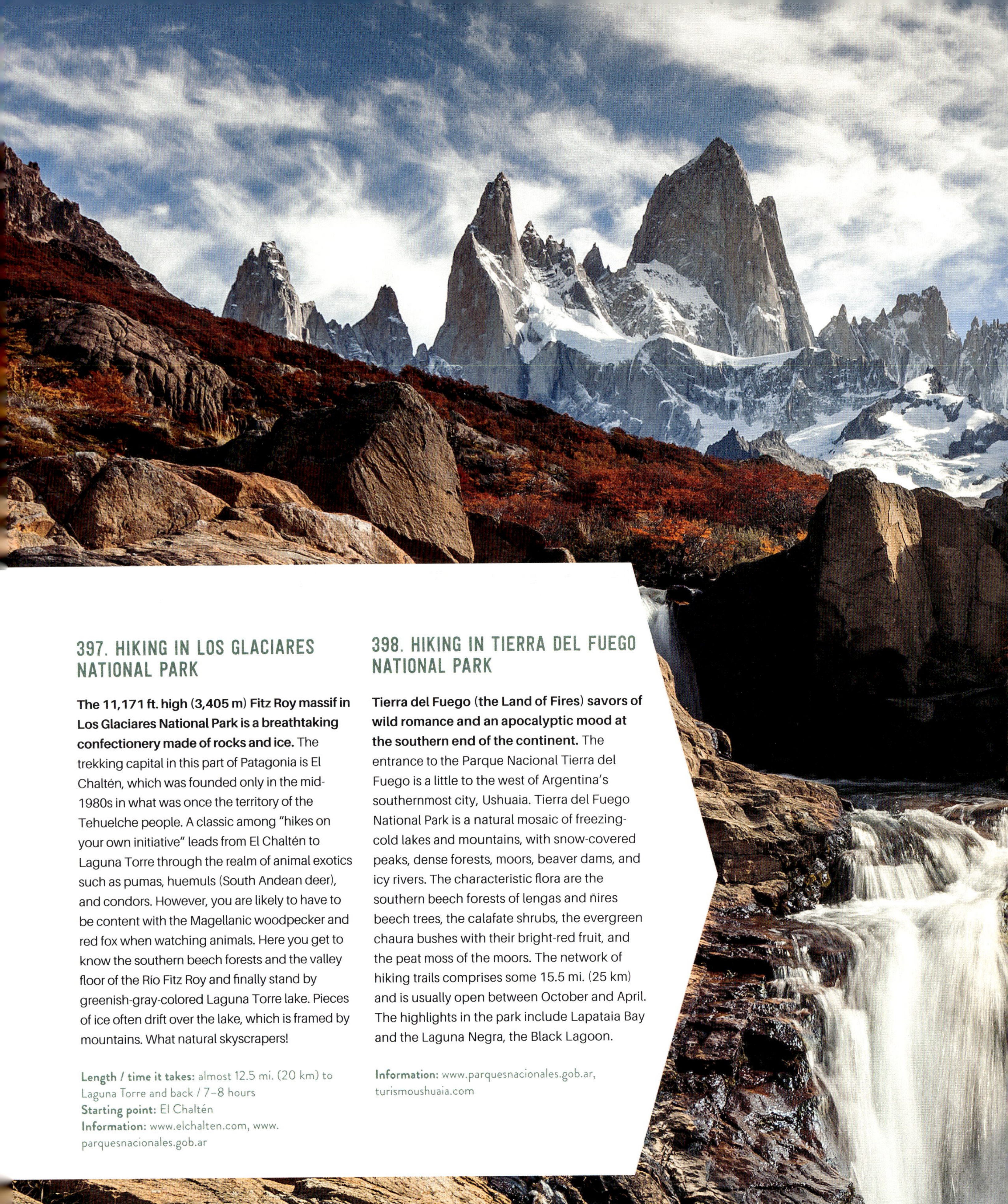

397. HIKING IN LOS GLACIARES NATIONAL PARK

The 11,171 ft. high (3,405 m) Fitz Roy massif in Los Glaciares National Park is a breathtaking confectionery made of rocks and ice. The trekking capital in this part of Patagonia is El Chaltén, which was founded only in the mid-1980s in what was once the territory of the Tehuelche people. A classic among "hikes on your own initiative" leads from El Chaltén to Laguna Torre through the realm of animal exotics such as pumas, huemuls (South Andean deer), and condors. However, you are likely to have to be content with the Magellanic woodpecker and red fox when watching animals. Here you get to know the southern beech forests and the valley floor of the Río Fitz Roy and finally stand by greenish-gray-colored Laguna Torre lake. Pieces of ice often drift over the lake, which is framed by mountains. What natural skyscrapers!

Length / time it takes: almost 12.5 mi. (20 km) to Laguna Torre and back / 7–8 hours
Starting point: El Chaltén
Information: www.elchalten.com, www.parquesnacionales.gob.ar

398. HIKING IN TIERRA DEL FUEGO NATIONAL PARK

Tierra del Fuego (the Land of Fires) savors of wild romance and an apocalyptic mood at the southern end of the continent. The entrance to the Parque Nacional Tierra del Fuego is a little to the west of Argentina's southernmost city, Ushuaia. Tierra del Fuego National Park is a natural mosaic of freezing-cold lakes and mountains, with snow-covered peaks, dense forests, moors, beaver dams, and icy rivers. The characteristic flora are the southern beech forests of lengas and ñires beech trees, the calafate shrubs, the evergreen chaura bushes with their bright-red fruit, and the peat moss of the moors. The network of hiking trails comprises some 15.5 mi. (25 km) and is usually open between October and April. The highlights in the park include Lapataia Bay and the Laguna Negra, the Black Lagoon.

Information: www.parquesnacionales.gob.ar, turismoushuaia.com

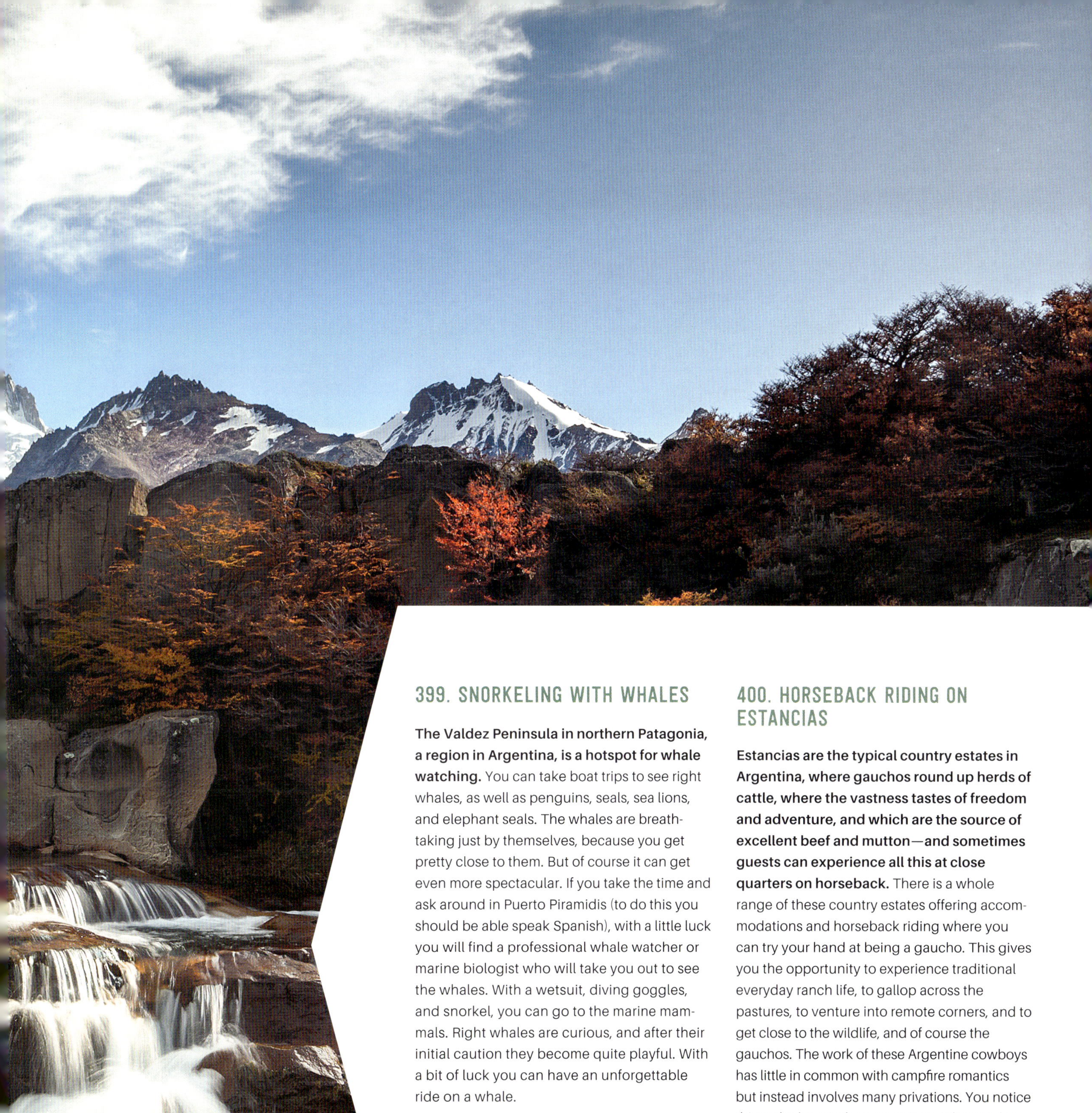

399. SNORKELING WITH WHALES

The Valdez Peninsula in northern Patagonia, a region in Argentina, is a hotspot for whale watching. You can take boat trips to see right whales, as well as penguins, seals, sea lions, and elephant seals. The whales are breathtaking just by themselves, because you get pretty close to them. But of course it can get even more spectacular. If you take the time and ask around in Puerto Piramidis (to do this you should be able speak Spanish), with a little luck you will find a professional whale watcher or marine biologist who will take you out to see the whales. With a wetsuit, diving goggles, and snorkel, you can go to the marine mammals. Right whales are curious, and after their initial caution they become quite playful. With a bit of luck you can have an unforgettable ride on a whale.

Time it takes: 2–3 hours
Starting point: Puerto Piramidis
Best time of year: mid-June and mid-December
Water temperature: icy cold
Information: www.puertopiramides.gov.ar

400. HORSEBACK RIDING ON ESTANCIAS

Estancias are the typical country estates in Argentina, where gauchos round up herds of cattle, where the vastness tastes of freedom and adventure, and which are the source of excellent beef and mutton—and sometimes guests can experience all this at close quarters on horseback. There is a whole range of these country estates offering accommodations and horseback riding where you can try your hand at being a gaucho. This gives you the opportunity to experience traditional everyday ranch life, to gallop across the pastures, to venture into remote corners, and to get close to the wildlife, and of course the gauchos. The work of these Argentine cowboys has little in common with campfire romantics but instead involves many privations. You notice this at the latest when you return exhausted to the estancia in the evening.

Information: www.estanciabuenavista.com.ar; www.nibepoaike.com.ar

ACTIVITIES

PHOTO CREDITS

Haafke, Udo: pages 67c, 67b
Seissenbacher, Priska: pages 152, 153b, 157, 162-163, 164, 180-181

Bildagentur LOOK
Pages 18-19 (Dave Derbis), 52 (Brigitte Merz), 59 (robertharding), 126 (Aurora Photos), 130-131 (age fotostock), 135a (robertharding), 235 (age fotostock)

mauritius images
Pages 22-23 (Ingram Premium Collection), 24 (Andreas Vitting), 27 below (Andreas Vitting), 28-29 (imageBROKER-Armin Floreth), 30 (Westend61-noonland), 34-35 (Westend61-Holger Spiering), 44 (Westend61-Günter Flegar), 45a (age fotostock-David & Micha Sheldon), 45c, (Christina Blum) 45 below (Buiten-Beeld-Misja Smits), 46a (imageBROKER-Mara Brandl), 47 (Bernd Römmelt), 54 (Ilkka Uusitalo-skiing-Alamy), 56-57 (Masterfile RM-F. Lukasseck), 75 (Westend61-Petra Silie), 80-81 (Loop Images-RJB Photograhic), 96-97 (Novarc-Annett Schmitz), 100 (Picfair-Andrei), 101a (Frank Fleischmann), 101b (All Canada Photos-Alamy), 102-103 (Steffen Rothammel), 104-105 (Cubolmages-Massimiliano Maddanu), 106a (ClickAlps), 107 (John G. Wilbanks-Alamy), 108-109 (Rasmus Kaessmann), 110 (James Hadley-Alamy), 111c (Westend61-Josep Rovirosa), 111 below (Joana Kruse-Alamy), 112 (Westend61-VITTA GALLERY), 113 (Chris Bull-Alamy), 116 (Cultura-William Perugini), 117c (Klaus Neuner), 117b (imageBROKER-Valentin Wolf), 118a (Ingram Premium Collection), 118 below (imageBROKER-Albrecht Weißer), 119 (Cro Magnon-Alamy), 156 below (Iain Masterton-Alamy), 206 (Patti McConville-Alamy)

Picture Alliance
Page 207a (DON EMMERT-AFP Creative)

Shutterstock
Page 12-13 (P. Eing), 14 (emka74), 15a (Animaflora PicsStock), 15c (emka74), 15b (Animaflora PicsStock), 16 (Heide Pinkall), 20-21 (rphstock), 25 (Tanja Esser), 26 (rdonar), 27a (Mike Mareen), 27c (FloriO), 31 (travelpeter), 32-33 (canadastock), 37 (Jenny Sturm), 38-39 (patrickdifeliciantonio), 40-41 (mRGB), 42-43 (Denis Belitsky), 46b (canadastock), 48-49 (cristo95), 50 (Globe Guide Media Inc.), 51 (milosk50), 55 (Angelo Cordeschi), 58 (Roger Nichol), 61 (lesmcluckie), 62 (Piotr Orlinski), 64-65 (Mauritius Martin), 66 (rphstock), 67a (Giorgio Caracciolo), 69 (Barnabas Davoti), 70 (Colorshadow), 71a (Ivan Chudakov), 71c (Glenn Pettersen), 71b (Jan Faukner), 72 (Kjetil Kolbjornsrud), 73 (solarseven), 74a (Jarno Holappa), 74c (Alexander Erdbeer), 74b (GUDKOV ANDREY), 76-77 (Eriks Z), 78 (Felix Lipov), 79a (Sander van der Werf), 79c (Coatesy), 79b (Fotokon), 82-83 (Evgeny Subbotsky), 84 (Evgeny Eremeev), 85 (CCat82), 86-87 (Alexander Pink), 88 (Olga Danylenko), 89 (Gergely Zsolnai), 90 (misha reme), 92 (Porojnicu Stelian), 93 (Calin Stan), 94-95 (Damian Pankowiec), 98 (Ales Krivec), 99 (JohannesS), 101c (makasana photo), 106c (Michal Zieba), 106b (Romas_Photo), 111a (bepsy), 114-115 (Krzysztof Wieprow), 117a (David Gonzalez Rebollo), 120-121 (Delbars), 122 (Natalya Erofeeva), 123 (apstockphoto), 124-125 (DongDongdog), 127a (EitaZul), 127b (Luke Wait), 127c (Martin Mwaura), 128-129 (Martin Mwaura), 133 (soft_light), 134 (Nadezda Murmakova), 135c (Fabio Lamanna), 135b (Tanguy de Saint-Cyr), 136-137 (sirtravelalot), 138-139 (Picture.Perfect), 140 (Zamir Popat), 141a (Fedor Selivanov), 141c (infografick), 141b (mbrand85), 142a (Kochneva Tetyana), 142b (Benny Marty), 142c (Oleg Trima), 143 (Kochneva Tetyana), 144-145 (Anton_Ivanov), 146-147 (Photonell_DD2017), 148 (Federica Violin), 149 (Luca Rei), 150-151 (Alexander Ingermann), 153a (Rafal Cichawa), 153c (Marcin Szymczak), 154-155 (Michal Knitl), 156a (BlueOrange Studio), 158-159 (chuyuss), 160 (Alexander Piragis), 161 (Katvic), 165 (Vasca), 166 (Christian Kornacker), 167 (zhu difeng), 168-169 (Jixin YU), 170 (Wantanee Chantasilp), 171 (Almazoff), 172 (Daniel Prudek), 174-175 (happystock), 176-177 (Dudarev Mikhail), 178-179 (Perfect Lazybones), 182-183 (Sarinee58), 184-185 (worldroadtrip), 186-187 (Neale Cousland), 188 (Tetyana Dotsenko), 189 (Luke Wait), 190 (lassedesignen), 191a (jamesteohart), 191b (Naruedom Yaempongsa), 191c (ChameleonsEye), 192-193 (Alizada Studios), 195 (Konrad Mostert), 196 (Maridav), 197a (Umomos), 197b (DZiegler), 197c (MNStudio), 198-199 (Michael Carni), 200 (Janice Chen), 201 (Ronnie Chua), 202-203 (James William Smith), 204-205 (stefk), 207b (Kristi Blokhin), 207c (Cheri Alguire), 208 (Stephen Moehle), 209 (wonrin), 210 (Elena Arrigo), 211 (Matthew79), 212 (My Good Images), 214-215 (Bojan Milinkov), 216-217 (stacyarturogi), 218 (Stephen N Haynes), 219 (Michael Bogner), 220 (LaGente.do), 221 (Rafal Cichawa), 222-223 (Christian Hartmann), 224 (Kimberly Shavender), 225 a. (Jess Kraft), 225 c. (Don Mammoser), 225 b. (Maridav), 226-227 (Henner Damke), 228 a. (Scott Biales), 228b (Michel Kluyskens), 229 (PeterBatarseh), 230-231 (windwalk), 232 (sharptoyou), 233a (Christian Vinces), 233b (PixieMe), 233c (Elzbieta Sekowska), 234a (sunsinger), 234b (Guaxinim), 234c (Nektarstock), 236-237 (Anton Petrus)

Front cover: Christa Thieser; back cover: lassedesignen a. l., wonrin a. r., Natalya Erofeeva c. LaGente.do b. l., Luca Rei b. r.

AUTHORS

Jörg Berghoff: 67, 68, 99, 100, 135, 136, 142
Norbert Blank: 32, 34, 36, 38, 45, 46, 48, 80, 101, 162, 168, 170-172, 175-178, 180-184, 208, 210, 241, 242, 244, 249, 251, 252, 256, 259, 261, 262, 265, 266, 269, 270, 277, 286, 290, 351, 367, 368, 369, 371, 372, 377-379, 385, 386, 388, 399
Astrid Därr: 207, 209, 211-240, 243, 247, 248
Andreas Drouve: 69-76, 79, 185-199, 201-206, 363, 365, 366, 373-376, 381-384, 387, 389-398, 400
Sarah Fischer: 155-158
Oliver Fülling: 11, 12, 15, 16, 39, 40, 51, 52, 85, 86, 147-150, 179, 311, 312
Udo Haafke: 13, 14, 30, 49, 50, 103-108
Jürgen Haberbauer: 57, 58, 77, 78, 89-96, 97, 119-122, 131, 132, 151-154, 161, 165-167, 173, 245, 246
Gunnar Habitz: 41, 42, 63-66, 87, 88, 118, 127-130, 133, 134, 137-141, 143-146, 255, 279
Roland F. Karl: 5, 6, 278, 298, 313-328
Thomas Krämer: 109, 110, 125, 126
Marion Landwehr: 81-84, 98, 331- 333, 335, 337-350, 352-357, 359-362
Kay Maeritz: 17, 23, 25, 31, 33, 43, 53, 54, 123, 124, 280, 283-285, 287-289, 291-297, 299, 300, 305-310, 364
Michael Nathan: 1, 7, 200, 254, 353
Priska Seisenbacher: 257-260, 267, 268, 271-274, 281, 282, 301-304
Hans J. Spitzensberger: 111-117, 263, 264
Anette Späth: 2, 3, 4, 18-22, 24, 27-29, 37, 44, 47, 59-62, 102, 159, 160, 329, 330, 334, 336, 358, 370
Thomas Stankiewicz: 55, 56
Bruckmann Editorial Department: 8, 9, 10

ABOUT THE AUTHORS

Jörg Berghoff, born in 1954, studied art history and ethnology and is a master winemaker and publishing book dealer. As a freelance author and journalist he has been running a press office since 1998. He makes regular trips to Australia and Tasmania, Ireland, and the United Kingdom. As a travel journalist, he is familiar with many countries around the world. He lives near Ansbach, Germany. More at www.prberghoff.

The photojournalist, world adventurer, and multimedia consultant **Norbert Blank**, who has a strong social commitment, loves authentic pictures in radical realism. But he also likes to take beautiful and promotionally effective photos of products and services for successful corporate marketing. A mechanical engineer, he assembles his mobile homes himself: www.bilder-botschaften.de.

Graduate geographer **Astrid Därr** was born in 1977 and has been traveling all over Africa since she was one year old. She has toured Morocco many times by car, on trekking tours, and by public transport. A freelance travel journalist and tour guide for Hauser excursions (www.hauser-exkursionen.de), she is the author of a number of illustrated books and travel guides about Africa. She lives near Munich, Germany. More at www.daerr.net.

Andreas Drouve, born in 1964, has a PhD and is one of the most successful German travel and cultural book authors, with well more than 120 published titles. He is represented at Bruckmann-Verlag with articles in several anthologies. He is also the author of the books *Costa Rica, South America, and Argentina-Chile*. The countries of Latin America have had a magical attraction for Drouve since he was a student; Argentina and Chile are among his favorites of all the most beautiful countries on earth. The author, from Düren, North Rhine-Westphalia, Germany, has lived in Spain as a freelance journalist and author for years. More at www.andreas-drouve.de.

Sarah Fischer, born in 1972, works as a travel journalist, lecturer, and author. Besides this, she researches articles for TV and film (ZDF, ARD, arte, Kika, Constantin Film, Endmol, VOX, BR, and many other broadcasters) and provides support for television and film teams in various countries. In addition to an extensive image archive, she plans the Mongolian trips for DAV Summit Club. Publications: *Heimatroulette—durch 160 Länder* ("Homeland roulette—through 160 countries") (Droemer & Knaur Verlag), and *Ein Mongolei Reiseführer* ("A Mongolian travel guide") (Reise Know How Verlag; ITB Book Prize 2015).

Oliver Fülling, who completed studies in Sinology, politics, and history in 1984, has traveled in the "Middle Kingdom" repeatedly since 1985. Author of numerous travel guides on Chinese cities and regions for renowned publishers and magazines. Developed travel concepts and programs for China and Japan and lived in Shanghai from 1996 to 1999 as a tour operator and author. Freelance author since 2004.

Udo Haafke studied visual communication at Fachhochschule (University of Applied Sciences), Dortmund, Germany. After completing a graduate photo designer course, initially did freelance artistic work and exhibition projects. After this, photography for travel guides, illustrated books, columns, etc., then more work with a journalistic background for magazines and daily newspapers, as well as online editorial departments. Supervision of the web portal content for www.schottlandberater.de.

Jürgen Haberhauer came into contact with wanderlust at an early age and has since been traveling the world together with his wife, Ruth, for more than two decades. Regardless of the travel destination, curiosity and interest in foreign cultures and respect for other ideas about life are always at the forefront for him. His work focuses on live reports and country portraits.

Gunnar Habitz worked in Zurich for many years as an international sales manager for Central and Eastern Europe. For years he guided travelers in his second home in Prague and has followed the development of the Golden City since the booming 1990s. The Swiss author, who has his roots in Bremen, Germany, has published many travel guides, hotel guides, and travelogues about the Czech Republic, Switzerland, and Lake Constance. Since 2016, this travel blogger with wanderlust has been exploring the southern hemisphere. He resides in Sydney, where he publishes writings on the themes of sales and leadership.

As a freelance author and photographer, **Roland F. Karl** has been producing travel reports for print media for 35 years, including for *Die Zeit, Stern, Handelsblatt*, and various travel magazines. In addition, he is involved in numerous book publications with his texts and images.

Author and photographer **Thomas Krämer** has had close ties to the North since his first trip to Scandinavia in 1983 and is now editor in chief of the Scandinavian magazine *Nordis*. He also likes to ride a motorcycle through the Alps and other interesting regions but is also out and about on foot, by bike, or on skis. So far, Bruckmann Verlag has published his *Highlights Schweden* ("Swedish highlights"), *Bildschönes Norwegen* ("Beautiful Norway"), and *Zeit für das Beste—Südschweden* ("Time for the best—southern Sweden"). For him, traveling means "discovering new things and finding beauty in the detail."

The author **Marion Landwehr** is a graduate journalist and, after many years of experience in the newspaper and magazine sector, has discovered the travel guide market, with a focus on the United States. Due to many trips to all parts of the country, she not only is an expert authority on the country and its people but also knows what is important when traveling to the United States and how best to take the reader in hand to make every vacation an adventure.

Kay Maeritz, born in 1958, is a graduate designer, photographer, and book author. An Asian specialist with a focus on the Himalayas, India, and Southeast Asia, he is one of the best-known German lecturers and reports on his travels in many multivision lectures in German-speaking countries. Many book and magazine publications.

Michael K. Nathan was born and raised in Israel and studied in Jerusalem and Hamburg. He was the correspondent for Germany and Europe for an Israeli news magazine from 1960 to 1974. As a freelance journalist he worked for *Die Zeit, Der Spiegel*, Stern, and ZDF. He is involved in the Israeli peace movement. After years in the management of large companies and after a six-year stay in Dhaka, Bangladesh, as a program manager for German development aid, he lives in Hamburg and works as a freelance author, journalist, and translator.

Priska Seisenbacher works as a travel photographer with a focus on Iran, central Asia, and India and has published a range of travel reports, books, and calendars. She lost her heart to Iran, not least because of the warmth of its people. Since her first trip, she has meticulously tried to discover more and more aspects of the country and is always enthusiastic about the cultural richness of Iran.

Anette Späth caught the travel bug at an early age and not only likes to roam distant countries but also often hikes and travels through her "home country" of Germany—inspiration for 20 years of work as a freelance lecturer, editor, and author in the travel and outdoor sector for book and magazine publishers. She currently lives in the beautiful Bavarian foothills of the Alps.

Hans-Joachim Spitzenberger holds a PhD in biology and has, among other things, spent time researching the wilderness of Norway's Svalbard archipelago several times. This made him enthusiastic about the polar region until he also discovered the whole of Norway as well as Antarctica, Africa, India, the South Seas, and South America as travel destinations. He travels to regions that cannot be reached by land as an leader on expedition cruise ships.

Thomas Stankiewicz was a camera assistant at Bayerischer Rundfunk and a film cameraman. He has been working as a photographer for newspapers and agencies since the mid-1980s, and as a travel photographer since the 1990s. His archive of travel pictures includes pictures of 87 countries. Nevertheless, he first traveled to New Zealand, his long-term dream destination, only last year, but then for almost two months.

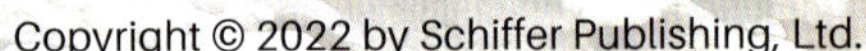

Originally published as *Rauszeit: 400 Beste Outdoorerlebnisse Weltweit* by Bruckmann, Munich © 2019 Bruckmann Verlag
Translated from the German by Simulingua, Inc.

Responsible: Joachim Hellmuth; editor: Mareike Weber; picture editor: Silwen Randebrock; layout: Leeloo Molnar; typesetting: satz & repro Grieb, Munich; cover design: Christa Thieser; repro: Ludwig:media; cartography: Verlagsservice Herbrecht; production: Bettina Schippel

Library of Congress Control Number: 2021936101

Edited by Ian Robertson
Production design by Danielle Farmer
Cover design by Jack Chappell
Type set in The Sans Extra Bold & Light/Aileron (OTF)/Festivo/ Brandon Grotesque
ISBN: 978-0-7643-6291-0
Printed in India

Published by Schiffer Publishing, Ltd.
4880 Lower Valley Road
Atglen, PA 19310
Phone: (610) 593-1777; Fax: (610) 593-2002
Email: Info@schifferbooks.com
Web: www.schifferbooks.com

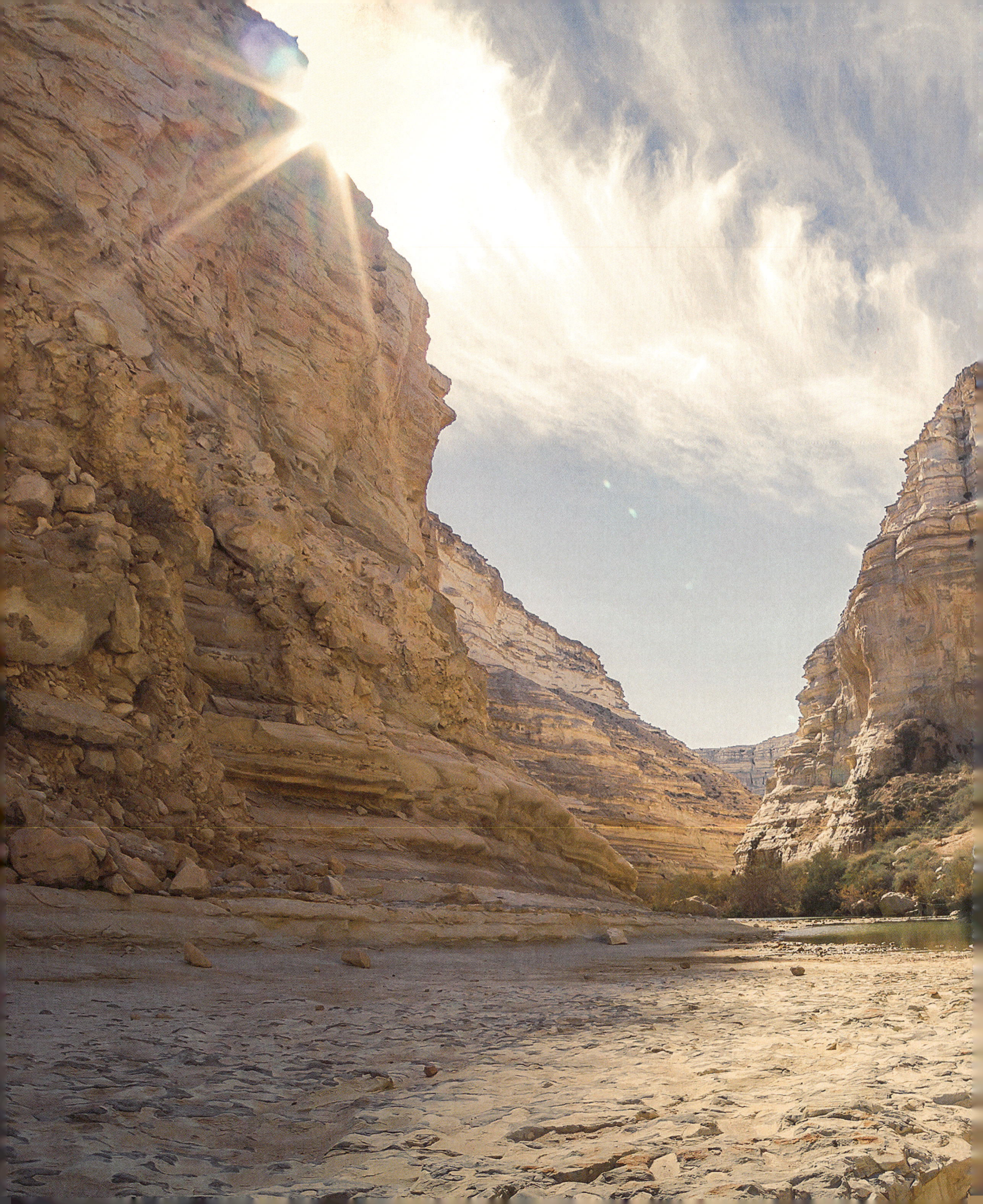